The Earth's Ionosphere
Second Edition

The Earth's Ionosphere
Plasma Physics and Electrodynamics
Second Edition

Michael C. Kelley

Cornell University
College of Engineering
School of Electrical and Computer Engineering
Ithaca, NY

ELSEVIER

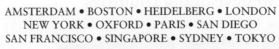

AMSTERDAM • BOSTON • HEIDELBERG • LONDON
NEW YORK • OXFORD • PARIS • SAN DIEGO
SAN FRANCISCO • SINGAPORE • SYDNEY • TOKYO

Academic Press is an imprint of Elsevier

Academic Press is an imprint of Elsevier
30 Corporate Drive, Suite 400, Burlington, MA 01803, USA
525 B Street Suite 1900, San Diego, California 92101-4495, USA
84 Theobald's Road, London WC1X 8RR, UK

This book is printed on acid-free paper. ♾

Library of Congress Cataloging-in-Publication Data
Kelley, Michael C.
    The earth's ionosphere : plasma physics and electrodynamics/Michael C. Kelley.
        p. cm. – (International geophysics series; v. 96)
    ISBN 978-0-12-088425-4 (hardcover : alk.paper) 1. Space plasmas. 2. Plasma
electrodynamics. 3. Ionosphere. I. Title.
    QC809.P5K45 2009
    551.51'45–dc22
                                                                    2009006467

British Library Cataloguing in Publication Data
A catalogue record for this book is available from the British Library

ISBN 13: 978-0-12-088425-4

For information on all Academic Press publications
visit our Web site at *www.elsevierdirect.com*

Typeset by: diacriTech, India

Printed and bound by CPI Group (UK) Ltd, Croydon, CR0 4YY

Transferred to digital print 2012

To my beloved family
Aidan, Brian, Elizabeth, Erica, Patricia, Scott, and Varykina

Contents

The companion Website containing Appendices, problem sets and solutions, images, and other supplemental materials can be found at http://www.elsevierdirect.com/companions/9780120884254.

Preface

During the past five decades, the underlying physical principles upon which ionospheric research is based have undergone a drastic change. Prior to this period, the main emphasis was on production and loss of electrons and ions by photoionization and particle beams. Ion chemistry, geomagnetic field variations, and other "aeronomical" processes formed primary subfields as well. With the development of incoherent scatter radar techniques as well as scientific satellites, a whole new realm of ionospheric processes became amenable to scientific study. In particular, space scientists were able to reliably probe the dynamics of the ionosphere and to realize the fundamental role played by electric fields. In a parallel development, the physics of plasmas became an important branch of science due to its role in controlled fusion research.

In this book, study of the plasma physics and electrodynamics of the ionosphere is given the highest priority, and some classical aeronomical processes are deemphasized. This is an unfortunate necessity, but the text by Rishbeth and Garriott (1969), published in the same International Geophysics Series, remains a valuable treatise on these topics, and a new publication by Schunk and Nagy (2000) is excellent in this area.

The ionosphere is somewhat of a battleground between the earth's neutral atmosphere and the sun's fully ionized atmosphere in which the earth is imbedded. One of the challenges of ionosphere research is to know enough about these two vast fields of research to make sense out of ionospheric phenomena. We try to give the reader some insights into how these competing sources of mass, momentum, and energy vie for control of the ionosphere. Unlike the first edition, we now include the D region of the ionosphere. Because the neutral atmosphere so dominates the physics, we include discussions of turbulence throughout the text where appropriate. The huge increases in dusty/icy plasma physics and the role polar mesospheric clouds may play in Global Change are reasons to include this region.

After some introductory material, the study is begun in earnest with the equatorial ionosphere. Here the earth's magnetic field is horizontal and many unique and fascinating phenomena occur. It is interesting to note also that the Coriolis force vanishes at the geographic equator, which makes for some unusual ocean and atmospheric dynamics as well. In Chapter 3 the electrodynamics of this zone are studied, and plasma instabilities are discussed in Chapter 4. This particular chapter is extended greatly from the first edition.

In Chapter 5 the midlatitude zone is studied, and, in particular, the role of gravity waves and neutral wind dynamics is emphasized. This is the ionospheric footprint of that portion of the near-space region of the earth that is dominated by the planet's rotation. Due to the explosion of information on midlatitude instabilities, a new Chapter 6 has been written. The role of the neutral wind in these instabilities simply was not recognized in the 1980s.

An entirely new chapter dealing with the polar summer mesosphere has been written due to the huge increase in interest and our understanding of this region. The role of charged ice in an icy (or dusty) plasma has led to this new interest, as has the possibility of global change being detected early in this region.

In Chapter 8 the interaction of the solar wind with the magnetic field of the earth is reviewed, with particular regard to the imposition of electric fields and field-aligned currents on the ionosphere. At high latitudes the imposed electric field from this source overcomes the field due to the rotation of the earth. The influence of the interplanetary magnetic field is described, as are the flow patterns that arise in the ionosphere. The effect of these electric fields on ionospheric plasma as well as on the neutral atmosphere is the topic of Chapter 9. In the final chapter, a number of processes that create structure in the high-latitude ionosphere are presented, including, but not limited to, the plasma instabilities that occur there.

As is already evident, the D region and mesosphere are included in this new addition. Problems are added on the companion Web site to facilitate using the book as a textbook. A solutions manual is available to teachers.

I am very much indebted to legions of students who have read and constructively criticized the text as it evolved, as well as to my many colleagues at Cornell and throughout the world. The study of space plasma physics is by definition an international discipline and epitomizes the way in which the human race can cooperate perfectly well at the highest technical and interpersonal levels. I have learned a great deal about both space physics and my fellow passengers on *Spaceship Earth* while traveling around to various rocket ranges, radar observatories, and scientific laboratories. Special thanks go to Rod Heelis, who helped considerably with Chapters 7 and 8 in the first edition, and to my editor, Laurie Shelton, for her skill and endless patience.

Michael C. Kelley
August 19, 2008

1 Introductory and Background Material

In this introductory chapter we present a qualitative treatment of several topics that we hope is sufficient to proceed with our study of ionospheric physics. The chapter begins with historical comments and a description of the limitations we have set for the text. In particular, we do not repeat or significantly update the material published by Rishbeth and Garriott (1969) earlier in this same International Geophysics Series. Rather, our emphasis is on electrodynamics and plasma physics, so we refer the interested reader to Rishbeth and Garriott and to Banks and Kockarts (1973) for more information about formation of the ionosphere, its ion chemistry, heat balance, and other aeronomic properties. Another and more recent resource for classical ionospheric science is the text by Schunk and Nagy (2000). Lyons and Williams (1984) have published a text on magnetospheric science.

1.1 Scope and Goals of the Text

1.1.1 Historical Perspective

The earth's ionosphere is a partially ionized gas that envelops the earth and in some sense forms the interface between the atmosphere and space. Since the gas is ionized, it cannot be fully described by the equations of neutral fluid dynamics. In fact, a major revolution in ionospheric physics has occurred in the past decades as the language and concepts of plasma physics have played an increasing role in the discipline. On the other hand, the number density of the neutral gas exceeds that of the ionospheric plasma, and certainly neutral particles cannot be ignored. A student of the ionosphere must thus be familiar with both classical fluid dynamics and plasma physics. Even a working knowledge of these two "pure" branches of physics, however, is not sufficient. Since the ionosphere lies at the interface between two very different and dynamic media, we must understand enough of both atmospheric dynamics and deep space plasma physics to understand how the ionosphere is formed and buffeted by sources from above and below. Added to these two is the requirement for a sufficient knowledge of ion and neutral chemistry and photochemistry to deal with production and loss processes.

Ionospheric physics as a discipline grew out of a desire to understand the origin and effects of the ionized upper atmosphere on radio wave propagation. The very discovery of the ionosphere came from radio wave observations and the recognition that only a reflecting layer composed of electrons and positive ions could explain the characteristics of the data. Most of the early work was aimed at explaining the various layers and their variability with local time, latitude, season, and so forth. The ionosonde, a remote sensing device that yields electron density profiles up to but not above the altitude of the highest concentration of charged particles, was the primary research tool. Such measurements revealed a bewildering variety of ionospheric behavior ranging from quiet, reproducible profiles to totally chaotic ones. Furthermore, a variety of periodic and aperiodic variations were observed, with time scales ranging from the order of the solar cycle (11 years) to just a few seconds. As time passed, the emphasis of ionospheric research shifted from questions dealing with formation and loss of plasma toward the dynamics and plasma physics of ionospheric phenomena.

Ionospheric research has greatly benefited from the space program with the associated development of instruments for balloons, rockets, and satellites (see Appendix A[1]). The combination of remote ionospheric sensing and direct in situ measurements made from spacecraft has accelerated the pace of ionospheric research. Equally important has been the development of plasma physics as a theoretical framework around which to organize our understanding of ionospheric phenomena. Likewise, the continuing development of magnetospheric and atmospheric science has very much influenced our understanding of the ionosphere.

In parallel with these developments, the incoherent scatter radar technique was devised (see Appendix A), and several large facilities were built to implement it. The primary advantage of this method was the ability to make quantitative measurements of numerous ionospheric parameters as a function of altitude at heights that were inaccessible to ground-based ionosondes. Since interpretation of the scattered signal requires detailed understanding of the interaction between the electrodynamic waves and thermal fluctuations in a plasma, a working knowledge of plasma physics became necessary for understanding the diagnostics as well as the science.

Ionospheric science has thus evolved toward the point of view that is encompassed by the term *space plasma physics*. The central theme of this book is the treatment of ionospheric physics within this context, but many other important issues are not discussed, including the topside ionosphere, seasonal behavior, heat balance, and global morphology.

1.1.2 *Organization and Limitations*

As indicated in the previous section, our goal is to treat the electrodynamics and plasma physics of the ionosphere in some detail. We do not, therefore, have the

---

[1] Appendices A and B are located on the book's companion Web site at http://www.elsevierdirect.com/companions/9780120884254.

space to start from the first principles of ionospheric science, particularly with regard to photoproduction, ion chemistry, recombination, and related topics. In this first chapter, we attempt a broad general introduction to this area, which we hope will provide enough background to make the remainder of the text meaningful.

It is also necessary to introduce elements of neutral atmospheric and magnetospheric physics, since the ionosphere is very much affected by processes that originate in these two regions. The ionosphere coexists with the upper portion of the neutral atmosphere and receives considerable energy and momentum from the lower atmosphere as well as from the magnetosphere. The energy and momentum fluxes are carried by particles, electromagnetic fields, and atmospheric waves. We devote some time in this first chapter to a qualitative description of the atmosphere and magnetosphere. Although useful as a starting point, the descriptive approach is not sufficient in all aspects, and we will treat some particularly important processes, such as gravity waves, in detail as they arise in the text.

Some knowledge about experimental techniques is necessary for a student of ionospheric dynamics, if only to understand the sources of the data. Accordingly, Appendix A is included, describing some of the most important measurement methods. The choice of instruments so dealt with is not exhaustive and certainly reflects the bias and expertise of the author.

Finally, we need to limit the scope of the text. We have somewhat regretfully chosen not to include a detailed discussion of the acceleration of auroral particle beams. Part of the rationale for this decision concerns the height range where the electron acceleration usually takes place, some 2000 km or more above the earth's surface, which is the limiting height range of the ionosphere as we define it here. Thus, even though particle acceleration can and does occur much lower than the 2000 km height, we consider this topic to be outside the domain of this book, if not always outside the height range under detailed consideration.

In Chapter 2 some basic equations and concepts are developed for use throughout the text. Some of these concepts would be quite useful as background for reading the rest of this first chapter, particularly the magnetospheric section. We thus recommend that a reader unfamiliar with the field read Chapter 1 twice, once for background and once again after reading Chapter 2. Chapters 3 and 4 deal with the electrodynamics and plasma physics of the equatorial ionosphere. This region is singled out, since the earth's horizontal magnetic field in that area leads to a number of unique phenomena. An interesting analogy exists with atmospheric and ocean dynamics. Since the Coriolis force vanishes at the equator, there are some unique phenomena common to both meteorology and oceanography that make equatorial dynamics very unusual and interesting. Chapters 5 and 6 deal with the tropical and midlatitude ionosphere, where the earth's magnetic field has a sizable inclination but is not vertical and usually does not link the ionosphere with the hot, tenuous, flowing plasmas of the magnetosphere or solar wind. Here, a toroidal region of relatively cold, dense plasma, called the plasmasphere or protonosphere, exchanges plasma with the ionosphere. In the remainder of the text after Chapter 6, we study the high-latitude region. As previously mentioned, several experimental techniques are described in Appendix A. Appendix B provides formulas and tables that will help the reader describe ionospheric phenomena. Appendix B also defines various magnetic activity indices used in the text.

1.2   Structure of the Neutral Atmosphere and the Main Ionosphere

Owing to the pervasive influence of gravity, the atmosphere and ionosphere are to first order horizontally stratified. Atmospheric structure can be neatly organized by a representative temperature profile, while the ionosphere is more sensibly organized by the number density of the plasma. Typical midlatitude profiles of temperature and plasma density are given in Fig. 1.1. The atmospheric temperature initially decreases with altitude from the surface temperature, with a "lapse rate" of about 7 K/km in the troposphere. At about 10 km altitude this temperature trend reverses (at the tropopause) and the stratosphere begins. This increase is due primarily to the absorption, by ozone, of part of the ultraviolet portion

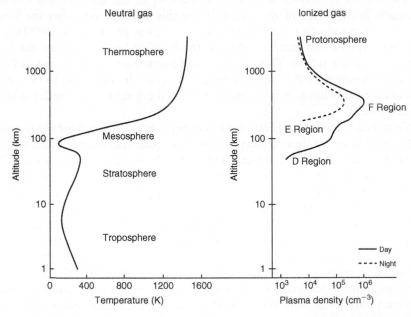

Figure 1.1   Typical profiles of neutral atmospheric temperature and ionospheric plasma density with the various layers designated.

of the solar radiation. This effect maximizes at 50 km, where the temperature trend again reverses at the stratopause. Radiative cooling creates a very sharp temperature decrease to a minimum in the range 130–190 K at about 90 km. For heights above the altitude of the temperature minimum (the mesopause), the temperature increases dramatically due to absorption of even higher energy solar photons to values that are quite variable but are often well above 1000 K. Not surprisingly, this region is termed the thermosphere. The atmosphere is relatively uniform in composition below about 100 km due to a variety of turbulent mixing phenomena. Above the "turbopause" the constituents begin to separate according to their various masses.

The temperature increase in the thermosphere is explained by absorption of UV and EUV radiation from the sun. The EUV radiation is also responsible for the production of plasma in the sunlit hemisphere, since these solar photons have sufficient energy to ionize the neutral atmosphere. Equal numbers of positive ions and electrons are produced in this ionization process. One requirement for a gas to be termed a plasma is that it very nearly satisfies the requirement of charge neutrality, which in turn implies that the number density of ions, n_i, must be nearly equal to the number density of electrons, n_e. Experimenters who try to measure n_i or n_e usually label their results with the corresponding title, ion, or electron density. In this text, we shall usually refer to the *plasma* density n, where $n \cong n_i \cong n_e$ is tacitly assumed to hold. Of course, if n_i exactly equaled n_e everywhere, there would be no electrostatic fields at all, which is not the case.

A note on units is in order here. Rationalized mks units will be used except in cases where tradition is too firmly entrenched. For example, measuring number density per cubic centimeter is so common that we shall often use it rather than per cubic meter. Likewise, the mho and the mho per meter will be used for conductivity. Some further discussion of units and of various parameters of interest to ionospheric physics is included in Appendix B.

Returning to Fig. 1.1, two plasma density profiles are given in the right-hand part of the figure, one typical of daytime midlatitude conditions and one typical of nighttime. In daytime, the solar spectrum is incident on a neutral atmosphere that is increasing exponentially in density with decreasing altitude. Since the photons are absorbed in the process of photoionization, the beam itself decreases in intensity as it penetrates. The combination of decreasing solar flux, increasing neutral density, and diffusion provides a simple explanation for the basic large-scale vertical layer of ionization shown in Fig. 1.1. The peak plasma density occurs in the so-called F layer and attains values as high as 10^6cm^{-3} near noontime. The factor that limits the peak density value is the recombination rate, the rate at which ions and electrons combine to form a neutral molecule or atom. This in turn very much depends on the type of ion that exists in the plasma and its corresponding interaction with the neutral gas.

Some experimental data on the ion and neutral composition above 100 km are reproduced in Fig. 1.2. Below and near that height, N_2 and O_2 have the same

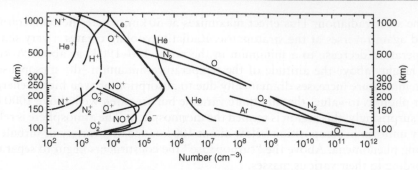

Figure 1.2 International Quiet Solar Year (IQSY) daytime atmospheric composition, based on mass spectrometer measurements above White Sands, New Mexico (32°N, 106°W). The helium distribution is from a nighttime measurement. Distributions above 250 km are from the *Elektron 11* satellite results of Istomin (1966) and *Explorer XVII* results of Reber and Nicolet (1965). [C. Y. Johnson, U.S. Naval Research Laboratory, Washington, D.C. Reprinted from Johnson (1969) by permission of the MIT Press, Cambridge, Massachusetts. Copyright 1969 by MIT.]

ratio as in the lower atmospheric regions—about 4:1—and dominate the gas. Near 120 km the amount of atomic oxygen reaches that of O_2, and above about 250 km the atomic oxygen density also exceeds that of N_2. This trend is due to the photodissociation of O_2 by solar UV radiation coupled with molecular diffusion and the absence of turbulent mixing above the turbopause. The dominance of atomic oxygen in the neutrals is mirrored by the plasma composition. The curve labeled e^- is similar to the right-hand side of Fig. 1.1 and represents the electron density (thus labeled with e^-). Near the peak in the plasma density, the ions are nearly all O^+, corresponding to the high concentration of atomic oxygen in the neutral gas. The altitude range 150–500 km is termed the F region, and the maximum density there is termed the F peak. (The F region is often separated into F1 and F2 during daytime due to the role of molecular ions.) Below the peak, NO^+ and O_2^+ become more important, dominating the plasma below about 150 km. The altitude range 90–150 km is called the E region, and the ionization below 90 km is, not surprisingly, termed the D region. These rather pedantic names have a curious history. The E region received its name from the electric field in the radio wave reflected by the "Heavyside" layer (the first name for the ionosphere). The other layers were simply alphabetical extensions. It was assumed initially that a plasma is absent between the layers. Unfortunately, many phenomena have been named for the instrument used to measure them or some other obscure parameter.

At the highest altitudes shown in Fig. 1.2, hydrogen becomes the dominant ion in a height regime referred to as the protonosphere. Helium ions are quite variable but sometimes reach 50% of the total ions at the base of the protonosphere. The hydrogen atom density is not shown in this early plot but exceeds helium above about 600 km. The hydrogen ions shown in Fig. 1.2 are produced by charge exchange with O^+, $O^+ + H \rightarrow H^+ + O$.

Turning to the nighttime profile in Fig. 1.1, the plasma density near the F peak is reduced in magnitude but not nearly so drastically as is the density at the lower altitudes. With reference to Fig. 1.2, this difference mirrors the change in composition. That is, in lower altitude regions where molecular ions dominate, the density is sharply curtailed at night. The O^+ plasma density, on the other hand, is sustained through the night. This trend is dramatically illustrated in Fig. 1.3 by the series of plasma frequency (f_p) profiles that were obtained by the incoherent scatter radar facility in Arecibo, Puerto Rico ($18.5°N, 66.8°W$). The electron plasma frequency (Hz) is related to the plasma density (cm^{-3}) by the equation $f_p = 8980\sqrt{n}$. The sunrise and sunset effects are very strong at the lower altitudes.

This substantial difference in ion behavior is due to the fact that molecular ions have a much higher recombination rate with electrons than do atomic ions. The two fast reactions that occur in E-region recombination are of the type

$$NO^+ + e^- \rightarrow N + O$$

and

$$O_2^+ + e^- \rightarrow O + O$$

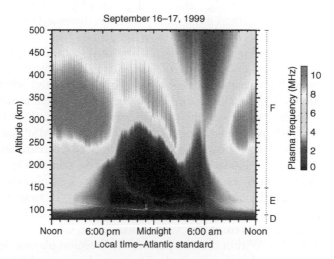

Figure 1.3 Plasma frequency contours during a typical night over Arecibo, Puerto Rico. (Figure courtesy of S. Collins.)

At higher altitudes one might expect

$$O^+ + e^- \rightarrow O + photon$$

to dominate. The former processes are called dissociative recombination, since the molecule ion breaks apart, while the latter is termed radiative recombination, since emission of a photon is required to conserve energy and momentum. The former two processes have a reaction rate much higher than the latter, which results in a much shorter lifetime for molecular ions than for atomic ions. Since the molecular ions are much shorter lived, when their production is curtailed at night, a recombination quickly reduces the plasma concentration. The O^+ plasma at higher altitudes, however, often survives the night at concentrations between 10^4 and $10^5 cm^{-3}$. The O^+ ions actually are lost through two-step processes (see Chapter 5) below the F peak. First, one of the following charge exchange reactions occurs:

$$O^+ + O_2 \rightarrow O + O_2^+$$

$$O^+ + NO \rightarrow NO^+ + O$$

and then the dissociative reactions given above complete the recombination process. In addition, at night, plasma can flow back into the F region from the high altitude region called plasmasphere.

The long-lived property of atomic ions also explains the numerous sharp layers of enhanced plasma density seen at low altitudes in Fig. 1.3. These contain heavy atomic ions such as Fe^+ and Mg^+ that are deposited by meteors in this height range. Just why they are gathered into sharp layers is discussed in some detail in Chapter 6, but once gathered, they last for a long time.

Photoionization by solar radiation is not the only source of plasma in the ionosphere. Ionization by energetic particle impact on the neutral gas is particularly important at high latitudes. Visible light is also emitted when particles strike the atmosphere. These light emissions create the visible aurora. A view of the aurora looking down on the earth is offered in Fig. 1.4. The picture was taken on the Defense Meteorology Satellite Program (DMSP) satellite from about 800 km altitude and shows some of the complex structure and intricate detail that the auroral emission patterns can have. This complexity is mirrored in the ionization that also results from particle impact. The photograph shows how much this ionization can vary in space. Variability in time is illustrated in Fig. 1.5, where consecutive plasma density profiles taken 20 s apart are displayed. The data were taken at the Chatanika Radar Observatory in Alaska (65.06°N, 147.39°W). Within the 40 s period, the E-region plasma density peak varied by almost one order of magnitude. The particle energies determine their penetration depth, and the distribution in their energies determines the resulting

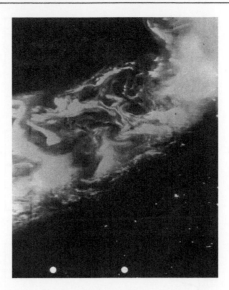

Figure 1.4 Photograph of an active auroral display taken from a high-altitude meteorology satellite over Scandinavia. The center-to-center distance of the two fiducial dots located on the lower edge is 1220 km, and the diameter of each dot is 100 km. [After Kelley and Kintner (1978). Reproduced by permission of the University of Chicago Press.]

ionization profile. The typical energy of electrons in the nighttime aurora is 3–10 keV, and electron impact is the dominant ionization source. This energy range results in large plasma production in the E region, as shown in the profiles in Fig. 1.5. The production rates are more than 10 times those provided by noontime photoionization in the E region. Lower-energy electrons can characterize the auroral precipitation pattern at other local times. These particles deposit their energy at higher altitudes, creating enhanced and highly variable F-region plasma concentrations.

A more global view of the aurora is provided by the Dynamics *Explorer-1* satellite imager, which has provided photographs looking down on the earth from an altitude of several thousand kilometers. The photo in Fig. 1.6 encompasses the entire polar region and shows that the band of auroral light extends completely around the polar zone. The auroral emissions wax and wane in a complicated manner related to the rate of energy input from the solar wind, as well as to the storage and release rate of this energy in the earth's magnetosphere and its ultimate release into the earth's upper atmosphere and ionosphere. The lower portion of the figure shows data from a scan through the ionosphere using an incoherent scatter radar simultaneously with the DE photograph. Contour plots of the electron density in the insert show a striking correlation with the light emissions.

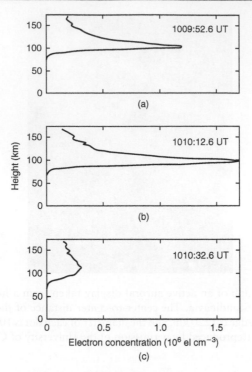

Figure 1.5 Sequence of E-region electron concentration profiles obtained using the Chatanika, Alaska, radar on September 27, 1971, near 1010 UT; azimuth = 209.04°, elevation = 76.58°. [After Baron (1974). Reproduced with permission of the American Geophysical Union.]

1.3 D-Region Fundamentals

The D region of the ionosphere was once thought to be of little plasma physics interest. The level of ionization is so low that the plasma physics is quite different from that of weakly collisional or collisionless plasmas, which captured the most attention of theorists. Furthermore, the region is too high for balloons and too low for satellites. Rockets often pass through it without even removing nosecones or windows. But as is usual in geophysics, as soon as better instruments were developed, interest rose steeply. In this section we discuss the basic source of the D region, which is fairly similar at equatorial and midlatitudes. We postpone discussion of its structure to future chapters.

A summary of rocket electron density measurements is presented in Fig. 1.7a for quiet conditions and 1.7b for an active sun. The flights were all made for solar zenith angles near 60°, and all were flown from the NASA base at Wallops Island, Virginia (37.9°N, 284.5°E). The electron density increases by almost five orders

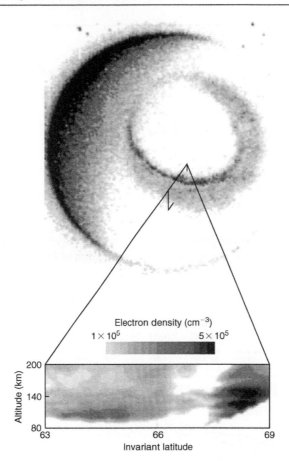

Figure 1.6 A view of the auroral oval from the *DE-1* satellite, along with simultaneous contour plots of electron density for a small portion of the photograph. [After Robinson et al. (1989). Reproduced with permission of the American Geophysical Union.]

of magnitude from 50 to 100 km. Any given profile exhibits at least one region of rapid increase in density, termed a ledge, and sometimes a series of ledges during the course of this rise. The largest difference between quiet and active solar conditions occurs between 65 and 75 km, where it exceeds an order of magnitude.

The ion composition is presented in Fig. 1.8. Ion mass spectrometers have much lower height resolution than electron instruments, so the profiles are very smooth looking. Water cluster ions begin to form below 86 km in this case. At night, except for the metal sporadic ion layers discussed in Chapter 5, the D region virtually disappears, dropping to levels of 100–$1000 \, cm^{-3}$ all the way to the base of the F region, where O^+ has a long lifetime.

The dominant source of ionization is photoionization of NO by atomic hydrogen Lyman $\alpha \, (\lambda = 122 \, nm)$. This low-energy photon has a very high flux (see

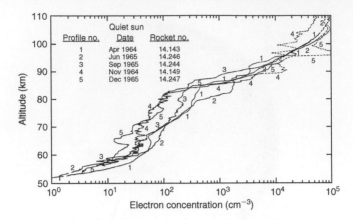

Figure 1.7a   Electron-density profiles in the lower ionosphere for quiet-sun conditions. [After Mechtly et al. (1972). Reproduced with permission of Pergamon Press.]

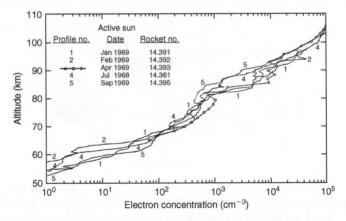

Figure 1.7b   Electron-density profiles in the lower ionosphere for active conditions. [After Mechtly et al. (1972). Reproduced with permission of Pergamon Press.]

Fig. 2.1b), but only the low ionization potential of NO allows a noticeable effect by this source. The ionization profile is very smooth, so the ledge effect cannot be explained by the source. The only other viable candidate is a height-dependent recombination effect. The processes are complex due to the many species involved (see Section 2.6).

Although the D region ionization is low compared to the E and F regions, it does have a huge effect on radiowaves. This holds because radiowave absorption is particularly high when even modest electron content exists at a height where the wave frequency equals the electron-ion (ν_{en}) collision frequency. In the D region $\nu_{en} \approx 10^6 \text{s}^{-1}$ (see Appendix B), which is comparable to AM radio

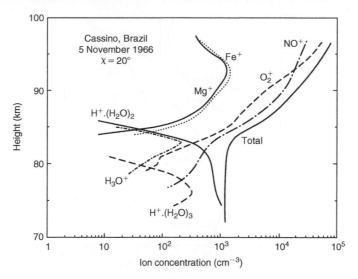

Figure 1.8 Positive-ion composition measurements at Cassino, Brazil, for full sun conditions before a solar eclipse. [After Narcisi et al. (1972). Reproduced with permission of Pergamon Press.]

frequencies. Thus, AM radio signals are highly absorbed during the day but are almost unattenuated at night. FM radio and TV are much higher frequency and almost entirely pass through the whole system into space. The ionization rates of various high-energy photons and particles are plotted in Fig. 1.9. These rates correspond to the P_j in (2.7). Table 1.1 provides some of the important reactions.

Ignoring transport, three continuity equations must be considered, all of the form,

$$\frac{\partial n_j}{\partial t} = P_j - L_j$$

where j is electrons, negative ions, or positive ions. At the time these coupled equations were first solved, reaction 2 in Table 1.1 was not well understood. But to create the ledge visible in all the rocket profiles in Figs. 1.7a and 1.7b, Reid (1970) needed the high effective rate of recombination plotted in Fig. 1.10. He concluded that such a high rate required the existence of water cluster ions in the 75–85 km height range, which were found later by Narcisi et al. (1972), as illustrated in Figure 1.8. Notice that the rate for reaction 2 is nearly 100 times that of reaction 1. A strong variation of α_d with altitude shown in Fig. 1.10 is caused by the change in the cluster-ion composition. Below 75 km, reaction 3 becomes important. The incoherent scatter radar at Arecibo has been used in the D region and verified most of the rocket results, albeit with less spatial resolution (Ioannidis and Farley, 1974).

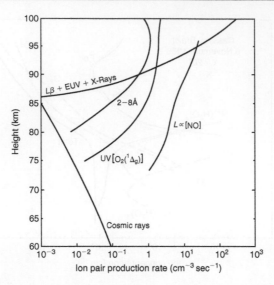

Figure 1.9 Ionization rates in the quiet daytime D region for solar minimum conditions and solar zenith angles near 60°. [After Thomas (1974). Reproduced with permission of the American Geophysical Union.]

Table 1.1 Important Ion Reactions in the D Region

	Reaction Type	Name	Rate
1	$xy^+ + e \to x + y$	Dissociative recombination	$\alpha_D^* = 4 \times 10^{-7} \text{cm}^3\text{s}^{-1}$
2	$H(H_2O)_n^+ + e \to$		
	$H(H_2O)_{n-1} + H_2O$	Cluster-ion recombination	$\alpha_d = 5 \times 10^{-5} \text{cm}^3\text{s}^{-1}$
3	$xy^+ + z^- \to xy + z$	Ion-ion mutual neutralization	$\alpha_i = 10^{-6} \text{cm}^3\text{s}^{-1}$
4	$z + z + e \to z^- + z$	Three-body attachment	$\alpha = 10^{-30} \text{cm}^6\text{s}^{-1}$
5	$z^- + h\nu \to z + e$	Photo detachment	$d = 0.4 \text{s}^{-1}$
6	$z^- + m \to mz + e$	Associative detachment	$f = 10^{-10} \text{cm}^3\text{s}^{-1}$

The simplest version of a D region model (Banks and Kockarts, 1973), which contains most of the chemistry, is thus

$$\frac{dn^+}{dt} = In - \alpha_d n_e n^+ - \alpha_i n^+ n^-$$

$$\frac{dn^-}{dt} = \alpha n^2 n_e - n^-(d + f n_m + \alpha_i n^+)$$

$$\frac{dn_e}{dt} = In - \alpha_d n_e n^+ - \alpha n^2 n_e + n^-(d + f n_m)$$

where I is the photoionization rate (mostly Lyman α on [NO]), n is the NO density, n^+ is the dominant positive ion density, n^- is the dominant negative ion

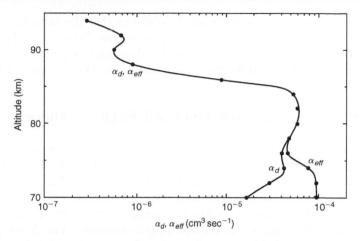

Figure 1.10 The inferred recombination coefficient α_d and the effective recombination coefficient α_{eff}. [After Reid (1970). Reproduced with permission of the American Geophysical Union.]

density, the α terms are defined in Table 1.1, and n_m is the chemically dominant minor constituent density (usually atomic oxygen). These equations are often solved after introduction of the parameter λ, where

$$\lambda = n^-/n_e$$

and

$$n^+ = n^- + n_e = (1 + \lambda)n_e$$

Following Banks and Kockarts, the differential equation for the electron density resulting from combining these equations is

$$\frac{dn_e}{dt} = \frac{In}{1 + \lambda} - (\alpha_d + \lambda\alpha_i)n_e^2 - \frac{n_e}{1 + \lambda}\frac{d\lambda}{dt}$$

Under the stationary condition ($dn_e/dt = 0$ and $d\lambda/dt = 0$), this equation reduces to the formula

$$In = (1 + \lambda)(\alpha_d + \lambda\alpha_i)n_e^2 = \alpha_{eff}n_e^2$$

or

$$n_e^2 = In/\alpha_{eff}$$

When the concentration of the negative ions is small (λ is small), the loss of charged particles in the D region is controlled by recombination of the positive ions with electrons ($\alpha_{eff} = \alpha_d$). Figure 1.10 shows that the influence of negative ions on recombination becomes significant below 75 km. Not shown, and only recently appreciated, are charged smoke, dust, and ice particles that create interesting dusty and icy plasma physics behavior in the D region (see Chapter 7).

1.4   The Earth's Magnetic Field and Magnetosphere

To first order the earth's magnetic field is that of a dipole whose axis is tilted with respect to the spin axis of the earth by about 11°. This offset, which is common to several planetary magnetic fields, is presently such that the dipole axis in the northern hemisphere is tilted toward the North American continent. The magnetic field **B** points down toward the surface of the earth in the Northern hemisphere and away from it in the Southern hemisphere. The dipole position wanders with time, and paleomagnetic studies show that it flips over with irregular time differences. The field is created by currents in the molten, electrically conducting core of the earth, currents that are in turn driven by thermal convection in the core. This convection is certainly quite complex, but the magnetic field contributions that are of higher order than the dipole term fall off faster with distance, leaving the dipole term dominant at the surface.

Some useful equations for a dipole field are gathered in Appendix B. The field at the earth's surface varies from about 0.25×10^{-4} tesla (0.25 gauss) near the magnetic equator to about 0.6×10^{-4} tesla near the poles. Sketches of magnetic field lines are useful when considering the ionosphere and magnetosphere, since plasma particles move very freely along field lines. A dipole field is sketched in Fig. 1.11. The equation for the magnetic field lines can be written in a modified spherical coordinate system as

$$r = L \cos^2 \theta$$

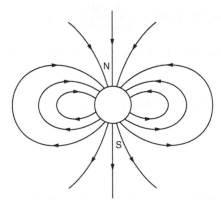

Figure 1.11   Sketch of dipole magnetic field lines extending into a vacuum.

where r is measured in units of earth radii ($R_e = 6371$ km), θ is the magnetic latitude, and L is the radius of the equatorial crossing point of the field line (also measured in earth radii). A field line that crosses the equatorial plane at $L = 4$ thus exits the earth's surface ($r = 1$) at a magnetic latitude of $60°$. Owing to the dipole tilt, this field line is located at about $49°$ geographic latitude in the North American sector and about $71°$ geographic latitude in the Euro-Asian region. As indicated in Fig. 1.11, if the earth were surrounded by a vacuum with no external sources of electrical currents, the dipole field lines would extend in loops of ever increasing dimension with the magnitude of **B** decreasing as $1/r^3$.

The earth is, however, immersed in the atmosphere of the sun. Like the earth's thermosphere, the upper atmosphere of the sun is very hot, so hot that the hydrogen and helium can escape gravitational attraction and form a steadily streaming outflow of material called the solar wind. Because of its high temperature and constant illumination by the sun, the solar wind is a fully ionized plasma (unlike the ionosphere, which contains more neutral particles than plasma). One surprising feature of the solar wind is that it is supersonic. Simply heating a gas cannot create supersonic flow, but a combination of heating, compression, and subsequent expansion can create this condition, as it does, for example, in a rocket exhaust nozzle. In the solar wind case, the solar gravitational field acts analogously to the rocket nozzle, and the solar wind becomes supersonic above a few solar radii. The sun has a very complex surface magnetic field created by convective flow of the electrically conducting solar material. Sunspots, in particular, have associated high magnetic fields. The expanding wind drags the solar magnetic field outward, with the result that the earth's magnetic field is continually bathed in a hot, magnetized, supersonic, collisionless plasma capable of conducting electrical current and carrying a large amount of kinetic and electrical energy.

The solar wind is a magnetohydrodynamic (MHD) generator, and some of the solar wind energy flowing by the earth finds its way into the ionosphere and the upper atmosphere. There it powers the aurora, creates magnetic storms and substorms that affect power grids and communication systems, heats the polar upper atmosphere, drives large neutral atmospheric winds, energizes much of the plasma on the earth's magnetic field lines, and creates a vast circulating system of hot plasma in and around the earth's nearby space environment. We cannot come close to exploring these topics in detail, but we need a framework about which to organize the ionospheric effects of these phenomena.

Suppose we first ignore the supersonic nature of the solar wind as well as the interplanetary magnetic field (IMF) and surround the earth and its magnetic field with a subsonic, streaming plasma. The magnetic force on a particle of charge q moving at velocity **V** is

$$\mathbf{F} = q\mathbf{V} \times \mathbf{B} \tag{1.1}$$

As illustrated in Fig. 1.12, and because of the polarity of the earth's field, this force will deflect solar wind ions to the right and electrons to the left as they

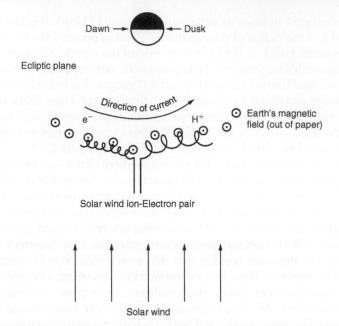

Figure 1.12   Schematic diagram showing deflection of solar wind particles by the earth's magnetic field. The view is in the ecliptic plane.

approach the earth. Once deflected, they will spiral around the magnetic field lines and drift around the earth. Since the magnetic field strength increases as the plasma approaches the earth, a distance is eventually reached where the force is sufficient to keep the particles from penetrating any farther, and they flow around the obstacle. Notice that since ions are deflected toward dusk and electrons toward dawn, a net duskward current exists in a thin sheet (extending also out of the plane of the diagram) where the force balance occurs. The secondary magnetic field generated by this current sheet is parallel to the earth's field in the region between the earth and the current sheet and is antiparallel to the earth's magnetic field in the solar wind. This magnetic field cancels the earth's field on the sunward side of the boundary and increases the value of the magnetic field inside the current sheet. If this were the entire interaction, the resulting configuration of flow and field would look something like the sketch in Fig. 1.13a. The volume inside the elongated region is termed the magnetosphere—the region dominated by the earth's magnetic field. Experiments show that the magnetosphere is indeed terminated by a very sharp boundary on the sunward side, which is called the magnetopause, but is very much more elongated in the antisunward direction than the sketch shows. (We return to this point later when we include the interplanetary magnetic field in the discussion.) A configuration such as that in Fig. 1.13a is called a closed magnetosphere, since there is almost no direct

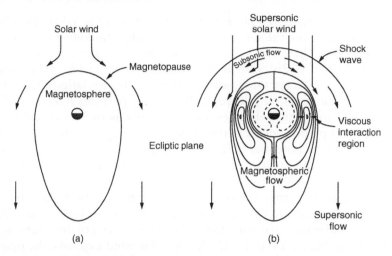

Figure 1.13 (a) Schematic diagram of a closed magnetosphere. The bow shock wave and internal flow pattern are sketched in (b).

access to particles from the solar wind and all of the magnetic field lines in the magnetosphere have two ends on the earth. The sketch in Fig. 1.13b shows the complication that arises when the solar wind is supersonic. A shock wave must form, across which the solar wind density and temperature rise abruptly. The solar wind velocity decreases, allowing for subsonic flow around the obstacle. The sketch in Fig. 1.13b also shows that a "closed" magnetosphere need not be static even though particles cannot readily enter. A "viscous" interaction may occur (Axford and Hines, 1961) in which plasma inside and near the flanks of the magnetosphere is weakly coupled to the solar wind flow, possibly by waves that propagate across the boundary. Some magnetospheric plasma is then dragged along in the antisunward direction, and because of the resulting pressure buildup in the nightside of the magnetosphere, there must be a flow back toward the sun in the center of the magnetosphere. The circulation that results (sketched in Fig. 1.13b) is in good qualitative agreement with observations but is now thought to be less important than the magnetospheric flow driven by a process termed "reconnection," which involves the interplanetary magnetic field and results in a partially open magnetosphere. This process is described next.

Before discussing reconnection, we must point out that both the single-particle dynamic and magnetofluid dynamic viewpoints discussed in Chapter 2 are such that the relationships

$$\mathbf{V}_\perp = \mathbf{E} \times \mathbf{B}/B^2 \tag{1.2a}$$

and

$$\mathbf{E}_\perp = -\mathbf{V} \times \mathbf{B} \tag{1.2b}$$

hold in a tenuous, magnetized plasma, where E is the electric field in the plasma, V is its flow velocity, and the subscript $\perp$ indicates the component perpendicular to B (Chen, 1984). These two expressions are, in fact, equivalent, since taking the cross product of (1.2b) with B and dividing by B^2 yields (1.2a). For our present purposes we accept these expressions as accurate relationships between $E_\perp$ and $V_\perp$. In the direction parallel to B, charged particles move very freely. This means that magnetic field lines usually act like perfect electrical conductors, transmitting perpendicular electric fields and voltages across vast distances with no change in the potential in the direction parallel to B. Thus, any flowing magnetized plasma can act as a source of voltage if there is a component of V perpendicular to B.

These electrical properties become very important factors in the interaction between the solar wind and the magnetosphere, once we include the interplanetary magnetic field in our solar wind model. Schematic views of the solar magnetic field are given in Figs. 1.14 and 1.15. As the solar wind expands, the magnetic field is stretched into a disklike geometry that has flutes much like a ballerina's skirt. Consideration of the Maxwell equation

$$\nabla \times B = \mu_0 J$$

shows that a geometry with adjacent magnetic fields that are antiparallel must have a current sheet separating them (indicated by the crosshatched surface in Fig. 1.14). The entire pattern rotates with the 27-day rotation period of the sun. As the pattern rotates past the earth, the position of the current sheet is alternately above and below the ecliptic plane. As illustrated in Fig. 1.15, the interplanetary magnetic field (IMF) can be considered approximately as a series of spirals emanating from the sun. Near the earth the magnetic field in the spiral is directed toward or away from the sun depending on whether the current sheet is above or below the plane. In addition, the inclination of the current sheet with respect to the ecliptic plane produces a northward or southward component of the IMF relative to an axis normal to the plane. Considerable small-scale turbulence structure exists in both the solar wind velocity and magnetic fields, and the interplanetary medium is also greatly affected by shock waves and solar flares. Taken together, these properties of the solar wind result in a highly variable buffeting of the earth's magnetosphere.

It is customary to specify the solar wind parameters in terms of three mutually perpendicular components with respect to fixed axes. In these geocentric solar ecliptic (GSE) coordinates, the x-axis points from the center of the earth to the sun, while the z-axis is positive parallel to the earth's spin axis (to the north) and is perpendicular to the ecliptic plane. The y-axis makes the third mutually perpendicular right-handed system. Figure 1.15 shows the situation looking in the negative z direction down onto the ecliptic plane. The interplanetary magnetic field is stretched out by the radially flowing solar wind but remains anchored to the rotating sun. The resulting geometry is similar to that of the water stream

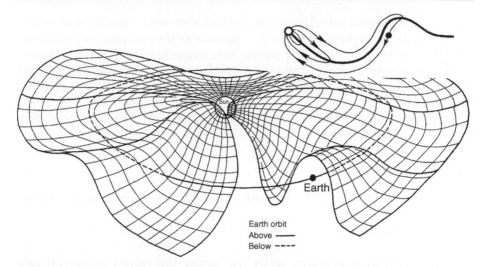

Figure 1.14   Three-dimensional sketch of the solar equatorial current sheet and associated magnetic field lines. The current sheet is shown as lying near the solar equator with
spiraled, outward-pointing magnetic fields lying above it and inward-pointing fields lying
below it. The current sheet contains folds or flutes. When the sun rotates, an observer
near the ecliptic will alternately lie above and below the current and will see a changing
sector pattern. The inset shows a meridional cross section with the earth below the current
sheet. (Figure courtesy of S.-I. Akasofu.)

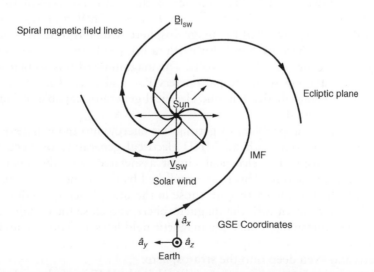

Figure 1.15   The spiral magnetic field viewed in the ecliptic plane as it is stretched out
from the solar surface by the solar wind. The illustrated coordinate system is the GSE
system. In this sketch there is one "away" sector and one "toward" sector in the magnetic
structure of the solar wind as it sweeps by the earth.

from a rotating garden sprinkler. At the earth's location this "garden hose angle" is such that the y component and the x component of the interplanetary magnetic field ($\mathbf{B}_{sw}$) are roughly equal and usually have opposite signs. The z component may be positive or negative, and the terms "B_z positive" or "northward" and "B_z negative" or "southward" are used to describe the orientation of the IMF with respect to the z-axis.

We now briefly describe the interplanetary electric field. To first order the solar wind blows radially outward from the sun, so that in the ecliptic plane

$$\mathbf{V}_{sw} = -|\mathbf{V}_{sw}|\mathbf{a}_x$$

In a reference frame fixed to the earth, there will be an electric field given by

$$\mathbf{E}_{sw} = -\mathbf{V}_{sw} \times \mathbf{B}_{sw}$$

This will generate a potential difference across the earth's magnetosphere given by

$$\mathbf{V}_m = -\mathbf{E}_{sw} \cdot L\mathbf{a}_y$$

where L is the effective size of the magnetosphere perpendicular to $\mathbf{V}_{sw}$. Note that when $\mathbf{B}_{sw}$ has a positive z component, $\mathbf{E}_{sw}$ has a dusk-to-dawn direction (negative y component). When $\mathbf{B}_{sw}$ has a negative z component, the interplanetary electric field is in the dawn-to-dusk direction across the earth. For typical values of $V_{sw} = 500$ km/s and $B_{sw_z} = 5 \times 10^{-9}$ tesla, $E_{sw_y} = 2.5$ mV/m. Integrating this electric field across the front of the magnetosphere, which is roughly a distance $L = 20\,R_e$ across (where R_e stands for earth radii), yields an available potential difference of the order of 300,000 volts. The magnitude of this voltage and its polarity fluctuate along with the parameters V_{sw} and B_{sw_z}. The earth is thus immersed in a magnetohydrodynamic electrical generator capable of hundreds of kilovolts of potential.

We now turn to the problem of getting this energy into the magnetosphere. Continuing our analogy with an MHD electrical generator, we need to tap this power by closing the electrical circuit associated with the solar wind-magnetosphere interaction. This is accomplished by the connection of magnetic field lines attached to the earth with those in the interplanetary medium. Current can then flow down into the magnetosphere and close the circuit through the conducting ionosphere. Since the magnetic field lines are equipotentials, the interplanetary electric field maps into the near-space region of the earth, into the ionosphere, and even deep into the stratosphere.

This "connection" can occur most easily when the IMF is southward, as illustrated in Fig. 1.16, where the cross-sectional view is now in the noon-midnight plane. This viewpoint was first suggested by Dungey (1961). The numbers

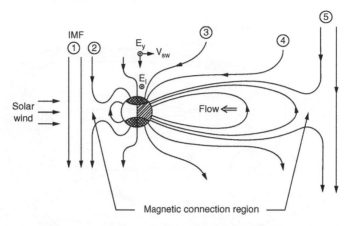

Figure 1.16 A "connected" and "reconnected" magnetic field topology for B_z south. The view is in the noon-midnight plane. $\mathbf{E}_y$ is the interplanetary electric field, and $\mathbf{E}_I$ is the electric field mapped into the ionosphere down the magnetic field lines.

correspond to sequential times as the interplanetary magnetic field connects to the earth's field and is then swept antisunward at the solar wind speed.

The Northern and Southern Hemispheric field lines are stretched into a magnetic tail and eventually reconnect with each other deep in the antisunward region. We spend considerable time in Chapter 8 discussing the electric field and associated $\mathbf{E} \times \mathbf{B}/B^2$ circulation patterns that arise in the magnetosphere for various solar wind configurations. In the "B_z south" case shown here, the interplanetary dawn-dusk electric field shown at the top of the sketch maps directly into the polar ionosphere along the connected magnetic field lines. There, the electric field $\mathbf{E}_I$ is "out of the paper" and the $\mathbf{E} \times \mathbf{B}$ drift is antisunward. Once reconnection occurs in the tail, the forces associated with the stretched magnetic field lines bring the plasma rapidly back toward the earth. This is called the region of sunward convection.

A cutaway view of the magnetosphere is given in Fig. 1.17, with the location of various regions and phenomena indicated. When the interplanetary magnetic field connects to the earth's magnetic field, it produces open field lines with only one foot on the earth. These field lines extend far into the magnetotail and form the northern and southern tail lobes. These open field lines also cross the magnetopause, where they connect to the interplanetary field in the magnetosheath. In so doing, they come into contact with plasma populations just inside the magnetopause, which are called the boundary layer and the mantle. The field lines that reach down the magnetotail have opposite directions on the northern and southern sides of the equatorial plane, and a current sheet must exist to separate the northern and southern tail lobes. This so-called neutral sheet where $|\mathbf{B}| \cong 0$ is also the region in which open field lines may reconnect to form closed

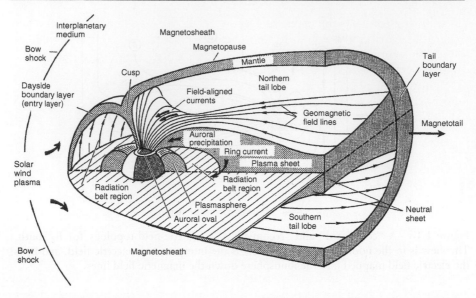

Figure 1.17   Schematic representation of the magnetosphere. (Courtesy of J. Roederer.)

field lines that convect back to the earth. As the hot plasma sheet flows toward the earth into an increasing magnetic field, the differential motion of the ions and electrons produces a ring current somewhat analogous to the interaction between the solar wind and the magnetosphere itself on the sunward side of the earth. Some of these hot plasma sheet particles move along the magnetic field lines and precipitate into the atmosphere in a ring around the polar regions called the auroral oval. This ring forms the light pattern seen in Fig. 1.6. Most of the solar wind and magnetospheric effects occur at high latitudes, although many interesting phenomena do extend to lower latitudes during active times. The active region is skewed in the antisunward direction, extending to 70° magnetic latitude on the sunward side and 60° on the nightside. In this text, some of the phenomena indicated in Fig. 1.17 will be treated in detail and others in a more sketchy fashion. The interested reader is referred to the text by Lyons and Williams (1984) and Kivelson and Russell (1995) for more complete treatises on magnetospheric processes.

Particle energization processes will not be discussed quantitatively at all in this text, but a brief sketch is in order here. Solar wind electrons have thermal and flow energies of only a few electron volts, while the bulk flow energy of the ions is about a kilovolt. The earth's bow shock wave, shown in Fig. 1.17, energizes the electrons to many tens or even hundreds of electron volts in the magnetosheath. Some of these hot electrons and ions directly penetrate into the ionosphere in the connection process or through the cusp in the magnetic field geometry. This precipitation of energetic particles contributes to the dayside part of the auroral oval. As the magnetic field lines reconnect in the tail and jet back toward the earth, the average electron energy is boosted to about a kilovolt in the region termed the plasma sheet. The plasma in this region provides a source of steady precipitation that corresponds visually to the widespread emissions called diffuse aurora. This precipitation also acts as an ionization source for the ionosphere. The hot flowing plasma has an associated electric field [see (1.2b)] that is impressed on the ionosphere. Electrical currents flow in response to this electric field. For nighttime conditions the precipitating plasma itself provides most of the conductivity, so an electrical feedback can develop between the plasma sheet and the ionosphere as currents flow back and forth between them.

This feedback is most complex in the region of the discrete aurora, which contains the very bright, active, auroral displays. Here, at some point along the magnetic field (usually above 3000 km), electric fields develop parallel to the magnetic field lines that accelerate electrons down into the atmosphere and ions out into the magnetosphere. Typical field-aligned potential drops associated with this parallel electric field range from 100 V to 10 kV. The electron density in the ionospheric E region can exceed 10^6cm^{-3} below a strong auroral display (see Fig. 1.5), and the horizontal electrical current in the ionosphere can create perturbations in the surface magnetic field of up to 3000 nanotesla—nearly 5% of the earth's surface field at high latitudes. Many other acceleration processes besides parallel electric fields operate in the magnetosphere and are beyond our scope to discuss in detail.

References

Axford, W. I., and Hines, C. O. (1961). A unifying theory of high-latitude geophysical phenomena and geomagnetic storms. *Can. J. Phys.* **39**, 1433.

Banks, P. M., and Kockarts, G. (1973). *Aeronomy*, Parts A and B. Academic Press, New York.

Baron, M. J. (1974). Electron densities within auroras and other auroral E-region characteristics. *Radio Sci.* **9**, 341.

Chen, F. F. (1984). *Introduction to Plasma Physics and Controlled Fusion*. 2nd ed., vol. 1: Plasma Physics. Plenum Press, New York.

Dungey, J. W. (1961). Interplanetary magnetic field and the aurorae zones. *Phys. Rev. Lett.* **6**, 47.

Iioannidis, G., and Farley, D. T. (1974). High resolution D-region measurements at Arecibo. *Radio Sci.* **9** (2), 151–157.

Istomin, V. G. (1966). Observational results on atmospheric ions in the region of the outer ionosphere. *Ann. Geophys.* **22**, 255.

Johnson, C. Y. (1969). Ion and neutral composition of the ionosphere. *Ann. IQSY* **5**.

Kelley, M. C., and Kintner, P. M. (1978). Evidence for two-dimensional inertial turbulence in a cosmic-scale low β–plasma. *Astrophys. J.* **220**, 339.

Kivelson, M. G., and Russell, C. T. (1995). *Introduction to Space Physics*. Cambridge University Press, Cambridge, UK.

Lyons, L. R., and Williams, D. J. (1984). *Quantitative Aspects of Magnetospheric Physics*. Reidel, Boston.

Mechtly, E. A., Bowhill, S. A., and Smith, L. G. (1972). Changes of lower ionosphere electron concentrations with solar activity. *J. Atmos. Terr. Phys.* **34**, 1899–1907.

Narcisi, R. S., Bailey, A. D., Wlodyka, L. E., and Philbrick, C. R. (1972). Ion composition measurements in the lower thermosphere during the November 1966 and March 1970 solar eclipses. *J. Atmos. Terr. Phys.* **34**, 647.

Reber, C. A., and Nicolet, M. (1965). Investigation of the major constituents of the April–May 1963 heterosphere by the Explorer XVII satellite. *Planet. Space Sci.* **13**, 617.

Reid, G. C. (1970). Production and loss of electrons in the quiet daytime D region of the ionosphere. *J. Geophys. Res.* **75**, 2551–2562.

Rishbeth, H., and Garriott, O. K. (1969). *Introduction to Ionospheric Physics*. Int. Geophys. Set., vol. 14. Academic Press, New York.

Robinson, R. M., Vondrak, R., Craven, J., Frank L., and Miller, K. (1989). A comparison of ionospheric conductances and auroral luminosities observed simultaneously with the Chatanika radar and the de-1 auroral imagers. *J. Geophys. Res.* **94**(A5), 5382–5396.

Schunk, R. W., and Nagy, A. (2000). *Ionospheres, Physics, Plasma Physics and Chemistry*. Cambridge University Press, Cambridge, UK.

Thomas, L. (1974). Recent developments and outstanding problems in the theory of the D region. *Radio Sci.* **9**, 121–136.

2 Fundamentals of Atmospheric, Ionospheric, and Magnetospheric Plasma Dynamics

In this chapter we model the ionospheric plasma as three interpenetrating fluids, with the electron and ion fluids immersed in the neutral gas. At all heights of interest in the study of ionospheric phenomena, the neutral gas density exceeds that of the plasma. In fact, the plasma density does not become comparable to that of the neutrals until several thousand kilometers in altitude. The primary difference between ionospheric plasma dynamics and thermospheric neutral gas dynamics is the effect of electromagnetic forces. The various forces acting on charged particles drive electric currents that in turn create electric fields that modify the plasma dynamics. The electrical conductivity of the medium is thus extremely important and is derived in this chapter. We discuss briefly the generation of electric fields in the ionosphere and the transmission of electric fields along magnetic field lines between the ionosphere and the magnetosphere. At middle and low latitudes, the electric field is generated primarily by the neutral wind field. In later chapters electric fields impressed on the ionosphere by solar wind and magnetospheric processes will be taken into account, as will their occasional penetration into the middle- and low-latitude sectors. In the analysis that follows, we first obtain the equations for a neutral fluid and then extend them to ionized gases. Finally, we develop the equations needed to describe collisionless plasmas in the absence of a neutral fluid, which is the appropriate approximation for the magnetosphere. Although not the primary topic of this text, such a development is necessary, since we must be able to describe certain key magnetospheric phenomena in some detail.

2.1 The Basic Fluid Equations

The ions, electrons, and neutrals can be considered as three interpenetrating fluids coupled by collisions and, in the case of ions and electrons, their self-generated electric and magnetic fields. The gas of ions and electrons, taken together, is often referred to as a plasma in the text. As usual in a fluid description, we assume that an element small enough to be treated as a differential volume in the mathematical sense still contains a sufficient number of atoms, ions, electrons, and/or

The Earth's Ionosphere: Plasma Physics and Electrodynamics

molecules that statistical techniques can be used to define such quantities as the temperature, density, and mean flow velocity per each species. Such a volume element is macroscopically small, even though microscopically it contains many particles colliding in a random fashion. We use this approach almost exclusively in the text. Schunk and Nagy (2000) discuss this and other approximations for tenuous fluid behaviors.

2.1.1 Conservation of Mass

Conservation of mass dictates that the flux of material into or out of a volume through its surface must be equal to the rate of increase or decrease of mass inside the volume. The mathematical statement in integral form relates the mass density ρ and the fluid velocity $\mathbf{U}$ through

$$\iiint_{v} \partial \rho / \partial t \, dV = - \iint_{\Sigma} \rho \mathbf{U} \cdot d\mathbf{a} \tag{2.1}$$

where the surface Σ encloses the volume V, and $d\mathbf{a}$ is normal to the surface and directed outward. The differential form of (2.1) can be derived from Gauss's theorem, which relates the surface integral on the right-hand side to a volume integral of the divergence of $\rho \mathbf{U}$—that is,

$$\iiint_{v} \partial \rho / \partial t \, dV = - \iint_{\Sigma} \rho \mathbf{U} \cdot d\mathbf{a} = - \iiint_{v} \nabla \cdot (\rho \mathbf{U}) \, dV$$

and thus,

$$\partial \rho / \partial t = -\nabla \cdot (\rho \mathbf{U}) \tag{2.2}$$

This constitutes the continuity equation for the neutral atmosphere with mass density ρ and velocity $\mathbf{U}$.

Expanding the divergence operator, (2.2) can also be written

$$\partial \rho / \partial t + \mathbf{U} \cdot \nabla \rho + \rho (\nabla \cdot \mathbf{U}) = 0 \tag{2.3}$$

or, equivalently,

$$d\rho / dt + \rho (\nabla \cdot \mathbf{U}) = 0 \tag{2.4}$$

where the total time derivative

$$d/dt = \partial / \partial t + \mathbf{U} \cdot \nabla \tag{2.5}$$

has been used. This operation yields the time rate of change of a quantity—ρ in this case—moving with the flow. Equation (2.4) states that the rate of change of ρ moving with the flow is determined only by the divergence of the velocity field. For an incompressible fluid the mass density of a parcel cannot change as it moves $(d\rho/dt = 0)$, and so from (2.4) it follows that $\nabla \cdot \mathbf{U} = 0$. In an incompressible fluid the velocity field is divergence free. The advective derivative $(\mathbf{U} \cdot \nabla)$ in (2.5) is important, since it describes the temporal variation of a quantity, at a point in space, due to the transport of that quantity into the region. For example, in the case of incompressible flow, the mass density can only change with time at a fixed point in space via this term, since

$$\partial\rho/\partial t = -(\mathbf{U} \cdot \nabla\rho) \tag{2.6}$$

In the case of a partially ionized medium, ion and electron pairs may be produced by impact of a photon or energetic particle or lost through recombination between positively and negatively charged particles. These processes are very important for the ionospheric plasma. If P_j denotes the rate of production of ions (and electrons) per cubic meter per second and L_j is the rate of loss, then the mass conservation equation for each of the ionized species is

$$\partial\rho_j/\partial t + \nabla \cdot \left(\rho_j\mathbf{V}_j\right) = \left(P_j - L_j\right)M_j \tag{2.7}$$

where M_j is the mass of each species. Notice that we use the notation $\mathbf{V}_j$ to represent the velocity of the charged species, reserving $\mathbf{U}$ for the velocity of the neutral gas. Since electric charge is a conserved quantity, it must be the case that the total number of electrons (e) gained or lost equals the sum of all the different types of ions gained or lost—that is,

$$\sum_{j=1}^{N} \left(P_j - L_j\right) = P_e - L_e$$

We ignore negative ions because their formation is unimportant above about 80 km altitude. Because the number density of the neutral gas far exceeds that of the ion and electron gas for heights below several thousand kilometers, we can ignore the loss of neutral particles when ion-electron pairs are formed. Furthermore, if we are not interested in the possible neutral mass density changes due to composition changes of the neutrals (e.g., formation of atomic oxygen by photodissociation), we may use (2.2) for the neutral atmosphere, while for the ionized particles we need (2.7).

The most important photoionization sources are illustrated in Figs. 2.1a and 2.1b in two different styles. In Fig. 2.1a the penetration depth of solar radiation is plotted versus wavelength. Except for absorption bands and some scattering, visible light reaches the ground, but shorter wavelengths are absorbed by the

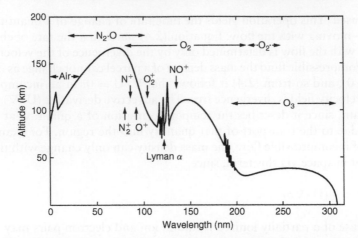

Figure 2.1a   Depth of penetration of solar radiation as a function of wavelength. Altitudes correspond to an attenuation of $1/e$. The principal absorbers and ionization limits are indicated. [After Smith and Gottlieb (1974). Reprinted with permission of Kluwer Academic Publishers.]

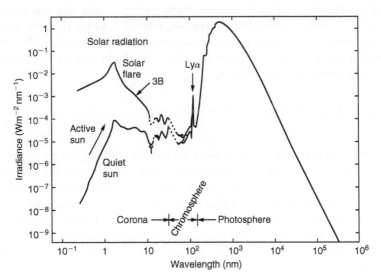

Figure 2.1b   Spectral distribution of solar irradiance and its variation with solar activity. The logarithmic representation emphasizes the contribution of x-rays and extreme ultraviolet radiation. [After Smith and Gottlieb (1974). Reprinted with permission of Kluwer Academic Publishers.]

indicated species. Figure 2.1b shows the variability in the solar photon emission at short wavelengths for quiet and active solar conditions, as well as for solar flare events. EUV radiation is absorbed at high altitudes, both heating and ionizing the E and F regions. X-rays penetrated quite deeply and, along with Lyman α (ionizing NO), control the D region ion densities. UV is absorbed by ozone, creating the temperature rise in the stratosphere.

2.1.2 Equation of State

For an ideal gas, the mass density and pressure, p_j, are related by

$$p_j = \rho_j k_B T_j / M_j = n_j k_B T_j \tag{2.8}$$

This is the equation of state for each of the fluids we consider (ions, electrons, neutrals), and we relate the mass density ρ_j to the number density n_j through $\rho_j = n_j M_j$. In this text we use k_B to represent Boltzmann's constant. The letter k used alone or with any other subscript represents a wave number.

2.1.3 Momentum Equation for the Neutral Fluid

The continuity equations and the equation of state must be supplemented by a dynamical equation relating the fluid velocity to the forces acting on the fluid. This is derived from the principle of conservation of momentum, which requires that the change of momentum per unit time within a volume be equal to the pressure gradient force and the total external force field F acting on the material inside the volume plus the momentum flux carried across the surface bounding the volume by viscosity, advection, or wave flux. We first treat the neutral gas. Here, as in most ionospheric and upper atmospheric studies, the wind direction is indicated by where the wind is going—in other words, an eastward wind is a wind toward the east. (In meteorology an easterly wind comes *from* the east.)

Advection of a vector quantity such as momentum across a boundary by flow is conceptually no more difficult than previously discussed for the scalar mass density. However, the mathematical description is more complex and is most readily accomplished by the use of tensor notation as employed here. The equation equivalent to (2.1) for the time rate of momentum change in a volume is

$$\iiint\limits_V \frac{\partial}{\partial t}(\rho \mathbf{U}) dV = \iiint\limits_V (-\nabla p)\, dV + \iiint\limits_V \mathbf{F} dV$$

$$- \iint\limits_\Sigma \pi_\mathrm{m} \cdot d\mathbf{a} - \iint\limits_\Sigma \pi_\mathrm{w} \cdot d\mathbf{a} \tag{2.9}$$

where $\mathbf{F}$ is the external force, p the pressure, π_m the momentum flux density tensor due to material motions, and π_w the momentum flux density tensor due to waves in the medium. Applying Gauss's theorem again, (2.9) becomes

$$\partial(\rho \mathbf{U})/\partial t = -\nabla p + \mathbf{F} - \nabla \cdot \pi_m - \nabla \cdot \pi_w \tag{2.10}$$

The pressure gradient term should be familiar. The external force $\mathbf{F}$ can be of many different types, which will be treated as they arise in the text.

To understand the material momentum tensor, consider a single particle of mass m moving at velocity $\mathbf{v}$. The momentum, $m\mathbf{v}$, is carried along by the velocity $\mathbf{v}$, resulting in a momentum flux tensor given by

$$\pi_m = \mathbf{v} m \mathbf{v}$$

Written as a 3×3 matrix, this becomes

$$(\pi_m)_{jk} = m v_j v_k$$

An analogous form for a fluid of mass density ρ characterized by a mean flow velocity $\mathbf{U}$ is given by

$$(\pi_m)_{jk} = \rho U_j U_k \tag{2.11}$$

This part of the momentum flux tensor describes how momentum is transferred within a fluid by the motion of the fluid. If there is any net divergence of this momentum flux, a net force will occur as indicated in (2.10). The divergence of a tensor is a vector and may be written

$$(\nabla \cdot \pi)_j = \sum_k \frac{\partial \pi_{jk}}{\partial x_k}$$

where x_k is the kth Cartesian coordinate. Applying this to the tensor in (2.11),

$$(\nabla \cdot \pi_m)_j = \sum_k \frac{\partial}{\partial x_k}(\rho U_j U_k)$$

$$= \sum_k \left[U_k \frac{\partial}{\partial x_k}(\rho U_j) + \rho U_j \frac{\partial}{\partial x_k}(U_k) \right]$$

which in vector form is

$$\nabla \cdot \pi_m = \mathbf{U} \cdot \nabla(\rho \mathbf{U}) + \rho \mathbf{U}(\nabla \cdot \mathbf{U})$$

The first term is the advective derivative of the momentum and can be combined with the partial derivative to form the total time derivative. In this text we follow the notation $(u, v, w) = $ (east,north,up) for the neutral wind.

We are not quite finished, since the term given in (2.11) is not the only contribution to the momentum flux tensor. Consider the situation illustrated in Fig. 2.2, where the x component of the mean fluid velocity increases in the z direction. Particles randomly crossing the plane $z = z_0$ from above that collide with particles below will, on average, contribute more x momentum to the fluid below $z = z_0$ than particles crossing in the other direction will contribute above. This momentum transfer results in forces on the fluid such that the velocity gradient in this direction is reduced. Unlike the momentum flux described by (2.11), this "viscous" force depends on collisions and is a dissipative process. Energy in the mean flow is converted into heat when viscosity is important. The exact form of the viscous momentum flux density tensor is quite complicated, and we refer the reader to a discussion by Landau and Lifshitz (1959) for details. In its simplest form, the momentum flux density depends on the first derivatives of the velocity with two associated proportionality constants η and η'. When the divergence of the momentum flux tensor is taken, as required by (2.10), the viscous force F_v is given by

$$\mathbf{F}_v = \eta \nabla^2 \mathbf{U} + \eta' \nabla(\nabla \cdot \mathbf{U}) \tag{2.12}$$

For incompressible flow $(\nabla \cdot \mathbf{U}) = 0$ and (2.12) reduces to the more familiar form

$$\mathbf{F}_v = \eta \nabla^2 \mathbf{U} \tag{2.13}$$

where η is called the dynamic viscosity coefficient. Equation (2.10) becomes

$$\partial(\rho \mathbf{U})/\partial t + \mathbf{U} \cdot \nabla(\rho \mathbf{U}) + \rho \mathbf{U}(\nabla \cdot \mathbf{U}) = - \nabla p + \mathbf{F}$$
$$- \nabla \cdot \pi_w + \text{viscous term} \tag{2.14}$$

Using the continuity equation, the left-hand side is just $\rho d\mathbf{U}/dt$. Finally, using (2.13)

$$\rho d\mathbf{U}/dt = -\nabla p + \mathbf{F} - \nabla \cdot \pi_w + \eta \nabla^2 \mathbf{U} \tag{2.15}$$

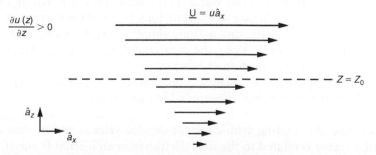

Figure 2.2   If the wind increases with increasing z, the viscous force will tend to reduce the velocity for $z > z_0$ and to increase it for $z < z_0$.

where as before d/dt is the total time derivative. Notice that the Laplacian operator in (2.15) acts on each velocity component. That is, the x component of the viscous force is given by

$$(\mathbf{F}_\nu)_x = \eta \frac{\partial^2 u}{\partial x^2}$$

For large-scale flow patterns $\nabla \cdot \mathbf{U} = 0$ is a good approximation, and (2.13) is used almost exclusively for viscous forces. We must note, however, that in the lower atmosphere, the coefficient η is determined not by molecular collisions but by the interaction between eddies in the flow. In other words, the effective viscosity coefficient is much larger than the molecular coefficient calculated from kinetic theory. Above the turbopause (near 100 km), where turbulent mixing ceases, the classical molecular viscosity coefficient is appropriate.

We now consider the forces that can contribute to the external force $\mathbf{F}$ in (2.15). The body force $\mathbf{F}$ includes the important gravitational term $\rho\mathbf{g}$. Another important body force for ionized species is due to electromagnetic effects. As is well known, when a current $\mathbf{I}$ passes through a wire in a magnetic field, the electromagnetic $\mathbf{I} \times \mathbf{B}$ force is transferred from the charged particles to the material in the wire. In continuous media the force density equivalent to this force is given by $\mathbf{J} \times \mathbf{B}$,

$$\mathbf{F}_{\text{EM}} = \mathbf{J} \times \mathbf{B} \tag{2.16}$$

This force density transfers electromagnetic energy to kinetic energy, accelerating or decelerating the fluid.

The upward flux of momentum π_w due to waves can be very important in the upper atmosphere. Waves from the dense lower regions tend to grow in amplitude as they propagate upward, since the mass density decreases exponentially with altitude. If the waves are absorbed at any given height, then $(\nabla \cdot \pi_w) \neq 0$ and the local atmosphere will be accelerated. Ocean swimmers often experience a phenomenon of this type when water waves break on the beach at an angle that is not exactly perpendicular to the shoreline. Some of the wave momentum is absorbed in the surf region, and a "long-shore" current results. This current transports sand (and people) along the coast and is an important factor in the formation, structure, and location of beaches.

This would complete the momentum equation for the neutrals except for the fact that the earth is rotating. Newton's laws as expressed in (2.15) refer to a frame of reference at rest or one moving with constant linear velocity. In a reference frame (R) rotating with constant angular velocity Ω, the time rate of change of a vector is related to the time derivative in an inertial frame (I) by

$$(d\mathbf{A}/dt)_{\text{I}} = (d\mathbf{A}/dt)_{\text{R}} + \Omega \times \mathbf{A} \tag{2.17}$$

An excellent discussion of this result is given by Goldstein (1950). In the rotating frame where we live and take measurements, the time derivative of the velocity in (2.15) must thus be replaced by

$$(d\mathbf{U}_I/dt)_I = (d\mathbf{U}_I/dt)_R + \Omega \times \mathbf{U}_I \tag{2.18}$$

We have included a subscript on the velocity vector to show specifically that even though we now have the proper time derivative in the rotating frame, the vector $\mathbf{U}_I$ is still the velocity in inertial space. Since the goal is to describe dynamics totally in the rotating coordinates, $\mathbf{U}_I$ must be expressed in that frame also. Viewed from inertial space, an object moving on the earth's surface with velocity $\mathbf{U}_R$ has an additional velocity $\Omega \times \mathbf{r}$, where $\mathbf{r}$ is the position vector drawn from the center of the earth—that is,

$$\mathbf{U}_I = \mathbf{U}_R + \Omega \times \mathbf{r} \tag{2.19}$$

Since Ω is constant, the time derivative of (2.19) in the inertial frame is

$$(d\mathbf{U}_I/dt)_I = (d\mathbf{U}_R/dt)_I + \Omega \times (d\mathbf{r}/dt)_I$$

Now each of the derivatives on the right-hand side must be replaced by the operation (2.18) in order to have the expression given entirely by quantities measured in the rotating coordinates, and, thus,

$$(d\mathbf{U}_I/dt)_I = (d\mathbf{U}_R/dt)_R + \Omega \times \mathbf{U}_R + \Omega \times \left[(d\mathbf{r}/dt)_R + \Omega \times \mathbf{r}\right]$$

For a parcel of fluid moving across the surface, $(d\mathbf{r}/dt)_R = \mathbf{U}_R$, so we have

$$(d\mathbf{U}_I/dt)_I = (d\mathbf{U}_R/dt)_R + 2\Omega \times \mathbf{U}_R + \Omega \times (\Omega \times \mathbf{r})$$

The second term is known as the Coriolis force. The last term is equal to $r\Omega^2 \cos\theta$, where θ is the latitude and has components both radially inward and equatorward. This term may be combined with $\mathbf{g}$ to describe an effective gravitational field. We will use the symbol $\mathbf{g}$ for these combined terms and move the Coriolis term to the right-hand side of (2.15), which yields the following equation of motion of the neutral atmosphere in a rotating frame:

$$\rho d\mathbf{U}/dt = - \nabla p + \rho\mathbf{g} + \eta\nabla^2\mathbf{U} - \nabla \cdot \pi_w$$
$$- 2\rho(\Omega \times \mathbf{U}) + \mathbf{J} \times \mathbf{B} \tag{2.20}$$

2.1.4 Momentum Equations for the Plasma

In Section 2.1.3 the only effect of a coexisting plasma embedded in the neutral atmosphere was the $\mathbf{J} \times \mathbf{B}$ force transferred to the neutrals. We now explore the fluid equations for the plasma species.

Each species will have its own momentum equation. Although neutral fluid dynamics on a rotating object such as the earth is strongly affected by the Coriolis force, this velocity-dependent force has little importance in geophysical plasma analysis, since the magnetic force (which is also velocity dependent) is much greater. The important body forces that do act on the ionospheric plasma and that we include in the force **F** are as follows:

$$\text{Gravitational:} \quad \rho_j \mathbf{g}$$
$$\text{Electric:} \quad n_j q_j \mathbf{E}$$
$$\text{Magnetic:} \quad n_j q_j (\mathbf{V}_j \times \mathbf{B})$$

where q_j is the charge of the jth species and **E** and **B** are the electric and magnetic fields. In addition, a frictional force is exerted on each species by collisions with all of the other species. For example, electrons will collide with neutrals as well as with the various ions. The force is proportional to the respective collision frequency and to the differential velocity between the particular fluid and the other fluids. The frictional force on each species may be written

$$\mathbf{F}_j = - \sum_{\substack{k \\ j \neq k}} \rho_j v_{jk} (\mathbf{V}_j - \mathbf{V}_k)$$

The v_{jk} are momentum transfer collision frequencies with units of s^{-1}. We leave any detailed discussion of the v_{jk} coefficients for other texts such as Banks and Kockarts (1973) and Schunk and Nagy (2000). The momentum equation we shall primarily use for each ionized species is then

$$\rho_j \frac{d\mathbf{V}_j}{dt} = -\nabla p_j + \rho_j \mathbf{g} + n_j q_j (\mathbf{E} + \mathbf{V}_j \times \mathbf{B}) - \sum_{\substack{k \\ j \neq k}} \rho_j v_{jk} (\mathbf{V}_j - \mathbf{V}_k)$$

Viscosity and momentum transfer by waves are ignored in this equation and will be discussed only briefly in the text where appropriate.

2.1.5 The Complete Equation Sets

The equations we have thus far obtained for the fluids making up the ionosphere may be summarized as follows. The neutral atmospheric equations of continuity, momentum, and state are

$$\partial \rho / \partial t = -\nabla \cdot (\rho \mathbf{U}) \tag{2.21a}$$

$$\rho d\mathbf{U}/dt = -\nabla p + \rho \mathbf{g} + \eta \nabla^2 \mathbf{U} - \nabla \cdot \pi_w$$
$$- 2\rho \mathbf{\Omega} \times \mathbf{U} + \mathbf{J} \times \mathbf{B} \tag{2.21b}$$
$$p = \rho k_B T_n / m_n = n_n k_B T_n \tag{2.21c}$$

where the subscript "n" stands for neutrals.

The corresponding equations for the ionized species are

$$\partial \rho_j / \partial t + \nabla \cdot (\rho_j \mathbf{V}_j) = (P_j - L_j) M_j \tag{2.22a}$$

$$\rho_j \, d\mathbf{V}_j / dt = -\nabla p_j + \rho_j \mathbf{g} + n_j q_j (\mathbf{E} + \mathbf{V}_j \times \mathbf{B})$$

$$- \sum_{\substack{k \\ j \neq k}} \rho_j \nu_{jk} (\mathbf{V}_j - \mathbf{V}_k) \tag{2.22b}$$

$$p_j = \rho_j k_B T_j / M_j = n_j k_B T_j \tag{2.22c}$$

Owing to the complexity of the equation sets (to which Maxwell's equations must yet be added), we will not yet consider the heat equations. This is equivalent to treating temperature profiles as given quantities. In this book we do not attempt to include a thermal analysis. (We refer the reader to the text by Schunk and Nagy, 2002.) This simplification would be very poor if our main interest was thermospheric neutral gas dynamics or the topside ionosphere. However, many interesting phenomena may be studied without including temperature changes self-consistently.

To treat the electric and magnetic fields, Maxwell's electrodynamic equations are needed, which, in their full form, are given by

$$\nabla \times \mathbf{E} = -\partial \mathbf{B} / \partial t \tag{2.23a}$$

$$\nabla \cdot \mathbf{E} = \rho_c / \varepsilon_0 \tag{2.23b}$$

$$\nabla \times \mathbf{B} = \mu_0 \left(\mathbf{J} + \varepsilon_0 \partial \mathbf{E} / \partial t \right) \tag{2.23c}$$

$$\nabla \cdot \mathbf{B} = 0 \tag{2.23d}$$

where ρ_c is the charge density ($\rho_c = \Sigma_j n_j q_j$) and $\mathbf{J}$ is the current density ($\mathbf{J} = \Sigma_j n_j q_j \mathbf{V}_j$). To this must be added the principle of conservation of charge

$$\frac{\partial}{\partial t} \iiint_v \rho_c dV = - \iint_S \mathbf{J} \cdot d\mathbf{a}$$

This equation states that the buildup or decay of charge inside a volume V is determined by the net electric conduction current across the surface Σ. Note the similarity to (2.1) for the conservation of mass. In differential form this may be written

$$\nabla \cdot \mathbf{J} = -\partial \rho_c / \partial t \tag{2.24}$$

Neither production nor recombination affects this equation because the net charge does not change in either process. In the ionosphere the conduction current is larger than the vacuum displacement current $\varepsilon_0 \partial \mathbf{E} / \partial t$ in (2.23c) for all wave frequencies of interest here, and the displacement current is there-fore dropped. Furthermore, since the largest changes in the magnetic field at ionospheric heights are such that $\delta B/B \leq 5\%$, we take (2.23c) as a diagnostic equation only. That is, given some current $\mathbf{J}$, we can find $\delta \mathbf{B}$, the perturbation to the geomagnetic field, from (2.23c), but the $\mathbf{B}$ we use in the rest of the dynamical equations is the earth's field, so we can take $\partial \mathbf{B}/\partial t = 0$. This approximation thus rules out induced electric fields from magnetic field changes. Of course, when we include electromagnetic waves such as Alfvén waves, we include both electric and magnetic field fluctuations. We are thus left with the set

$$\nabla \times \mathbf{E} = 0 \tag{2.25a}$$

$$\nabla \cdot \mathbf{E} = \rho_c/\varepsilon_0 \tag{2.25b}$$

$$\nabla \times \mathbf{B} = \mu_0 \mathbf{J} \tag{2.25c}$$

$$\nabla \cdot \mathbf{B} = 0 \tag{2.25d}$$

$$\nabla \cdot \mathbf{J} = -\partial \rho_c/\partial t \tag{2.25e}$$

Equation (2.25a) shows that the electric field is derivable from a potential function $\phi(\mathbf{r}, t)$ through $\mathbf{E} = -\nabla\phi$. Finally, one further important simplification is possible. In an ionized medium, very small charge differences create large electric fields. Thus, a plasma must exhibit nearly perfect charge neutrality. This implies

$$\nabla \cdot \mathbf{J} = -\partial \rho_c/\partial t \cong 0 \tag{2.26}$$

That is, the divergence of the electric current on any macroscopic time scale must be zero. As we shall see in Section 2.4, in most cases we use $\nabla \cdot \mathbf{J} = 0$ to calculate the electric field rather than (2.25b). Poisson's equation [i.e., (2.25b)] is, of course, still valid, but it is not very useful, since the charge difference associated with geophysically important electric fields is very small. This situation is similar to the problem of calculating the vertical velocity in atmospheric dynamics. The vertical component of the momentum equation is not very useful, since the vertical pressure gradient and the gravitational terms nearly cancel each other out. In practice, the vertical velocity is often found from the divergence of the horizontal wind, much as we use $\nabla \cdot \mathbf{J} = 0$ to find the electric field. The following combined set of dynamic and electrodynamic equations for the plasma (plus the equation of state) remains after these simplifications:

$$\rho_j \, d\mathbf{V}_j/dt = -\nabla p_j + \rho_j \mathbf{g} + \frac{q_j \rho_j}{M_j}(\mathbf{E} + \mathbf{V}_j \times \mathbf{B})$$
$$- \sum_{\substack{k \\ j \neq k}} \rho_j \nu_{jk}(\mathbf{V}_j - \mathbf{V}_k) \tag{2.27a}$$

$$\frac{\partial \rho_j}{\partial t} + \nabla \cdot (\rho_j \mathbf{V}_j) = (P_j - L_j)M_j \tag{2.27b}$$

$$p_j = \rho_j k_B T_j/M_j = n_j k_B T_j \tag{2.27c}$$

$$\mathbf{E} = -\nabla \phi \tag{2.27d}$$

$$\nabla \cdot \mathbf{J} = 0 = \nabla \cdot \mathbf{B} \tag{2.27e}$$

$$\nabla \times \mathbf{B} = \mu_0 \mathbf{J} \tag{2.27f}$$

Note that from (2.26) we have the important result that the number of electrons per unit volume must be almost equal to the number of positive ions of all types

$$n_e \cong \sum_{\text{ions}} n_j$$

This means we can define a plasma density n, which is equal to both $n_e n_e$ and $\Sigma_j n_j$.

2.2 Steady-State Ionospheric Plasma Motions Due to Applied Forces

In this section we start with (2.27a) and derive the equations that determine the electrodynamic response of a partially ionized plasma to applied steady forces. For the present we specify the distribution of plasma density and the wind field. Considering (2.27a), with the plasma pressure distribution specified, we can argue that the response of the plasma constituents to changing forces occurs very quickly (i.e., $d\mathbf{V}_j/dt \approx 0$). This can be seen by comparing the terms in (2.27a) that include the velocity. The acceleration terms on the left-hand side are of order V_j/τ and, V_j^2/L, where τ is the response time to a new set of forces and L is a distance scale for velocity change. The Lorentz term (the third term on the right-hand side) is of order $\mathbf{V}_j\Omega_j$, where Ω_j is the gyrofrequency ($q_j\mathbf{B}/M_j$), and the frictional term is of order $\mathbf{V}_j\nu_j$ where ν_j is the collision frequency. As long as $\tau \gg \Omega_j^{-1}$ or ν_j^{-1}, the acceleration term can be neglected. Collision frequencies and gyrofrequencies are sufficiently high that, in most problems of interest to macroscopic dynamics, the acceleration term can be ignored. Also, since fluid velocities are usually subsonic, if L is greater than the gyro radius and the mean free path, the advective term is small. The plasma constituents are thus assumed to be in velocity equilibrium with the existing force fields.

The equilibrium fluid velocity of each species may now be found from (2.27a) by setting the total time derivative equal to zero and specifying the force fields and pressure distributions

$$0 = -\nabla(n_j k_B T_j) + n_j M_j \mathbf{g} + q_j n_j (\mathbf{E} + \mathbf{V}_j \times \mathbf{B})$$

$$- \sum_{\substack{k \\ k \neq j}} n_j M_j \nu_{jk} (\mathbf{V}_j - \mathbf{V}_k) \tag{2.28}$$

The collision frequencies v_{jk} play a crucial role in a partially ionized plasma. Texts by Rishbeth and Garriott (1969), Banks and Kockarts (1973), and Schunk and Nagy (2000) discuss the physics in some detail. Here it suffices to plot and discuss representative collision profiles such as those given in Fig. 2.3, which are valid for high sunspot conditions (e.g., see Johnson, 1961). The numerical values of various ionospheric and atmospheric parameters used in making this plot are given in Appendix B. A useful approximate formula for the ion-neutral collision frequency is

$$v_{in} = 2.6 \times 10^{-9}(n_n + n_i)A^{-1/2} \tag{2.29a}$$

where A denotes the mean neutral molecular mass in atomic mass units and the electron neutral plus the electron ion collision frequency is

$$v_e \equiv v_{en} + v_{ei} = 5.4 \times 10^{-10}n_n T_e^{1/2}$$
$$+ \left[34 + 4.18 \ln\left(T_e^3/n_e\right)\right] n_e T_e^{-3/2} \tag{2.29b}$$

where T_e is measured in Kelvin and all the densities are expressed per cubic centimeter. The ion-electron collision frequency is not included because it is negligible over the altitude range of interest. At night the electron-neutral collision frequency equals the electron-ion collision frequency near about 280 km. As we shall see, the equatorial F-region plasma is often driven to very high altitudes in the evening hours, and in such a case the transition height can be much higher. During the daytime, transition between v_{en} and v_{ei} occurs at about 200 km. For the moment we consider the case that electron-neutral collisions are more common than electron-ion collisions, which simplifies the algebra considerably.

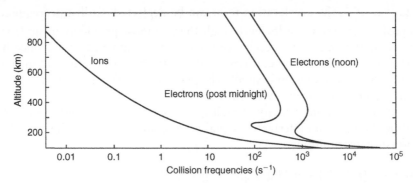

Figure 2.3 Typical electron neutral plus electron ion collision frequency along with the ion-neutral collision frequency at a high sunspot number.

It is instructive and useful to solve (2.28) for the ion and electron velocities in terms of the driving forces. For simplicity we consider a single ion species of mass M, using the symbol m for the electron mass and the symbol e for the elemental charge, which is taken to be positive. For spatially uniform ion and electron temperatures, we have the two equations

$$0 = -k_B T_i \nabla n + n M \mathbf{g} + n e (\mathbf{E} + \mathbf{V}_i \times \mathbf{B}) - n M v_{in} (\mathbf{V}_i - \mathbf{U}) \tag{2.30a}$$

$$0 = -k_B T_e \nabla n + n m \mathbf{g} - n e (\mathbf{E} + \mathbf{V}_e \times \mathbf{B}) - n m v_{en} (\mathbf{V}_e - \mathbf{U}) \tag{2.30b}$$

where we have used the fact that $n_i = n_e = n$, the plasma density. The electric field in this equation is the one that would be measured in an earth-fixed coordinate system. This is usually the electric field that is measured in ionospheric experiments. It is nevertheless instructive to express these equations in a reference frame moving with the neutral flow velocity $\mathbf{U}$. Transformation between two coordinate systems moving at a relative velocity $\mathbf{U}$ does not leave the electric field invariant even if $|\mathbf{U}| \ll c$, where c is the speed of light. Jackson (1975) discusses transformation of the electromagnetic fields $\mathbf{E}$ and $\mathbf{B}$ between two coordinate systems and shows that in the moving frame,

$$\mathbf{E}' = (\mathbf{E} + \mathbf{U} \times \mathbf{B}) \Big/ \left(1 - U^2/c^2\right)^{1/2} \tag{2.31a}$$

$$\mathbf{B}' = \left(\mathbf{B} - \mathbf{U} \times \mathbf{E}/c^2\right) \Big/ \left(1 - U^2/c^2\right)^{1/2} \tag{2.31b}$$

where the primed variables are those measured in the moving frame and the unprimed variables are measured in the earth-fixed frame.

It is easy to show that the $\mathbf{U} \times \mathbf{E}/c^2$ term in (2.31b) is small compared to $\mathbf{B}$ for any reasonable values of $\mathbf{U}$ and $\mathbf{E}$ in the earth's atmosphere or ionosphere. However, $\mathbf{U} \times \mathbf{B}$ is the same order of magnitude as $\mathbf{E}$ and must be retained. Thus, for $|\mathbf{U}| \ll c$,

$$\mathbf{E}' = \mathbf{E} + \mathbf{U} \times \mathbf{B} \tag{2.32a}$$

$$\mathbf{B}' = \mathbf{B} \tag{2.32b}$$

Another way to interpret these equations is that in a nonrelativistic transformation the current density is not significantly changed, $\mathbf{J}' \cong \mathbf{J}$, but the charge density is, $\rho_c' \neq \rho_c$. The explanation for this "asymmetry" between the electric and magnetic fields lies in the fact that very small charge densities can produce significant electric fields in a plasma.

Following the notation of Haerendel (personal communication, 1973), we may now transform the terms in (2.28) to a reference frame moving with the neutral wind $\mathbf{U}$. We use the subscript notation again for brevity, where j stands for i (the single-ion gas) or e (the electron gas). Since $\mathbf{V}'_j = \mathbf{V}_j - \mathbf{U}$, (2.28) becomes

$$0 = -k_B T_j \nabla n + nM_j \mathbf{g} + nq_j \mathbf{E}' + nq_j \left(\mathbf{V}'_j \times \mathbf{B}\right) - nM_j v_{jn} \mathbf{V}'_j \qquad (2.33)$$

where everything is expressed in the moving reference frame (note that ∇n and $\mathbf{g}$ are unchanged in a nonrelativistic transformation). If we divide through by $nM_j v_{jn}$ and gather terms, this can be written

$$\mathbf{V}'_j - \kappa_j \left(\mathbf{V}'_j \times \hat{\mathbf{B}}\right) = -D_j \nabla n/n + b_j \mathbf{E}' + \left(D_j/H_j\right) \hat{\mathbf{g}} \equiv \mathbf{W}'_j \qquad (2.34)$$

where $\hat{\mathbf{B}}$ is a unit vector in the $\mathbf{B}$ direction, $\hat{\mathbf{g}}$ is a unit vector in the $\mathbf{g}$ direction, κ_j is the ratio of gyrofrequency to collision frequency $(q_j B/M_j v_{jn})$, which has the same sign as the particle charge, D_j is the diffusion coefficient $(k_B T_j/M_j v_{jn})$, b_j is the mobility $(q_j/M_j v_{jn})$, which also has the algebraic sign of q_j, and H_j is the scale height $(k_B T_j/M_j g)$. Notice that the velocity $\mathbf{W}'_j$ is the fluid velocity that would arise in an unmagnetized plasma subject to the same forces.

The absolute value of κ_j determines whether a particle does or does not make a cycle about the magnetic field before a collision takes place. For a small absolute value of κ_j, many collisions occur and the particle basically moves parallel to the applied forces as if there were no magnetic field. This is illustrated in Fig. 2.4a for ions and electrons subject to an electric field. The collisionless case (κ infinite) is shown in Fig. 2.4b for particles initially at rest. After about one gyroperiod the particles are moving at right angles to the electric field. In this important case the final velocity is identical for ions and electrons and equal to $\mathbf{E} \times \mathbf{B}/B^2$. For an absolute value of $\kappa = 1$ (Fig. 2.4c), the net motion is at a $45°$ angle to the electric field.

These results can be seen analytically from (2.34). Consider first the case of a very high collision frequency ($\kappa_j \ll 1$). Then the first term on the left-hand side dominates and

$$\mathbf{V}'_j = \mathbf{W}'_j = b_j \mathbf{E}' - D_j \nabla n/n + \left(D_j/H_j\right) \hat{\mathbf{g}} \qquad (2.35)$$

which is the same as the fluid velocity that would arise in the case of an unmagnetized plasma. The velocity is parallel to the forces. For κj very large (a "collisionless" plasma), the component of (2.34) parallel to $\mathbf{B}$ is unchanged:

$$\left(\mathbf{V}'_j\right)_{\parallel} = [b_j \mathbf{E}' - D_j \nabla n/n + \left(D_j/H_j\right) \hat{\mathbf{g}}] \cdot \hat{\mathbf{B}} \qquad (2.36a)$$

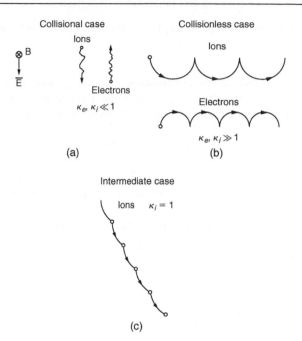

Figure 2.4 Ion and electron trajectories for various values of κ for particles initially at rest.

and thus collisions are still very important. But in the perpendicular direction, the second term on the left-hand side of (2.34) dominates and the components perpendicular to **B** become

$$\left(\mathbf{V}_j'\right)_\perp = \kappa_j^{-1}\left[b_j\mathbf{E}' - D_j\nabla n/n + \left(D_j/H_j\right)\hat{\mathbf{g}}\right] \times \hat{\mathbf{B}} \qquad (2.36b)$$

or, equivalently, when κ_j is evaluated we have

$$\left(\mathbf{V}_j'\right)_\perp = \left(1/B^2\right)\left[\mathbf{E}' - \left(k_BT_j/q_j\right)\nabla n/n + \left(M_j/q_j\right)\mathbf{g}\right] \times \mathbf{B} \qquad (2.36c)$$

The individual terms on the right-hand side of (2.36c) are all perpendicular to the forces that drive them. Furthermore, since the first term does not depend on the charge, it is identical for ions and electrons. The ions and electrons move together at the "$\mathbf{E} \times \mathbf{B}$" velocity in a collisionless plasm, and no net current flows in response to an applied electric field. Thus, Ohm's law in its usual form is of little use in a collisionless magnetized plasma. There is another interesting subtlety in (2.36c). Ignoring ∇n and **g**, we express all the remaining variables in earth-fixed coordinates:

$$\left(\mathbf{V}_j - \mathbf{U}\right)_\perp = \left[\left(\mathbf{E} + \mathbf{U}\times\mathbf{B}\right)\times\mathbf{B}\right]/B^2$$

Carrying out the triple cross product, $[(U \times B) \times B]/B^2 = -U_\perp$ and so

$$\left(V_j\right)_\perp = E \times B/B^2$$

This equation shows that in the collisionless case the plasma moves at the $E \times B$ velocity in any reference frame, provided the electric field and velocity are expressed in that reference frame. In the earth-fixed frame, then,

$$\left(V_j\right)_\perp = \left(1/B^2\right)\left[E - \left(k_B T_j/q_j\right)\nabla n/n + \left(M_j/q_j\right)g\right] \times B \qquad (2.36d)$$

which is identical in form to (2.36c). In (2.36a) we have left the prime on E', but it should be noted that the transformation (2.32a) leaves the component of E parallel to B invariant.

The solutions of (2.34) for intermediate values of κ are given by

$$\left(V'_j\right)_\parallel = \left(W'_j\right)_\parallel \qquad (2.37a)$$

and

$$V'_{j\perp} = \frac{W'_{j\perp}}{1+\kappa_j^2} + \frac{\kappa_j W'_{j\perp}}{1+\kappa_j^2} \times \hat{B} \qquad (2.37b)$$

where again we have expressed the result in terms of the steady-state unmagnetized velocity solution $W'_{j\perp}$. These expressions show explicitly that for small κ, V'_j tends toward $W'_{j\perp}$, while for large κ the motions tend to be perpendicular to the forces. The absolute values of κ_e and κ_i are plotted in Fig. 2.5 for an

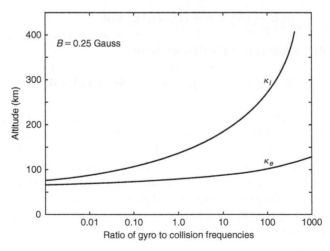

Figure 2.5 Typical values for κ_e and κ_i in the equatorial ionosphere for a magnetic field of $0.25\,G = 2.5 \times 10^{-5}$ tesla.

equatorial ionosphere with $B = 2.5 \times 10^{-5}$ tesla (0.25 gauss). The transition from a molecular ion plasma (NO^+ and O_2^+) to an atomic ion plasma (O^+) has been included in the calculation of κ. The absolute value of κ_e passes through unity near 75 km, while κ_i does so at 130 km. In making the plot of κ_e, we have used the total electron collision frequency $\nu_e = \nu_{en} + \nu_{ei}$. This is not entirely consistent with the preceding discussion, which assumes $\nu_e = \nu_{en}$. However, the modification is of little importance, since the absolute value of κ_e is very large above 100 km.

The relationship between $\mathbf{J}'$ and $\mathbf{E}'$ may now be determined from the definition $\mathbf{J}' = ne(\mathbf{V}_i' - \mathbf{V}_e')$, with $\mathbf{V}_j'$ given by (2.37a,b) and (2.35). The result may be expressed through a tensor relationship $\mathbf{J}' = \sigma \cdot \mathbf{E}'$, where

$$\sigma = \begin{pmatrix} \sigma_P & -\sigma_H & 0 \\ \sigma_H & \sigma_P & 0 \\ 0 & 0 & \sigma_0 \end{pmatrix} \tag{2.38}$$

To obtain this form, $\mathbf{B}$ has been taken to be parallel to the z-axis, and we have defined

$$\sigma_0 = ne(b_i - b_e) \tag{2.39a}$$

$$\sigma_P = ne\left[b_i\Big/\left(1 + \kappa_i^2\right) - b_e\Big/\left(1 + \kappa_e^2\right)\right] \tag{2.39b}$$

$$\sigma_H = (ne/B)\left[\kappa_e^2\Big/\left(1 + \kappa_e^2\right) - \kappa_i^2\Big/\left(1 + \kappa_i^2\right)\right] \tag{2.39c}$$

The three conductivity parameters, σ_0, σ_P, and σ_H, are called the specific, Pedersen, and Hall conductivities, respectively. (Remember that b_e is negative.)

Plots of σ_0, σ_P, and σ_H for a typical daytime midlatitude ionosphere are given in Fig. 2.6. These plots correspond to the daytime collision frequencies in Fig. 2.3 and a magnetic field of 5×10^{-5} tesla (0.5 gauss). The specific or parallel conductivity σ_0 is dominated by the high electron mobility and is equal to $ne^2/m\nu_e$ to a good approximation. At high altitudes when electron-neutral collisions become rare, the plasma density factor in ν_{ei} cancels the same factor in the numerator of σ_0, and therefore σ_0 is independent of density above 400 km. The variation above that height displayed in Fig. 2.6 is related to the electron temperature, since according to (2.29b), ν_{ei} is very nearly proportional to $(T_e)^{-3/2}$. The parallel conductivity is so high that the ratio σ_0/σ_P is greater than 1×10^4 above 130 km. Above about 75 km, κ_e is very large, and in the plane perpendicular to $\mathbf{B}_0$, the electrons only move perpendicular to the forces that act on them. Then the Pedersen conductivity may be written in the form

$$\sigma_P = ne^2\Big/\left[M\nu_{in}\left(1 + \kappa_i^2\right)\right] \tag{2.40a}$$

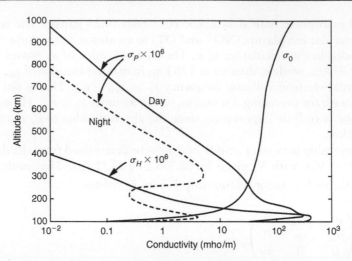

Figure 2.6 Typical conductivity values for the midlatitude daytime ionosphere. Notice the change of scale for σ_P and σ_H. The dashed curve is a typical nighttime profile of σ_P also multiplied by 10^6.

For $\kappa_i \gg 1$ (above 130 km) this expression becomes even simpler,

$$\sigma_P = ne^2 v_{in}/M\Omega_i^2 = nMv_{in}/B^2 \qquad (2.40b)$$

The Hall conductivity σ_H falls off more rapidly with height than does $\sum_P$ and is important only in a narrow height range where three conditions are met: $\kappa_e \gg 1$, $\kappa_i \lesssim 1$, and n is large. A typical nighttime curve for σ_P is also given.

Finally, we remind the reader that the calculations have thus far been performed in the neutral reference frame, where $J' = \sigma \cdot E'$. More usually we measure the neutral wind U and electric field E in the earth-fixed frame. However, since $E' = E + U \times B$ and $J = J'$ for nonrelativistic transformations, we have the important and most usual form of the current equation

$$J = \sigma \cdot (E + U \times B) \qquad (2.41)$$

where all parameters are measured in the earth-fixed coordinates. Earth-fixed measurements of electric fields, of course, can determine only the E in (2.41), not the entire quantity $E' = E + U \times B$.

To summarize, we note that the ionospheric plasma is subject to electromagnetic forces in addition to those felt by the neutral atmosphere. The dipole nature of the magnetic field is not greatly affected by ionospheric currents. The result is that the magnetic field creates geometric constraints on the plasma behavior, constraints that are quite different at different magnetic latitudes. Electric fields, on the other hand, come and go quite rapidly and put the plasma constituents in motion perpendicular to B. Electric fields thus play a dominant role in the dynamics of the ionosphere. In the next sections we briefly discuss the generation and mapping of electric fields. Specific electric field sources are discussed as they arise in subsequent chapters.

2.3   Generation of Electric Fields

Although we have allowed for the possibility of an electric field in the earth's upper atmosphere, we have yet to show that such fields exist. The other forces in (2.28) are, for our present purposes, given quantities—that is, the forces due to the plasma pressure gradients to the magnetic and gravitational fields, and to the atmospheric winds. Electric fields arise as a result of these forces when the ions and electrons respond differently to them. This is expressed quantitatively via the current divergence equation

$$\nabla \cdot \mathbf{J} = -\partial \rho_c / \partial t \tag{2.42}$$

Any charge density, ρ_c, of course, must create electric fields through Poisson's equation

$$\nabla \cdot \mathbf{E} = \rho_c / \varepsilon_0 \tag{2.43}$$

Given the complexity of the forces in (2.28), it is not surprising that the electric current associated with the difference between the ion and electron velocities has a finite divergence. However, this divergence creates a charge density via (2.42) that, via (2.43), creates an electric field that forces the divergence to zero. In other words, if the complex forces acting on the ion and electron fluids create a divergence in $\mathbf{J}$, an electric field builds up quickly to modify the fluid velocities so that once again $\nabla \cdot \mathbf{J} = 0$.

For example, consider an electric field of 10 mV/m, which is large by equatorial and midlatitude standards. Assuming a scale length of 1 km in (2.43), we find $\rho_c = 8.85 \times 10^{-17} \text{C/m}^3$, which amounts to an excess of ions or electrons of a few thousand per cubic meter compared to a total of at least 10^9 m^{-3}. The time scale for buildup of such a charge density can be estimated from (2.42) and (2.43). From (2.42)

$$\tau \sim \rho_c / \nabla \cdot \mathbf{J} = \varepsilon_0 \nabla \cdot \mathbf{E} / \nabla \cdot \mathbf{J}$$

Assuming for the moment that σ is uniform and isotropic, then $\mathbf{J} = \sigma \mathbf{E}$ and

$$\tau \cong \varepsilon_0 / \sigma \tag{2.44}$$

Using the lowest value of any component of σ in the ionosphere, we find the largest value for $\tau = 10^{-6}$s. Electric fields thus build up very quickly indeed in response to any divergence of $\mathbf{J}$. Such divergences arise whenever there are spatially varying forces on the plasma or when the conductivity changes in space. In practice, it is not possible to calculate ρ_c from (2.42) and then $\mathbf{E}$ from (2.43). Rather, the electric field is treated as a free parameter that adjusts in magnitude and direction to fit the requirement that $\nabla \cdot \mathbf{J} = 0$. In the next chapters we show in detail how electric fields arise in the earth's ionosphere.

When an electric field is created by a wind, the process is often called a dynamo in analogy to a motor-driven electric generator in which a conductor is moved across a magnetic field. In this case, or any case in which electrical energy is created, the quantity $(\mathbf{J} \cdot \mathbf{E})$ should be negative, and in addition the electrical forces must act in opposition to the source of the charge separation.

When electric fields are applied from an external source, such as occurs at high latitudes due to the solar wind–magnetosphere interaction, it is usually the case that $\mathbf{J} \cdot \mathbf{E} > 0$ in the ionosphere. In this case electrical energy is converted into mechanical energy in the ionosphere and is released in the form of heat. Such Joule heating is a very important process at high latitudes and may greatly affect the thermospheric winds. In this case the ionosphere acts like an electrical load on some external generator. Likewise, momentum may be transferred to the thermospheric gas through the "ion drag" term if the ions are driven very strongly by an externally applied electric field. In such a case the ionosphere-magnetosphere system acts like a motor with electrical energy converted into mechanical energy.

2.4 Electric Field Mapping

The high conductivity parallel to the earth's magnetic field, σ_0, has important implications concerning the transmission of electric fields for long distances along $\mathbf{B}$. In fact, if σ_0 were infinite, there would be zero potential drop along the magnetic field, and the potential difference between any two field lines would be constant. In such a case any electric field generated at ionospheric heights would be transmitted along the magnetic field lines to very high altitudes. For example, an electric field generated at $60°$ magnetic latitude would be communicated to the equatorial plane at an altitude over $25,000$ km. Likewise, electric fields of solar wind or magnetospheric origin could be transmitted to ionospheric heights.

This phenomenon can be studied quantitatively as follows (following Farley, 1959, 1960). Suppose first that the conductivity is anisotropic but uniform and that the neutral wind is absent. If the electric field perpendicular to $\mathbf{B}$ is $\mathbf{E}_\perp$ and the field parallel to $\mathbf{B}$ is $\mathbf{E}_{||}$ the total current is

$$\mathbf{J} = \sigma_P \mathbf{E}_\perp - \sigma_H \left(\mathbf{E}_\perp \times \hat{\mathbf{B}} \right) + \sigma_0 \mathbf{E}_{||} \qquad (2.45)$$

For an electrostatic field $\mathbf{E} = -\nabla\phi$. Substituting this expression for $\mathbf{E}$ into (2.45), taking the divergence, setting $\nabla \cdot \mathbf{J} = 0$, and taking $\mathbf{B}$ to be in the z direction yields

$$- (\sigma_P)\, \partial^2\phi/\partial x^2 - (\sigma_H)\, \partial^2\phi/\partial x\partial y - (\sigma_P)\, \partial^2\phi/\partial y^2$$
$$+ (\sigma_H)\, \partial^2\phi/\partial y\partial x - (\sigma_0)\, \partial^2\phi/\partial z^2 = 0$$

The terms containing σ_H cancel, leaving

$$\partial^2\phi/\partial x^2 + \partial^2\phi/\partial y^2 + (\sigma_0/\sigma_P)\, \partial^2\phi/\partial z^2 = 0 \tag{2.46}$$

Making the change of variables

$$dz' = (\sigma_P/\sigma_0)^{1/2}\, dz$$
$$dx' = dx$$
$$dy' = dy$$

converts (2.46) to

$$\left(\nabla'\right)^2 \phi = 0 \tag{2.47}$$

which is Laplace's equation in the "reduced" coordinate system. That is, the substitution has transformed the real medium into an equivalent isotropic medium with a greatly reduced depth parallel to the magnetic field (the z direction in the calculation). The ratio $(\sigma_0/\sigma_P)^{1/2}$ is plotted in Fig. 2.7 for a typical ionospheric profile. Above 130 km the ratio exceeds 100, reaching 1000 at 300 km. At high altitudes, σ_0 becomes independent of density. The ratio σ_0/σ_P continues to increase as the ion-neutral collision frequency and the plasma density, which determine σ_P, continue to decrease. One of the basic approximations of magnetohydrodynamics (MHD) is that if the conductivity parallel to the magnetic field

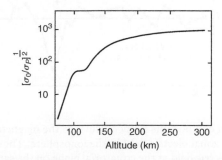

Figure 2.7   The mapping ratio $(\sigma_0/\sigma_P)^{1/2}$ plotted as a function of height for a typical mid- to high-latitude ionosphere.

becomes very large, then the parallel electric field component vanishes. For many applications, MHD theory applies on the high-altitude portions of the field lines that contact the ionosphere. This theory is discussed in the next section in the context of the "frozen-in" condition.

The implication of these calculations is that large-scale electrical features perpendicular to **B** map for long distances along the earth's magnetic field lines. This has been verified experimentally via the simultaneous measurements shown in Fig. 2.8. In this experiment the zonal electric field component was measured in the Northern Hemispheric ionosphere with the Millstone Hill Radar (see Appendix A) and in the inner magnetosphere at a point very close to where the same magnetic field line crossed the equatorial plane. The latter measurement was accomplished using the whistler technique (Carpenter et al., 1972). The two measurements clearly have the same temporal form, but the magnetospheric component is 10 times smaller. This difference may be explained as a geometric effect arising from the spreading of the magnetic field lines as follows. First we take σ_0 to be infinite so that the field lines act like perfect conductors. This implies that there is no potential difference along them and in turn that the voltage difference between two lines is conserved. The magnetic flux density (measured in tesla) decreases along the field line as a function of distance from the mid- or high-latitude ionosphere to the equatorial plane in the magnetosphere. Since the voltage between adjacent field lines is constant, the perpendicular

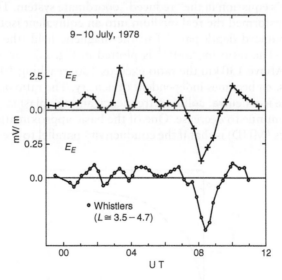

Figure 2.8 Electric field components perpendicular to the magnetic field. The top panel is a measurement of the zonal electric field in the ionosphere. The lower panel is a measurement of the zonal electric field at the equatorial plane on the same magnetic field line. [After Gonzales et al. (1980). Reproduced with permission of the American Geophysical Union.]

electric field must also decrease along the field lines. For a dipole field, Mozer (1970) has shown that the two electric field components (meridional E_{MI} and zonal E_{ZI}) map from the ionosphere to the magnetosphere in the equatorial plane as

$$E_{MI} = 2L\left(L - \frac{3}{4}\right)^{1/2} E_{RM} \qquad (2.48a)$$

$$E_{ZI} = L^{3/2} E_{ZM} \qquad (2.48b)$$

where the L value is the distance from the center of the earth to the equatorial crossing point measured in earth radii (R_e), E_{RM} is the radial magnetospheric component at the equatorial plane, and E_{ZM} is the zonal magnetospheric component there. The equation for the zonal component (2.48b) is in excellent agreement with the corresponding data in Fig. 2.8. Notice that the zonal ionospheric electric field component maps to a zonal field in the equatorial plane but that the meridional component in the ionosphere, E_{MI}, becomes radial in the equatorial plane (E_{RM}). In particular, a poleward ionospheric electric field points radially outward at the equatorial plane. Thus, large-scale electric fields generated in the E and F regions of the ionosphere can map upward to the magnetosphere and create motions there. Likewise, electric fields of magnetospheric and solar wind origin can map from deep space to ionospheric heights and have even been detected by balloons at stratospheric heights (Mozer and Serlin, 1969). For perfectly conducting field lines, the ratio E^2/B is conserved, since B varies as the area, while E varies as the linear distance along the magnetic field.

Farley (1959) studied the upward mapping process realistically by including the z dependence of the conductivities in his analysis. The basic equations $\nabla \cdot \mathbf{J} = 0$, $\mathbf{E} = -\nabla\phi$, and $\mathbf{J} = \sigma \cdot \mathbf{E}$ are the same used in deriving (2.47), and, with the assumption that variations of σ occur only in the z direction, they yield

$$\partial^2 \phi/\partial x^2 + \partial^2 \phi/\partial y^2 + (1/\sigma_P)\partial/\partial z(\sigma_0 \partial\phi/\partial z) = 0$$

The same change of variables now yields

$$(\nabla')^2 \phi + (\partial\phi/\partial z')(\partial/\partial z')\left[\ln(\sigma_0\sigma_P)^{1/2}\right] = 0 \qquad (2.49a)$$

where $(\sigma_0\sigma_P)^{1/2} = \sigma_m$ is termed the geometric mean conductivity. Furthermore, if σ_m can be modeled in the form $\sigma_m = c \exp(c_0 z')$, then the equation simplifies to

$$(\nabla')^2 \phi + c_0(\partial\phi/\partial z') = 0 \qquad (2.49b)$$

This differential equation has a straightforward analytical solution. By considering the solutions with different Fourier wave numbers in the source field,

Farley showed that (1) larger-scale features map more efficiently to the F region from the E region than small-scale features; (2) the height of the source field is very important, with upper E-region structures very much favored as F-region sources compared to lower E-region sources; and (3) roughly speaking, perpendicular structures with scale sizes greater than a few kilometers map unattenuated to F-region heights. The implication here is that if very large-scale electric fields are generated in the E region, the potential differences thereby created map up into the F-region ionosphere and beyond along the magnetic field lines deep into space. As we shall see, these low-altitude electric fields dominate motions of the plasma throughout the dense plasma region around the earth, termed the plasmasphere.

Considering sources at F region and even higher altitudes, in the magnetosphere and solar wind, for example, the previous analysis can be used to show that the mapping efficiency to the E region is even greater. In fact, within the magnetospheric and solar wind plasmas, the parallel conductivity is often taken to be infinite, and thus the parallel electric field vanishes even when finite field-aligned currents flow. As we shall see later, this assumption that $E_{||} = 0$ is a powerful analytical device, allowing great conceptual simplifications in the understanding of magnetospheric electric field and flow patterns. On the other hand, it is exactly in the regions where the assumption of infinite conductivity breaks down that very interesting phenomena occur. The generation of the aurora is an example.

Since large-scale electric fields map along the magnetic field lines, we may consider them to be independent of z in the magnetic coordinate system just used (B in the z direction). This has some interesting consequences for ionospheric electrodynamics. Consider the current divergence equation separated into its perpendicular and parallel parts:

$$\nabla_\perp \cdot \mathbf{J}_\perp = -\partial J_z/\partial z$$

where $\mathbf{J}_\perp = \boldsymbol{\sigma}_\perp \cdot (\mathbf{E}_\perp + \mathbf{U} \times \mathbf{B})$ and $\boldsymbol{\sigma}_\perp$ is the 2×2 perpendicular conductivity matrix. Ignoring the neutral wind for the moment,

$$\nabla_\perp \cdot (\boldsymbol{\sigma}_\perp \cdot \mathbf{E}_\perp) = -\partial J_z/\partial z$$

Remembering that the z-axis is parallel to the magnetic field and integrating from the top of the Northern Hemisphere ionosphere ($z = 0$) to a value z_0 below which no significant perpendicular currents flow yields

$$J_z(0) - J_z(z_0) = \int_0^{z_0} [\nabla \cdot (\boldsymbol{\sigma}_\perp \cdot \mathbf{E}_\perp)]dz$$

Although small currents do exit the base of the ionosphere and link up with atmospheric electrical currents to complete the global atmospheric electrical circuit, we ignore this interesting phenomenon by setting $J_z(z_0) = 0$. $J_z(0)$ is then the field-aligned current entering the ionosphere from above and is related to the divergence of the perpendicular current density. Now, since $\mathbf{E}_\perp$ is independent of z, we may move it and the divergence operator through the integral to yield

$$J_z = \nabla \cdot (\Sigma \cdot \mathbf{E}_\perp) \tag{2.50}$$

where Σ is the perpendicular height-integrated conductivity tensor

$$\Sigma = \begin{pmatrix} \Sigma_P & -\Sigma_H \\ \Sigma_H & \Sigma_P \end{pmatrix}$$

and we have dropped the argument of J_z and assume that we measure J_z at a sufficiently high altitude that all perpendicular ionospheric currents are below it. Since conductivity has units of mhos per meter, Σ has units of mhos. The E-region values vary from a few tenths of a mho in the nighttime midlatitude region to several tens of mhos during a strong auroral precipitation event. F-region values maximize at the magnetic equator where the field lines are very long and rapidly decrease at midlatitudes and higher.

Some insight can be achieved by letting Σ be uniform in the horizontal directions. Then the perpendicular divergence operator does not operate on it and

$$J_z = \underset{P}{\Sigma} \left(\partial E_x/\partial x \right) - \underset{H}{\Sigma} \left(\partial E_y/\partial x \right) + \underset{H}{\Sigma} \left(\partial E_x/\partial y \right) + \underset{P}{\Sigma} \left(\partial E_y/\partial y \right)$$

which can be written

$$J_z = \underset{P}{\Sigma} \left(\nabla_\perp \cdot \mathbf{E}_\perp \right) + \underset{H}{\Sigma} \left[\left(\partial E_x/\partial y - \partial E_y/\partial x \right) \right]$$

The last term is the z component of $\nabla \times \mathbf{E}$, which vanishes for steady magnetic fields, and we have

$$J_z = \underset{P}{\Sigma} \left(\nabla_\perp \cdot \mathbf{E}_\perp \right) \tag{2.51}$$

In the auroral zone $\Sigma_P \approx 10$ mho, $\mathbf{E}_\perp \sim 50$ mV/m, and $\nabla_\perp \sim 1/L \sim 10^{-5}$ m, which yields a typical large-scale J_z of order $5 \ \mu\text{A/m}^2$. Small-scale currents can be more than an order of magnitude higher. Field-aligned currents at mid- and low latitudes are much smaller but still play a very important role in the physics. Equation (2.51) plus Poisson's equation (2.25b) show that current flows downward in the Northern Hemisphere where the net charge density in the ionosphere is positive.

This seems counterintuitive but corresponds to the fact that the ionosphere is an electrical load at higher latitudes.

Finally, we turn to the question of downward mapping of magnetospheric and ionospheric electric fields into the region below the ionosphere—that is, into the lower atmosphere. In the late 1960s it was realized that these fields could be detected at balloon altitudes ($\sim$ 30 km), since most of the atmospheric resistivity is in the last atmospheric scale height, which has a value of about 8 km near the surface of the earth (Mozer and Serlin, 1969). At the surface of the earth, of course, horizontal fields must vanish, since the earth is a good conductor compared with the atmosphere. However, at balloon height the horizontal ionospheric/magnetospheric electric field is almost unattenuated. Some theories of solar activity effects on climate appeal to this downward penetrating electric field as a causative factor.

2.5 Elements of Magnetospheric Physics

One may choose between (at least) two philosophies when describing a collisionless magnetized plasma. In the magnetohydrodynamic approach, the plasma is considered as a single conducting fluid with a certain bulk flow velocity determined by the forces acting on the fluid. These forces must include the electromagnetic force as well as the gravitational, pressure, and viscous forces acting on a hydrodynamic system. Another viewpoint considers the microscopic motions of particles subject to the various forces. Both descriptions are used in space plasma physics and are discussed in the next two sections.

2.5.1 The Guiding Center Equations and the Adiabatic Invariants

One way to view particle dynamics is to consider the motion of the so-called guiding centers. In this picture the near-circular gyromotion about the magnetic field is considered to be the dominant motion with a time scale (Ω^{-1}) much shorter than any other dynamical time scale τ (i.e., $\Omega^{-1} \ll \tau$).

To illustrate this, consider the single-particle motion associated with a general force $\mathbf{F}$ and the magnetic field $\mathbf{B}$. Ignoring spatial derivatives,

$$(M\partial\mathbf{V}/\partial t) = q(\mathbf{V} \times \mathbf{B}) + \mathbf{F}$$

Taking the component parallel to $\mathbf{B}$ gives

$$M\left(\partial\mathbf{V}_{\parallel}/\partial t\right) = \mathbf{F}_{\parallel}$$

which is straightforward. Perpendicular to $\mathbf{B}$, this leaves

$$(M\partial\mathbf{V}_{\perp}/\partial t) = q(\mathbf{V}_{\perp} \times \mathbf{B}) + \mathbf{F}_{\perp} \tag{2.52}$$

Now let

$$V_\perp = W_D + u$$

where

$$W_D = (F_\perp \times B)/qB^2 \tag{2.53}$$

This substitution is motivated by the result we already have that $W_D = (E \times B)/B^2$ when $F_\perp = qE$. Substituting (2.53) into (2.52) yields

$$M\partial W_D/\partial t + M\,\partial u/\partial t = \left(F_\perp \times B/B^2\right) \times B + qu \times B + F_\perp$$

The first term on the left vanishes, while the first term on the right-hand side equals $-F_\perp$ and thus cancels the other $F_\perp$ term. This leaves

$$M\partial u/\partial t = q(u \times B)$$

The solution to this, of course, is just the gyromotion at frequency

$$\Omega = qB/M$$

The interpretation we make is that in a frame moving at W_D, the particle motion is pure gyration. This yields the concept of a guiding center motion, since W_D gives the velocity of the center of gyromotion. Some examples of guiding center drifts due to various forces are as follows. For an electric field, $F = qE$ and

$$W_D = qE \times B/qB^2 = E \times B/B^2$$

For the gravitational field, $F = Mg$ and

$$W_D = Mg \times B/qB^2$$

For the inertial force $F = -M\,\partial W_D/\partial t$ and, letting $W_D = E \times B/B^2$,

$$W_D = (F \times B)/qB^2 = \left(-1/qB^2\right)\left[M\,\partial/\partial t\left(E \times B/B^2\right) \times B\right]$$

$$W_D = -\left(M/qB^2\right)(\partial E/\partial t \times B) \times B$$

$$W_D = \left(M/qB^2\right)(\partial E/\partial t)$$

Notice that this expression can be related to a displacement current in the plasma, since using $J = (neW_{Di} - neW_{De})$ and $M \gg m$ yields

$$J = ne\left(M/eB^2\right)\partial E/\partial t = \left(nM/B^2\right)\partial E/\partial t$$

Substituting this into one of Maxwell's equations,

$$\nabla \times \mathbf{B} = \mu_0 \mathbf{J} + \mu_0 \varepsilon_0 \partial \mathbf{E}/\partial t$$

gives

$$\nabla \times \mathbf{B} = \mu_0 \left(nM/B^2 + \varepsilon_0 \right) \partial \mathbf{E}/\partial t$$

so we can define a dielectric constant via the expression

$$\varepsilon = \left(nM/B^2 \right) + \varepsilon_0 \cong \left(nM/B^2 \right)$$

Now a required assumption necessary for using the guiding center approximation is that the time scale $\tau > \Omega_i^{-1}$. Hence, this dielectric constant should be valid for electromagnetic waves with frequencies $f < \Omega_i$. Indeed, the expression for the phase velocity of an electromagnetic wave is

$$V_{ph} = (1/\mu_0 \varepsilon)^{1/2}$$

Substituting the dielectric constant derived above yields

$$V_{ph} = B/(\mu_0 \rho)^{1/2}$$

This velocity is the Alfvén speed, which is the velocity of an electromagnetic wave propagating parallel to $\mathbf{B}$ in a magnetized plasma when its frequency satisfies $f \ll \Omega_i$ (e.g., Spitzer (1962) or any elementary plasma text).

If the magnetic field is curved as in a dipole field, particles moving along $\mathbf{B}$ will feel a force given by

$$\mathbf{F} = -\left(MV_{\parallel}^2/R \right) \hat{n}$$

where $\hat{n}$ is a unit vector pointed inward, $V_{\parallel}$ is the particle velocity parallel to $\mathbf{B}$, and R is the radius of curvature (see Fig. 2.13 for a sketch of the coordinate system). Substituting this force into (2.55), we find that the particles drift perpendicular to the field with velocity

$$\mathbf{W}_D = \left(MV_{\parallel}^2/R \right) \left(\mathbf{B} \times \hat{n}/qB^2 \right)$$

Similarly, if there is a gradient in the magnetic field with a gradient-scale length large compared to a gyroradius, we can consider the force on a magnetic dipole of moment μ, which is given by

$$\mathbf{F} = -\mu \nabla B$$

A gyrating particle has a magnetic moment since it carries a current I and surrounds an area πr_g^2, where r_g is the gyroradius, so

$$\mu = IA = I\pi r_g^2$$

The current equals the charge divided by the gyro period, τ_g, and using

$$\tau_g = \left(2\pi r_g/V_\perp\right) \text{ and } r_g = \left(MV_\perp/qB\right)$$

yields the magnetic moment

$$\mu = \left(MV_\perp^2/2B\right) = K_\perp/B$$

where $K_\perp$ is the perpendicular kinetic energy of the particle. Thus, the force due to the magnetic field gradient is

$$\mathbf{F} = -(K_\perp/B)\,\nabla B$$

and the corresponding "gradient drift" is

$$\mathbf{W}_D = -\mu\left(\nabla B \times \mathbf{B}/qB^2\right) = \left(MV_\perp^2/2B\right)\left(\mathbf{B} \times \nabla B/qB^2\right)$$

Notice that the gradient and curvature drifts are proportional to the particle perpendicular and parallel energies, respectively, and are in the same direction in a dipole field. The gradient-driven motion can be visualized easily with reference to Fig. 2.9. If ∇B is downward, as in the figure, a gyrating particle will have a slightly smaller radius of curvature in one portion of its cycle than in the other. A net drift results to the right for positively charged particles and to the left for negatively charged particles.

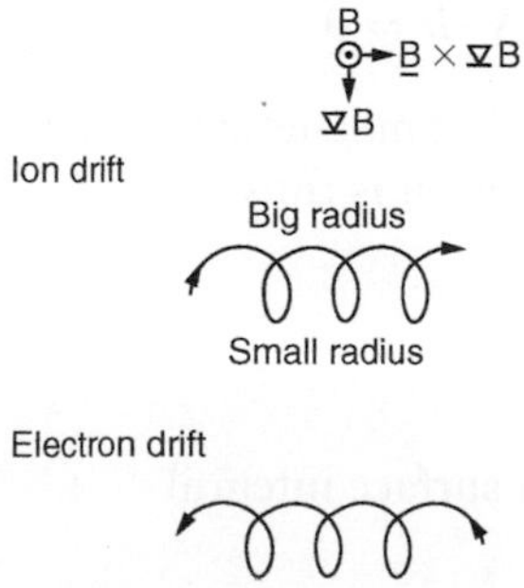

Figure 2.9   Gyromotion in a magnetic field with a gradient pointing downward.

To summarize, for time scales greater than Ω^{-1} and length scales greater than r_g the guiding center perpendicular drift equation for particle motion in a magnetic field is given by

$$\mathbf{W}_D = \left(\mathbf{E} \times \mathbf{B}/B^2\right) + \left(M/qB^2\right) d\mathbf{E}/dt + \left(M/qB^2\right)(\mathbf{g} \times \mathbf{B})$$

$$+ \left(1/qB^2\right)\left(MV_\perp^2/2B\right)(\mathbf{B} \times \nabla B) + \left(MV_\parallel^2/qRB^2\right)(\mathbf{B} \times \hat{n}) \qquad (2.54)$$

Since the particle motions just described are quite complex, it has proved useful to develop an intuition based on parameters that are nearly conserved in the motion. The most "rugged" of these adiabatic invariants is the particle magnetic moment μ. Here, we prove that if the time scale for field changes is larger than $\tau_g = \Omega^{-1}$ (the gyroperiod), then μ is conserved. We refer the reader to Schmidt (1966) for proof that conservation also holds if the length scale for changes in $\mathbf{B}$ is much greater than r_{gi} (the ion gyroradius).

If $\mathbf{B}$ is uniform in space but time varying,

$$\partial\mu/\partial t = \partial/\partial t(K_\perp/B) = (\partial K_\perp/\partial t)/B - (\partial B/\partial t)\left(K_\perp/B^2\right)$$

The rate of change of perpendicular energy in the guiding center approximation can be estimated by the energy gained in one gyration divided by the time the gyration takes. Thus,

$$\Delta K_\perp/\Delta t = \left(\Delta K_g/\tau_g\right)$$

and

$$\Delta K_g = \int \mathbf{F} \cdot d\mathbf{l} = q \int (\mathbf{E} + \mathbf{V} \times \mathbf{B}) \cdot d\mathbf{l}$$

where the integral is around one gyroloop of the particle motion. We note that $d\mathbf{l} = \mathbf{V}dt$ so

$$(\mathbf{V} \times \mathbf{B}) \cdot d\mathbf{l} = (\mathbf{V} \times \mathbf{B}) \cdot \mathbf{V} \, dt = 0$$

This is an example of the fact that magnetic forces do not change particle energy. The energy change in one gyration is thus

$$\Delta K_g = \int q\mathbf{E} \cdot d\mathbf{l}$$

We can transform this to a surface integral

$$\Delta K_g = q \iint (\nabla \times \mathbf{E}) \cdot d\mathbf{a}$$

and using (2.23a)

$$\Delta K_g = -q \iint (\partial \mathbf{B}/dt) \cdot d\mathbf{a}$$

Note that the vectors $q\mathbf{B}$ and $d\mathbf{a}$ are in opposite directions. Now, since $\mathbf{B}$ is uniform in space, we can move it outside the integral and

$$\Delta K_\perp/\Delta t = \Delta K_g/\tau_g = \left(q\pi r_g^2/\tau_g\right)(\partial B/\partial t) = (K_\perp/B)(\partial B/\partial t)$$

and finally

$$\partial \mu/\partial t = (K_\perp/B)(\partial B/\partial t)/B - (\partial B/\partial t)\left(K_\perp/B^2\right) = 0$$

and so μ is conserved.

For illustrative purposes, we can apply these findings to a "magnetic bottle" geometry similar to the earth's magnetic field as illustrated in Fig. 2.10. Suppose a particle starts on the symmetry axis at the place where the field strength is B_0 and the velocity vector of the particle makes an angle α (the pitch angle) with respect to the magnetic field line. At the magnetic equator, then,

$$V_\| = V \cos\alpha, \quad V_\perp = V \sin\alpha$$

The velocity $V_\|$ carries the particle into a region of larger field strength, as shown in Fig. 2.10b. However, if μ is conserved, as $|\mathbf{B}|$ increases $MV_\perp^2/2$ must also increase. Since the particle energy must also be conserved, this can only

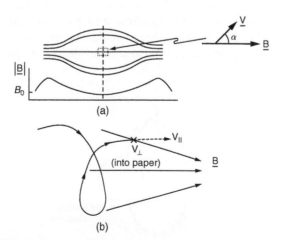

(a)

(b)

Figure 2.10 (a) Sketch of the magnetic field lines and the corresponding magnetic field magnitude along the axis of a magnetic bottle. (b) The gyromotion associated with a converging magnetic field.

come at the expense of $V_{\parallel}$ and the particle slows down in its parallel motion. That the particle eventually bounces off the increasing magnetic field and returns toward the equatorial plane and the other hemisphere can be seen as follows. The magnitude of the Lorentz force is $|\mathbf{F}_L| = qV_{\perp}B$. From Fig. 2.10b it is clear that the direction of the force is to the left, which is opposite to the particle parallel velocity. Even at the right-hand side of the figure where the parallel velocity has gone to zero, we see that $\mathbf{F}_L = q\mathbf{V}_{\perp} \times \mathbf{B}$ still has a small component of force toward the magnetic equator due to the convergence of the magnetic field lines. The parallel velocity of the particle will increase until the equatorial plane is crossed. Then the axial component of the Lorentz force reverses sign and the particle again slows down and is eventually reflected. If the initial pitch angle is too small, the particle will penetrate so deeply into the atmosphere at the end of the "bottle" that it will be lost by collisions. A "loss cone" then develops in the distribution function such that particles with pitch angles less than a certain value α_0 escape and particles with $\alpha > \alpha_0$ are trapped for many bounce cycles.

This bounce motion leads to another time scale with a time constant τ_{bounce}, which can be considered in the particle dynamics. A second adiabatic invariant is related to this dynamical time scale. The second invariant holds only for time variations much longer than τ_{bounce}. Finally, the gradient and curvature drifts take particles entirely around the earth, creating another time scale τ_{drift} that is related to a third adiabatic invariant. This has an even longer time scale, so it is the easiest condition to break down. In general, we have

$$\tau_{\text{drift}} \gg \tau_{\text{bounce}} \gg \tau_{\text{gyration}}$$

and the particles gyrate many times between mirror points and bounce between the mirrors many times, while the gradient and curvature drifts move them around the earth. Sketches of these three oscillatory motions are presented in Fig. 2.11. Neglecting scattering and plasma instabilities, the particles could be

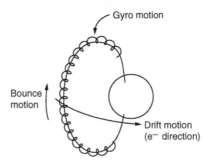

Figure 2.11 The three oscillatory motions in the earth's magnetic field that are associated with the three adiabatic invariants.

trapped in the earth's magnetic field forever. In practice, some species with particular energies (e.g., protons with energies $\sim$100 MeV) are trapped for $\sim$100 years. These particles travel $\sim$1 $\times$ 10^{10} cm/s $\times$ 3 $\times$ 10^9 s = 3 $\times$ 10^{19} cm. Since the scale size of the system is $\sim$$10^9$ cm, such particles traverse the system $\sim$$10^{10}$ times before escaping.

We now raise the interesting question "Can the current in some complicated plasma problem be computed using only the guiding center equations?" As pointed out by Spitzer (1962), the answer is no, since the guiding center current is only part of the total current. The particle gyromotion produces an additional current if there is a population gradient, exactly analogous to the magnetization current, that arises from edge effects in solids. When this contribution is calculated exactly, it contains a term that can even cancel the gradient and curvature drift terms. The actual currents have nothing to do with the values of $\mathbf{B}$ and ∇B, and we must use the macroscopic equations given in Section 2.5.2. The sign of the current is generally given correctly by the sign of the gradient and curvature drifts, but the magnitudes must be derived from other equations.

2.5.2   Magnetohydrodynamics

We now take up a second viewpoint on the dynamics of a collisionless magnetized plasma. In magnetohydrodynamics it is conjectured that the electrical conductivity is so high parallel to $\mathbf{B}$ that in a reference frame moving with velocity

$$\mathbf{V} = \mathbf{E} \times \mathbf{B}/B^2$$

the electric field vanishes both parallel and perpendicular to $\mathbf{B}$. Since this is the plasma reference frame, the plasma can be considered as a single fluid having infinite conductivity. Study of such a one-fluid model is referred to as magnetohydrodynamics. Its range of applicability is particularly great in astrophysics due to the large distance scales involved. Thus, plasmas in the solar wind, solar flares, sunspots, the earth's magnetosphere, and interstellar regions can all be treated to first order with a magnetohydrodynamic approach.

A close analogy exists with hydrodynamics, in which, as we have noted, a single fluid is subject to a variety of forces: pressure, viscosity, gravity, and so forth. The presence of an electrical conductivity requires the inclusion of Maxwell's equations and a volume force on the fluid given by $\mathbf{J} \times \mathbf{B}$. We thus have the continuity equation

$$\partial \rho / \partial t + \nabla \cdot (\rho \mathbf{V}) = 0 \tag{2.55a}$$

the equation of motion,

$$\rho (d\mathbf{V}\, dt) = -\nabla p + \rho \mathbf{g} + \mathbf{J} \times \mathbf{B} \tag{2.55b}$$

and Maxwell's equations,

$$\nabla \times \mathbf{B} = \mu_0 \mathbf{J}$$

$$\nabla \times \mathbf{E} = -\partial \mathbf{B}/\partial t$$

$$\nabla \cdot \mathbf{B} = 0$$

$$\nabla \cdot \mathbf{E} = \rho_c/\varepsilon_0$$

where the vacuum displacement current has been ignored due to the high conductivity and low frequency. These equations constitute the single-fluid description of a magnetized plasma or of conducting fluids such as the element mercury and the earth's molten core.

If we neglect the displacement current, we can also set $\nabla \cdot \mathbf{E} = 0$, since $\rho_c = 0$. Whatever electric fields are present, they change in space only when $\partial \mathbf{B}/\partial t$ is nonzero or the reference frame is changed. These equations constitute an MHD approximation, and to them must be added a relationship between the fields and currents. For a true conducting fluid this is just

$$\mathbf{J}' = \sigma \mathbf{E}'$$

since the conductivity is isotropic. As before, $\mathbf{E}'$ is the electric field in the frame moving with the fluid. In some other frame in which the plasma velocity is $\mathbf{V}$ and the electric field is $\mathbf{E}$, we have $\mathbf{E} = \mathbf{E}' - \mathbf{V} \times \mathbf{B}$ and $\mathbf{J} = \mathbf{J}', \mathbf{B} = \mathbf{B}'$ for $V \ll c$. Thus, $\mathbf{J} = \mathbf{J}' = \sigma \mathbf{E}'$ and

$$\mathbf{J} = \sigma(\mathbf{E} + \mathbf{V} \times \mathbf{B}) \tag{2.55c}$$

with $\mathbf{J}, \mathbf{E}, \mathbf{V},$ and $\mathbf{B}$ all measured in the second frame. The conductivity of a plasma is not isotropic, so it is not clear that the MHD fluid approach should work. However, for a collisionless plasma, σ_0 is so high parallel to $\mathbf{B}$ that in the frame of reference moving with the plasma the electric field is zero. This is also true in an infinitely conducting isotropic fluid. Thus, in the limit that σ_0 goes to infinity, the plasma will behave like an infinitely conducting fluid, even though it is an anisotropic material.

For σ large, $\mathbf{E}' \approx 0$ and (2.55c) is not very useful. Instead, for the perpendicular component of $\mathbf{J}$, we solve (2.55b) for $\mathbf{J}_\perp$ by taking the cross product with the magnetic field. Then,

$$\mathbf{J}_\perp = \frac{\mathbf{B} \times \nabla \rho}{B^2} + \frac{\rho(\mathbf{g} \times \mathbf{B})}{B^2} - \rho\left(\frac{d\mathbf{V}}{dt} \times \mathbf{B}\right)\bigg/B^2 \tag{2.56}$$

which is independent of $\mathbf{E}$.

Finally, we need an equation of state. Several possibilities exist depending on the properties of the fluid. For a true conducting metallic fluid like mercury or the earth's core, we could use the incompressibility condition

$$\nabla \cdot \mathbf{V} = 0$$

For a fluid with a high heat conductivity, we could use isothermal conditions

$$dT/dt = d(p/\rho)/dt = 0$$

or for an adiabatic fluid

$$d(p^\gamma/\rho^\gamma)/dt = 0$$

The particular equation of state to be used depends on the application.

To reduce the number of equations and gain physical insight, the current density and electric field can be eliminated. Returning to the equation of motion and ignoring viscosity and gravity, we have

$$\rho(d\mathbf{V}/dt) = -\nabla p + \mathbf{J} \times \mathbf{B} = -\nabla p + (1/\mu_0)(\nabla \times \mathbf{B}) \times \mathbf{B} \tag{2.57}$$

Using the vector identity

$$\nabla(\mathbf{X} \cdot \mathbf{Y}) = (\mathbf{X} \cdot \nabla)\mathbf{Y} + (\mathbf{Y} \cdot \nabla)\mathbf{X} + \mathbf{X} \times (\nabla \times \mathbf{Y}) + \mathbf{Y} \times (\nabla \times \mathbf{X})$$

with $\mathbf{X} = \mathbf{Y} = \mathbf{B}$ yields

$$\nabla\left(B^2\right) = 2(\mathbf{B} \cdot \nabla)\mathbf{B} + 2\mathbf{B} \times (\nabla \times \mathbf{B})$$

and thus the $\mathbf{J} \times \mathbf{B}$ term in (2.57) becomes

$$f_B = (1/\mu_0)(\nabla \times \mathbf{B}) \times \mathbf{B} = -\nabla\left(B^2/2\mu_0\right) + (\mathbf{B} \cdot \nabla)\mathbf{B}/\mu_0$$

and the equation of motion becomes

$$\rho(d\mathbf{V}/dt) = -\nabla\left(p + B^2/2\mu_0\right) + (\mathbf{B} \cdot \nabla)\mathbf{B}/\mu_0 \tag{2.58}$$

We study this equation in more detail following. In a similar fashion we may eliminate E from Maxwell's magnetic field equation,

$$\partial \mathbf{B}/\partial t = -(\nabla \times \mathbf{E}) = -\nabla \times \left(\mathbf{J}/\sigma - \mathbf{V} \times \mathbf{B}\right)$$
$$= \nabla \times (\mathbf{V} \times \mathbf{B}) - (1/\sigma)(\nabla \times \mathbf{J})$$
$$= \nabla \times (\mathbf{V} \times \mathbf{B}) - \nabla \times (\nabla \times \mathbf{B})/\sigma\mu_0$$

But using $\nabla \times (\mathbf{V} \times \mathbf{B}) = \nabla (\nabla \cdot \mathbf{B}) - \nabla^2 \mathbf{B} = -\nabla^2 \mathbf{B}$ yields finally

$$\partial \mathbf{B}/\partial t = \nabla \times (\mathbf{V} \times \mathbf{B}) + (1/\sigma\mu_0)\, \nabla^2 \mathbf{B} \tag{2.59}$$

The MHD equations can now be written

$$\partial \rho/\partial t + \nabla \cdot (\rho \mathbf{V}) = 0 \tag{2.60a}$$

$$\nabla \cdot \mathbf{B} = 0 = \nabla \cdot \mathbf{E} \tag{2.60b}$$

$$\rho \left(d\mathbf{V}/dt \right) = -\nabla \left(p + B^2/2\mu_0 \right) + (\mathbf{B} \cdot \nabla)\, \mathbf{B}/\mu_0 \tag{2.60c}$$

$$\partial \mathbf{B}/\partial t = \nabla \times (\mathbf{V} \times \mathbf{B}) + (1/\sigma\mu_0)\, \nabla^2 \mathbf{B} \tag{2.60d}$$

One of the equations of state completes the system.

The concepts of magnetic pressure and tension are derivable from (2.60c). One should remember that the terms involving the magnetic field **B** all stem from the $\mathbf{J} \times \mathbf{B}$ force but that **J** and **B** are inextricably related through Maxwell's equations. Thus, any deviation of **B** from a force-free configuration must be balanced by fluid pressure when the fluid is in equilibrium. Fluid acceleration occurs in the case of a nonequilibrium condition. For example, it is straightforward to show that the dipole magnetic field in the upper portion of Fig. 2.12 is such that the magnetic terms in (2.60c) cancel everywhere outside the core of the earth. The distorted dipole field shown following, however, has forces that will cause the plasma to flow back toward the earth. Since $\nabla \times \mathbf{B} \neq 0$ in this distorted field, there must be a current in the region, as indicated. Such a current is associated with a $\mathbf{J} \times \mathbf{B}$ force on the plasma in the direction toward the earth, since there is a small component of **B** upward while **J** is out of the page. This $\mathbf{J} \times \mathbf{B}$ force is equivalent to the magnetic pressure and magnetic tension forces. In a steady state, then, there must be a particle pressure gradient pointing toward the earth.

The meaning of the magnetic terms can be understood in a local coordinate system (see Fig. 2.13) defined by a unit vector $\hat{s}$ parallel to **B**, a unit vector $\hat{n}$

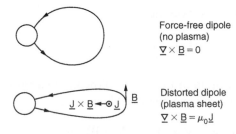

Figure 2.12 Illustration of the difference between a force-free dipole and a distorted field configuration. The magnetic tension force is related to the $\mathbf{J} \times \mathbf{B}$ force shown.

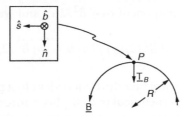

Figure 2.13 Local coordinate system used to describe a curved magnetic field. The inset gives the directions of the unit vectors at the point P. In a stretched magnetic field the tension force $\mathbf{T}_B$ is parallel to and inversely proportional to R.

normal to $\mathbf{B}$ (and antiparallel to the radius of curvature), and a unit vector $\hat{b}$ given by $\hat{b} = \hat{n} \times \hat{s}$. We have just shown that $\mathbf{J} \times \mathbf{B}$ may be written in the form of the magnetic force $\mathbf{f}_B = \mathbf{J} \times \mathbf{B}$, or

$$\mathbf{f}_B = (1/\mu_0)(\mathbf{B} \cdot \nabla)\mathbf{B} - \nabla\left(B^2/2\mu_0\right) \tag{2.61}$$

where

$$\mathbf{B} = B\hat{s}$$

In terms of $\hat{s}$, $\hat{n}$, and $\hat{b}$,

$$(\mathbf{B} \cdot \nabla)\mathbf{B} = B(\partial/\partial s)(B\hat{s}) = B(\partial B/\partial s)\hat{s} + B^2(\partial\hat{s}/\partial s)$$

Now

$$\partial\hat{s}/\partial s = \hat{n}d\theta/Rd\theta = \hat{n}/R$$

so

$$(\mathbf{B} \cdot \nabla)\mathbf{B} = B(\partial B/\partial s)\,\hat{s} + \left(B^2/R\right)\hat{n}$$

Substituting into (2.61),

$$\mathbf{f}_B = \left(B^2/2\mu_0 R\right)\hat{n} + (B/\mu_0)\,(\partial B/\partial s)\,\hat{s} - \nabla\left(B^2/2\mu_0\right)$$

But

$$(1/\mu_0)(B\partial B/\partial s) = \partial\left(B^2/2\mu_0\right)/\partial s$$

so this term cancels the $\hat{s}$ component of $\nabla B^2/2\mu_0$ and we have finally

$$\mathbf{f}_B = \left(B^2/2\mu_0 R\right)\hat{n} - \left[\partial\left(B^2/2\mu_0\right)\Big/\partial n\right]\hat{n} - \left[\partial\left(B^2/2\mu_0\right)\Big/\partial b\right]\hat{b}$$

The magnetic $\mathbf{J} \times \mathbf{B}$ forces are therefore equivalent to a pressure $B^2/2\mu_0$ acting isotropically in the plane perpendicular to $\mathbf{B}$ plus a force $\mathbf{T}_B$ also normal to $\mathbf{B}$, which acts in the plane of curvature of the magnetic field. The sense of $\mathbf{T}_B$ is similar to that of forces due to tension in a stretched string, which are parallel to $\hat{n}$ and proportional to $1/R$ (see Fig. 2.13). As noted above, for a dipole field the tension and pressure forces in the $\hat{n}$ direction exactly cancel (as they must when $\nabla \times \mathbf{B} = 0$).

The case of an equilibrium requires $dV/dt = 0$, which in turn implies from (2.57)

$$0 = -\nabla p + \mathbf{J} \times \mathbf{B}$$

which is equivalent to

$$\nabla\left(p + B^2/2\mu_0\right) = (\mathbf{B} \cdot \nabla)\,\mathbf{B}/\mu_0$$

The magnetic pressure $(B^2/2\mu_0)$ enters just like the particle pressure, while the curvature (tension) term appears on the right-hand side of this force balance equation. These concepts are quite useful in discussing magnetospheric dynamics and equilibria.

Equation (2.60d) also has an interesting and useful interpretation. Assuming first that σ is infinite,

$$\partial\mathbf{B}/\partial t - \nabla \times (\mathbf{V} \times \mathbf{B}) = 0$$

Consider the magnetic flux ϕ across an arbitrary surface Σ moving with velocity $\mathbf{V}$:

$$d\phi/dt = \iint\limits_S (\partial\mathbf{B}/\partial t) \cdot d\mathbf{a} + \int \mathbf{B} \cdot (\mathbf{V} \times d\mathbf{l})$$

The first integral yields the change in ϕ due to the time variation of $\mathbf{B}$ and the second integral yields the change due to the motion of the surface. Working on this second term,

$$\int \mathbf{B} \cdot (\mathbf{V} \times d\mathbf{l}) = -\int (\mathbf{V} \times \mathbf{B}) \cdot d\mathbf{l} = -\iint\limits_S \nabla \times (\mathbf{V} \times \mathbf{B}) \cdot d\mathbf{a}$$

so

$$d\phi/dt = \iint\limits_{S} [\partial\mathbf{B}/\partial t - \nabla \times (\mathbf{V} \times \mathbf{B})] \cdot d\mathbf{a}$$

But for σ infinite we have shown that the bracket vanishes so

$$d\phi/dt = 0$$

Thus, the flux is constant through the surface. We say that the magnetic fluid is "frozen in" and can be considered to move with the fluid. In this sense, if the magnetic field line is labeled by particles on the line at time t, they still label the line at any other time. The concept of frozen-in field lines is very useful and allows visualization of complex flow if we know the magnetic field geometry. Conversely, if the motion is known, the field geometry can be deduced.

If $\sigma \neq \infty$, the field can slip through the fluid, and we have

$$\partial\mathbf{B}/\partial t - \nabla \times (\mathbf{V} \times \mathbf{B}) = (1/\mu_0\sigma)\nabla^2\mathbf{B}$$

For a stationary case ($\mathbf{V} = 0$),

$$\partial\mathbf{B}/\partial t = (1/\mu_0\sigma)\nabla^2\mathbf{B}$$

which is a diffusion equation. If the diffusion scale length is L, the diffusion time constant is given by

$$\tau = \mu_0\sigma L^2$$

For typical laboratory dimensions, L is small and so, even for good conductors, τ is short. In cosmic plasmas or conducting fluids, however, τ is large, and the concept of frozen-in fields is correspondingly important.

Before leaving this section, it is useful to derive an energy relationship based on MHD principles. The particle pressure $p = nk_BT$ is equivalent to the particle energy density, and this is also true for magnetic pressure and magnetic energy density. That is, $B^2/2\mu_0$ yields the energy stored in a magnetic field per unit volume. The total stored magnetic energy in a system is then

$$W_B = (1/2\mu_0) \int B^2 dV$$

where dV is the volume element and we have used a single integral sign to designate a triple integral. Changes of this quantity with time can be written

$$\partial W_B/\partial t = (1/\mu_0) \int (\mathbf{B} \cdot \partial\mathbf{B}/\partial t)\, dV$$

Then using $\partial \mathbf{B}/\partial t = -\nabla \times \mathbf{E}$, $\mathbf{B} = \mu_0 \mathbf{H}$, and the vector identity $\nabla \cdot (\mathbf{E} \times \mathbf{B}) = \mathbf{B} \cdot (\nabla \times \mathbf{E}) - \mathbf{E} \cdot (\nabla \times \mathbf{B})$, we have

$$\partial W_B/\partial t = - \int \nabla \cdot (\mathbf{E} \times \mathbf{H}) dV - \int \mathbf{E} \cdot (\nabla \times \mathbf{H}) dV$$

Using $\mathbf{J} = \sigma(\mathbf{E} + \mathbf{V} \times \mathbf{B})$ and $\mathbf{J} = \nabla \times \mathbf{H}$ and applying the divergence theorem to convert the first volume integral to a surface integral gives

$$\partial W_B/\partial t = - \iint_S (\mathbf{E} \times \mathbf{H}) \cdot d\mathbf{a} - \int [(\mathbf{J}/\sigma - \mathbf{V} \times \mathbf{B}) \cdot \mathbf{J}] dV$$

where the area element $d\mathbf{a}$ points outward from the surface of the volume. Finally, rearranging this equation yields

$$\partial W_B/\partial t = - \iint_S (\mathbf{E} \times \mathbf{H}) \cdot d\mathbf{a} - \int \left(J^2/\sigma \right) dV - \int [\mathbf{V} \cdot (\mathbf{J} \times \mathbf{B})] dV \quad (2.62)$$

In words, the change in stored magnetic energy in a volume equals the energy flux across the surface into the volume in the form of the Poynting flux $(\mathbf{E} \times \mathbf{H})$, minus the resistive energy loss inside the volume, minus the mechanical work done against the $\mathbf{J} \times \mathbf{B}$ force inside the volume.

In a truly closed magnetosphere with the surface an equipotential, no energy crosses the surface ($\mathbf{E}$ is everywhere normal to an equipotential), and there could be no internal circulation (convection), no dissipation by ionospheric currents, and no storage of magnetic energy for later release (e.g., in what are called magnetic substorms). Two sources for generating a component of $\mathbf{E}$ parallel to the magnetopause, and thus a net Poynting flux inward, are viscous interaction and reconnection. Both of these processes thus can result in a net flow of energy into the magnetosphere. We return to these matters in Chapter 8.

2.6   Are Ionospheric Electric Fields Real?

Vasyliunas and Song (2005) have argued that the paradigm used in ionospheric physics is based on a false premise: that ionospheric electric fields cause the plasma motions. They argue that the plasma velocity is fundamental and that it generates the electric field rather than the other way around. Such an argument is clearly valid in the case of the solar wind, which is driven by intense pressure and has no electric field in the wind frame. In the frame of the earth, however, there is an electric field equal to $-\mathbf{V} \times \mathbf{B}$ in a region that is connected to the polar ionosphere by magnetic field lines. Is this field real, or is it generated by the motion of the polar cap plasma?

Information is transmitted from one region to the other in the solar wind–magnetosphere–ionosphere system by waves of various types. For example, the solar wind generates Alfvén waves, which travel down magnetic field lines carrying energy and momentum. This system can be described as a transmission line with the ionosphere as the load. If the load impedance (the field line-integrated Pedersen conductivity (Σ_P)) does not match the transmission line impedance (η_A), reflection will occur, but after a few bounce times this electrodynamic description may be replaced by what is termed electroquasistatics. Most of ionospheric physics is carried out in this context.

Vasyliunas and Song (2005) have argued that the use of quasistatics hides the fundamental physics and that electric fields cannot be thought of as acting on the plasma. Rather, only the plasma velocity matters, and the electric field is an illusion of one's frame of reference. In this approach, the waves put the ionosphere in motion and the electric field follows. However, decades of successful application of electroquasistatics in the ionosphere should not be replaced when it is applicable. This approach is valid on time scales longer than the scale of the system divided by the Alfvén speed or equivalently, for the Alfvén wavelength larger than the system scale. We argue that if you live on a resistor—for example, the ionosphere—then the electric field and Joule heating are very real and can be used as a proper language. In this text we use electroquasistatics whenever possible. At times we do use the full electrodynamic approach, such as for the solar wind–generated Poynting flux input to the polar cap.

2.7 Coordinate Systems

In standard meteorological practice, a local coordinate system has the x-axis eastward, the y-axis to the north, and the z-axis vertically upward. The three components of the neutral wind vector U are usually denoted by (u, v, w) in those coordinates. We will use this notation here as well.

Some complication arises in ionospheric plasma studies due to the importance of the magnetic field direction and the fact that it varies from horizontal at the magnetic equator to vertical at the poles. The reader should be alert to this and to the fact that the magnetic coordinate systems vary somewhat in the text. For example, in Chapters 3 and 4 we use a coordinate system in which the magnetic field is in the direction of the y-axis. The conductivity tensor does not, then, have the form given in (2.38). In deriving (2.38) and in Chapters 5–10 we have taken B parallel to the z-axis. The reader should be aware of the possible confusion caused by the use of different coordinate systems. Furthermore, in the Northern Hemisphere the "z-axis" associated with meteorology is nearly antiparallel to the magnetic field-aligned z-axis. In addition, researchers sometimes define a $\hat{z}'$ axis antiparallel to B in the Northern Hemisphere.

References

Banks, P. M., and Kockarts, G. (1973). *Aeronomy*, Parts A and B. Academic Press, New York.

Carpenter, D. L., Stone, K., Siren, J. C., and Crystal, T. L. (1972). Magnetospheric electric fields deduced from drifting whistler paths. *J. Geophys. Res.* 77, 2819.

Farley, D. T. (1959). A theory of electrostatic fields in a horizontally stratified ionosphere subject to a vertical magnetic field. *J. Geophys. Res.* 64, 1225.

———. (1960). A theory of electrostatic fields in the ionosphere at nonpolar geomagnetic latitudes. *J. Geophys. Res.* 65, 869.

Goldstein, H. (1950). *Classical Mechanics*. Addison-Wesley, Reading, MA.

Gonzales, C. A., Kelley, M. C., Carpenter, D. L., Miller, T. R., and Wand, R. H. (1980). Simultaneous measurements of ionospheric and magnetospheric electric fields in the outer plasmasphere. *Geophys. Res. Lett.* 7, 517.

Jackson, J. D. (1975). *Classical Electrodynamics*. Wiley, New York.

Johnson, F., ed. (1961). *Satellite Environment Handbook*. Stanford University Press, Stanford, CA.

Landau, L. D., and Lifshitz, E. M. (1959). *Fluid Mechanics*. Pergamon, Oxford.

Mozer, F. S. (1970). Electric field mapping from the ionosphere to the equatorial plane. *Planet. Space Sci.* 18, 259.

Mozer, F. S., and Serlin, R. (1969). Magnetospheric electric field measurements with balloons. *J. Geophys. Res.* 74, 4739.

Rishbeth, H., and Garriott, O. K. (1969). *Introduction to Ionospheric Physics*. Int. Geophys. Ser., Vol. 14. Academic Press, New York.

Schmidt, G. (1966). *Physics of High Temperature Plasmas: An Introduction*. Academic Press, New York.

Schunk, R. W., and Nagy, A. (2000). *Ionospheres, Physics, Plasma Physics and Chemistry*. Cambridge University Press, Cambridge, UK.

Smith, E. V. P., and Gottlieb, D. M. (1974). Solar flux and its variations. *Space Sci. Rev.* 16(5–6), 771–802.

Spitzer, L. (1962). *Physics of Fully Ionized Gases*. Wiley (Interscience), New York.

Vasyliunas, V. M., and Song, P. (2005). Meaning of ionospheric Joule heating. *J. Geophys. Res.* 110, A02301, doi:10.1029/2004JA010615.

3 Dynamics and Electrodynamics of the Equatorial Zone

In this chapter we study the dynamics and electrodynamics of the magnetic equatorial zone. To a great extent our knowledge of the electrical structure of this region comes from measurements made at the Jicamarca Radio Observatory, located just east of Lima, Peru. This incoherent scatter radar facility was designed to optimize measurements of plasma flow perpendicular to the earth's magnetic field, which is nearly horizontal over the site. In the F region of the ionosphere, κ_i and κ_e are very large, so the ion and electron velocities perpendicular to $\mathbf{B}$ are very nearly equal to each other. This means that a plasma flow velocity can be uniquely defined and related to the electric field. We deal first with the generation of electric fields by thermospheric winds in the F region and follow with an analysis of the E-region dynamo and the equatorial electrojet. The latter is an intense current jet that flows in the E region at the magnetic equator. These dynamos are the primary sources of the low-latitude electric field, but high-latitude processes also contribute and are discussed as well. An introduction to D-region dynamics is also presented.

3.1 Motions of the Equatorial F Region: The Database

We choose first to study the equatorial F-layer dynamo, since it seems conceptually simpler, although it was not discovered first. Most of the data concerning equatorial electrodynamics come from incoherent scatter radar observations near the magnetic equator over Jicamarca, Peru. Details concerning Jicamarca and several other observatories are given in Appendix A, along with a discussion of the measurement method. The Jicamarca system is capable of determining the plasma temperature, density, composition, and ion drift velocity as functions of altitude and time from the backscatter due to thermal fluctuations in the plasma. The radar can be directed perpendicular to the magnetic field, where the frequency width of the backscatter spectrum is very narrow—meaning that even small mean Doppler shifts can be detected and converted to very accurate ionospheric drift velocities. In practice, the radar is split into two beams, one oriented 3° to the geomagnetic east and one 3° to the geomagnetic west. The difference

The Earth's Ionosphere: Plasma Physics and Electrodynamics

of the Doppler shifts detected by the two beams yields the zonal eastward drift speed of the ion gas, and their average yields the vertical drift speed of the ions. In the F region, κ_i is very large, and the ion velocity perpendicular to **B** is given by (2.36d)—that is,

$$(\mathbf{V}_i)_\perp = \left[\mathbf{E} - (k_B T_i/q_i)\nabla n/n + (M/q_i)\mathbf{g}\right] \times \left[\mathbf{B}/B^2\right] \tag{3.1}$$

where all quantities are measured in the earth-fixed frame. Notice that the neutral wind velocity does not appear in this equation, so the radar measurements cannot be used directly to measure any F-region wind components perpendicular to **B**. At typical measurement heights, we can estimate the drift velocities due to the pressure and gravity terms. The former is the order of $k_B T_i/q_i L B$, where L is the density gradient scale length. Taking $L = 10$ km and $T_i = 1000$ K yields a drift velocity of roughly 0.4 m/s. The gravitational drift is even smaller, less than one-tenth of this value. Since the observed drift velocities are much greater than these values, they must be due to the electric field.

These results hold for both daytime and nighttime conditions; that is, incoherent scatter radar measurements in the F region can, in principle and often in practice, yield the ion drift and thus the electric field at all local times (Fejer, 1991). There are some limitations of the technique, however. A plasma instability termed equatorial spread F often occurs in the evening and, when present, precludes incoherent scatter measurements, even in the upper F region. Also, the incoherent scatter method requires a minimum plasma density in the scattering volume determined by the system noise, antenna size, transmitter power, integration time, and so on. For the Jicamarca Radar Observatory, this minimum is about 10^4 cm^{-3}, which often precludes measurements at night in the altitude range below the F peak. Barium ion cloud releases, rocket probes, and radar interferometric methods have been used successfully in this height range.

A compilation of the Jicamarca data set spanning nearly an entire solar cycle is presented in Figs. 3.1 and 3.2. In Fig. 3.1 the zonal eastward drift measured near the F peak is plotted, whereas in Fig. 3.2 the vertical drift component is shown. Both solar and seasonal dependencies are illustrated, and both components have a strong diurnal modulation. To first order, the drifts are up and to the west during the day and down and to the east at night.

Some aspects we wish to emphasize are as follows:

1. The peak eastward drift at night is twice as great as the peak westward drift during the day.
2. The zonal drifts are much larger than the vertical velocities.
3. The vertical drift is often strongly enhanced just after sunset but shows no comparable feature near sunrise. This is termed the prereversal enhancement of the vertical drift or, equivalently, of the eastward electric field component.
4. There are strong solar cycle effects in the vertical drifts and moderate seasonal effects in both data sets.

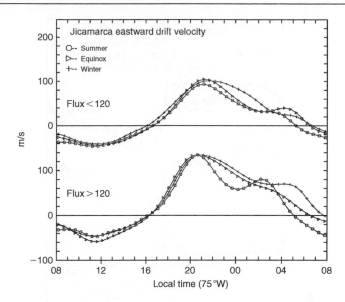

Figure 3.1 Seasonal variations of the zonal plasma drifts during periods of low and high solar flux. [After Fejer et al. (1991). Reproduced with permission of the American Geophysical Union.]

As just discussed, these drifts are directly related to the ambient perpendicular electric field (where, as usual, we mean the component perpendicular to **B**) through

$$\mathbf{E}_\perp = -\mathbf{V}_i \times \mathbf{B} \tag{3.2}$$

The magnetic field over Jicamarca is about 2.5×10^{-5} tesla. Thus, a zonal eastward drift is due to a vertically downward electric field component, with a 100 m/s drift corresponding to a 2.5 mV/m electric field. Likewise, an upward drift of 40 m/s corresponds to an eastward electric field of about 1 mV/m.

Evidence for the repeatability of the average drifts just plotted is given in Fig. 3.3, where a large number of individual 24 h measurement sets are superimposed. Except for the July 1968 event, which was associated with changes in the interplanetary magnetic field, the data are remarkably well behaved. Such a plot suggests that the temporal sequences plotted in these figures can be interpreted equally well in terms of Universal Time measurements, at least in an average sense. That is, if we take a snapshot of the instantaneous electric field pattern around the earth's equatorial zone, it should look very much like the Jicamarca equatorial measurements as a function of local time plotted here. One test of this hypothesis uses the fact that for electrostatic fields, the Maxwell equation $\nabla \times \mathbf{E} = 0$ implies

$$\oint \mathbf{E} \cdot d\mathbf{l} = 0 \tag{3.3}$$

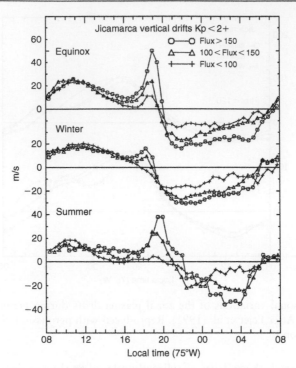

Figure 3.2 Average vertical plasma drifts measured at Jicamarca during equinox (March–April, September–October), winter (May–August), and summer (November–February) for three levels of solar flux. [After Fejer et al. (1991). Reproduced with permission of the American Geophysical Union.]

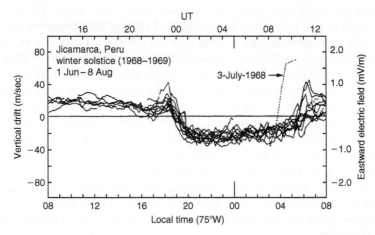

Figure 3.3 Superimposed vertical drifts (eastward electric fields) measured at Jicamarca, Peru. [After Woodman (1970). Reproduced with permission of the American Geophysical Union.]

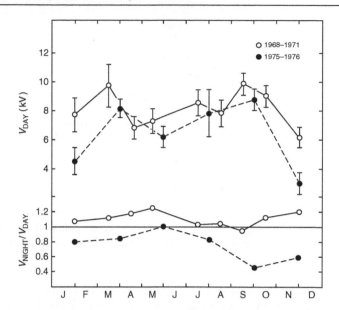

Figure 3.4 Monthly variation of the local daytime voltage drop between the reversals and of the nighttime-to-daytime ratio. Standard deviation of the latter is about 15–20%. [After Fejer et al. (1979). Reproduced with permission of the American Geophysical Union.]

That is, the line integral of the zonal electric field entirely around the earth must vanish at any one time. An equivalent statement is that the voltage difference between the dusk and dawn terminators (see Fig. 3.5) when the electric field is westward (at night) should equal the voltage drop when the field is eastward (during the day). Fejer et al. (1979) performed this calculation for a number of days using the Jicamarca time series, and the results are plotted in Fig. 3.4. To first order, the voltage drop across the dayside is close to the drop across the night, both equaling about 8 kV. The agreement is not perfect, however, and the ratio $V_{\text{NIGHT}}/V_{\text{DAY}}$ dropped as low as 0.5 in October 1975–1976, indicating that local conditions may affect the Jicamarca measurements and that the 24-hour plots may not represent an instantaneous pattern in some seasons. The prereversal enhancement may also have a different seasonal behavior in different longitude sectors. Evidence for this comes from the different seasonal behavior of the equatorial spread F around the globe (see Chapter 4). Since the prereversal enhancement (17–19 LT) accounts for 40% of the dayside voltage, its variability is important. Pingree and Fejer (1987) showed that the region of space around the prereversal enhancement is consistent with $\nabla \times \mathbf{E} = 0$, using the assumption that local time and longitude are equivalent.

From Poisson's equation,

$$\nabla \cdot \mathbf{E} = \rho_c/\varepsilon_0$$

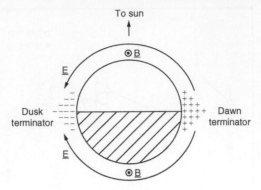

Figure 3.5 Schematic diagram showing the zonal electric field component and its relationship to the charge densities at the terminators. Since the zonal electric field varies slowly with height, the charge density is also a weak function of height from 100 to 500 km.

we expect the observed dayside and nightside fields to be the result of charge buildup at the terminators, with the dusk terminator charged negatively and the dawn terminator positively. Assuming a change of the electric field equal to 2 mV/m over a 2 h local time interval corresponding to a distance of about 3×10^6 m at the equator, $\rho_c \approx 7 \times 10^{-21}$ C/m^3. This is less than one excess elemental charge per cubic meter, again illustrating how very small net charge densities can yield significant electric fields in a plasma. A schematic diagram based on the "charged terminator model" is given in Fig. 3.5 from the viewpoint of looking down on the earth from above the Northern Hemisphere.

3.2   The Equatorial F-Region Dynamo

Before discussing the F-region dynamo, we need to understand a little about the thermospheric winds in the equatorial region, since they provide the source of energy that maintains the electric field. Early observational data on thermospheric winds came from studies of the drag exerted on artificial satellites by the neutral atmosphere (King-Hele, 1970). Analysis of such data shows that the change in satellite inclination is directly related to the angular velocity of the atmosphere with respect to the rotating earth. The surprising conclusion was that, on average, the low-latitude thermosphere superrotates; that is, there is a net eastward average zonal flow of about 150 m/s near 350 km altitude and about 50 m/s at 200 km altitude. The effect is most pronounced in the 2100–2400 local time period. Direct measurements of thermospheric winds are now available at night using the Fabry-Perot technique to determine the Doppler shift of airglow emissions (Sipler and Biondi, 1978; Biondi et al., 1999). Some of these data (which were taken over Kwajalein in the Pacific Sector) are reproduced in Fig. 3.6a in the form of azimuth plots of the tip of a vector in the direction toward which the wind is blowing (note that at Kwajalein 8 UT is equal to 2000 LT).

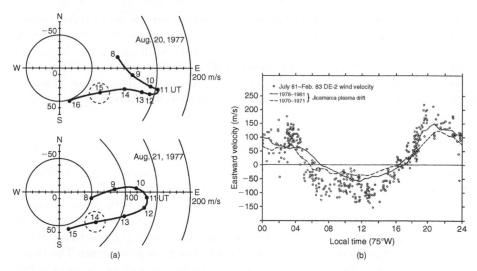

Figure 3.6 (a) Position of the tip of the neutral wind vector measured over Kwajalein as a function of universal time on August 20 and 21, 1977. The dashed circle indicates the 15 m/s uncertainty in the determination. [After Sipler and Biondi (1978). Reproduced with permission of the American Geophysical Union.] (b) Comparison of the Jicamarca average drifts with thermospheric wind data from DE-2. [After Fejer et al. (1985). Reproduced with permission of the American Geophysical Union.]

Although the winds display a high degree of variability from day to day, to a first approximation the winds are eastward and quite strong (~ 150 m/s) in the post-sunset period, decaying in amplitude to less than 50 m/s after midnight. More recent airglow observations in the Peruvian sector are in good agreement with these data (Meriwether et al., 1986).

Airglow observations are restricted to nighttime, but satellite observations have yielded full 24 h coverage of the thermospheric winds (Wharton et al., 1984). Some of these data are presented in Fig. 3.6b, superimposed on the Jicamarca zonal plasma drift pattern. Both are clearly diurnal, with larger winds and plasma drifts occurring at night. There is a lot of scatter in the winds measured by the satellite, but the pattern is unmistakable, as is its correlation with the plasma drifts.

Setting aside the superrotation phenomenon for the moment, we can ask whether forcing due to solar heating of the thermosphere can explain the high winds observed in the postsunset period. Since the Coriolis force vanishes at the equator, in a steady state the winds should blow in the $-\nabla p$ direction from west to east across the sunset terminator. This pressure gradient is due to the subsolar neutral atmospheric temperature bulge that occurs on the sunward side of the earth. The problem is not so much the direction of the wind, which agrees with the data, but the magnitude of the theoretically calculated winds – the initial estimates of the wind speed were too high (Lindzen, 1966).

Without a Coriolis term, the primary limitation on the wind speed comes from the various drag terms. If the plasma is ignored, as the early modelers naturally did, viscosity is the only frictional factor. As discussed in Chapter 2, in the lower atmosphere viscosity is provided by the interaction of eddies in the atmospheric flow—that is, by eddy viscosity. Above the turbopause, classical molecular viscosity becomes the dominant factor. An expression for the viscosity coefficient is given by

$$\eta = 4.5 \times 10^{-5}(T/1000)^{+0.69} \text{ kg/m} \cdot \text{s} \qquad (3.4)$$

where atomic oxygen has been assumed to be the dominant neutral (Dalgarno and Smith, 1962). This is a good approximation in the F region. The vertical temperature gradient is small in the upper thermosphere, and η is therefore virtually independent of height. However, when comparing viscosity with the acceleration terms such as $\mathbf{F} = -(1/\rho)\nabla p$, the kinematic viscosity (η/ρ) is what matters. Unlike η, this ratio increases exponentially with height as ρ decreases. This drastic increase of the ratio η/ρ is responsible for the importance of viscosity in the thermosphere. That a tenuous gas has a high viscosity seems counterintuitive, but what matters is the mean free path that carries velocity shear information from one height to another.

A quantitative comparison between the various terms in the neutral momentum equation (2.21b) has been provided by Rishbeth (1972) and is reproduced in Fig. 3.7. The graph was constructed for 45° latitude, but all parameters (except for the Coriolis term) are applicable to the equatorial case. For now, we are interested in the curves labeled F and $(\eta\tilde{u}/\rho H^2)$. The former is the magnitude of the driving horizontal pressure gradient acceleration term and is deduced from neutral atmospheric measurements. The upper scale corresponds to F and is given in meters per second squared. Since F is inversely proportional to ρ, it increases with height. Surprisingly, however, F increases only roughly linearly with height. To plot the viscous term, we need to scale it appropriately according to the form it takes in the momentum equation, $(\eta/\rho)\nabla^2\mathbf{U}$. We would not expect the winds to vary more quickly with altitude than the scale height H. Suppose the wind changed by $\tilde{u} = 60$ m/s over a scale height. The associated force is the order of $\eta\tilde{u}/\rho H^2$ and is plotted in Fig. 3.7. By 200 km it exceeds all of the other terms. Since the viscous term increases rapidly with height, it will eventually become large enough to suppress all variations of $\mathbf{U}$ with altitude. Therefore, most numerical models require that $\partial\mathbf{U}/\partial z = 0$ as an upper boundary condition.

In his early study of thermospheric winds, Lindzen (1966) took the observed diurnal component of the pressure gradient given experimentally by Harris and Priester (1965) and solved the simplest form of the resulting zonal wind equation at equatorial latitudes:

$$\partial u/\partial t = -(1/\rho)(\partial p/\partial x) + \frac{\eta}{\rho}\frac{\partial^2 u}{\partial z^2} \qquad (3.5)$$

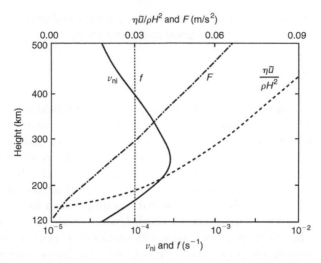

Figure 3.7 Graphs of neutral acceleration time constants as well as viscous and pressure accelerations for midday at sunspot minimum. The upper scale applies to the acceleration due to the pressure gradient, $F(- \cdot - \cdot -)$, and the normalized kinematic viscosity parameter, $\eta \tilde{u}/\rho H^2 (- - -)$. The lower scale refers to the ion drag parameter, $\nu_{ni}(—)$, and the Coriolis parameter for a latitude of $45°$, $f(----)$. [After Rishbeth (1972). Reproduced with permission of Pergamon Press.]

where u is the zonal component of the wind. He found an analytic solution to this equation by generating reasonable models for the functional variations of p and ρ with x and z. Even with a low estimate of the pressure term, however, zonal velocities of about 300 m/s were found in this analysis. In later work, Lindzen (1967) used a numerical solution with a more realistic pressure variation and found winds as high as 550 m/s! We leave this dilemma for the moment to determine the electrodynamic effect of the observed and predicted postsunset zonally eastward thermospheric wind. Later we will show that including frictional drag on the neutrals by the ionospheric plasma explains the lower wind values that are actually observed.

In the remainder of this chapter and in most of the next, we will slightly redefine our coordinate system to preserve the conventional notation that the a_z axis is upward. At the equator, we thus take $\mathbf{B} = |\mathbf{B}|a_y$, which is horizontal and northward, and take a_x toward the east. The conductivity tensor is then

$$\sigma = \begin{pmatrix} \sigma_P & 0 & \sigma_H \\ 0 & \sigma_0 & 0 \\ -\sigma_H & 0 & \sigma_P \end{pmatrix} \tag{3.6}$$

In the F region $\sigma_P \gg \sigma_H$ and the conductivity tensor is diagonal, although it still holds that $\sigma_P \ll \sigma_0$. To a very good approximation,

$$\sigma_P = \frac{ne^2 \nu_{in}}{M\Omega_i^2}$$

in the F region. Assume first that the horizontal magnetic field lines extend forever or, equivalently, that they terminate at both ends in an insulating layer. In fact, the field lines bend and enter the E region, which has a finite conductivity that varies with local time, season, and solar activity. (We return to this point later.) The vertical component of the large-scale neutral wind field in the atmosphere is always small, so we consider a simple model in which the thermospheric wind is eastward and has a magnitude u that is uniform with height. In the text we use the meteorological notation in which $\mathbf{U} = (u, v, w)$, where u, v, and w correspond to the zonal (positive eastward), meridional (positive northward), and upward components, respectively. From (2.41), an electric current will flow with magnitude and direction given by

$$\mathbf{J} \simeq \sigma \cdot (u\mathbf{a}_x \times \mathbf{B})$$

The wind-driven current is therefore vertically upward with magnitude $J_z = \sigma_P u B$. The current J_z is quite small with a peak value of the order of $0.01\ \mu A/m^2$. However, σ_P varies considerably with altitude due to its dependence on the product $n\nu_{in}$. The zonal wind component, u, may also vary with height, but we assume for now that viscosity keeps this variation small. At any rate, $d(\sigma_P u B)/dz \neq 0$, and an electric field must build up in the z direction to produce a divergence-free current. However, the "insulating plates" we have assumed at the ends of the magnetic field lines do not allow a magnetic field-aligned current to flow at all (i.e., $J_y = 0$), so in this first approximation a stronger condition on $\mathbf{J}$ holds than the expression $\nabla \cdot \mathbf{J} = 0$, namely, $\mathbf{J} = 0$.

A plasma density profile for the postsunset equatorial F layer is shown in Fig. 3.8a. As we noted in Section 2.2, gravitational forces do not cause the plasma (with large κ_i, κ_e) to fall at the magnetic equator, since the velocity due to gravity is perpendicular to the gravitational force. Recombination "eats away" at the molecular ions on the bottomside, forming a steep upward-density gradient. The result is a dense O^+ plasma with a well-defined lower boundary. To study the electrodynamics of this region in a little more detail, we approximate the actual situation shown in Fig. 3.8a with the configuration illustrated in Fig. 3.8b, which shows a slab geometry with σ_P constant inside the slab and zero elsewhere and with zonal wind u constant everywhere. Since the current is upward inside the layer and zero outside, charges pile up at the two boundaries as shown in the figure. The magnitude of the electric field that builds up as a result of these charges is such that

$$J_z = \sigma_P E_z + \sigma_P u B = 0$$

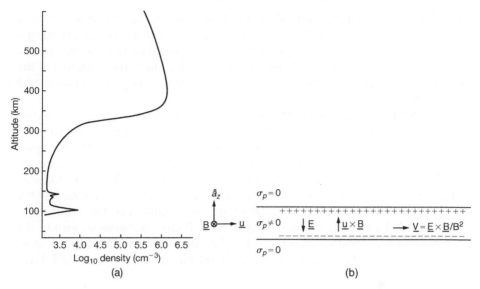

Figure 3.8 (a) Schematic equatorial plasma density profile in the evening local time period. (b) Electrodynamics of the equatorial F region in which the density and conductivity profiles are modeled with a slab geometry, subject to a constant zonal eastward neutral wind. Even though the plasma density does not fall off quickly with height alone, the peak, ν_{in} and thus σ_p, falls off exponentially.

which yields

$$E_z = -uB \tag{3.7}$$

Note that the plasma inside the slab will drift with an $\mathbf{E} \times \mathbf{B}/B^2$ velocity equal in magnitude and direction to the zonal wind speed. Furthermore, the electric field in the reference frame of the neutral wind, $\mathbf{E}' = \mathbf{E} + \mathbf{U} \times \mathbf{B}$, vanishes. This must be true because the current is independent of reference frame and we have set the current equal to zero—that is, since $\mathbf{J} = \mathbf{J}' = \sigma_P \mathbf{E}' = 0$, $\mathbf{E}'$ must be zero.

In this simple model there is a very strong shear in the plasma flow velocity at the two interfaces that is not shared by the driving neutral wind; that is, at the interfaces, the plasma velocity changes abruptly from u to 0. This "prediction" of a sheared plasma flow is intriguing, since such shears have been observed in the equatorial F-region and are discussed following. However, in this model the shear is created somewhat artificially by the slab conductivity assumption.

The insulating end plate assumption made here is most nearly valid at night when the E-region molecular ion and electron pairs rapidly recombine with no offsetting production by sunlight (see, for example, Rishbeth and Garriott, 1969). The dominant O^+ (atomic) ions in the F layer are much longer lived and support the F-layer dynamo. As shown in Fig. 3.1, the maximum nighttime zonal drift and maximum vertical electric field (downward) are about 150 m/s and 4.5 mV/m, respectively. This observed nighttime plasma velocity is consistent in magnitude and direction with the eastward zonal neutral thermospheric wind found experimentally and discussed previously. The vertical pattern of the nighttime electric field is thus consistent with the simple F-region dynamo model we have presented.

To understand the diurnal variation in more detail, however, we need to consider the role of the "end" plates in the E layer, which, contrary to the approximation used previously, are good conductors during the day. First, consider again an idealized slab geometry that ignores the magnetic field line curvature and dip angle but includes conductivity variations with distance (y) along the magnetic field direction. The actual geometry is shown schematically in Fig. 3.9a, and the slab model is shown in Fig. 3.9b. There are no variations in the x direction. In this model the wind is a function of z in the F layer but goes to zero in the E layer. The finite density in the E layer and its attendant conductivity act as an electric load on the dynamo. (In this three-level slab model we take the E-layer

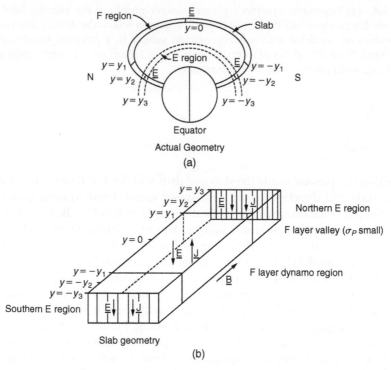

Figure 3.9   (a) Side view of the dipole magnetic field geometry near the magnetic equator. The curves are exaggerated to show the coupling geometry between the F region at the equator and the off-equatorial E region. (b) F-layer slab geometry including conducting end plates in the northern and southern hemispheres.

conductivity to be constant as a function of y.) The current parallel to **B** now is nonzero and must be supplied by the dynamo and completed in the E layer. The full divergence equation must be used. First, as discussed in Chapter 2, we note that for large-scale dynamo sources, the magnetic field lines are nearly equipotentials due to the high ratio of σ_0 to $\sigma_P (\geq 10^5)$. The electric field is thus mapped down to the E-region altitudes, where we have assumed that the neutral wind vanishes in order to study just the F-region dynamo. The $\nabla \cdot \mathbf{J} = 0$ condition in the F region now yields

$$\frac{d\,[\sigma_P(E_z + uB)]}{dz} = \frac{-dJ_y}{dy}$$

We now integrate this equation along the y direction from $y = 0$ to $y = y_1$, which corresponds to integrating from the equatorial plane to the base of the F layer in the Northern Hemisphere. By symmetry the contribution to the integral from the Southern Hemisphere is identical. Symmetry also requires that the field-aligned current vanish directly at the equator $[J_y(0) = 0]$, and the integral yields

$$\int_0^{y_1} \frac{d}{dz}\,[\sigma_P(E_z + uB)]\,dy = -J_y(y_1) \tag{3.8}$$

Due to the low perpendicular conductivity in the "valley" between the E and F regions, the current we have just calculated, which leaves the F-region dynamo at $y = y_1$, must be equal to the field-aligned current that enters the E region at $y = y_2$. In the E layer we have only the (mapped) electric field, E_z, to contend with because we have set the wind speed as $u = 0$ for now. Applying $\nabla \cdot \mathbf{J} = 0$ yields

$$d(\sigma_P E_z)/dz = -dJ_y/dy$$

Integrating from y_2 to y_3 and requiring $J_y(y_3) = 0$—that is, an insulating atmosphere below y_3—we have

$$\int_{y_2}^{y_3} \frac{d}{dz}(\sigma_P E_z)\,dy = J_y(y_2) = J_y(y_1) \tag{3.9}$$

and, finally, using (3.8) and (3.9),

$$\int_0^{y_1} \frac{d}{dz}\left[\sigma_P^F(E_z + uB)\right]dy = -\int_{y_2}^{y_3} \frac{d}{dz}\left(\sigma_{PN}^E E_z\right)dy \tag{3.10}$$

Now E_z is not a function of y (equipotential magnetic field lines), so if as a first approximation we take uB constant with y until the field line enters the F-layer valley, we can perform the y integrals in (3.10), giving

$$\frac{d}{dz}\left[\left(\frac{1}{2}\Sigma_P^F + \Sigma_{PN}^E\right)E_z + \frac{1}{2}\Sigma_P^F uB\right] = 0$$

where Σ_P^F and Σ_P^E are the field line–integrated conductivities discussed in Chapter 2. A more realistic model would include the full y dependence of u and Σ, but some insights are possible, even in this simple case. Including the Southern Hemisphere conductivity, Σ_{PS}^E, and assuming $\Sigma_{PN}^E = \Sigma_{PS}^E = \Sigma_P^E$, the solution for $E_z(z)$ is thus

$$E_z(z) = -\left[u(z)B\Sigma_P^F(z)\right]\Big/\left[\Sigma_P^F(z) + 2\Sigma_P^E(z)\right] \tag{3.11}$$

In (3.11) we have dropped a constant of integration, since the slab geometry is not physically accurate and would be replaced by a smooth transition to a two-dimensional (J_x and J_z) equatorial electrojet current in a more accurate model (see Section 3.4).

Equation (3.11) clearly shows that if the off-equatorial E region is insulating ($\Sigma_P^E = 0$), $E(z)$ will be equal to $-u(z)B$ everywhere. We can now use (3.11) to explain some aspects of the diurnal variation in the vertical electric field. During the nighttime, Σ_P^E becomes quite small and therefore $E_z \approx -uB$ in regions where Σ_P^F is large—for example, near the peak in the F-region plasma density. During the daytime, however, the E-region conductivity is comparable to or larger than the magnetic field line–integrated F-region conductivity. For large Σ_P^E the electrodynamic control of the ionosphere is vested in the E region. As discussed following, a dynamo operates there as well, but it is driven by daytime tidal winds, which are smaller than the thermospheric winds in the F region. This explains the diurnal variation in the F-region zonal plasma drifts plotted in Fig. 3.1. At night the E-region conductivity is low, and the high zonal winds at several hundred kilometers altitude determine the vertical electric field and thus the horizontal plasma flow. For $\Sigma_P^F \gg \Sigma_P^E$ the electric field will be almost equal to uB. The eastward plasma velocity thus nearly matches the neutral wind speed. During the day, however, the F-region dynamo loses control of the electrodynamics, and the resulting electric fields are determined by winds in the E region. Since these winds tend to be weaker, the plasma drift is smaller during the day. Curiously, nighttime E-region winds measured by chemical releases are comparable to the F-region winds (see Chapter 5), but since Σ_P^E is small, they do not seem to create large E-region dynamo electric fields.

An analogous electric circuit is shown schematically in Fig. 3.10. The three batteries correspond to the two (north and south) E-region dynamos and the F-region dynamo. Each battery has a finite internal resistance given by

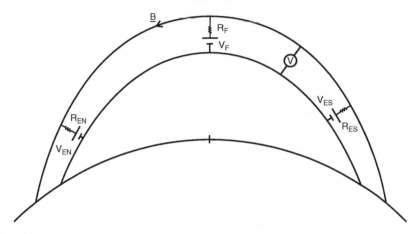

Figure 3.10   Electric circuit analogy to the voltage sources in the equatorial F region. Off-equatorial E-region wind dynamo competes with the F-region dynamo at the equator to determine the voltage differences between magnetic field lines.

$R = (\Sigma_P)^{-1}$. The voltage measured by the meter corresponds to the electric field times the distance between the two magnetic field lines (the conducting wires). The battery with the lowest internal resistance determines the voltage and thus the electric field as well. In the F layer during the nighttime Σ_P^F is larger, and the F region dominates. During the day, E-region sources determine the electric field, and the F layer acts as a load.

A model calculation of the field line–integrated Pedersen conductivity and the vertical electric field is presented in Fig. 3.11 as a function of height measured at the magnetic equator. The local time is 1900, and the driving eastward neutral wind used in the calculation is taken to be constant with a value of 160 m/s. The model used for the ionospheric plasma density is similar to the data plotted in Fig. 3.8a except that the E-region density below 150 km was taken to be constant at a value of 2×10^4 cm^{-3}. The parenthetical expressions in the figure yield the percentage of the total Σ_P found below 300 km along that particular magnetic field line. At first glance the Σ_P values seem high, since the value of σ_P at 300 km in Fig. 2.6 is only about 5×10^{-5} mho/m. However, near the equator the magnetic field lines are nearly horizontal and are the order of several thousand kilometers long. Large values of Σ_P then result. Notice that Σ_P continues to increase with height above the F peak at about 350 km, even though when measured at the magnetic equator, σ_P decreases with height above 350 km. This is due to the length of the field line and to the fact that the field lines crossing the equator "reenter" the F layer at off-equatorial latitudes where the equatorial anomaly occurs (see Chapter 5). Above 400 km altitude, most of the conductivity is located in the F layer, and the local thermospheric winds drive the electrical system. However, as altitude decreases, the E layer begins to short out the electric field

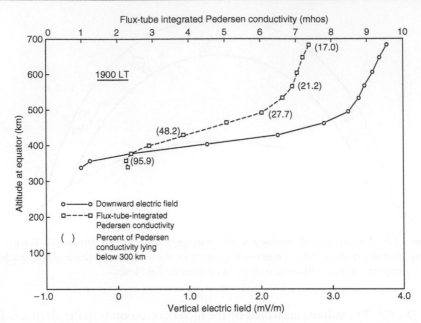

Figure 3.11 Equatorial F-region electric field and conductivity calculations using realistic ionospheric parameters, a 160 m/s zonally eastward neutral wind in the F layer, and a diurnal tidal mode in the off-equatorial E region. [After Anderson and Mendillo (1983). Reproduced with permission of the American Geophysical Union.]

and it decreases rapidly. A considerable shear thus occurs in the plasma flow due to the rapid change in the driving electric field, even though the thermospheric wind was taken to be uniform with height. This detailed calculation is totally consistent with the simple model presented above. (Note that in Fig. 3.11 the electric field reverses sign at low altitude, since a tidal wind field was included in the E region.)

Historically such a shear has been difficult to measure with an incoherent scatter radar, since it occurs where the plasma density decreases to values below the signal-to-noise threshold for plasma velocity measurements. The barium cloud technique can be used, however, and a number of experiments have been performed using sounding rockets. The results are summarized in Fig. 3.12a as a function of altitude and local time. The dashed line roughly indicates the base of the F layer. Below the F layer the plasma velocity is strongly westward. Thus, the downward electric field is not only reduced but also reversed in sign. Simultaneous neutral wind measurements displayed velocities that were very different from the plasma velocities. These surprising results were verified by measuring the velocity of plasma irregularities, using the Jicamarca radar system in an interferometer mode. In this technique coherent scatter echoes are used to trace the position of localized nonthermal scatterers in the radar beam. Four consecutive

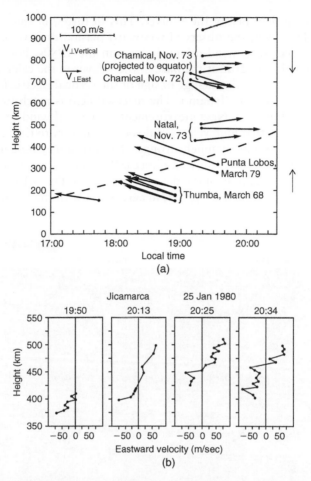

Figure 3.12 (a) Summary plot of plasma velocity derived from a number of barium ion cloud experiments in the equatorial zone. The arrows show the electric field direction. [After Fejer (1981). Reproduced with permission of Pergamon Press.] (b) Zonal drifts of plasma irregularities as a function of height and time using an interferometric technique. Note that the shear point where the velocity is zero moves up as the scattering layer rises. [After Kudeki et al. (1981). Reproduced with permission of the American Geophysical Union.]

altitude profiles of the eastward irregularity velocity are plotted in Fig. 3.12b. The plasma velocities plotted here also change sign with height in agreement with the results from the barium experiments.

A possible explanation for the shear is a variation on the shorting effect just discussed. If the off-equatorial E-region dynamo creates a meridional electric field that is in the poleward direction, and if the E-region dynamo internal resistance is small enough, then the electric field will map to the equatorial plane as a vertically upward field below the F layer. This in turn will cause a westward plasma drift. However, the observed westward plasma drift velocity is higher than was expected for a wind-driven process in the 1980s. For example, the weak westward plasma drift at the base of the F layer in the calculation shown in Fig. 3.11 (indicated by the change in sign of the vertical electric field) was due to the theoretical E-region dynamo. The reversed field is much lower than the measured values, however. New measurements of neutral winds in the E region (see Chapter 5) show that high velocities do occur contrary to tidal theory. Thus, the E-region dynamo is likely much stronger than previously thought. Improved programs at Jicamarca have allowed direct ISR observations of the postsunset equatorial zone. An example of the data is presented in Fig. 3.13. A remarkable vortex is seen in the postsunset region, in agreement with the barium cloud and

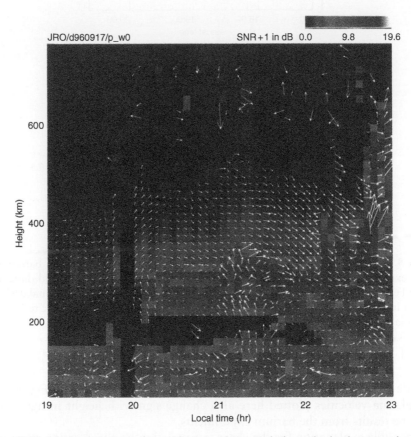

Figure 3.13 A combination of irregularity and $\mathbf{E} \times \mathbf{B}$ drifts and a backscattered power map for September 17, 1996 (sunset time 18:03 LT, Kp = 3−). A data gap exists at ∼20:00 LT. [After Kudeki and Bhattacharyya (1999). Reproduced with permission of the American Geophysical Union.] See Color Plate 1.

interferometer data. This shear most likely has an important role in the plasma physics described in Chapter 4.

3.3   E-Region Dynamo Theory and the Daytime Equatorial Electrojet

To this point, we have discussed only the nighttime vertical equatorial electric field component associated with the F-region dynamo. During the daytime, the vertical field is controlled by the meridional electric field, which is generated by the global E-region dynamo we discuss next. The zonal electric field is generated in both daytime and nighttime by the E-region dynamo. Although the zonal component is small, it is very important because it causes the plasma to move vertically. This motion greatly affects the plasma density, since it causes the F-region plasma to interact with quite different neutral densities as it changes altitude, strongly affecting the recombination rate and, in turn, the plasma content.

The zonal component of the electric field and the daytime vertical component are due primarily to winds in the E region. This E-region dynamo is driven by tidal oscillations of the atmosphere. An excellent book on this topic has been written by Chapman and Lindzen (1970), and we only touch on some aspects of tidal theory here. The largest atmospheric tides are the diurnal and semidiurnal tides driven by solar heating. The semidiurnal lunar gravitational tide is next in strength in the upper atmosphere. It is interesting to note that the lunar semidiurnal tide is the strongest in the case of ocean tides.

One may legitimately question a terminology in which we discuss diurnal "tides" in the lower thermosphere (E region) but refer to a diurnal "wind" in the upper thermosphere (F region). The difference is that the tidal modes propagate into the lower thermosphere from below, whereas the upper thermospheric winds are driven by absorption of energy in the thermosphere itself. We could thus refer to the thermospheric response as being due to an in situ diurnal tide as opposed to a propagating tide, but we will stick with the traditional usage here.

Tidal theory is quite complex. The equations of the neutral atmosphere must be solved on a rotating spherical shell subject to the earth's gravitational field. Considerable insight is obtained by studying the free oscillations of the atmosphere—that is, the normal modes of the system. This is accomplished by reducing the set of equations to one second-order partial differential equation that is often written in terms of the divergence of the wind field. The resulting equation is separable in terms of functions of latitude (θ), longitude (ϕ), altitude (z), and time (t). The longitude/time dependence is of the form $\exp[i(s\phi+ft)]$, where s must be an integer. For $s = 0$, the temporal behavior does not propagate with respect to the earth. For $s = 1$, the disturbance has one oscillation in longitude and propagates westward following the sun; this is the diurnal tide. The θ dependence can be expressed in terms of so-called Hough functions, which may be related to spherical harmonic functions.

The height dependence of these normal modes is a crucial factor in tidal theory, since when forcing (e.g., solar heating) is included in the equations, the only modes excited are those having a vertical structure that matches the vertical structure of the forcing. Although the mathematical models were well developed by 1900, controversy raged over the actual nature of atmospheric tides until definitive measurements of the temperature structure of the earth's upper atmosphere became available from sounding rockets. The features of these temperature profiles, which are of most relevance to tidal theory, are due to the absorption of sunlight by ozone in the stratosphere and by water vapor in the troposphere.

When both the θ and z dependences of forcing functions and the normal modes are taken into account, many of the dominant features of tidal oscillations can be explained. For our purposes we summarize these features as follows:

1. As tidal oscillations propagate upward, the associated wind speed amplitude also grows up to F-region heights, where the amplitude becomes constant. (This amplification is due to the decreasing density of the atmosphere and is a consequence of energy conservation. This important feature of vertical wave propagation is discussed in detail in Chapter 5, where gravity waves are studied.)
2. Diurnal tides can propagate vertically only below 30° latitude. At higher latitudes they remain trapped in the stratosphere.
3. With the decreasing importance of the diurnal tide, the semidiurnal tide becomes dominant at latitudes higher than 30°.

Armed with this modest understanding of tidal theory, we may now investigate some aspects of the E-region dynamo.

In the E region of the ionosphere, the conductivity tensor is not diagonal, and the Hall terms must also be considered. Furthermore, the electric field cannot be taken as being entirely self-generated by local wind fields, as we assumed for the nighttime F-layer dynamo. This can be understood as follows. The entire dayside ionosphere is a good electrical conductor in which currents are driven by lower thermospheric tides. The tidal E-region wind field $U(r, t)$ is global in nature and will drive a global current system given by $J_w = \sigma(r, t) \cdot [U(r, t) \times B]$. Now, because both $\sigma(r, t)$ and $U(r, t)$ depend on r, the current J_w is not, in general, divergence free. Thus, an electric field $E(r, t)$ must build up such that the divergence of the total current is zero and

$$\nabla \cdot [\sigma(E(r, t) + U(r, t) \times B)] = 0 \qquad (3.12)$$

The resulting $E(r, t)$ is as rich and complex as the driving wind field and the conductivity pattern that produce it. The latter has a primarily diurnal variation over the earth's surface, whereas the dominant tidal winds change from diurnal to semidiurnal, depending on latitude and altitude. As a first approximation we might then expect a diurnal electric field pattern at low latitudes and a mixture of diurnal and semidiurnal electric fields at higher latitudes. This crude analysis actually describes the situation fairly well.

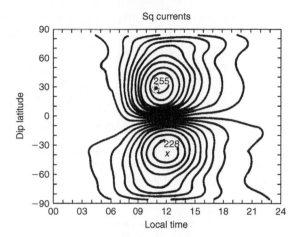

Figure 3.14   Average contours of vertical magnetic field in nT due to the Sq system measured during the International Geophysical Year. [After Matsushita (1969). Reproduced with permission of the American Geophysical Union.]

The diurnal variations of the electric field components measured at Jicamarca have already been pointed out in Section 3.1. The global vertical magnetic perturbation pattern measured by ground-based magnetometers is shown in Fig. 3.14 and gives further evidence for this simple picture of the electric field. Such data have been used to construct the pattern of electrical currents in the entire ionosphere that is referred to as the Sq current system. (S stands for solar and q for quiet in this notation.) As noted in Chapter 2, ground magnetometers respond primarily to horizontal Hall currents, since the magnetic field due to field-aligned currents is canceled out by the Pedersen currents that link the field lines horizontally through the ionosphere. Since the electric field must be nearly the same at both ends of a magnetic field line, any differences between the two hemispheres create currents that flow between the hemispheres to maintain the field lines as equipotentials. However, to first order, at equinox the tidal modes are symmetrical and the field-aligned currents are therefore minimal. In this case the Sq pattern due to Hall currents represents the true total currents in the ionosphere quite well.

Although the Sq current deduced from data such as those shown in Fig. 3.14 flows primarily on the dayside (where the conductivity is high), this does not mean that the electric field is confined to the dayside. In fact, the charge density that builds up at the terminators to force $\nabla \cdot \mathbf{J} = 0$ creates the diurnal zonal electric field pattern observed at Jicamarca. In this chapter we consider the effect of the zonal electric field at the equator. The effect of the E-region dynamo at midlatitudes is discussed in Chapter 5.

Typical plasma densities and conductivities at the dayside equator are shown in Fig. 3.15. The zonal electric field component will drive a small Pedersen current

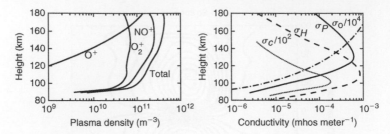

Figure 3.15 Vertical profiles of daytime composition and plasma density (left) and conductivities (right) for average solar conditions. [After Forbes and Lindzen (1976). Reproduced with permission of Pergamon Press.]

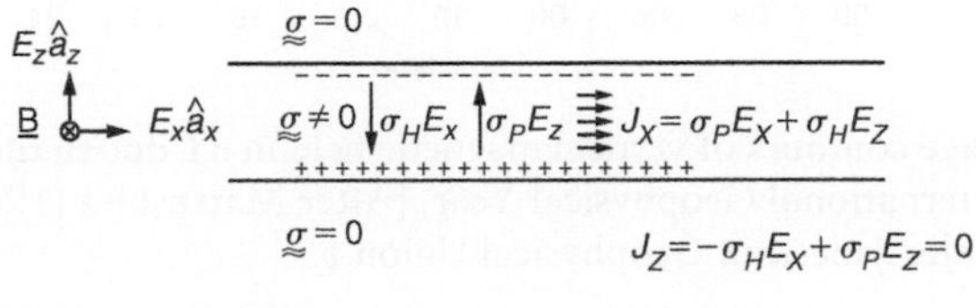

$$E_z\hat{a}_z$$

$$\underline{\underline{\sigma}} = 0$$

$$B \quad E_x\hat{a}_x \qquad \underline{\underline{\sigma}} \neq 0 \quad \downarrow\sigma_H E_x \quad \uparrow\sigma_P E_z \quad J_x = \sigma_P E_x + \sigma_H E_z$$

$$\underline{\underline{\sigma}} = 0 \qquad J_z = -\sigma_H E_x + \sigma_P E_z = 0$$

Figure 3.16 The equatorial electrojet in a slab geometry.

eastward along the dayside equator. More importantly, a vertically downward Hall current will also flow as shown in Fig. 3.16. Now from $\nabla \cdot \mathbf{E} = 0$, we can deduce that

$$\partial E_x/\partial z = \partial E_z/\partial x$$

This means that the variation of the zonal component with altitude can be estimated by the ratio

$$\delta E_x = \delta E_z(\delta z/\delta x)$$

The scale size (δx) of the horizontal conductivity pattern is 100 times that of the vertical (δz) variations of conductivity, whereas both experiments and theory indicate E_z is at most 10–20 times E_x. Taken together, this means that the zonal electric field can change only slightly in the E region. The data presented in Fig. 3.17b verify this argument, and we will thus assume that conductivity gradients rather than variations in E_x dominate the divergence of the vertical Hall current. Figure 3.15 shows that the Hall conductivity dominates below 120 km and, indeed, is highly altitude dependent (i.e., a slablike geometry is appropriate).

As a first approximation to the physics of the E-region dynamo, we again consider a slab conductivity geometry such as that illustrated in Fig. 3.16, subject to a constant zonal electric field. The Hall current cannot flow across the boundary, and charge layers must build up, generating an upward-directed electric

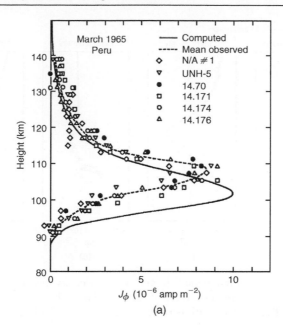

Figure 3.17a Observed and computed eastward current density profiles near noon at the dip equator off the coast of Peru in March 1965, normalized to a magnetic field perturbation of 100 nT at Huancayo. Measured profiles are from Shuman (1970) (flight N/A #1), Maynard (1967) (flight UNH-5), and Davis et al. (1967) (flights 14.170, 14.171, 14.174, and 14.176). The theoretical profile is from Richmond's (1973a) theory. [After Richmond (1973b). Reproduced with permission of Pergamon Press.]

field. In a steady state in this slab model, no vertical current may flow, and the vertical Pedersen current must exactly cancel the Hall current. This implies that

$$\sigma_H E_x = \sigma_P E_z$$

and hence that

$$E_z = (\sigma_H/\sigma_P)E_x \tag{3.13}$$

Since $\sigma_H > \sigma_P$, the vertical electric field component considerably exceeds the zonal electric field component. In addition, $E_z(z)$ has the same z dependence as the function $\sigma_P(z)/\sigma_H(z)$. The zonal current is now given by

$$J_x = \sigma_H E_z + \sigma_P E_x \tag{3.14a}$$

$$J_x = \left[(\sigma_H/\sigma_P)^2 + 1\right]\sigma_P E_x = \sigma_c E_x \tag{3.14b}$$

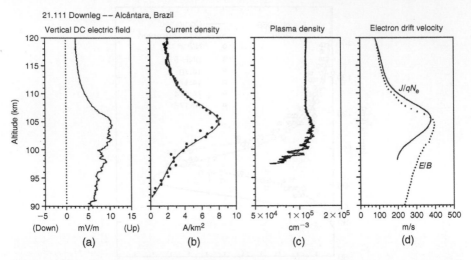

Figure 3.17b Electric and magnetic field and plasma density measurements along with the derived current density and electron drift measured over the magnetic equator. [After Pfaff et al. (1997). Reproduced with permission of the American Geophysical Union.]

where σ_c is the so-called Cowling conductivity. Notice that the local neutral wind does not enter this calculation at all; the electrojet is set up by the global tidal winds that create the diurnal zonal electric field component measured at the equator. In a more complete theory, complications due to the zonal neutral wind may be included, since it will drive an ion current equal to u. Note that the meridional wind component does not enter at the magnetic equator as a local dynamo, since for that component the cross product with the magnetic field vanishes.

In effect, (3.14ab) shows that the zonal conductivity is enhanced by the large factor $1 + \sigma_H^2/\sigma_P^2$, the Cowling conductivity factor, which leads to the intense current jet at the magnetic equator. This can be seen in Fig. 3.14, in which the magnetic field contours become very close together at the magnetic equator. This channel of electrical current is termed the equatorial electrojet. The Cowling conductivity is also plotted in Fig. 3.15 (divided by 100) and displays a peak at 102 km with a half-width of 8 km.

There have been many rocket measurements of the magnetic field due to the electrojet current. Some of the data are reproduced here in Fig. 3.17a. The plot displays the altitude variation of the jet over Peru with each profile normalized to a 100 nT variation of the magnetic field measured on the ground at Huancayo, Peru; that is, the actual perturbed magnetic field at Huancayo during each rocket flight was used to scale all data to a common ground perturbation of 100 nT.

Of more importance to the present text is the generation of large vertical electric fields in the electrojet, an example of which is presented in Fig. 3.17b, along with the horizontal current and the electron drift (Pfaff et al., 1997). As discussed in detail in the next chapter, these fields create electron drifts that are unstable to the generation of plasma waves. Peak vertical electric fields in the range 10–20 mV/m have been predicted theoretically and would create electron drift speeds of 350–700 m/s. This exceeds the acoustic speed C_s, which is about 360 m/s, and leads to the generation of intense plasma waves via the two-stream instability. Indeed, as we shall see, very large Doppler shifts occur when radar signals are scattered from the electrojet waves. These observations show that the vertical electric field reaches values at least as large as that given by $E_z = C_s B \simeq$ 9 mV/m at the very threshold of the two-stream instability, and it is very likely that the field reaches values half again as large. This is the largest electric field found in the ionosphere below subauroral zone latitudes. Remarkably, the data in Fig. 3.17a are the *only* vertical electric field measurements ever made.

The measurements in panels (a–c) have been combined in the right-hand panel to show the differential (net current) velocity as well as the $\mathbf{E} \times \mathbf{B}$ drift speed. At high altitudes the agreement is quite good, but below 104 km there is a systematic difference. As discussed in Chapter 2, neutral winds in the E region carry the ions with them but barely affect the electrons at all. The systematic difference between the two curves in panel d suggests that a high, westward neutral wind may have been present, a result in agreement with TMA measurements made in the evening two days later (Larsen and Odom, 1997).

Notice that the measured current density in Fig. 3.17a field peaks almost 5 km above the theoretical curve. This has been a mystery ever since Gagnepain et al. (1977). A fix can be made by arbitrarily quadrupling the electron-ion collision frequency. This approach works for horizontal magnetic field measurements as well (P. Alken, personal communication, 2008). However, Ilma and Kelley (2009) have shown that merely by evaluating the field line–integrated conductivity and using it to calibrate the current, the problem goes away. As in the F region, this is due to the increasing length of the line integral as a function of altitude. This effect is classical but for zonal electric fields larger than 1 mV/m, plasma waves "anomalously" decrease the conductivity (Kelley et al., 2009b).

An electrojet theory suitable for comparison with such detailed data must include realistic altitude and latitude variations of σ. Away from the equator, σ is often expressed in geographic rather than geomagnetic coordinates, and the resulting conductivity tensor is not as simple as (3.6). If $\mathbf{R}$ is the rotation matrix relating geographic and geomagnetic coordinates, we have

$$\mathbf{R} \cdot \mathbf{J} = \mathbf{R} \cdot (\sigma \cdot \mathbf{E}')$$

where $\mathbf{E}' = \mathbf{E} + \mathbf{U} \times \mathbf{B}$. Inserting the identity matrix $\mathbf{I} = \mathbf{R}^{-1} \cdot \mathbf{R}$, we have

$$\mathbf{R} \cdot \mathbf{J} = (\mathbf{R} \cdot \sigma \cdot \mathbf{R}^{-1}) \cdot \mathbf{R} \cdot \mathbf{E}'$$

and we find a new tensor given by

$$\sigma^R = R \cdot \sigma \cdot R^{-1}$$

For the simple case when geomagnetic and geographic north coincide and the dip angle is I, R corresponds to a rotation counterclockwise by the angle I about the x axis and

$$R = \begin{pmatrix} 1 & 0 & 0 \\ 0 & \cos I & -\sin I \\ 0 & +\sin I & \cos I \end{pmatrix}$$

Application of σ^R to E' determines J in the geographic coordinate system. Following Forbes (1981), we denote the eastward direction as the λ axis, the northward as the θ axis, and the vertical as the z-axis. Then

$$\sigma^R = \begin{pmatrix} \sigma_{\lambda\lambda} & \sigma_{\lambda\theta} & \sigma_{\lambda z} \\ \sigma_{\theta\lambda} & \sigma_{\theta\theta} & \sigma_{\theta z} \\ \sigma_{z\lambda} & \sigma_{z\theta} & \sigma_{zz} \end{pmatrix} \tag{3.15}$$

with

$$\sigma_{\lambda\lambda} = \sigma_P$$
$$\sigma_{\lambda\theta} = -\sigma_{\theta\lambda} = -\sigma_H \sin I$$
$$\sigma_{\lambda z} = -\sigma_{z\lambda} = +\sigma_H \cos I$$
$$\sigma_{\theta\theta} = \sigma_P \sin^2 I + \sigma_0 \cos^2 I \tag{3.16}$$
$$\sigma_{\theta z} = \sigma_{z\theta} = (\sigma_0 - \sigma_P) \sin I \cos I$$
$$\sigma_{zz} = \sigma_P \cos^2 I + \sigma_0 \sin^2 I$$

Note in this form that if we set $I = 0$, we recover a matrix identical to σ in (3.6). Taking the divergence of J and substituting $E = -\nabla\phi$, we have the dynamo equation

$$\nabla \cdot \left[\sigma^R \cdot (-\nabla\phi + U \times B) \right] = 0$$

where the quantities are all measured in the earth-fixed frame and expressed in the rotated coordinate system. This may be written as,

$$\nabla \cdot \left[\sigma^R \cdot \nabla\phi \right] = \nabla \cdot \left[\sigma^R \cdot (U \times B) \right] \tag{3.17}$$

Even taking U and σ^R as known functions, this equation is a complicated partial differential equation with nonconstant coefficients. A common simplifying assumption is that no vertical current flows anywhere in the system; that is, all currents flow in a thin ionospheric shell. Setting J_z equal to zero (in geographic coordinates now) yields

$$\sigma_{z\lambda} E'_\lambda + \sigma_{z\theta} E'_\theta + \sigma_{zz} E'_z = 0$$

or

$$E'_z = -\left(\sigma_{z\lambda} E'_y + \sigma_{z\theta} E'_\theta\right)/\sigma_{zz}$$

For $I = 0$ this reduces to (3.13) at the equator. Using this expression to eliminate E'_z, $\mathbf{J}^R$ can now be written as a two-dimensional vector and the dynamo equation as a function of θ and λ only (Forbes, 1981),

$$\begin{pmatrix} J_\lambda \\ J_\theta \end{pmatrix} = \begin{pmatrix} \xi_{\lambda\lambda} & \xi_{\lambda\theta} \\ \xi_{\theta\lambda} & \xi_{\theta\theta} \end{pmatrix} \cdot \begin{pmatrix} E'_\lambda \\ E'_\theta \end{pmatrix}$$

where

$$\xi_{\lambda\lambda} = \sigma_P + \left(\sigma_H \cos I\right)^2 / \left(\sigma_P \cos^2 I + \sigma_0 \sin^2 I\right)$$

$$\xi_{\lambda\theta} = -\xi_{\theta\lambda} = \left(-\sigma_0 \sigma_H \sin I\right)/\left(\sigma_P \cos^2 I + \sigma_0 \sin^2 I\right)$$

$$\xi_{\theta\theta} = \left(\sigma_0 \sigma_P\right)/\left(\sigma_P \cos^2 I + \sigma_0 \sin^2 I\right)$$

are the so-called layer conductivities. Insertion into $\nabla \cdot \mathbf{J} = 0$ in geographic coordinates yields

$$(\partial/\partial\lambda)\left[(\xi_{\lambda\lambda}/R_e \sin\theta)(\partial\phi/\partial\lambda) + (\xi_{\lambda\theta}/R_e)(\partial\phi/\partial\theta)\right]$$
$$+ (\partial/\partial\theta)\left\{\sin\theta\left[(\xi_{\theta\lambda}/R_e \sin\theta)(\partial\phi/\partial\lambda) + (\xi_{\theta\theta}/R_e)(\partial\phi/\partial\theta)\right]\right\} \qquad (3.18)$$
$$= (\partial/\partial\lambda)\left[-\xi_{\lambda\lambda} u B_z + \xi_{\lambda\theta} v B_z\right] + (\partial/\partial\theta)\left\{\sin\theta\left[-\xi_{\theta\lambda} u B_z + \xi_{\theta\theta} v B_z\right]\right\}$$

Assuming E'_λ and E'_θ to be independent of height, (3.18) can be integrated over height to give the thin-shell dynamo equation

$$(\partial/\partial\lambda)\left[\left(\textstyle\sum_{\lambda\lambda}/R_e \sin\theta\right)\left(\partial\phi/\partial\lambda\right) + \left(\textstyle\sum_{\lambda\theta}/R_e\right)\left(\partial\phi/\partial\theta\right)\right]$$
$$+ \left(\partial/\partial\theta\right)\sin\theta\left[\left(\textstyle\sum_{\theta\lambda}/R_e \sin\theta\right)\left(\partial\phi/\partial\lambda\right) + \left(\textstyle\sum_{\theta\theta}/R_e\right)\left(\partial\phi/\partial\theta\right)\right] \qquad (3.19)$$
$$= \frac{\partial}{\partial\lambda}\int\left[-\xi_{\lambda\lambda} u + \xi_{\lambda\theta} v\right] B_z dh + \frac{\partial}{\partial\theta}\left[\sin\theta\int\left[-\xi_{\theta\lambda} u + \xi_{\theta\theta} v\right] B_z dh\right]$$

where the Σ denote height-integrated layer conductivities, u and v are the zonal and meridional wind components, and R_e is the radius of the earth. It is important to note that assuming $\mathbf{E}' = \mathbf{E} + \mathbf{U} \times \mathbf{B}$ to be independent of height implies that u, v, and $-\nabla\phi$ are all independent of height, which is not very realistic. Nonetheless, comparisons between the electrojet currents obtained using thin-shell theory and experimentally measured currents have been made by treating the zonal electric field as a free parameter (e.g., Sugiura and Cain, 1966; Untiedt, 1967) and by solving the dynamo equation by setting the zonal derivatives equal to zero. However, the peak current density seems to be underestimated in these models, and details concerning the vertical and latitudinal extent of the electrojet are not reproduced. More seriously, without vertical currents at the equator, which are suppressed by the thin-shell model, it is very hard to satisfy the divergence requirement

$$\frac{\partial J_\lambda}{\partial \lambda} + \frac{\partial J_\theta}{\partial \theta} = 0$$

since the electrojet varies rapidly in latitude but only slowly in longitude. Forbes and Lindzen (1976) pointed out that allowing a vertical current could actually increase the electrojet intensity predicted by the models. For example, returning to the current equation at the equator and again ignoring the neutral wind,

$$J_x = \sigma_P E_x + \sigma_H E_z$$
$$J_z = -\sigma_H E_x + \sigma_P E_z$$

Eliminating E_z yields

$$J_x = \sigma_c E_x + (\sigma_H/\sigma_P) J_z \tag{3.20}$$

Thus, if J_z is nonzero and positive at the equator, J_x would exceed the Cowling current and the electrojet current would be stronger than $\sigma_c E_x$.

More realistic models that include such possibilities have been developed by Richmond (1973a, b; see Fig. 3.17a) and Forbes and Lindzen (1976). Richmond's improvement over the thin-shell model involved integrating the divergence equation along a magnetic field line as discussed above for the F-layer dynamo. The zonal derivatives were all set to zero. The vertical electric field at the equator then becomes $(\Sigma_H/\Sigma_P) E_x$ rather than $(\sigma_H/\sigma_P) E_x$; that is, the integrated conductivities along the field line are used rather than the local values. Above 100 km the integrated conductivity ratio exceeds the local value so $J_z \neq 0$ and both the eastward current and the vertical electric field are greater than the thin-shell model. The result seems to overestimate the current (see Fig. 3.17a). Forbes and Lindzen (1976) relaxed the requirement of zonal invariance in their solution of the three-dimensional $\nabla \cdot \mathbf{J} = 0$ equation, but they did not include integration along $\mathbf{B}$.

The excellent review by Forbes (1981) details these and other electrojet studies, including day-to-day variability, neutral wind, and lunar tidal effects that are beyond the scope of the present text.

3.4 Further Complexities of Equatorial Electrodynamics

The simple E- and F-region dynamos just described explain some of the observations presented earlier in Fig. 3.1. Since the E region controls the physics during the day, the small daytime vertical electric fields in the F region presumably mirror the meridional E-region polarization fields in the latitude ranges just north and south of the equator. The zonal field, both daytime and nighttime, is global in nature and is driven by the large-scale tides and winds in the sunlit hemisphere. In the postsunset period, the F-layer vertical field is enhanced due to the local F-layer dynamo previously discussed. In the next paragraphs we point out some of the features not explained by these simple models.

3.4.1 The Prereversal Enhancement

The postsunset enhancement or prereversal enhancement of the zonal field occurs during all epochs and seasons studied except for the solar minimum solstices. The effect of this brief duration, large eastward electric field can be quite significant since the F-layer plasma often is driven to very high altitudes where recombination is slight and collisions are rare. Heelis et al. (1974) successfully predicted the postsunset effect in their model, which included *horizontal* conductivity gradients near sunset in the F-layer dynamo mechanism (in addition to the vertical gradients we have studied). In effect, near such a sharp east-west gradient, an enhanced zonal electric field is established to keep $\nabla \cdot \mathbf{J} = 0$. Experimental data support this explanation since the enhancement begins when the sun sets on either of the E layers in contact along $\mathbf{B}$ with the equatorial F region.

The evidence from their numerical simulation is presented in Fig. 3.18. The solid line shows the calculated vertical ion velocity in the F region (equivalent to a zonal electric field) when only a diurnal E-region tidal mode is considered. The basic day-night features are reproduced, but no prereversal enhancement occurs. The dashed line shows a calculation including the F-region dynamo in the physics. The latter agrees quite well with the dotted line, which is a typical vertical drift profile for the plasma over Jicamarca.

The prereversal enhancement has been successfully simulated for the first time by a general circulation model, the National Center for Atmospheric Research thermosphere/ionosphere/electrodynamic general circulation model (TIEGCM; Fesen et al., 2000). The TIEGCM reproduces the zonal and vertical plasma drifts for equinox, June, and December for low, medium, and high solar activity. The crucial parameter in the model needed to produce the prereversal enhancement (PRE) is the nighttime E-region electron densities; densities greater than 10^4 cm^{-3}

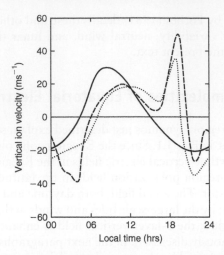

Figure 3.18 Calculated vertical ion drift velocities for several driving wind components. The solid line includes only the tidal-driven E-region dynamo, while the dashed line includes the F-region dynamo as well. Typical measured vertical plasma drifts are indicated by the dotted line. [After Heelis et al. (1974). Reproduced with permission of Pergamon Press.]

preclude the PRE development by short-circuiting the F-region dynamo. The E-region semidiurnal 2,2 tidal wave largely determines the magnitude and phase of the daytime F region drifts in this model (see Fig. 3.19).

A simple sketch that may explain the prereversal enhancement is given in Fig. 3.20. The equatorial plane is shown here as a vertical plane, and its projection onto the Southern Hemisphere along **B** is also shown. The wind blows across the terminator, generating a vertical electric field, E_z, which is downward on both sides of the terminator. E_z is much smaller on the dayside than on the nightside due to the shorting effect, but it is never zero. Thus, E_z maps along **B** to an equatorward electric field component off the equator. This field drives a westward Hall current $J_{\theta\phi}$ on both sides of the terminator. However, even though E_z is smaller on the dayside by up to 90%, the Hall conductivity is more than 10 times the nighttime value. The result is that a negative charge density builds up near the terminator, creating the localized zonal E_ϕ perturbation (the PRE) shown in the figure. Using the notation of Farley et al. (1986), the current $J_{\phi\phi}$ cancels $J_{\theta\phi}$ in a steady state. The absence of a sunrise enhancement may be due to the lower value of E_z, which occurs on the nightside of the dawn terminator relative to the nightside of the dusk terminator as well as the lower conductivities that occur in the predawn F region. Another idea is that the equatorial electrojet partially closes with a current through the postsunset F-layer valley. With the low conductivity in this region, both an eastward and an upward E field would result. This could explain both the prereversal enhancement and westward flows seen

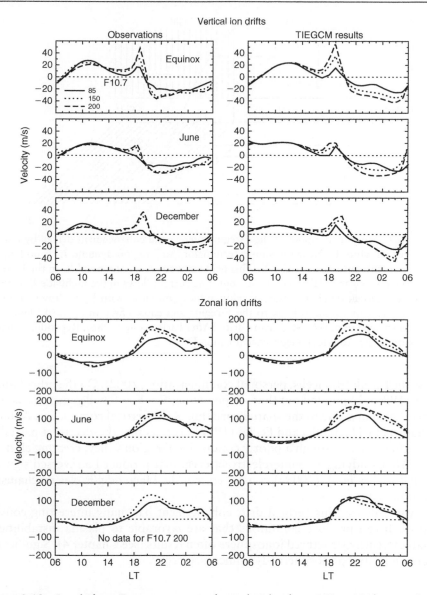

Figure 3.19 Ion drifts at Jicamarca averaged over heights from 300 to 400 km as a function of season and solar activity. Left side, observations; right side, TIEGCM simulations. Top three panels, vertical ion drifts; bottom three panels, zonal ion drifts. Within each panel, the different curves illustrate results for different levels of solar activity corresponding to F10.7 indices of 85, 150, and 200. [After Fesen et al. (2000). Reproduced with permission of the American Geophysical Union.]

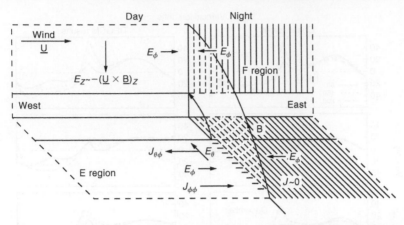

Figure 3.20 Simplified model of the F-region prereversal enhancement driven by a uniform F-region wind, **U**. Near the sunset terminator the F-region dynamo E_z is no longer shorted out and approaches $-\mathbf{U} \times \mathbf{B}$. This field maps to an equatorward E_θ in the E layer and drives a westward Hall current, $J_{\theta\phi}$. But if no current flows in the nightside E region, a negative polarization charge must develop at the terminator, with E_ϕ as shown and $J_{\phi\phi}$ canceling $J_{\theta\phi}$. This E_ϕ maps back to the F region and causes first an upward (day) and then a downward (night) $\mathbf{E} \times \mathbf{B}$ plasma drift. [After Farley et al. (1986). Reproduced with permission of the American Geophysical Union.]

below the F peak (R. Heelis, personal communication, 2002). On the other hand, Eccles (1998a, b) pointed out that simply setting $\nabla \times \mathbf{E} = 0$ will generate a zonal electric field pulse due to the spatial variation of the vertical component. A third mechanism by Haerendel and Eccles (1992) suggests that the electrojet partially closes after sunset through a poorly conducting F-region valley, creating electric fields. After studying a counterelectrojet event that produced a reversed sign of the PRE, Kelley et al. (2009a) concluded that the Haerendel/Eccles mechanism must dominate.

As we shall see, this vertical drift enhancement has many interesting consequences. It can produce conditions that are favorable to plasma instabilities, and an extremely structured ionosphere often results (see Chapter 4). Significant effects on neutral atmospheric dynamics also occur.

3.4.2 High-Latitude Effects on the Equatorial Electric Field

The equatorial plasma is not always as well behaved as indicated thus far. A number of interesting processes can affect equatorial electrodynamics, in addition to the neutral wind-generated electric fields already discussed, which are driven by the solar heat input carried by photons. Energy also comes from the sun via particles in the solar wind and via the electromagnetic (Poynting) flux in the solar wind. These energy sources affect primarily the high-latitude polar cap and auroral zones but can also be detected at the equator. The sensitivity of radars to electric fields has, in fact, increased our knowledge of how the solar wind affects the magnetosphere/ionosphere system (Gonzales et al., 1979, 1983; Somayajulu et al., 1985, 1987; Earle and Kelley, 1987).

A classical event (Nishida, 1968) illustrating a clear relationship between the interplanetary magnetic field (IMF), which is of solar origin, and the magnetic field created by the equatorial electrojet is reproduced in Fig. 3.21. The two magnetic field measurements are clearly correlated, although the magnetometers were separated by 10^5 km! Thus, although the magnetosphere, ring current, and plasmasphere all act to shield out interplanetary and solar wind effects from low latitudes, the protection is not perfect. Some insight into the complex relationship between interplanetary and equatorial ionospheric phenomena has come from simultaneous measurements of electric fields at Jicamarca and in the auroral zone. Since, as discussed elsewhere in this text, the much shorter chain of cause and effect between the IMF and auroral fields is itself not yet entirely understood, we can be satisfied here with some plausible explanations for the equatorial effects.

Forty-eight hours of Jicamarca electric field data are presented in the lower panel of Fig. 3.22. The eastward electric field component is plotted along with a lighter line that shows the average quiet-day value for the same solar cycle conditions and season. The top trace is a superposition of magnetic field data from a number of auroral zone magnetometer stations, which are used to generate the AU and AL auroral zone magnetic indices (see Appendix B). A series of six bursts of auroral substorm activity are clear in this data set. The northward component of the interplanetary magnetic field is also plotted. The latter, unfortunately, has

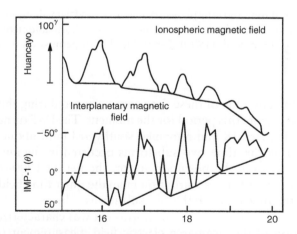

Figure 3.21 Correlation of horizontal geomagnetic fluctuations observed at Huancayo on December 3, 1963, with changes in the direction of the interplanetary magnetic field component perpendicular to the sun-earth line observed from the satellite IMP-1 versus UT. [After Nishida (1968). Reproduced with permission of the American Geophysical Union.]

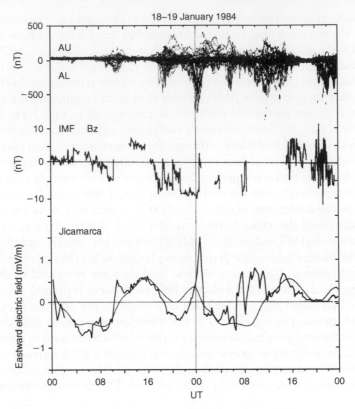

Figure 3.22 Auroral zone and interplanetary magnetic field data along with zonal electric field measurements over Jicamarca, Peru, for a 48 h period. Local midnight occurs at 5 UT at Jicamarca. [After Fejer et al. (1990). Reproduced with permission of Terra Scientific Publishing Co.]

a number of data gaps. One of these data gaps occurred during the first isolated substorm, and we ignore this period for the moment. The IMF turned southward at 1700 UT on January 18 and remained southward until about 0100 UT on January 19. During this time two substorms occurred in the auroral oval. At Jicamarca, which was in the postnoon sector, the period 1900–2300 UT was characterized by a westward perturbation from the quiet-time field. This period corresponds to the time between the two substorms. The period 0700–1200 UT on January 19 also fell between two substorms but was characterized by an east-ward perturbation of the Jicamarca electric field measurement (postmidnight sector). Notice that there was almost exactly 12 h between these events, which means that the Jicamarca data were taken on the opposite side of the system when viewed in a magnetospheric or solar wind reference frame. The burst of eastward perturbation field at 1000–1200 UT on January 18 is now seen to be consistent with the postmidnight "substorm recovery" characteristics seen 24 h later (i.e., 0700–1200 UT on January 19). These eastward perturbations in the postmidnight sector are very commonly reported in the various studies of disturbance fields using the Jicamarca data base. In part, this is due to the fact that they are easily recognizable in that sector since the quiet-time field is usually steady and westward. (Notice, for example, that the July 3, 1968, event identified in Fig. 3.3 is of this type.) A local time dependence may exist due to the ion conductivity postmidnight, which leads to enhanced perturbations in this sector. The westward perturbation in the daytime Jicamarca data (1700–2400 UT on Jan. 18) is now also interpreted as the same (between substorm) type of event viewed on the opposite side of the system. The transition in sign is clearly seen at 0 UT (1900 LT) when B_z turned north: substorm activity ceased and Jicamarca recorded an eastward perturbation.

Another event of this class is shown in Fig. 3.23. The shaded period is characterized as follows:

1. The IMF turned northward (panel 2) and magnetic activity decreased (panel 1).
2. The auroral zone electric field measured over Alaska (Chatanika data) and eastern Canada (Millstone Hill) decreased dramatically.
3. An eastward electric field perturbation occurred in the postmidnight sector over Peru (Jicamarca data).

A qualitative explanation for such events was proposed by Kelley et al. (1979). As we will discuss in more detail in subsequent chapters, it is known that if a steady magnetospheric electric field exists, divergence of the ring current near $L = 4$ will eventually create a charge separation in such a way that the magnetospheric electric field will be shielded from low latitudes. This charge separation region is called the Alfvén layer (Vasyliunas, 1972). Now if the magnetospheric field rapidly increases or decreases, the charges will be temporarily out of balance with the new configuration, and a brief low-latitude disturbance will result. An example of the effect of a rapid decrease in the external electric field is illustrated in Fig. 3.24a. The transient state inside the plasmasphere when the external field vanishes is an eastward perturbation on the nightside and a westward perturbation during the day, just as observed in the data shown in Figs. 3.22 and 3.23. In other words, the low-latitude electric field may be due to Alfvén layer charges that still exist, even though no external field is left in the outer regions to shield out! In circuit terminology, the finite inductance of the ring current maintains the voltage in the inner region when the source is turned off. This effect is also called overshielding.

Perhaps a better way to understand these penetrating electric fields is illustrated in Fig. 3.24b. Here the physics is described by a current source rather than a voltage source. In the middle panel we illustrate the dayside voltage response to enhanced Region 1 (high latitude) currents (increasing magnetic activity) and at the bottom we illustrate the opposite effect when Region 2 (low latitude) currents exceed those in Region 1, which occurs when the solar wind–magnetosphere

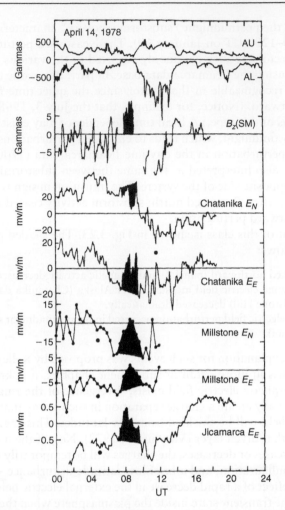

Figure 3.23 Auroral (AU, AL) and interplanetary (B_z) magnetic field data along with auroral (Chatanika and Millstone) and equatorial (Jicamarca) electric field measurements. The shaded region corresponds to a rapid decrease in magnetospheric convection apparently triggered by a northward turning of the interplanetary field. [After Kelley et al. (1979). Reproduced with permission of the American Geophysical Union.]

interaction abruptly decreases (see Chapter 8). The increasing activity currents (Region 1 > Region 2) yield an eastward perturbation on the dayside and a westward perturbation at night. Decreasing activity current unbalance (Region 2 > Region 1) yields the opposite changes.

An event similar to Nishida's original study shown in Fig. 3.21 was studied on February 17, 1976, using electric field data as well as magnetic measurements. The IMF, the auroral zone electric field, and the equatorial electric field all

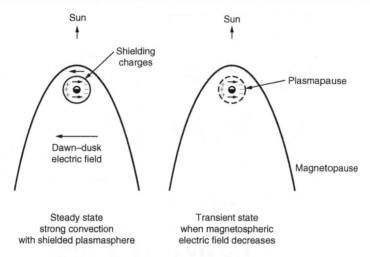

Figure 3.24a A schematic diagram drawn in the equatorial plane showing how a rapid decrease in magnetospheric convection could create electric field changes inside the plasmasphere.

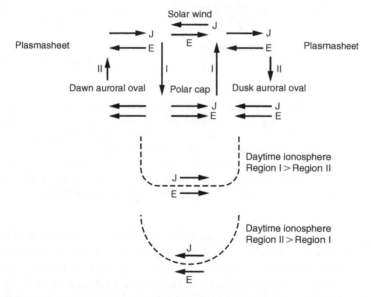

Figure 3.24b Circuit diagram looking from the sun. Polarities of closure currents are shown for two different cases of field-aligned current imbalance.

display similar oscillations, as illustrated in Fig. 3.25a. These time variations are too quick for the charges in the Alfvén layer to change and thus seem to be consistent with the concept of very effective direct electric field penetration into the plasmasphere. Power spectra of the electric field (from Alaska and Peru)

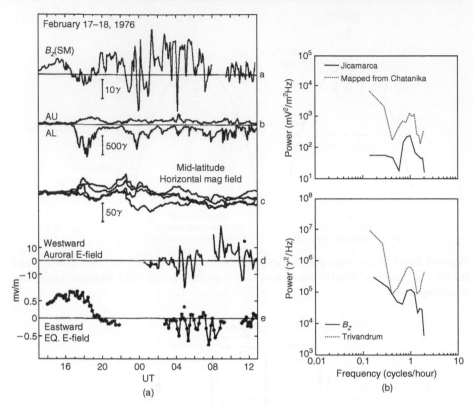

Figure 3.25   (a) Interplanetary (B_z), auroral (AU, AL), and midlatitude magnetic field data along with auroral and equatorial electric field data during a series of rapid interplanetary magnetic field changes. [After Gonzales et al. (1979). Reproduced with permission of the American Geophysical Union.] (b) Power spectra of the data plotted in (a). The Chatanika electric field data were reduced by the factor $L^{3/2}$ before analysis to show the value that would arise in the equatorial plane at $L = 5.5R_e$. [After Earle and Kelley (1987). Reproduced with permission of the American Geophysical Union.]

and magnetic field data (from India and the interplanetary region) are plotted in Fig. 3.25b and are very similar. The daytime poleward magnetic field data from India were 180° out of phase with the nighttime eastward electric field data. The same anticorrelation on opposite sides of the earth was shown recently by Kelley et al. (2007). The anticorrelation of Indian poleward magnetic field and Peruvian eastward electric field data shows that when the electric field perturbation is eastward at night it is westward during the day. Control of the earth's low-latitude electric field from hundreds of thousands of kilometers away in the interplanetary medium is indeed remarkable. The auroral zone data (from Chatanika, Alaska) were mapped to the equatorial plane before the Fourier analysis was done so that they could be directly compared to the Jicamarca data (which are also in the equatorial plane). The power is down by only a factor of four at Jicamarca, which means the field penetrating to the equator in Fig. 3.25b is reduced by only 50% from the magnetospheric (auroral zone) value in the equatorial plane. This is typical of the electric field values in the few-hour frequency range studied by Earle and Kelley (1987), who also have reported experimental evidence that the shielding process acts like a high-pass filter, allowing signals with periods shorter than about 8 h to penetrate to the equatorial ionosphere. This is in good agreement with the time constant predicted by ring current shielding theory (Vasyliunas, 1972). Furthermore, they show that for the zonal electric field component at Jicamarca, the signal in the frequency range corresponding to a period of a few hours exceeds the "geophysical noise" due to gravity waves at a moderate level of magnetic activity ($K_p = 3$). In other words, at frequencies above the ring current shielding frequency, high-latitude effects may always be present at all latitudes but cannot be detected unless they are above the fluctuation level (geophysical noise) due to neutral wind-driven electric fields.

Examples of the low-latitude effect of increasing magnetospheric electric fields are rare for reasons that are not altogether clear. One good example on March 23–24, 1971, is shown in Fig. 3.26a. The IMF turned southward abruptly at 0700 UT on March 24 after 5 hours of northward field. A strong *westward* perturbation occurred over Jicamarca, which was in the postmidnight sector. Notice that on the next day the more common eastward perturbation was seen in conjunction with a northward turning of the IMF at 0900 UT. There is an unfortunate data gap, but it does seem that a westward electric field perturbation occurred in conjunction with the B_z south event at 0500 UT on the 25th. One of the best documented events (Kelley et al., 2003a) occurred on April 17, 2002. For example, Fig. 3.26b directly compares the interplanetary electric field (IEF) in the upper panel to the electric field measured with the incoherent scatter radar (middle panel). The IEF has been divided by 15 before plotting in the middle panel and is highly correlated with the equatorial field at Jicamarca. Gonzales et al. (1979) and more recently Anderson et al. (2002) developed a technique using magnetic field data from two sites—one on and one off the equator—which has promise for greatly improving our knowledge of the electric field. A direct comparison of the Anderson-Gonzales method with the measured electric field is shown in the lower panel of Fig. 3.26b and is very convincing. Not shown is a good correlation between the Jicamarca electric field and that measured at Sondre Stromfjord (Kelley et al., 2003b).

Fejer and Scherliess (1995) created a semi-empirical model and compared it with the Rice Convection model, which includes both Region 1 and Region 2 currents. The former corresponds to the highest latitude currents, which are closely controlled by and linked to the solar wind generator. The Region 2 currents correspond to the inner magnetospheric portion of the auroral current system and link up with the ring current system. Their results are shown in Fig. 3.27 for an increase in the Region 1 current without a corresponding Region 2 current, corresponding to increased polar cap convection. For a rapid decrease in convection

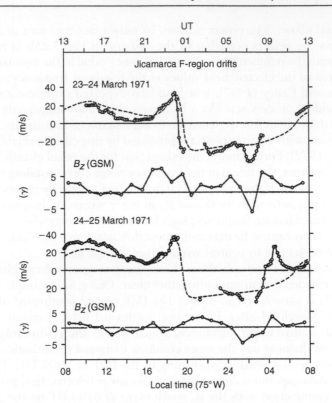

Figure 3.26a F-region vertical drifts measured over Jicamarca (corresponding to an eastward electric field component) and IMF data. Note that the large nighttime drift perturbations are well correlated with southward and northward IMF changes. Quiet-time patterns are shown by dashed lines. [After Fejer (1986). Reproduced with permission of Terra Scientific Publishing Co.]

(e.g., if B_z returned northward), the Region 2 currents exceed Region 1 and the opposite equatorial perturbation exists. These conditions are illustrated in Fig. 3.24b. This work has continued, as described by Scherliess and Fejer (1997) and Fejer and Scherliess (1997).

Another mechanism that creates low-latitude electrodynamic changes is called a disturbance dynamo. If auroral zone heating and momentum sources are sufficiently strong, the low-latitude neutral atmospheric winds can also be affected. In turn, these winds will create different electric field patterns through the usual E- and F-region dynamo process. Unlike the virtually instantaneous changes shown previously, the disturbance dynamo takes more time to develop and lags the auroral inputs by many hours. Gravity waves can also be generated by high-latitude processes and propagate to low latitudes. It seems likely that the wind fields associated with gravity waves will also create dynamo electric fields

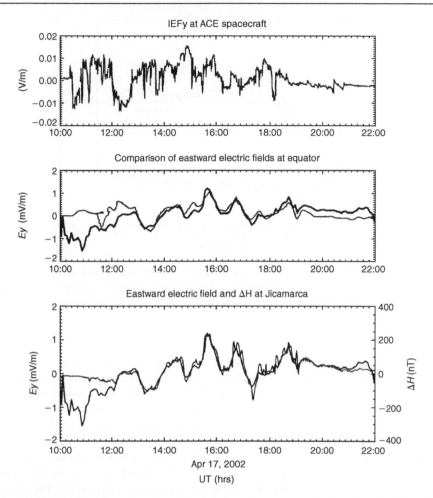

Figure 3.26b Direct comparison of the IEF and the equatorial electric field (using ISR and the Anderson technique) after adjusting for the time delay from the ACE satellite to the magnetopause. In the middle and bottom panels the dark line is the eastward component over Jicamarca that corresponds to the dawn-to-dusk direction. The light line in the middle panel is the dawn-to-dusk component of the IEF. In the lower panel the light line is the predicted eastward field from ΔH values. [Adapted from Kelley et al. (2003a). Reproduced with permission of the American Geophysical Union.]

that vary in time with the wave period. Again, a time delay of many hours is expected if the gravity wave source is in the auroral zone (see Fig. 3.30a, b and associated text).

Very clear examples of both instantaneous and delayed equatorial perturbations are illustrated in Fig. 3.28. In the upper two panels, the auroral AU and AL magnetic fields are plotted for a 48 h period on August 8–10, 1972. Below, the

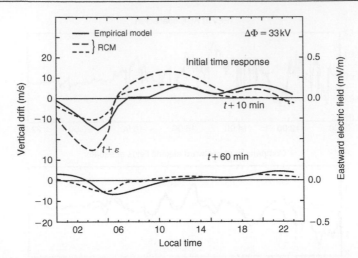

Figure 3.27 Equatorial ionospheric perturbation electric field patterns calculated for three times after an increase in the Region 1 field-aligned currents. Ionospheric dynamo effects were not included. [After Fejer and Scherliess (1995). Reproduced with permission of the American Geophysical Union.]

corresponding zonal electric field data measured at Jicamarca are given. In each of the latter plots the mean field pattern is shown as the solid line. The abrupt change in the Jicamarca field between 2300 and 0300 local time on August 8/9 has been studied by Gonzales et al. (1979) and has a waveform identical to that of the electric field measured simultaneously at Chatanika, Alaska. (See also Fig. 3.25a for identical auroral and equatorial waveforms.) This is a clear case of rapid penetration of high-latitude electric fields to low latitudes. On the next day the Jicamarca zonal field component displayed a long-lived deviation from the average, commencing at 2200 local time and lasting for 8 h. Fejer et al. (1983) argue that such slowly developing, delayed electric field changes are due to a worldwide change in the thermosphere winds. These changes alter the E region and equatorial F-region dynamo and thereby change the electric field at low latitudes from the normal pattern.

Taken together, we see that these and other high-latitude effects contribute to the "weather" in the low-latitude electric field and neutral wind patterns. Their study therefore yields information on both high- and low-latitude phenomena. Much can be learned about global electrodynamics and thermospheric physics through events of these types. The true test of global electrodynamic model calculations may be their ability to predict the low-latitude ionospheric effects of such diverse external influences. Such studies are well under way (Maruyama et al., 2005). In Chapter 5 we discuss further the penetration of high-latitude electric fields to midlatitudes. The origins of the high-latitude electric fields themselves are considered in some detail in Chapter 8.

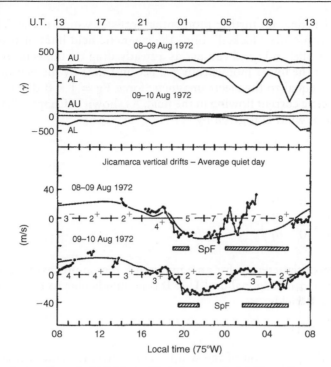

Figure 3.28 Auroral magnetic fields and F-region vertical drifts at Jicamarca on August 8–10, 1972. The solid curves in the lower panel show the average quiet-time diurnal variation. Deviations from this pattern beginning at 2300 LT on August 8 are due to direct penetration effects, whereas the slower deviations starting at 2200 LT on August 9 are due to the disturbance dynamo. [After Fejer et al. (1983). Reproduced with permission of the American Geophysical Union.]

3.5 Feedback Between Electrodynamics and Thermospheric Winds

We now return briefly to the problem of the large calculated thermospheric winds mentioned in Section 3.2. Part of the key to this dilemma lies in the crucial role of the plasma in controlling the thermospheric wind, even though the ionized component is only a very minor constituent (one part in 10^3 at 300 km). When a moving neutral particle strikes an ionized particle at rest in the thermosphere, some of the momentum imparted is converted into motion of the particle about the magnetic field and yields a net deflection in the $q(\mathbf{U} \times \mathbf{B})$ direction, where q includes the sign of the particle charge. Thus, unlike a collision with a neutral particle, the momentum transfer does not appear as linear momentum parallel to $\mathbf{U}$ in the ionized gas. A similarly dense minor neutral species would, after one or two collisions, merely accelerate to a velocity equal to the background wind speed, at which time no further momentum transfer would occur. The ionized particles, however, are "locked" onto the magnetic field lines instead of being accelerated parallel to U and can act as a steady drag on the neutral wind. We now show that the linear momentum lost per unit time by the neutral fluid due to this drag is equal to the electromagnetic force $F_E = J \times B$ due to the wind-generated electric current flowing in the fluid. As shown in Chapter 2, the current due to a neutral wind U is given by

$$J = \sigma \cdot (U \times B)$$

with

$$\sigma = \begin{pmatrix} \sigma_P & 0 & 0 \\ 0 & \sigma_0 & 0 \\ 0 & 0 & \sigma_P \end{pmatrix}$$

being the appropriate tensor for an equatorial F-region process with B in the $\hat{a}_y$ direction. Since σ is diagonal and $U \times B$ is perpendicular to B, the triple cross product in $J \times B$ yields

$$F_E = -\sigma_P B^2 U_\perp \tag{3.21}$$

where $U_\perp$ is the projection of the wind vector onto the plane perpendicular to B. The plasma thus clearly acts as a drag on the neutrals, since the force acts in the opposite direction to $U_\perp$. We have shown previously that in the F region we can approximate σ_P by the expression

$$\sigma_P = \left(ne^2 v_{in}/M\Omega_i^2 \right) = nv_{in}M/B^2 \tag{3.22}$$

where n is the plasma density, M the ion mass, Ω_i the ion gyrofrequency, and v_{in} the ion-neutral collision frequency. Substituting (3.22) into (3.21) yields

$$F_E = -nMv_{in}U_\perp \tag{3.23}$$

which even more clearly shows the draglike nature of the $J \times B$ force. At first glance this expression has the form of a frictional or drag force, F_D, on the wind, but it is not quite right, since it should include the *neutral* density, not the plasma density, if it is to be inserted into the Navier Stokes equation; that is, it should be of the form

$$F_D = -n_n Mv_{ni}U_\perp$$

However, for equal-mass ions and neutrals, it follows that the collision frequencies for momentum transfer are related by

$$n_n v_{ni} = n_i v_{in} = nv_{in}$$

and the two expressions for $\mathbf{F_E}$ and $\mathbf{F_D}$ are seen to agree with each other. In the preceding last term, we have simply noted that $n_i = n$, the plasma density.

Referring back to Fig. 3.7, we may now interpret the parameters ν_{ni} and f. The solid curve is ν_{ni}, the daytime neutral-ion collision frequency, and the vertical dotted line is the Coriolis parameter, $f = \Omega \sin \theta$, evaluated for 45° latitude. (Here Ω is the rotation frequency of the earth, 7.35×10^{-4} rad/s.) The latter plays no role at the equator but the drag term does. For example, consider the steady-state solution to

$$dU/dt = -(1/\rho) \nabla p - \nu_{ni} U = \mathbf{F} - \nu_{ni} U$$

which is given by

$$U = -\mathbf{F}/\nu_{ni}$$

where typical values for F are also plotted in Fig. 3.7. Near the F-layer plasma density peak at 250 km the equation yields a magnitude for U of only about 70 m/s. At 300 km the velocity rises to about 130 m/s. These values are much more reasonable than those found by Lindzen (1966) without ion drag and show that the ionosphere has great control over the neutral atmosphere. In summary, the $\mathbf{J} \times \mathbf{B}$ force on the plasma is transferred to the neutral gas by collisions. As discussed earlier, vertical wind variations are suppressed by viscosity in the middle and upper thermosphere.

There is an interesting and important day-night asymmetry to the drag term due to the nighttime F-region dynamo. During the day, the F-region dynamo is shorted out by the E layer and the plasma cannot move along with the wind. Thus, the thermospheric wind blows through ions that are tied to the magnetic field lines. Collisions between neutrals and the "immovable" ions create the drag. But at night, the plasma tends to catch up to the neutrals, and the drag becomes small. Alternatively, we could argue that $\mathbf{J} = \mathbf{E} + \mathbf{U} \times \mathbf{B}$ becomes small and, in any case, the wind can accelerate.

Given the control exhibited by the ionospheric plasma on the thermosphere, it is not surprising that the winds are so variable. In addition to the F-layer dynamo and the effect of the vertical electric fields, another factor is the altitude of the nighttime ionospheric plasma layer, which is determined by the magnitude of the zonal component of the electric field. Anderson and Roble (1974) have calculated zonal neutral wind contours in conjunction with an empirical model of the electric field pattern for the two cases illustrated in Fig. 3.29. In the upper plot a diurnally varying electric field pattern is used that is typical of sunspot minimum conditions, while the lower plot includes the prereversal enhancement of the electric field, which is largest during sunspot maximum. In the latter case the ionospheric plasma is driven to very high altitudes in the 2000–2400 LT sector. With the plasma out of the way, the ion drag is greatly reduced and the thermospheric wind speed nearly doubles. Since changes in the zonal electric field

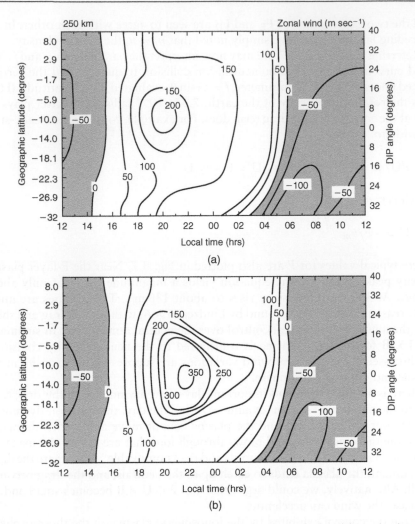

Figure 3.29 Contours of the zonal neutral wind at 250 km, calculated from the thermospheric model as a function of latitude and local time: (a) without the postsunset enhancement of the $\mathbf{E} \times \mathbf{B}$ drift velocity and (b) including the postsunset enhancement. [After Anderson and Roble (1974). Reproduced with permission of the American Geophysical Union.]

occur due to auroral disturbances, it is easy to see that weather in the equatorial thermosphere is also related to weather in the solar wind.

Another variable is the effect of enhanced high-latitude atmospheric heating during auroral storms. Models have shown that large-amplitude gravity waves can be created that propagate throughout the thermosphere, forming so-called traveling ionospheric disturbances. In extreme cases the worldwide wind patterns

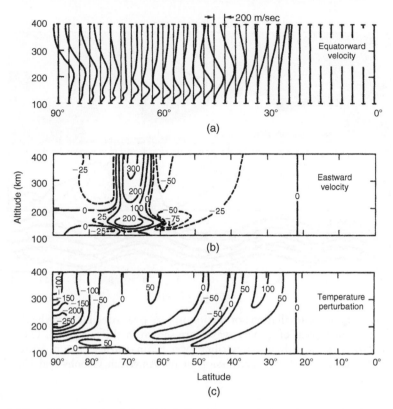

Figure 3.30a "Snapshot" of the global disturbance wind at time = 2 h after a strong auroral event. (a) Equatorward neutral wind velocity profiles every 3° in latitude. (b) Eastward neutral wind velocity. The contour spacing is 100 m/s for the solid lines and 25 m/s for the dashed lines. (c) Neutral temperature perturbation. The contour spacing is 50 K. [After Richmond and Matsushita (1975). Reproduced with permission of the American Geophysical Union.]

can be affected by heat input during such storms. Richmond and Matsushita (1975) have modeled the effect of an isolated magnetospheric substorm lasting for two hours near 70° latitude. At the end of this time they calculated the patterns of wind and temperature variation shown in Fig. 3.30a. A large disturbance is seen to propagate equatorward at a speed of about 750 m/s. The simulation shows that the effect reaches the equator with very little attenuation but requires many hours to arrive. The effect is larger in the meridional wind component than in the zonal component. An example of a set of waves passing Arecibo, PR, and later creating an intense convective ionospheric storm (aka equatorial spread F) at the equator is shown in Fig. 3.30b. This set of waves was caused by a series of magnetic substorms. Notice the downward phase progression in the lowest panel, a signature of gravity wave propagation.

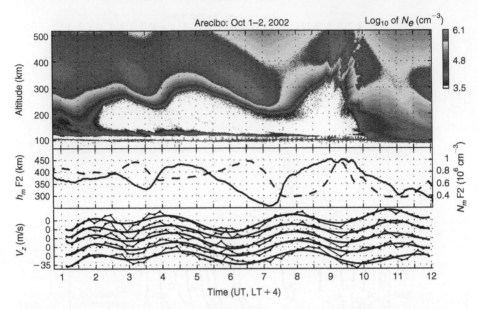

Figure 3.30b Periodic oscillations of the ionosphere over Arecibo with the downward phase progression typical of gravity waves are shown in the top panel. One of these oscillations triggered a turbulent upwelling over Jicamarca near dawn. [After Nicolls et al. (2004). Reproduced with permission of the American Geophysical Union.] See Color Plate 2.

A thermospheric model has been run by Sipler et al. (1983), who used it as a diagnostic tool in a numerical experiment to determine which processes dominate the wind variability. The measurements and model calculations of the wind are plotted in Fig. 3.31. In this study, the thermospheric general circulation model (TGCM) was run for August 21, 1978, and compared to the measured winds indicated by the heavy dashed line in Fig. 3.31. Then the lower atmospheric tides were added along with the high-latitude influence (light dashed line). Finally, ionospheric electric fields were added (dotted line). The conclusion was that the day-to-day variability was tied most closely to the tidal and electric field effects but that even in relatively quiet conditions the high-latitude plasma circulation played a role in the equatorial winds.

Finally, we return to the curious fact that the earth's upper atmosphere super-rotates—that is, the mean zonal wind in the rotating frame is eastward in the equatorial zone. A number of explanations have been put forth concerning this effect (e.g., the review by Rishbeth, 1972). Since a high eastward velocity is most common in the sunset-to-midnight period, theories that control the thermospheric winds via the ionospheric drag effect look very promising, since in this period the winds are the least opposed by the ionosphere. This happens for two reasons. First, due to the prereversal enhancement, the ionosphere rises and

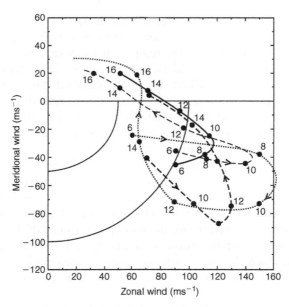

Figure 3.31 Position of the tip of the measured wind vector as a function of universal time for August 21, 1978. The heavy dashed line gives the measured values and the solid line represents the TGCM calculations. Two other TGCM calculations are also presented: the fine dashed curve is a numerical experiment that includes tides from the lower atmosphere and enhanced (60 kV cross-tail potential) magnetospheric convection at high latitudes; the dotted line further includes enhanced $\mathbf{E} \times \mathbf{B}$ drifts representative of solar maximum conditions. [After Sipler et al. (1983). Reproduced with permission of the American Geophysical Union.]

"moves out of the way," Second, the downward electric field builds up due to the F-region dynamo (with little E-region shorting), and ion drag becomes small.

3.6   Mesospheric and Lower Thermospheric Dynamics

3.6.1   *Atmospheric Winds in the Mesosphere and Lower Thermosphere*

The mesosphere and lower thermosphere are collocated with the E region. Most of what we know about dynamics comes from rocket and radio wave techniques but more recently, lidars have become practiced. Meteor radars use the drift of meteor trails to accumulate line-of-sight velocities over some suitable period of time and convert them to vector winds. MF (medium frequency) radars use partial reflections from the D-layer structures seen in Fig. 1.7a, b to determine the winds. Large incoherent scatter radars can be used in the daytime, but it is not practical to use this technique for climatological studies.

In Chapter 1 a representative temperature profile was presented for the earth's atmosphere. Below about 60 km the temperature at a given height stays within a rather narrow range due to the huge thermal inertia in the system. Above this height the situation is quite different. In the extreme case of the upper atmosphere discussed already in this chapter, the temperature can vary by nearly 500 K in a single day. Naturally, huge diurnal tides surge across the system from night to day, leading to the 200 m/s (720 km/hr) winds observed. This type of wind field is termed an in situ tide, since it is generated and detected where it is created: in the thermosphere. Propagating tides usually are only easily detectable after they increase in amplitude by propagating upward into regions of low atmospheric density and are first evident in the lower thermosphere.

A sample analysis of two years of data from the MF radar on Kauai, Hawaii, is presented in Fig. 3.32. Here we see many interesting wave modes, including planetary waves at 2- to 10-day periods, diurnal (S1) and semidiurnal (S2) tides, and even evidence for S3 and S4 tides. Humphreys et al. (2005) have shown that S1 through S6 tides are generated in the troposphere using GPS techniques. There is concern that some contamination occurs in the MF data at high frequencies and high altitudes (> 85 km) due to gravity wave phase velocity detection (rather than winds). But for tidal frequencies and lower, there is general acceptance of these results.

We postpone for now a discussion of atmospheric waves in these high-frequency regimes, usually referred to as gravity waves, but we must point out

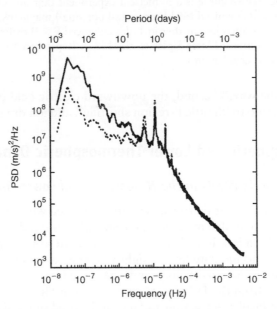

Figure 3.32 Frequency spectral of zonal (solid) and meridional (dashed) velocities in standard form for a two-year data set obtained with the Hawaii MF radar. [After Fritts and Isler (1994). Reproduced with permission of the American Meteorological Society.]

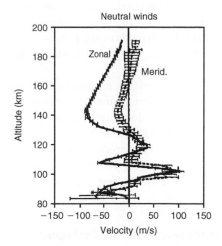

Figure 3.33 Neutral wind profile derived from one TMA rocket release. [After Larsen and Odom (1997). Reproduced with permission of the American Geophysical Union.]

that such waves increase in amplitude as they propagate upward. Upward-propagating tides have the same property, so quite substantial winds can occur in the upper atmosphere. A plot of the wind field determined from a rocket trail measurement is presented in Fig. 3.33. The data were obtained from one of the several hundred trail observations made since 1960. Tri-methyl aluminum (TMA) is the chemical of choice, since it burns in the rich atomic oxygen environment for several tens of minutes. Using triangulation, the winds can be measured as a function of altitude from 80 to 130 km. These large winds are typical of TMA observations over the years (Larsen, 2002) and represent a problem for theoreticians: Such large values are not predicted by the models.

Observations such as those presented in Fig. 3.33 indicating 100 m/s winds in the lower thermosphere are quite remarkable and are not as yet explainable by tidal theory. We postpone discussion of these ubiquitous high winds until our discussion of gravity waves in Chapter 6. These large winds are accompanied by large shears, shears large enough to be dynamically unstable to the Kelvin-Helmholtz instability. We investigate this neutral atmospheric process in Chapters 4 and 6 because it leads to turbulence and small-scale structure in both the atmosphere and the ionosphere. Such effects are likely to occur at all latitudes, but very few equatorial observations exist at this point. We thus postpone discussion of the winds to later chapters.

These high winds have, for the most part, been ignored by theorists and experimentalists, but the data in Fig. 3.33 suggest that this is not a good idea! Also, considering Fig. 3.17b, classic electrojet theory would yield $J/ne = |E/B|$, but this is clearly not the case. The difference can only be due to the current $nev_i \cong neu$ where u is the zonal neutral wind that, in this case, must be opposite to the Hall current observed. A westward wind of over 100 m/s is required by the data in Fig. 3.17b. The winds cannot be measured during the day using the TMA method, but two days after these data were taken, the wind speeds measured just after sunset were of this magnitude (Larsen and Odom, 1997).

At still lower levels, a very interesting feature of the mesospheric temperature field is shown in Fig. 3.34. This solstice view of the earth's atmospheric temperature shows the curious feature that the polar summer mesosphere, in full sunlight for 24 hours a day, is nearly 100 K colder than its winter counterpart (Geller, 1983). Radiative equilibrium theory predicts just the opposite, and the explanation must involve a dynamic effect. This too requires a discussion of gravity waves and their profound effect on the system, and we revisit the polar summer mesosphere in detail in Chapter 7.

3.6.2 A Primer on Turbulence and the Turbopause

Neutral turbulence has been studied for many years in both the atmosphere and the laboratory. The crucial parameter is the Reynolds number, which is a measure of the importance of the different terms in the momentum equation. The momentum equation, also called the Navier-Stokes equation, for a neutral gas (ignoring gravity) is

$$\rho \frac{(\partial U)}{\partial t} + (U \cdot \nabla)\rho U = -\nabla p + \eta \nabla^2 U$$

If we compare the second and fourth terms we get

$$\frac{\rho |U|^2}{L} \quad : \quad \frac{\eta |U|}{L^2}$$

Their ratio yields the Reynolds number

$$R = \frac{|U| L}{\nu}$$

where $\nu = \eta/\rho$ is the kinematic viscosity. In the atmosphere L is quite large (compared to the laboratory), and high Reynolds numbers are common ($R \geq 1000$).

In laboratory experiments energy is generated at some scale by, for example, flowing through a grid. The eddies generated at the grid scale (L) generate smaller-scale waves in the inertial subrange where energy is passed from scale to scale without losses. Ultimately, a scale is eventually reached at which viscosity damps out the structure. That this must occur at small scales can be understood from the viscosity term $(\eta/\rho)\nabla^2 U$. When viewed in the Fourier domain this becomes

$$(\eta/\rho)\nabla^2 U \approx k^2 (\eta/\rho) U$$

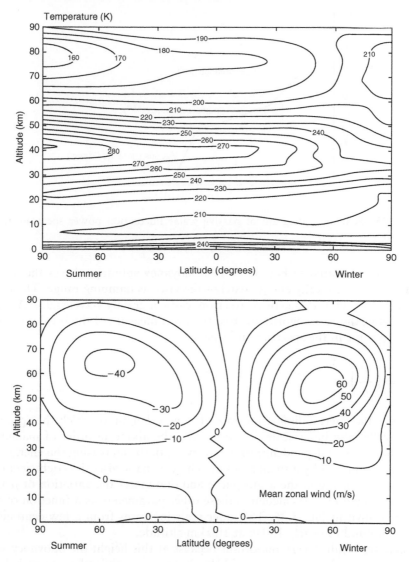

Figure 3.34   Calculated temperatures and mean zonal winds including a Rayleigh friction parameter. [After Geller (1983). Reproduced with permission of the Reidel Publishing Company.]

which increases dramatically at small scales. The theory on which this rests is Kolmogorov's turbulence theory. In the laboratory grid case, energy cascades from the grid scale through the inertial subrange to the viscous subrange. In the atmosphere the grid is replaced by other energy input processes due to gravity waves and other disturbances in the system. The resulting fluctuation power

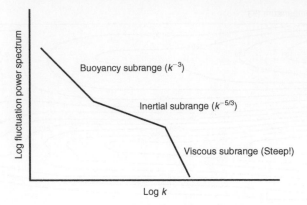

Figure 3.35 Schematic diagram of passive scalar fluctuation power spectrum in the atmosphere.

spectra are illustrated in Fig. 3.35. The buoyancy subrange drives the inertial subrange and eventually energy reaches the viscous damping range. The buoyancy scale L_B is determined by convection scales or gravity wavelengths. At small wavelengths, the Kolmogorov microscale μ is a measure of the breakpoint to a very steep slope in the spectrum at large k:

$$\mu = \left(\frac{\nu^3}{\varepsilon}\right)^{1/4}$$

where ε is the energy dissipation rate. The wavelength at which the spectrum breaks, called the inner scale, is estimated to be about an order of magnitude larger than μ. The dependence on ε is very weak, so increasing the energy cascade only changes μ by a small amount. The kinematic viscosity coefficient does change dramatically in the atmosphere and dominates the variation of μ with height. Hocking (1983) calculated these other parameters as a function of altitude, as shown in Fig. 3.36. The inner scale increases from a few centimeters near the ground to nearly 100 m at 100 km altitude.

This latter result is very important because at this height the buoyancy scale and the microscale are nearly equal. Without an inertial subrange turbulence cannot exist and the atmosphere is no longer mixed. This height is called the turbopause; above it the molecules separate by diffusion according to their mass. In the transition region, turbulence is very patchy, and for any given profile, a patch of turbulent atmosphere can actually be above a region of nonturbulent fluid. An observation illustrating this phenomenon comes from pictures of chemiluminescent meteor trails and is shown in Fig. 5.25b. The well-defined double trail portion of the trail is below the very puffy, turbulent piece. The puffy portion was found to expand at a rate consistent with turbulent diffusion, whereas the lower turbulent portion exhibited molecular diffusion (Kelley et al., 2003b).

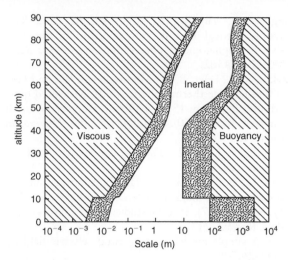

Figure 3.36 Various subranges in the earth's atmosphere. (Figure courtesy of W. Hocking.)

Since atmospheric radars can detect scatter from clear air turbulence, Fig. 3.36 shows why a given radar can only see up to a certain height. If the radar Bragg scale is too deeply into the viscous subrange, no echo can be detected. In the mesosphere it is necessary for electrons to be mixed by the turbulence to get a signal because the air is so tenuous. Thus, large long-wavelength (e.g., 50 MHz) radar systems can obtain mesospheric echoes in daytime and M-S-T (Mesosphere-Stratosphere-Troposphere) radars can be built to study virtually the entire atmosphere up to 100 km.

In summary:

1.   Below 110 km or so, on average, the atmosphere is mixed by turbulence, which feeds off buoyancy or shear forces. Such structures are roughly isotropic.
2.   Above this height the atmospheric constituents begin to separate according to multi-component molecular diffusion (Vlasov and Davydov, 1982).
3.   Above the turbopause the neutral atmosphere no longer has a turbulent structure, and any radar scatter must be due to plasma instabilities or incoherent scatter. The former is highly anisotropic.

References

Anderson, D. N., and Mendillo, M. (1983). Ionospheric conditions affecting the evolution of equatorial plasma depletions. *Geophys. Res. Lett.* 10, 541.

Anderson, D. N., and Roble, R. G. (1974). The effect of vertical $\mathbf{E} \times \mathbf{B}$ ionospheric drifts on F-region neutral winds in the low-latitude thermosphere. *J. Geophys. Res.* **79**, 5231.

Anderson, D., Anghel, A., Yumoto, K., Ishitsuka, M., and Kudeki, E. (2002). Estimating daytime vertical $E \times B$ drift velocities in the equatorial F region using ground-based magnetometer observations. *Geophys. Res. Lett.* **29**, 1596, doi:10.1029/2001GL014562.

Biondi, M. A., Sazykin, S. Y., Fejer, B. G., Meriwether, J. W., and Fesen, C. G. (1999). Equatorial and low latitude thermospheric winds: Measured quiet time variations with season and solar flux from 1980 to 1990. *J. Geophys. Res.* **104**(A8), 17,091.

Chapman, S., and Lindzen, R. S. (1970). *Atmospheric Tides: Thermal and Gravitational.* Gordon & Breach, New York.

Dalgarno, A., and Smith, F. J. (1962). Thermal conductivity and viscosity of atomic oxygen. *Planet. Space Sci.* **9**, 1.

Davis, T. N., Burrows, K., and Stolarik, J. D. (1967). A latitude survey of the equatorial electro-jet with rocket-borne magnetometers. *J. Geophys. Res.* **72**, 1845.

Earle, G., and Kelley, M. C. (1987). Spectral studies of the sources of ionospheric electric fields. *J. Geophys. Res.* **92**, 213.

Eccles, J. V. (1998a). A simple model of low-latitude electric fields. *J. Geophys. Res.* **103**(A11), 26,699–26,708.

―――. (1998b). Modeling investigation of the evening prereversal enhancement of the zonal electric field in the equatorial ionosphere. *J. Geophys. Res.* **103**(A11), 26,709.

Farley, D. T., Bonelli, E., Fejer, B. G., and Larsen, M. F. (1986). The prereversal enhancement of the zonal electric field in the equatorial ionosphere. *J. Geophys. Res.* **91**, 13,723.

Fejer, B. G. (1981). The equatorial ionospheric electric field. A review. *J. Atmos. Terr. Phys.* **43**, 377.

―――. (1986). Equatorial ionospheric electric fields associated with magnetospheric disturbances. In *Solar Wind-Magnetosphere Coupling.* Y. Kamide and J. Slavin (eds.). Terra Scientific Pub. Co., Tokyo, 519–545.

―――. (1991). Low latitude electrodynamic plasma drifts: A review. *J. Atmos. Terr. Phys.* **53**, 677.

Fejer, B. G., and Scherliess, L. (1995). Time dependent response of equatorial ionospheric electric fields to magnetospheric disturbances. *Geophys. Res. Lett.* **22**(7), 851.

―――. (1997). Empirical models of storm time equatorial zonal electric fields. *J. Geophys. Res.* **102**, 24,047.

Fejer, B. G., Farley, D. T., Woodman, R. F., and Calderon, C. (1979). Dependence of equatorial F-region vertical drifts on season and solar cycle. *J. Geophys. Res.* **84**, 5792.

Fejer, B. G., Kelley, M. C., Senior, C., de La Beaujardiere, O., Holt, J. A., Teplet, C. A., Burnside, R., Abdu, M. A., Sobral, J. H. A., Woodman, R. F., and Lepping, R. (1990). Low- and midlatitude ionospheric electric fields during the January 1984 GIMOS campaign. *J. Geophys. Res.* **95**, 2367.

Fejer, B. G., Kudeki, E., and Farley, D. T. (1985). Equatorial F-region zonal plasma drifts. *J. Geophys. Res.* **90**, 12,249.

Fejer, B. G., Larsen, M. F., and Farley, D. T. (1983). Equatorial disturbance dynamo electric fields. *Geophys. Res. Lett.* **10**, 537.

Fejer, B. G., de Paula, E. R., González, S. A., and Woodman, R. F. (1991). Average vertical and zonal F-region plasma drifts over Jicamarca. *J. Geophys. Res.* **96**(A8), 13,901.

Fesen, C. G., Crowley, G., Roble, R. G., Richmond, A. D., and Fejer, B. G. (2000). Simulation of the prereversal enhancement in the low latitude vertical in drifts. *Geophys. Res. Lett.* **27**(13), 1851.

Forbes, J. M. (1981). The equatorial electrojet. *Rev. Geophys. Space Phys.* **19**, 469.

Forbes, J. M., and Lindzen, R. S. (1976). Atmospheric solar tides and their electrodynamic effects. II. The equatorial electrojet. *J. Atmos. Terr. Phys.* **38**, 911.

Fritts, D. C., and Isler, J. R. (1994). Mean motions and tidal and two-day structure and variability in the mesosphere and lower thermosphere over Hawaii. *J. Atmos. Sci.* **51**, 2145.

Gagnepain, J., Crochet, M., and Richmond, A. D. (1977). Comparison of equatorial electrojet models. *J. Atmos. Terr. Phys.* **39**, 1119–1124.

Geller, M. A. (1983). Dynamics of the middle atmosphere. *Space Sci. Rev.* **34**, 359.

Gonzales, C. A., Kelley, M. C., Fejer, B. G., Vickrey, J. F., and Woodman, R. F. (1979). Equatorial electric fields during magnetically disturbed conditions, 2, Implications of simultaneous auroral and equatorial measurements. *J. Geophys. Res.* **84**, 5803.

Gonzales, C. A., Kelley, M. C., Behnke, R. A., Vickrey, J. F., Wand, R. H., and Holt, J. (1983). On the latitudinal variations of the ionospheric electric field during magnetospheric disturbances. *J. Geophys. Res.* **88**, 9135.

Haerendel, G., and Eccles, J. V. (1992). The role of the equatorial electrojet in the evening ionosphere. *J. Geophys. Res.* **97**, 1181–1192.

Harris, I., and Priester, W. (1965). On the diurnal variation of the upper atmosphere. *J. Atmos. Sci.* **22**, 3.

Heelis, R. A., Kendall, P. C., Moffett, R. J., Windle, D. W., and Rishbeth, H. (1974). Electrical coupling of the E and F-regions and its effect on F-region drifts and winds. *Planet. Space Sci.* **22**, 743.

Hocking, W. K. (1983). Mesospheric turbulence intensities measured with a HF radar at 35°S, II. *J. Atmos. Terr. Phys.* **45**, 103.

Humphreys, T. E., Kelley, M. C., Huber, N., and Kintner Jr., P. M. (2005). The semidiurnal variation in GPS-derived zenith neutral delay. *Geophys. Res. Lett.* **32**, L24801, doi:10.1029/2005GL024207.

Ilma, R., and Kelley, M. C. (2009). On explaining the altitude of the equatorial electrojet. *J. Atmos. Solar-Terr. Phys.*, in press.

Kelley, M. C., Fejer, B. G., and Gonzales, C. A. (1979). An explanation for anomalous ionospheric electric fields associated with a northward turning of the interplanetary magnetic field. *Geophys. Res. Lett.* **6**, 301.

Kelley, M. C., Kruschwitz, C. A., Gardner, C. S., Drummond, J. D., and Kane, T. J. (2003a). Mesospheric turbulence measurements from persistent Leonid meteor train observations. *J. Geophys. Res.* **108**, 8454, doi:10.1029/2002JD002392.

Kelley, M. C., Makela, J. J., Chau, J. L., and Nicolls, M. J. (2003b). Penetration of the solar wind electric field into the magnetosphere/ionosphere system. *Geophys. Res. Lett.* **30**, 1158, doi:10.1029/2002GL016321.

Kelley, M. C., Nicolls, M. J., Anderson, D., Anghel, A., Sekar, R., Subbarao, K. S. V., and Bhattacharyya, A. (2007). Multi-longitude case studies comparing the interplanetary and equatorial ionospheric electric fields using an empirical model. *J. Atmos. Solar-Terr. Phys.* **69**(10–11), 1160–1173.

Kelley, M. C., Ilma, R. R., and Crowley, G. (2009a). On the origin of pre-reversal enhancement of the zonal equatorial field. *Ann. Geophys.*, in press.

Kelley, M. C., Ilma, R. R., Alken, P., and Maus, S. (2009b). Evidence for an anomalous cowling conductivity in the strongly driven equatorial electrojet. *Geophys. Res. Lett.*, in press.

King-Hele, D. G. (1970). Average rotational speed of the upper atmosphere from changes in satellite orbits. *Space Res.* **10**, 537.

Kudeki, E., and Bhattacharyya, S. (1999). Postsunset vortex in equatorial F-region plasma drifts and implications for bottomside spread F. *J. Geophys. Res.* **104**, 28,163.

Kudeki, E., Fejer, B. G., Farley, D. T., and Ierkic, H. M. (1981). Interferometer studies of equatorial F-region irregularities and drifts. *Geophys. Res. Lett.* **8**, 377.

Larsen, M. F. (2002). Winds and shears in the mesosphere and lower thermosphere: Results from four decades of chemical release wind measurements. *J. Geophys. Res.* **107**, 1215, doi:10.1029/2001JA000218.

Larsen, M. F., and Odom, C. D. (1997). Observations of altitudinal and latitudinal E-region neutral wind gradients near sunset at the magnetic equator. *Geophys. Res. Lett.* **24**, 1711.

Lindzen, R. S. (1966). Crude estimate for the zonal velocity associated with the diurnal temperature oscillation in the thermosphere. *J. Geophys. Res.* **71**, 865.

———. (1967). Reconsideration of diurnal velocity oscillation in the thermosphere. *J. Geophys. Res.* **72**, 1591.

Maruyama, N., Richmond, A. D., Fuller-Rowell, T. J., Codrescu, M. V., Sazykin, S., Toffoletto, F. R., Spiro, R. W., Millward, G. H. (2005). Interaction between direct penetration and disturbance dynamo electric fields in the storm-time equatorial iono- sphere. *Geophys. Res. Lett.* **32**, L17105, doi:10.1029/2005GL023763.

Matsushita, S. (1969). Dynamo currents, winds, and electric fields. *Radio Sci.* **4**, 771.

Maynard, N. C. (1967). Measurements of ionospheric currents off the coast of Peru. *J. Geophys. Res.* **72**, 1863.

Meriwether, J. W., Moody, J. W., Biondi, M. A., and Roble, R. G. (1986). Optical inter- ferometric measurements of nighttime equatorial thermospheric winds at Arequipa, Peru. *J. Geophys. Res.* **91**, 5557.

Nicolls, M. J., Kelley, M. C., Coster, A. J., González, S. A., and Makela, J. J. (2004). Imaging the structure of a large-scale TID using ISR and TEC data. *Geophys. Res. Lett.* **31**, L09812, doi:10.1029/2004GL019797.

Nishida, A. (1968). Geomagnetic Dp2 fluctuations and associated magnetospheric phenomena. *J. Geophys. Res.* **73**, 1795.

Pfaff, R. F., Jr., Acuña, M. H., Marionni, P. A., and Trivedi, N. B. (1997). DC polariza- tion electric field, current density, and plasma density measurements in the daytime equatorial electrojet. *Geophys. Res. Lett.* **24**, 1667.

Pingree, J., and Fejer, B. G. (1987). On the height variation of the equatorial F-region vertical plasma drifts. *J. Geophys. Res.* **92**, 4763.

Richmond, A. D. (1973a). Equatorial electrojet. I. Development of a model including winds and instabilities. *J. Atmos. Terr. Phys.* **35**, 1083.

Richmond, A. D. (1973b). Equatorial electrojet. II. Use of the model to study the equa- torial ionosphere. *J. Atmos. Terr. Phys.* **35**, 1105.

Richmond, A. D., and Matsushita, S. (1975). Thermospheric response to a magnetic substorm. *J. Geophys. Res.* **80**, 2839.

Rishbeth, H. (1972). Superrotation of the upper atmosphere. *Rev. Geophys. Space Phys.* **10**, 799.

Rishbeth, H., and Garriott, O. K. (1969). *Introduction to Ionospheric Physics*. Int. Geo- phys. Ser., Vol. 14, Academic Press, New York.

Scherliess, L., and Fejer, B. G. (1997). Storm time dependence of equatorial disturbance dynamo zonal electric fields. *J. Geophys. Res.* **102**(A11), 24,037.

Shuman, B. M. (1970). Rocket measurement of the equatorial electrojet. *J. Geophys. Res.* **75**, 3889.

Sipler, D. P., and Biondi, M. A. (1978). Equatorial F-region neutral winds from nightglow OI 630.0 nm Doppler shifts. *Geophys. Res. Lett.* **5**, 373.

Sipler, D. P., Biondi, M. A., and Roble, R. G. (1983). F-region neutral winds and temperatures at equatorial latitudes: Measured and predicted behaviors during geomagnetically quiet conditions. *Planet. Space Sci.* **31**, 53.

Somayajulu, V. V., Reddy, C. A., and Viswanathan, K. S. (1985). Simultaneous electric field changes in the equatorial electrojet in phase with polar cusp latitude changes during a magnetic storm. *Geophys. Res. Lett.* **12**, 473.

Somayajulu, V. V., Reddy, C. A., and Viswanathan, K. S. (1987). Penetration of magnetospheric convective electric fields to the equatorial ionosphere during the substorm of March 22, 1979. *Geophys. Res. Lett.* **14**, 876.

Sugiura, M., and Cain, J. C. (1966). A model equatorial electrojet. *J. Geophys. Res.* **71**, 1869.

Untiedt, J. (1967). A model of the equatorial electrojet involving meridional currents. *J. Geophys. Res.* **72**, 5799.

Vasyliunas, V. M. (1972). The interrelationship of magnetospheric processes. In *Earth's Magnetospheric Processes*. B. M. McCormac (ed.). Reidel, Boston, 29.

Vlasov, M. N., and Davydov, V. E. (1982). Theoretical description of the main constituents in the earth's upper atmosphere. *J. Atmos. Terr., Phys.* **44**, 461.

Wharton, L. E., Spencer, N. W., and Mayr, H. G. (1984). The earth's thermospheric super-rotation from Dynamics Explorer 2. *Geophys. Res. Lett.* **11**, 531.

Woodman, R. F. (1970). Vertical drift velocities and east-west electric fields at the magnetic equator. *J. Geophys. Res.* **75**, 6249.

4 Equatorial Plasma Instabilities and Mesospheric Turbulence

In this chapter we study the plasma physics and aspects of neutral turbulence at low latitudes in the earth's ionosphere. Most of the information we have concerning these processes comes from radio wave scattering or reflection experiments conducted from the surface of the earth. In situ measurements made with rocket- and satellite-borne sensors have also contributed significantly to our present understanding. As in the previous chapter, instabilities occurring in the F region are addressed first, followed by consideration of lower-altitude phenomena. We also discuss mesospheric turbulence because plasma turbulence is often described in an analogous manner.

4.1   F-Region Plasma Instabilities: Observations

Plasma instability phenomena occurring in the equatorial F-region ionosphere are usually grouped under the generic name equatorial spread F (ESF). This stems from the earliest observations using ionosondes, which showed that on occasion the reflected echo did not display a well-behaved pattern but was "spread" in range or frequency (Berkner and Wells, 1934). A physics-based name for this phenomenon is "convective equatorial ionospheric storm" (CEIS). The phenomenon occurs primarily at night, although isolated daytime events occur (Woodman et al., 1985; Chau and Woodman, 2001). The modern era in CEIS (ESF) studies began on a "low" theoretical note in 1970 with the publication of the first compilation of measurements made by the Jicamarca Radar Observatory in Peru (Farley et al., 1970). The authors concluded that no theory published to date could explain the data! Fortunately, considerable progress has occurred since then, and the theory has kept pace with continuing additions and improvements in the data.

In the early 1970s, range-time-intensity (RTI) "weather" radar maps came into vogue as a method of following the position and intensity of CEIS plasma density irregularities. Since the Jicamarca radar is stationary and has a small field of view ($\sim 1°$), the RTI maps are similar to pictures from a slit camera. One of the more spectacular examples of the maps obtained at the Jicamarca Observatory is

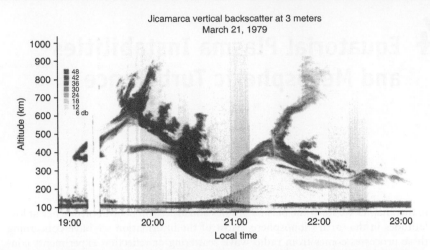

Figure 4.1 Range-time-intensity (RTI) map displaying the backscatter power at 3 m wavelengths measured at Jicamarca, Peru obtained on March 21, 1979. The gray scale is decibels above the thermal noise level. [After Kelley et al. (1981). Reproduced with permission of the American Geophysical Union.]

reproduced in Fig. 4.1, and a more typical map is reproduced in the top panel of Fig. 4.2. The Jicamarca radar transmits nearly vertically, and at its 50 MHz frequency it is sensitive to backscatter from waves that satisfy the Bragg matching condition. This condition requires that $k_r = k_s + k_m$, where k_r is the radar wave vector, k_s the scattered wave vector, and k_m the wave vector in the medium. (See Appendix A for more details.) Since $k_s = -k_r$ for backscatter, it follows that $k_m = 2 k_r$. Thus, the Jicamarca 50 MHz radar ($\lambda = 6$ m) detects only waves with vertical wave vectors corresponding to a 3 m wavelength. In the F region of the ionosphere (above 200 km), both figures display intense echoes over a 100 km altitude extent for much of the time. The gray scale shows the echo strength to be up to 50 dB higher than that caused by thermal fluctuations at the 3 m wavelength. This "thermal backscatter level" is the source of the incoherent scatter echoes often used to determine ionospheric parameters (see Appendix A). The thick irregularity layer moves up or down with time rather periodically in Fig. 4.1 and twice is interrupted by a period of intense backscatter signal that extends to very high altitudes. In Fig. 4.2 this happens only once at about 2110 local time. These towering echoing features have been termed "plumes." As we shall see, this name is quite apt for structures that are probably the result of a convective instability driven by buoyancy forces, much like tropospheric thunderstorms. The horizontal time scale is such that the mean horizontal plasma drift speed (125 m/s toward the east) multiplied by time yields the same horizontal distance scale as given in the vertical altitude scale. Thus, the picture yields an accurate description of the ionospheric scatterers (as viewed looking south

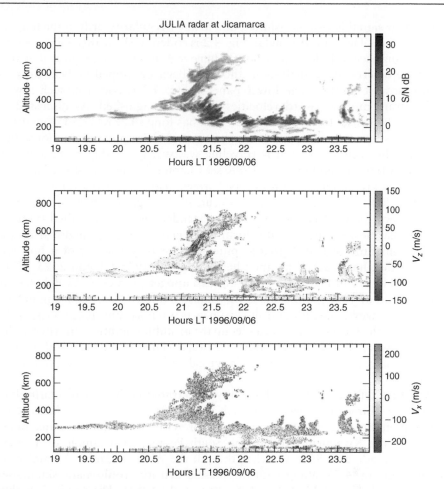

Figure 4.2 An RTI map for September 6, 1996, is reproduced in the top panel. In the central panel the map is color scaled to represent the vertical line-of-site velocity of the irregularities. In the lower panel the east-west velocity is represented. [After Hysell and Burcham (1998). Reproduced with permission of the American Geophysical Union.] See Color Plate 3.

antiparallel to the magnetic field) if the irregularities are frozen into a medium that moves with that constant velocity. Since the probing radar wavelength is very small compared to the scales involved in the phenomena, they can be considered as tracers, but care must be taken in the interpretation of the observed features in these maps. Nonetheless, the inspired interpretation made by Woodman and LaHoz (1976) that the plumes represent upwelling of plasma "bubbles" has stood the test of time and is consistent with in situ data from rockets and satellites.

The radar signal is Doppler-shifted due to any irregularity drift in the line-of-sight direction. The central panel in Fig. 4.2 is coded to show this with green to red, denoting ascent. Indeed, the central plume region is moving rapidly upward. An interferometric data analysis method has been developed at Jicamarca to reveal east-west motion. The lower panel in Fig. 4.2 is coded in such a way that eastward and westward velocities can be distinguished. As discussed in Chapter 3, by sunset the zonal F-region drift has long since reversed from westward to eastward, easily explaining the westward irregularity motions. But the initial thin echoing region is traveling westward, creating a shear in the plasma flow (Hysell and Burcham, 1998; see Chapter 3). These westward drifting irregularities are termed "bottom-type" irregularities. The thickening of the layer after 2040 (discounting plumes) occurs in the region of F-region dynamo control, since the drift is eastward. Such extended regions are called bottom-side irregularities. Plumes usually form just once on a given night, if at all, and are sometimes called "apogee" plumes when they rise out of a bottom-side layer that has reached its apex height (e.g., at 1935 LT and 2125 LT in Fig. 4.1; 2110 LT in Fig. 4.2). Detecting two apogee plumes (as in Fig. 4.1) at a limited field-of-view ground site is very unusual and is discussed again in Section 4.2.3. However, satellite and all-sky imagers often see multiple plumes, probably because the terminator, which sets up the unstable condition, moves rapidly westward, whereas the plumes move eastward. Thus, a narrow field-of-view ground site may only see one plume on a typical night but, on occasion, can detect several.

Several rockets and satellites have now penetrated both the irregularity layers and the plume structures. For such studies, it has proven useful to correlate the in situ probing with Jicamarca or with the Altair scanning radar located on the Kwajalein Atoll in the South Pacific. An example of a radar (RTI) map made at 0.96 m wavelength with the Altair radar on Kwajalein is presented in Fig. 4.3a, along with the plasma density profile made simultaneously onboard the PLUMEX I rocket (Rino et al., 1981). The profile is highly irregular throughout the rocket trajectory, but the most crucial observation is that the intensely scattering top, or "head," of the radar plume is collocated with a region of depleted plasma, a bubble. Another example that correlates AE-E satellite plasma density measurements with Altair data is presented in Fig. 4.3b. The plot shows that the radar plume is associated with a very structured region of plasma in which the mean density is much less than the value outside the volume of backscatter (Tsunoda et al., 1982). An Altair radar scan in the north-south plane is represented in Fig. 4.4 and shows that the depleted regions are elongated along the direction of the magnetic field for hundreds of kilometers.

Plasma density wave number spectra from several rocket flights through the unstable bottomside layers are presented in Fig. 4.5a. The spectra are similar but

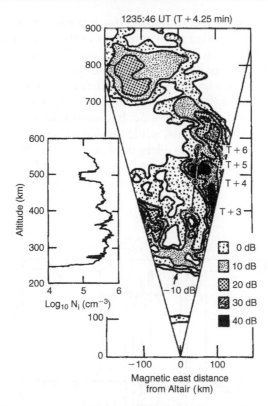

Figure 4.3a Simultaneous vertical rocket plasma density profile and backscatter map made with the Altair radar on the island of Kwajalein. Dots show the rocket trajectory. [After Rino et al. (1981). Reproduced with permission of the American Geophysical Union.]

not identical, and it is not possible to declare a universal spectral form with any certainty (unlike Kolmogorov turbulence: see Section 4.9). We can say that they are typically of the form k^{-n} with $2 \leq n \leq 3$ and that they break to a much steeper form somewhere in the range of 60 m. In Fig. 4.5b we contrast horizontal spectra from AE-E, which has access to longer wavelengths with the four rocket spectra obtained in the Kwajalein rocket campaign and the distribution of scales detected by an HF channel probe (see Fig. 4.12). The satellite data do not extend to small enough scales to detect the steep portion of the spectrum, which is more clearly evident here than in Fig. 4.5a. Figure 4.5b suggests that the two dimensions in the plane perpendicular to **B** are not equivalent. Larger perturbation scales exist horizontally due to its larger extent, of course. But also at intermediate scales, the rocket spectra are more intense and have steeper spectra. The two vertical lines correspond to the wave number responsible for Jicamarca and Altair backscatter. The satellite horizontal spectra show a peak at 200 km and a secondary (local) maximum near 50 km. Both of these features are detected in the HF data (see Fig. 4.12 and related discussion of the HF technique).

A new dimension to our understanding of CEIS came with the advent of electric field measurements in Project Condor. The first spectral data of this sort are shown on the right-hand side of Fig. 4.5c. It is important to note that the electric field spectrum is almost featureless, while the density spectrum changes form at 40 Hz, which corresponds to a wavelength of about 50 m. Even better data were obtained in the Kwajalein campaign, which we explore further in Section 4.4. It is clear, however, that the physics changes in the 50–100 m scale size regime.

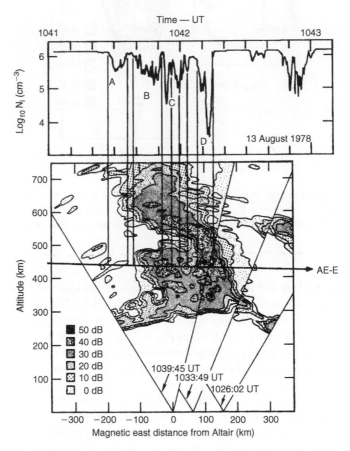

Figure 4.3b Simultaneous horizontal satellite plasma density profile and backscatter map made with the Altair radar. [After Tsunoda et al. (1982). Reproduced with permission of the American Geophysical Union.]

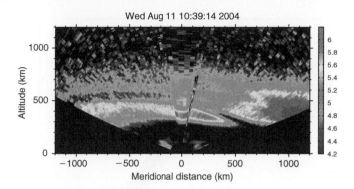

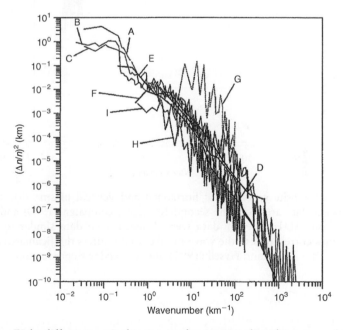

Figure 4.4 Incoherent backscatter map of the equatorial zone during a CEIS event. The depleted plasma region follows the contour of the magnetic field lines. The strong signals just off vertical were detected when the radar was pointed perpendicular to the magnetic field lines. The magnetic equator is located near 5 degrees south of the radar. The equatorial anomaly can be seen at the right. The color scale corresponds to $\log_{10} n$ (from 4.2 to 6.2). (Figure courtesy of D. Hysell.) See Color Plate 4.

Figure 4.5a Eight different vertical wave number spectra directly measured in unstable ionospheric conditions: A–C Kwajalein; D, G, H, I Peru; E Brasil; F Florida Ba cloud (see Chapter 6). The most unusual spectrum (G) was obtained on the wall of a plume and may be due to a wind-driven secondary instability (LaBelle et al., 1986). [After Kelley and Hysell (1991). Reproduced with permission of Pergamon Press.]

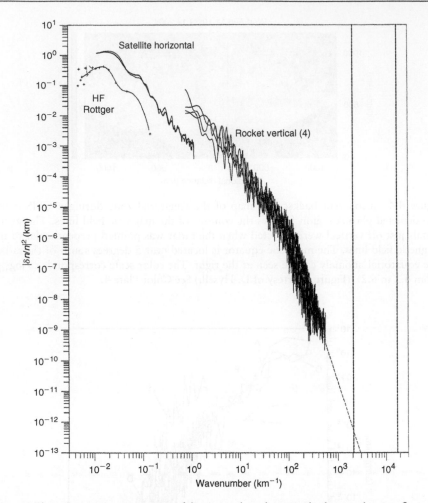

Figure 4.5b Composite spectrum of horizontal and vertical plasma density fluctuations. The rocket data are from the second Kwajalein campaign (1990) and the satellite data are from AE-E. The HF data were deduced from data similar to Fig. 4.12. The vertical lines correspond to the wave number that scatters the Jicamarca and Altair radar signals. [After Kelley and Hysell (1991). Reproduced with permission of Pergamon Press.]

Forward scatter, or "scintillation," measurements have provided information on structure in the intermediate wavelength range (100–1000 m), since, at the radio frequencies typically used for these experiments, the scale of the irregularities that do the scattering is in that range (Basu et al., 1978). In the example shown in Fig. 4.6, the VHF scintillation intensity (lower trace) continues for a considerable time after the most intense 3 m backscatter ends. This shows

March 14, 1983

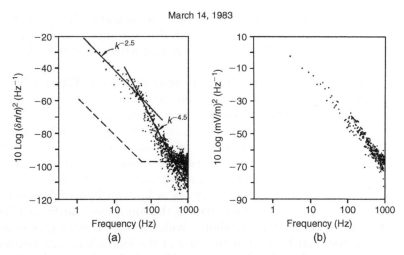

(a)

(b)

Figure 4.5c   Spectra of density and electric field irregularities in equatorial spread F as observed from a rocket launched March 14, 1983, from Punta Lobos, Peru: (a) spectrum of density irregularities; (b) spectrum of electric field irregularities. The dashed line indicates the background spectrum measured during a quiet period. [After LaBelle et al. (1986). Reproduced with permission of the American Geophysical Union.]

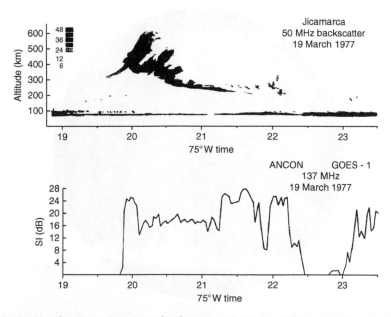

Figure 4.6   Simultaneous Jicamarca backscatter power and VHF scintillation intensity due to irregularities in the same scattering volume. [After Basu et al. (1978). Reproduced with permission of the American Geophysical Union.]

that the large-scale features remain in the medium long after the smallest structures disappear. Decaying neutral fluid turbulence also behaves in this manner; that is, the shortest-scale features decay first when the driving forces are
removed.

Airglow observations have also proven to be quite valuable. Figure 4.7a shows
an image taken in the 630 nm emission due to excited atomic oxygen in the
thermosphere. This is emitted in the two-step recombination of O^+ discussed
in Chapter 1, which also yields a 557.7 nm emission (e.g., Fig. 4.7b). The dark
bands indicate regions that are either low in plasma density or high in altitude;
very likely both attributes contribute based on all of the other data shown in this
section. Figure 4.7b shows that plumes can extend to well over 1500 km. Here,
two CCD camera outputs are combined to show both the base of the equatorial
ionosphere and the plumes as they map to the ionosphere north of Hawaii.
Many bifurcations occur, and the plumes widen as they reach high equatorial
altitudes. Structure can be seen at the base of the ionosphere (see arrows). It
is important to note that the airglow over Hawaii, for example, is generated
locally. Thus, the structure observed must be created by electric fields mapped
from the high-altitude equatorial plane acting on the vertical gradient in the local
ionosphere.

Sept 28, 1995

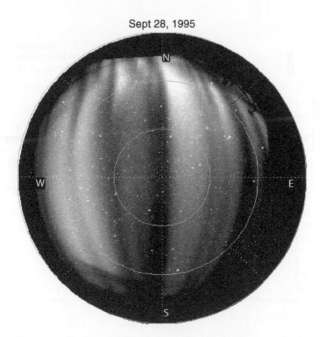

Figure 4.7a   All-sky airglow image at 630 nm taken on Christmas Island during CEIS
conditions. (Figure courtesy of M. Taylor.)

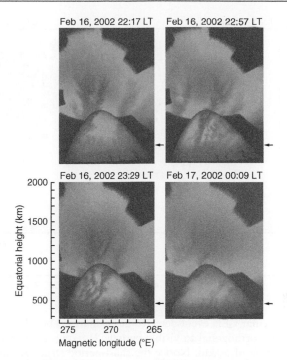

Figure 4.7b A narrow field-of-view camera looking south from Hawaii parallel to the magnetic field provides a two-dimensional view of the bottomside near the magnetic equator while the collocated all-sky camera shows that the plumes map to latitudes poleward of Hawaii. The image is in the 557.7 nm oxygen line. GPS satellite signal scintillations occurred when the line of sight penetrated a dark region. [After Kelley et al. (2002). Reproduced with permission of the American Geophysical Union.]

Fully developed equatorial storms often display remarkable internal structure, some of which is periodic. We have already seen long-period modulations of the whole scattering region (e.g., Figs. 4.1 and 4.7a), which certainly must reflect long-wavelength variations as well. Somewhat shorter-scale organization of radar echoes is seen in the "downleg" miniplumes seen in Figs. 4.1 and 4.2 and in the airglow pictures in Fig. 4.7b. The horizontal spectral data of Fig. 4.5b show a peak in the 400–600 km range and a secondary peak at 50 km. It is very tempting to associate the latter with the miniplume spacing. Some evidence for vertical periodic behavior is seen in Fig. 4.1—for example, at 2145 when three bands of irregularities are detected with vertical separations of about 70 km. A more striking example is presented in Fig. 4.8. Here, many examples of near-horizontal striations are seen. The various cuts shown have been Fourier analyzed and indicate a characteristic separation of 50 km. This event had three plumes between 2010 and 2110 LT.

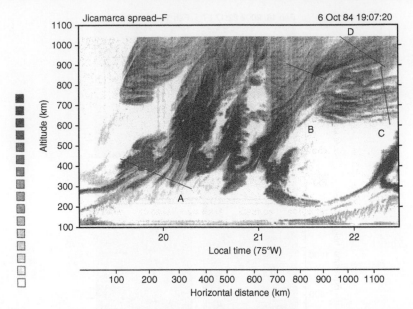

Figure 4.8 A spectacular CEIS event detected at Jicamarca. The labeled lines show cuts made though the data orthogonal to visible periodic layering. Fourier analysis shows a characteristic separation of 50 km. Notice the short-period structures drawn out of the 100–110 km region where the equatorial electrojet instabilities occur. [After Hysell et al. (1990). Reproduced with permission of the American Geophysical Union.]

As we shall see following, winds and waves in the neutral atmosphere may be responsible in part for some of these characteristics scales. But before we can explore these processes, we need to understand plasma instability theory as it applies to these equatorial ionospheric storms.

4.2 Development and Initiation of Convective Equatorial Ionospheric Storms (a.k.a. Equatorial Spread F)

4.2.1 Linear Theory of the Rayleigh-Taylor Instability

Dungey (1956) first proposed the Rayleigh-Taylor (RT) instability as the process driving CEIS. This mechanism was temporarily rejected along with all the other candidate theories by Farley et al. (1970), since, as we shall see, it seemed capable only of generating structure on the bottomside of the F-region plasma density profile. The manner in which the RT instability can cause irregularities to grow in the equatorial ionosphere is illustrated in Fig. 4.9a, using a two-dimensional model. Here the steep upward-directed gradient, which develops on the bottom-side of the nighttime F layer (see Chapter 3), is approximated by a step function.

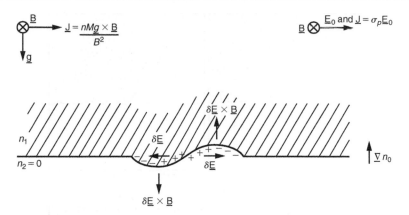

Figure 4.9a   Schematic diagram of the plasma analog of the Rayleigh-Taylor instability in the equatorial geometry.

The density is equal to n_1 above the interface and zero below. The gravitational force is downward, antiparallel to the density gradient, and the magnetic field is horizontal, into the paper. An initial small sinusoidal perturbation is also illustrated, and we assume that the plasma is nearly collisionless—that is, that κ_i and κ_e are large. From (2.36c) we can determine the electrical current by considering the ion and electron velocities due to the pressure gradients and gravity. First, we note that the pressure-driven current does not create any perturbation electric fields, since the current is everywhere perpendicular to the density gradient. The pressure-driven current thus flows parallel to the modulated density pattern and has no divergence.

Turning to the gravitational term in (2.36c), the species velocity is proportional to its mass, so the ion term dominates. A net current flows in the x direction with magnitude

$$J_x = nMg/B$$

Since the current is in the $\mathbf{g} \times \mathbf{B}$ direction, which is strictly horizontal, J_x will be large when n is large and small when n is small. There is thus a divergence, and charge will pile up on the edges of the small initial perturbation. As a result, perturbation electric fields (δE) build up in the directions shown. These fields in turn cause an upward $\delta \mathbf{E} \times \mathbf{B}$ drift of the plasma in the region of plasma depletion and a downward drift in the region where the density is high. Lower (higher) density plasma is therefore advected upward (downward), creating a larger perturbation, and the system is unstable. An analogous hydrodynamic phenomenon is illustrated in the series of sketches in Fig. 4.9b. These have been derived from photographs of the hydrodynamic Rayleigh-Taylor instability when a light fluid supports a heavier fluid against gravity. Initial small oscillations in

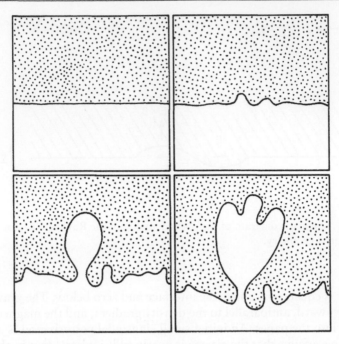

Figure 4.9b Sequential sketches made from photos of the hydrodynamic Rayleigh-Taylor instability. A heavy fluid is initially supported by a transparent lighter fluid.

the surface grow "in place," pushing the lighter fluid upward. In the ionospheric case the "light fluid" is the low-density plasma, which carries a gravity-driven current that provides the $\mathbf{J} \times \mathbf{B}$ force, preventing the plasma from freely falling. The system is unstable when $\mathbf{g}$ and ∇n are oppositely directed. We discuss the E_0 term shown at the top right of Fig. 4.9a following.

We now calculate more formally the growth rate of the RT instability, assuming a small initial perturbation in plasma density and electric field. Inertial terms are dropped, which corresponds to the condition $\nu_{in} \gg \partial/\partial t$. This approximation breaks down at high altitudes but is certainly valid on the bottomside and up to 450 km or so. Using (2.37b) and ignoring neutral winds for now, the steady velocity of each species under the influence of electric and gravitational fields is given by

$$\mathbf{V}_{j\perp} = \left[1 \Big/ \left(1 + \kappa_j^2\right)\right] \mathbf{W}_{j\perp} + \left[\kappa_j \Big/ \left(1 + \kappa_j^2\right)\right] \left[\mathbf{W}_{j\perp} \times \hat{B}\right] \qquad (4.1)$$

where

$$\mathbf{W}_j = b_j \mathbf{E} + \mathbf{g}/\nu_{jn} - D_j \nabla n/n \qquad (4.2)$$

For the electrons κ_e is so large that the second term dominates in (4.1) and

$$\mathbf{V}_e = \mathbf{E} \times \mathbf{B}/B^2 - \left(m/eB^2\right)(\mathbf{g} \times \mathbf{B}) + \left(k_B T_e/eB^2 n\right)(\nabla n \times \mathbf{B}) \qquad (4.3)$$

For the ions $\kappa_i \gg 1$ in the F region, but it is not so great that the first term in (4.1) can be ignored, as was the case for the electrons. For the F region that term can be written as $\mathbf{W}_{i\perp}/\kappa_i^2$, and we have

$$\mathbf{V}_{i\perp} = \left(b_i/\kappa_i^2\right)\mathbf{E} + \mathbf{g}/\left(\nu_{in}\kappa_i^2\right) - \left(D_i/\kappa_i^2\right)(\nabla n/n) + \mathbf{E} \times \mathbf{B}/B^2$$

$$+ \left(M/eB^2\right)(\mathbf{g} \times \mathbf{B}) - \left(k_B T_i/eB^2 n\right)(\nabla n \times \mathbf{B}) \qquad (4.4)$$

The electron mass is so small that the $\mathbf{g} \times \mathbf{B}$ term in (4.3) can be dropped. The second and third terms in the right-hand side of (4.4) are small due to the fact that $\kappa_i \gg 1$, and we take $n_i = n_e = n$. Now, since the $\mathbf{E} \times \mathbf{B}$ term is identical for ions and electrons, no current flows due to those terms. $\mathbf{J}$ is given by

$$\mathbf{J} = ne(\mathbf{V}_i - \mathbf{V}_e)$$

$$= \sigma_P \mathbf{E} + (ne/\Omega_i)\mathbf{g} \times \hat{B} - \left(k_B/B^2\right)(T_i + T_e)(\nabla n \times \mathbf{B}) \qquad (4.5)$$

Here we have used the fact that for large κ_i, $\sigma_P = neb_i/\kappa_i^2$. The gravity term also is rewritten in terms of Ω_i. Notice that the gravitational current flows even in a collisionless plasma, while the electric field term exists only if $\sigma_P \neq 0$—that is, in a collisional plasma.

We now study the linear stability of a vertically stratified equatorial F layer under only the influence of gravity—that is, the pure Rayleigh-Taylor case. We set $\mathbf{E}_0 = 0$ for now but retain a first-order electric field perturbation $\delta \mathbf{E}$. The continuity and current divergence equations from Chapter 2 will be used in the analysis. Ignoring production and loss, which is reasonable in the postsunset time period when the F layer is very high, the continuity equation is

$$\partial n/\partial t + \mathbf{V} \cdot \nabla n + n(\nabla \cdot \mathbf{V}) = 0 \qquad (4.6)$$

where for $M \gg m$ the plasma velocity $\mathbf{V}$ may be approximated by $\mathbf{V}_i$. First, consider the "compressibility" term $(\nabla \cdot \mathbf{V})$. From (4.4) with $\mathbf{E} = 0$ and κ_i large,

$$\nabla \cdot \mathbf{V} = \nabla \cdot \left\{\left(M/eB^2\right)(\mathbf{g} \times \mathbf{B}) - \left(k_B T_i/enB^2\right)(\nabla n \times \mathbf{B})\right\} \qquad (4.7)$$

Since $\mathbf{g}$ and $\mathbf{B}$ do not vary in the $\mathbf{g} \times \mathbf{B}$ direction, the first term vanishes. Since we also have $\nabla \cdot (\nabla n \times \mathbf{B}) = 0$ and $(\nabla n \times \mathbf{B}) \cdot \nabla n = 0$, Eq. (4.7) vanishes and the plasma flow is incompressible. This is usually a good approximation and is often taken to be valid a priori for any ionospheric F-region calculation. However, care must be taken in applying this result since it is not a fundamental principle and must be checked in each case. It is certainly not true in the E region, where compressibility plays an important role in the formation of images of F-region phenomena (see Chapter 10). Setting $(\nabla \cdot \mathbf{V}) = 0$, the equations we shall linearize are

$$\partial n/\partial t + \mathbf{V} \cdot \nabla n = 0 \tag{4.8a}$$

$$\nabla \cdot \mathbf{J} = 0 \tag{4.8b}$$

To study the electrostatic instability of these equations in the presence of a vertical zero-order density gradient, we can write the electric potential and the plasma density as

$$\phi = \delta\phi e^{i(\omega t - kx)} \tag{4.9a}$$

$$n = n_0(z) + \delta n e^{i(\omega t - kx)} \tag{4.9b}$$

where the initial perturbation propagates in the x direction. Note that we have already assumed charge neutrality so that $n_e = n_i = n$. Using (4.5) with E replaced by $\delta\mathbf{E}$, (4.8b) becomes

$$\nabla \cdot \left[(ne/\Omega_i)\mathbf{g} \times \hat{B} + \left(ne^2 \nu_{\text{in}}/M\Omega_i^2 \right)\delta\mathbf{E} \right] = 0 \tag{4.10}$$

where again we have used the fact that $\nabla \cdot (\nabla n \times \mathbf{B}) = 0$ to set the divergence of the pressure-driven current equal to zero. Here $\delta\mathbf{E}$ is the perturbation electric field associated with the potential ϕ, and we have substituted (2.40b) for σ_P in the F region. Since $\delta\mathbf{E} = -\nabla\phi$, the vector inside the square bracket in the preceding equation only has an x component and taking the x derivative yields

$$\left(eg/\Omega_i \right)(\partial n/\partial x) - \left(e^2 \nu_{\text{in}}/M\Omega_i^2 \right)(\partial n/\partial x)(\partial\phi/\partial x) - \left(ne^2 \nu_{\text{in}}/M\Omega_i^2 \right)\left(\partial^2\phi/\partial x^2 \right) = 0$$

The second term is of second order, and thus the linear form of this equation is

$$\left(eg/\Omega_i \right)(\partial n/\partial x) - \left(ne^2 \nu_{\text{in}}/M\Omega_i^2 \right)\left(\partial^2\phi/\partial x^2 \right) = 0 \tag{4.11a}$$

Making the substitutions

$$P = M\nu_{\text{in}}/B^2, \quad Q = Mg/B$$

(4.11a) becomes

$$Q(\partial n/\partial x) - Pn\left(\partial^2\phi/\partial x^2 \right) = 0 \tag{4.11b}$$

Turning now to the continuity equation (4.8a), we have

$$(\partial n/\partial t) + V_x(\partial n/\partial x) + V_z(\partial n/\partial z) = 0$$

The two velocity components may be obtained from (4.4) if we remember that the electric field δE is of first order, κ_i is large, and the pressure-driven velocity does not contribute to $\mathbf{V} \cdot \nabla n$. Since $\partial n/\partial x$ is of first order, only the zero-order V_x contributes to the second term—hence, $V_x = Mg/eB = Q/e$. In the third term, $\partial n/\partial z$ is of zero order due to the vertical density gradient in the plasma. We must then include the first-order vertical velocity given by $V_z = \delta E_x/B$. The linearized continuity equation is therefore

$$\partial n/\partial t + (Q/e)(\partial n/\partial x) - (1/B)(\partial \phi/\partial x)(\partial n/\partial z) = 0 \tag{4.12}$$

Using the plane wave solutions, (4.11b) and (4.12) may be written

$$-ikQ\delta n + n_0 k^2 P\delta\phi = 0 \tag{4.13a}$$

$$(i\omega - ikQ/e)\delta n + (ik/B)\,(\partial n_0/\partial z)\,\delta\phi = 0 \tag{4.13b}$$

These are two equations in two unknowns, δn and $\delta\phi$, and may be solved by setting the determinant of coefficients equal to zero. This yields the dispersion relation

$$\omega = (kQ/e) - i\big(g/\nu_{\text{in}}\big)\,(1/n_0)\,(\partial n_0/\partial z) \tag{4.14}$$

The real part of ω, ω_r, shows that the plane waves propagate eastward with phase velocity V_ϕ given by

$$V_\phi = \omega_\text{r}/k = Q/e = Mg/eB \tag{4.15a}$$

For an atomic oxygen plasma at the equator $V_\phi \approx 6\,\text{cm/s}$, which is quite small. The imaginary part of ω is

$$\omega_\text{i} = -\big(g/\nu_{\text{in}}\big)\,[(1/n_0)\,(\partial n_0/\partial z)] \tag{4.15b}$$

When $\partial n_0/\partial z$ is positive (corresponding here to the density gradient antiparallel to $\mathbf{g}$), ω_i is negative and

$$e^{i\omega t} = e^{i\omega_\text{r} t}e^{\gamma t}$$

where γ is positive and thus yields a growing solution. The parameter γ is the growth rate of the instability and is given by

$$\gamma = g/L\nu_{\text{in}} \tag{4.16}$$

where L is the gradient scale length ($L = [(1/n_0)\partial n_0/\partial z]^{-1}$).

Although quite a simple expression, this result offers explanations for a number of properties of CEIS. First, in the initial development of spread F, there is a strong tendency for a VHF radar to obtain echoes confined to the height range where the density gradient is upward. In fact, the early Jicamarca study showed that the onset of nonthermal backscatter usually began at a density level about 1% of the plasma density at the F peak. Several rockets have been flown during bottomside ESF at times when no radar echoes were obtained above the F peak, and, indeed, intense irregularities were found below the peak but a smooth profile was found above. These cases are thus in agreement with the linear theory in that the latter predicts instability only when $\mathbf{g}$ is antiparallel to ∇n. Another feature predicted by the theory is a height dependency for γ due to the collision frequency term in the growth rate denominator: the higher the layer, the lower ν_{in} and the larger the instability growth rate. As mentioned earlier, Farley et al. (1970) noted a strong tendency for irregularities to be generated when the layer was at a high altitude where the collision term is small. Notice that the plume structure in Fig. 4.2, the initial plume in Fig. 4.1, and the multiple plumes in Fig. 4.8 were all generated when the echoing layer was at its peak altitude. This also suggests that the most spectacular effects occur when the ionosphere is high.

Indeed, strong evidence for the control of CEIS by the height of the F layer was presented by Fejer et al. (1999). An example from their study is summarized in Fig. 4.10, based on Jicamarca radar data. In this plot the vertical lines indicate the presence of 3 m irregularities, which define active conditions. We qualify this since, as shown in Fig. 4.6, longer wavelength structures can exist long after the 3 m irregularities are gone. The circles indicate the vertical drift velocity and the x's indicate the altitude of maximum incoherent scatter backscatter power, which occurs just below the F peak. Magnetic activity (average K_p) for these September 1987 examples were, sequentially, 2.9, 2.9, and 4.5, and the solar decimetric flux, a measure of solar activity, was near 80 units (solar minimum conditions). Inspection shows that CEIS begins if and only if the peak reached 400 km at onset. The extended altitude on September 23 occurred on the night that the highest altitude was reached. Note that high K_p suppressed the zonal electric field and CEIS on September 25. This could be due to either a disturbance dynamo or the penetration electric fields discussed in Chapter 3.

The fundamental destabilizing source is the current density arising due to the differential drift of ions and electrons, which is dependent on mass. However, the growth rate of CEIS (4.16) is independent of mass. This is essentially due to the assumption of a single ion composition in the F region. Molecular ions are also found to exist in the F region, along with atomic oxygen, during evening (Sridharan et al., 1997) and nighttime (Narcisi and Szuszczewicz, 1981) in the equatorial ionosphere. The growth rate expression derived by including the ionic constituents (Sekar and Kherani, 1999) reveals that the growth rate of CEIS depends on the mass and number densities of both ions.

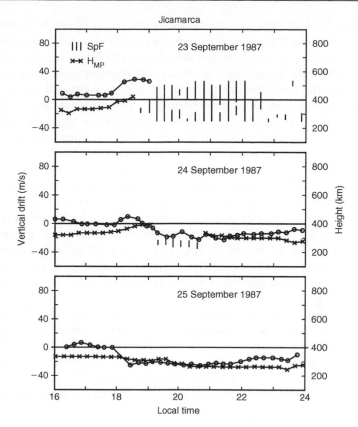

Figure 4.10 Vertical drift velocities (circles), heights of maximum backscattered power (x), and spread F scattering layers (vertical lines) during March 1985. The maximum height probed by the radar was 650 km. [After Fejer et al. (1999). Reproduced with permission of the American Geophysical Union.]

4.2.2 The Generalized Rayleigh-Taylor Process: Electric Fields, Neutral Winds, and Horizontal Gradients

Gravity is not the only destabilizing influence in the equatorial ionosphere. If we return to (4.5), we can include the effect of the ambient electric field $\mathbf{E}_0$ as well as the neutral wind. The latter can easily be included by remembering (see Chapter 3) that when electric fields and winds both exist, the current density is $\sigma \cdot \mathbf{E}_0'$ where $\mathbf{E}_0' = \mathbf{E}_0 + \mathbf{U} \times \mathbf{B}$. Since the fundamental destabilizing source is the current, $\sigma_P \mathbf{E}'$ is the correct quantity to investigate. First we note that for a zonally eastward electric field, the zero-order Pedersen current is in the same direction ($\mathbf{g} \times \mathbf{B}$) as the gravity-driven current. The derivation outlined previously, which considered only gravity, can be generalized to include the effect of an electric field by replacing g/ν_{in} with $g/\nu_{in} + E'_{x0}/B$ in the growth rate, where E'_{x0} is the zonal component of the electric field in the neutral frame of reference. A zonally eastward electric field drives a Pedersen current to the east. Any undulation of the boundary will intercept charge just as in the gravitational case and cause the perturbation to grow. Hence, an eastward E_{x0} (eastward E_x and/or downward wind) is destabilizing. A zonally westward field will be stabilizing on the bottomside. The general condition for instability is that the $\mathbf{E}'_{x0} \times \mathbf{B}$ direction be parallel to the plasma density gradient. As discussed in the previous chapter, the zonal electric field component at the equator often increases to a large eastward value just after sunset, driving the F layer to very high altitudes. This uplift contributes in two ways to the destabilization of the plasma. Not only is the electric field in the right direction for instability but also the gravitational term becomes large due to the high altitude of the layer. The growth rates of the gravitational and electric field–driven processes are plotted as a function of height in Fig. 4.11. For a 0.5 mV/m eastward electric field, the two sources of instability are equal at an altitude of 375 km. The gravitational term dominates above this height and increases exponentially with altitude.

Since a large-scale neutral wind is usually horizontal, $(\mathbf{U} \times \mathbf{B}) \times \mathbf{B}$ is also horizontal and thus usually has no component parallel to ∇n if the ionosphere is vertically stratified. Hence, $(\mathbf{E}'_{x0} \times \mathbf{B}) \cdot \nabla n$ above is due entirely to the zonal electric field. However, other terms can be added to the linear growth rate by considering the possibility of a horizontal component of ∇n and/or a vertical wind. In fact, since the layer does change height during the course of any given

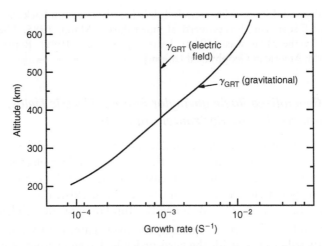

Figure 4.11   Linear growth rates for the gravitational and electric field-driven Rayleigh-Taylor instabilities in the equatorial ionosphere for typical conditions. [After Kelley et al. (1979). Reproduced with permission of the American Geophysical Union.]

night, there is every reason to expect that the layer must be tilted with respect to the vertical. In such a case, the linear growth rate may be written in the following form:

$$\gamma_{RT} = \frac{E_x - wB}{LB} \cos \Delta + \frac{g}{\nu_{in} L} \cos \Delta + \frac{E_z + uB}{LB} \sin \Delta \qquad (4.17)$$

where L is the gradient scale length on the bottomside, Δ is the tilt angle, the z-axis is upward, the x-axis is eastward, and we ignore any dip angle of the magnetic field. The tilt angle is defined such that if it represented an inclined plane a ball would roll down and west for positive Δ and down and east for negative Δ. This is the growth rate of the generalized Rayleigh-Taylor instability as it is applied to the equatorial ionosphere. Kelley et al. (1981), for example, used it to interpret Fig. 4.1 and concluded that the preference for plumes to be generated during negative slopes of the Jicamarca radar profile was due to the contribution of an eastward neutral wind ($u\hat{a}_x$) to the instability growth rate. That is, if the ionosphere is tilted such that the neutral wind blows toward the east into a region of increasing plasma density so that $((u\hat{a}_x \times \mathbf{B}) \times \mathbf{B})$ is antiparallel to the density gradient, a stable configuration occurs. However, if the wind blows antiparallel to the plasma gradient, which is believed to occur during the downward slopes in Fig. 4.1, the wind is destabilizing. Tsunoda (1981, 1983; see also Fig. 4.14), using the scanning radar at Kwajalein, has also shown that wind effects are important. Note, however, that if a perfect polarization field develops due to the F-region dynamo, $E_z + uB = 0$, and there would be no effect of the third term in (4.17). This is a reminder that these instabilities are current driven, not electric field driven. However, due to a finite E-region loading of the current, u usually exceeds E_z/B by 20% or so, and a vertical current does flow. Since the electric field is generated by both local and remote wind fields (see the previous chapter) the generalized growth rate represents quite a complicated set of phenomena (we have not yet mentioned the role of shear in the plasma flow).

Vertical neutral winds do occur when gravity waves propagate upward through the thermosphere-ionosphere system. In addition, there is some evidence that larger-scale, downward winds may occur near the sunset terminator. Raghavarao et al. (1987), for example, have reported just such conditions from rocket-borne chemical release experiments. Of course, it could have been the case that even these winds were due to a long-period gravity wave. For reference, a vertical density gradient pointed upward is unstable to a downward wind velocity (Sekar and Raghavarao, 1987). Hysell et al. (1990) conjectured that the stripes in Fig. 4.8 were due to the vertical velocity phase of a gravity wave.

Linear instability theory may also be used to explain the day-night asymmetry of CEIS, since the instability is influenced by the effects of the E region on the charge buildup that leads to the instability. The equatorial ionosphere is not two-dimensional, since electric currents can easily flow down the magnetic field lines if there is a conducting path through the E region. The physics is identical to that discussed in Chapter 3, which showed that the F-region dynamo was suppressed during the day, when the low-altitude conductivity dominates. During the daytime, any perturbation electric fields due to the instability are shorted out in the E region, which explains why spread F does not begin until well after sunset, even though g and ∇n are antiparallel on the bottomside during the day as well as at night. A full explanation of the morphology of CEIS thus must include the diurnal, seasonal, and solar cycle effects on the electric field, on the neutral density, temperature, and wind patterns, and on conductivities of the E and F regions. Developments in the ability to create realistic models of the equatorial ionosphere are occurring rapidly and many of the effects mentioned above now can be included (de La Beaujardiére et al., 2006).

Longitudinal differences occur due to the offset of the magnetic and geographic equators, the possible influences of orographic features, and the varying declination of the magnetic field. Fejer et al. (1999) suggest that all of these factors influence the prereversal enhancement electric field. This quantity, in turn, controls the height and tilt of the ionosphere and, thus, CEIS. We showed in Chapter 3 that the zonal electric field measured over Peru as a function of local time cannot be extrapolated to represent the zonal electric field as a function of longitude for a fixed local time. Thus, to ever predict CEIS, we need data-assimilative models that include information on the worldwide zonal electric field, the thermospheric wind patterns, and the conductivity of the E region.

The zonal electric field and neutral winds also influence the equatorial ionization anomaly (discussed in Chapter 5), which develops well before the sunset and thus possibly can be used to predict the onset of CEIS. Based on ionospheric sounding studies over the Indian zone, an empirical correlation between the occurrence of CEIS and the strength (ratio) between the electron densities at the crest and trough region of the equatorial ionization anomaly was found by Raghavarao et al. (1988). This same concept was followed by Sridharan et al. (1994) using a daytime optical technique to get a real time prediction. Along similar lines, Alex et al. (1989) reported a correlation between the occurrence of CEIS and the latitudinal gradient in electron densities at a fixed altitude over American longitudes during postsunset hours. Recent multi-instrumented studies (Mendillo et al., 2001) also suggested that a good precursor for premidnight CEIS is the late-afternoon strength of the equatorial ionization anomaly. Further investigations over various regions and seasons are required to consolidate this notion.

4.2.3   The Seeding of Convective Ionospheric Storms by Gravity Waves

In addition to the factors just mentioned, there is very likely an additional random factor in the occurrence of CEIS that may hinder attempts to develop predictive

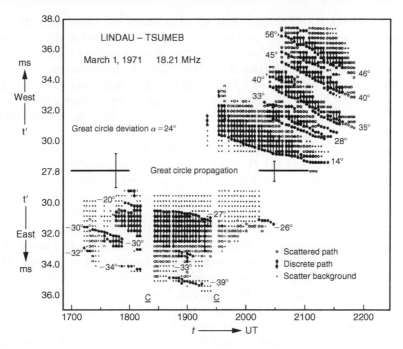

Figure 4.12   Forward scatter measurements in the African sector showing regular wave-like regions of enhanced plasma density. [After Röttger (1973). Reproduced with permission of Pergamon Press.]

capabilities: if an internal gravity wave (see Chapters 5 and 6) organizes equatorial plasma into high- and low-altitude contours with the same horizontal wavelength as the gravity wave. The generalized RT instability can take over and cause the oscillation to grow. In fact, the generalized R-T growth rate is so small that seeding may be essential, either by gravity waves or by the collisional Kelvin-Helmholtz process discussed following. Such seeding can greatly decrease the time needed to develop a large-amplitude disturbance. Strong evidence for the organization of bottomside plasma into structures with scales of several hundred kilometers is given in Fig. 4.12 (see also Figs. 4.1 and 4.5b). The data are from a transequatorial HF radio propagation experiment. If the ionosphere were uniform, refraction would yield a single "great circle" path with some minimum time delay. However, when the plasma density has an east-west structure, additional, larger time delays occur, which can be used to characterize the structure. The data show that paths other than the great circle route occur in regular intervals, which are found to be typical of internal gravity wave wavelengths. One explanation for this effect, proposed by Beer (1973), was that the temperature variation in the gravity wave would change the recombination rate of the ionospheric plasma and thereby create a density modulation. The effect would be particularly strong if the plasma drift speed matched the phase velocity of the wave, since in such a case the perturbation would always act in the same sense on a given parcel of plasma. This is a variation on the so-called spatial resonance theory suggested first by Whitehead (1971). Wave-like modulations of ordinary tropospheric clouds are created by this effect as the local temperature is raised above or lowered below the dew point by a gravity wave. Beer's idea has not passed the test of time, but other gravity wave effects have been proposed, as discussed next.

Klostermeyer (1978) also appealed to the spatial resonance effect but pointed out that the internal wind field, δU in a gravity wave must also drive electrical currents, δJ. Due to the finite wavelength of the gravity waves, the associated winds are not uniform in space. The divergence of the wind-driven electrical current is therefore not zero, and an electric field, δE, must build up with a wavelength equal to that of the gravity wave. This process is illustrated in Fig. 4.13. If the E region is a perfect insulator, no field-aligned currents can flow and the $\nabla \cdot \mathbf{J} = 0$ equation may be replaced by the more restrictive equation $\delta J = 0$, where δJ has components due to the gravity wave wind and the electric field. This requirement has been used to generate the diagram. The resulting electric field pattern has alternating eastward and westward components, which, due to the $\delta E \times B$ drift, will cause portions of the ionosphere to rise and portions to fall. Now, if the plasma has a zero-order vertical density gradient, the $\mathbf{V} \cdot \nabla n$ term in the continuity equation will lead to a sinusoidal density pattern with the same horizontal wavelength as the gravity wave. These east-west oscillations may then act as an initial perturbation, which is amplified by the Rayleigh-Taylor process (see Fig. 4.1). An additional attractive feature of the gravity wave theory is that after sunset the ionosphere begins to descend. This is also the direction of the phase velocity of a gravity wave carrying energy upward from below (an upward gravity wave energy flux has a downward phase velocity, as discussed in Chapter 6).

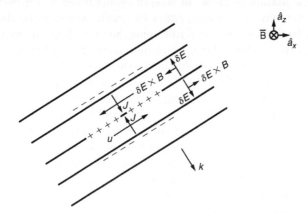

Figure 4.13a   Schematic diagram showing how the perturbation winds in a gravity wave generate electric fields.

Spatial resonance could thus occur with a zonally eastward-propagating wave, since it is well known that the plasma also drifts in the eastward direction (see Chapter 3). Furthermore, the magnitudes of the typical eastward and downward plasma drifts are roughly comparable to the typical phase velocities of large-scale gravity waves in the upper atmosphere.

Gravity wave–induced electric fields have now been observed (Kudeki et al., 1999; Varney et al., 2009). Figure 4.13b shows the classic downward-phase progression of the vertical velocity over Arecibo. If we assume a nonconducting E region and a gravity wave at F-region heights, then, as previously argued, the divergence of the wind-driven current will set up polarization charges and an electric field such that

$$\delta J = \sigma_P(\delta E + \delta U \times B) = 0$$

and thus,

$$\delta E = -(\delta U \times B)$$

where σ_P is the Pedersen conductivity, δE is the perturbation electric field in the earth-fixed frame, δU is the perturbation wind velocity due to the gravity wave, and B is the magnetic field. The net effect of the dynamo is to cause the plasma and the neutral gas to move together in the gravity wave wind field, since $\delta V = \delta E \times B/B^2 = \delta U$. To determine quantitatively the effect of such a perturbation, we first consider the case of perfect spatial resonance in which the mean unperturbed plasma is at rest in a fixed phase front pattern associated with the gravity wave; that is, the plasma drift matches the wave phase velocity

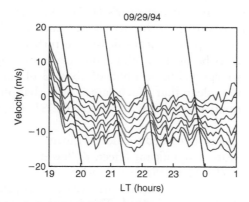

Figure 4.13b Stack plot of the vertical drift velocities at individual heights. The lowest height is 225 km and each subsequent line is 15 km higher and has been shifted by 2 m/s upward. The red portions of the plot are points inferred by interpolation. The slanted black lines are identical and have been aligned with peaks in the data to illustrate the downward-phase velocity. See Color Plate 5.

exactly. Since the plasma behaves as an incompressible fluid, the density variation with time at a fixed point in space is given by the advective derivative

$$\partial n / \partial t = -\delta \mathbf{V} \cdot \nabla n \tag{4.18}$$

where n is the plasma density and $\delta \mathbf{V}$ the perturbation plasma velocity. Here we again use a right-hand coordinate system with x positive in the eastward direction and y positive toward the north. For perfect spatial resonance the perturbation neutral velocity in the plasma reference frame is time independent (the wave frequency is Doppler shifted to zero) and (4.18) becomes

$$\partial n / \partial t = -w_0 e^{-ik_x x + ik_z z} (\partial n / \partial z) \tag{4.19}$$

For an initial density profile of the form $n_0 e^{z/L}$ the solution to (4.19) is

$$n(x, z, t) = n_0 e^{z/L} e^{\gamma(x,z)t} \tag{4.20a}$$

where

$$\gamma(x, z) = R \left\{ - (w_0/L) \, e^{-ik_x x + ik_z z} \right\} \tag{4.20b}$$

where R means "take the real part." These equations describe a spatial pattern of rising and falling density contours in the plasma reference frame. For $w_0 = 2$ m/s and $L = 20$ km, the peak value of γ is $10^{-4} \mathrm{s}^{-1}$. This "perfect" spatial resonance theory gives a perturbation of 5% in 550 s. Note, however, that the perturbation plasma ion velocity due to the presence of the wave *can never exceed* the amplitude of the wave-induced neutral velocity. This constraint severely limits the altitude modulation of the F layer due to a pure gravity wave-driven process. In Klostermeyer's (1978) nonlinear approach to this problem, an anomalous diffusion due to plasma microinstabilities was modeled, and the saturation amplitude of the density perturbation for perfect spatial resonance was found to be at most about one order of magnitude (e.g., see his Fig. 4.1). This corresponds to an uplift of about 3 plasma scale heights or about 50 km. It is now clear that much larger perturbations occur (e.g., see Fig. 4.1) and a pure gravity wave theory explaining CEIS height modulation is not tenable.

This simple model can be extended to address the question of how "resonant" a wave must be to create a significant effect. Kelley et al. (1981) showed that the horizontal phase velocity of the gravity wave only has to be within only about 100 m/s of the horizontal plasma drift velocity to produce a 5% seeding effect in one-half of a wave period. This is not a very severe constraint and makes it very plausible that gravity waves can be responsible for seeding spread F. Huang and Kelley (1996) performed nonlinear simulations (see Section 4.3.2) with and without spatial resonance and verified that it is not necessary.

Returning again to Fig. 4.1, Kelley et al. (1981) concluded that a gravity wave could have seeded the event but that the large uplifts (~80 m/s) observed must have resulted from an amplification of the induced perturbation by the Rayleigh-Taylor process. Sometimes plumes have velocities over 1000 m/s, which is completely out of the question for a neutral air speed (it is supersonic!). They also argued that the multiple plumes located on the descending phase of the oscillation were due to a neutral wind blowing eastward across a structure tilted such that it had a westward density gradient. Tsunoda (1983) has shown this effect convincingly, using the scanning Altair radar. In the data shown in Fig. 4.14, for example, several plumes grow from a structure tilted in just this fashion, while the east wall is stable.

Prakash and Pandy (1980) have pointed out that the perturbation electric fields that develop in response to the wave-driven electric current will be shorted out if magnetic field lines can link the regions of positive and negative charge. For pure (magnetic) zonal propagation this will not occur, but if a finite meridional component of the gravity wave vector exists, the seeding process will be limited to a small angular propagation zone near east-west. Those gravity waves generated as a result of auroral activity [large-scale traveling ionospheric disturbances (TIDs)] propagate from the poles to the equator, while medium-scale TIDs may propagate in any direction. Thus, gravity waves generated in the auroral zone may be less effective for seeding equatorial spread F than waves generated by tropospheric sources such as thunderstorms, frontal systems, and orographic forcing. Seed processes may thus add another longitudinal factor to CEIS occurrence probability.

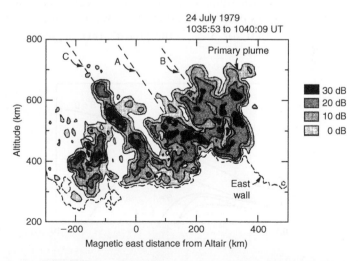

Figure 4.14 Backscatter map showing a primary plume and three other plume-like features on the west wall of a plasma upwelling. [After Tsunoda (1983). Reproduced with permission of the American Geophysical Union.]

4.2.4 Role of Velocity Shear in Convective Ionospheric Storms

The question of gravity wave seeding and the general problem of the very long wavelengths that occur in equatorial spread F must be viewed in a context, which includes the velocity shear that we now know exists in and below the F layer (e.g., see Chapter 3 and Fig. 4.2). Extending earlier work dealing with the $\mathbf{E} \times \mathbf{B}$ instability (Perkins and Doles, 1975), Satyanarayana et al. (1984) showed that velocity shear acts to stabilize the Rayleigh-Taylor instability. This is illustrated in Fig. 4.15, where the normalized growth rate $\hat{\gamma} = \gamma \nu_{in} L_N/g$ is plotted versus the normalized wave number $\hat{k} = 2\pi L_N/\lambda$ for various values of the shear parameter

$$\hat{S} = \frac{dV/dz}{\left(g/L_N\right)^{1/2}} \approx \frac{V_0/L}{\left(g/L\right)^{1/2}}$$

where L_N is the vertical density scale length, V_0 is the maximum in the plasma flow velocity, and the scale length L characterizes both the velocity shear and the density gradient scale length. $\hat{S} = 0$ corresponds to no shear, and the corresponding curve yields the dependence of the Rayleigh-Taylor growth rate on $\hat{k}$. This result differs from the simple analysis given earlier, for which $\hat{\gamma} = 1$ for all values of k (until diffusion sets in). This more accurate linear theory shows that $\hat{\gamma}$ equals unity only for $\hat{k} \geq 10$ or $\lambda \leq 2\pi L/10 \simeq 0.6L$. Since $L = 15$ km, the pure RT growth rate is small for $\lambda \geq 10$ km, where, experimentally, much of the largest perturbation occurs. For $\hat{S} > 0$ the growth rate decreases at all $\hat{k}$, *but a peak in $\hat{\gamma}$ begins to evolve for small values of $\hat{k}$*. This means that velocity shear is stabilizing but that it does push the most unstable waves to long wavelengths. It is interesting to note that for $\hat{S} > 2$ the growth rate begins to rise again,

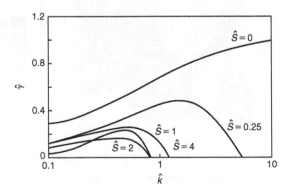

Figure 4.15 Normalized growth rate plotted versus normalized wave number for various shear strengths. [After Satyanarayana et al. (1984). Reproduced with permission of the American Geophysical Union.]

indicating that a Kelvin-Helmholtz instability has set in. The parameter $\hat{S}$ may be written

$$(\hat{S})^{-2} = \left[\frac{(g/n)dn/dz}{(dV/dz)^2} \right] = (R_{ip})$$

where we have defined a plasma Richardson number R_{ip}, which is analogous to the atmospheric Richardson number. In the atmospheric case the numerator is given by the Brunt-Vaisala frequency $\omega_B^2 = (g/\theta)(d\theta/dz)$ where θ is the potential temperature, so $R_i = \omega_B^2/(dV/dz)^2$. It is well known in atmospheric science that a shear is unstable for a Richardson number less than 1/4. This corresponds to $\hat{S} > 2$. However, very intense shears would be required to meet this criterion— for example, if $L = 15$ km and $\hat{S} = 2$, $V_0 = 2(gL)^{1/2} = 760$ m/s, which is much larger than the typical plasma velocity. Another way to look at this is to calculate the required shear if $\hat{S} = 2$. For $L = 15$ km the required shear is 50 m/s per km. This is considerably larger than the observed large-scale vertical shears reported in the literature, some of which are shown in Chapter 3.

Kelley et al. (1986) turned this argument around and asked, "Given the observed shears, what would $\hat{k}$ be for maximum growth?" They found the expected values of $\hat{k}$ to be in the range 0.5–2, which corresponds to λ values ranging from 45 to 180 km. These wavelengths are shorter than the outer scale limit for the horizontal spread F structure reported by Röttger (1973) and indicated by the studies of Kelley et al. (1981, 1986) and Tsunoda (1983). However, these values of λ are in good agreement with multiple plume spacings reported in the same three papers (see also Figs. 4.1, 4.7b, and 4.12). It thus may well be the case (Kelley et al., 1986) that gravity waves determine the outer horizontal scale by preferentially seeding a region of k space and creating perturbations with relatively low growth rates. Velocity shear and neutral winds then combine to structure the westward-directed horizontal density gradients, creating the observed multiple plumes.

Shear effects have been reconsidered in the nonlinear domain (Hysell and Kudeki, 2004), as discussed in Section 4.3.2, and this notion that shear-related seeding of long wavelength structures has resurfaced as a viable mechanism.

4.2.5   Summary of Linear Theory Results

To summarize, it is clear from linear theory that the generalized Rayleigh-Taylor instability plays a crucial role in the production of convective ionospheric storms and that each of the destabilizing terms can be important at times. In addition, the initial conditions that lead to an event on a given night depend in a complicated way on the state of the ionosphere, the neutral atmosphere, and the patterns of gravity waves in the upper atmosphere. Finally, the preferred scale for the generation of multiple plumes (30–80 km) may depend on the velocity shear that occurs in the bottomside of the F layer, while larger plume separations (100–600 km) may require gravity wave seeding. We turn now to nonlinear calculations.

4.3   Nonlinear Theories of Convective Ionospheric Storms

4.3.1   Two-Dimensional Computer Simulations

Considerable effort has gone into the development of computer simulations of the Rayleigh-Taylor instability. The set of equations, which are solved in the two dimensions perpendicular to **B**, is a subset of the full governing equations. They consist of two continuity equations for the electrons and ions,

$$(\partial n_j/\partial t) + \nabla \cdot \left(n_j \mathbf{V}_j\right) = 0 \tag{4.21a}$$

the electron velocity equation with $\kappa_e \gg 1$,

$$\mathbf{V}_e = \frac{\mathbf{E} \times \hat{a}_y}{B} \tag{4.21b}$$

the ion velocity equation for intermediate κ_i,

$$\mathbf{V}_i = (1/B)[(M/e)\mathbf{g} + \mathbf{E}] \times \hat{a}_y + \left(v_{in}M/eB^2\right)[(M/e)\mathbf{g} + \mathbf{E}] \tag{4.21c}$$

the charge continuity equation,

$$\nabla \cdot \mathbf{J} = 0 = \nabla \cdot (n_i e \mathbf{V}_i - n_e e \mathbf{V}_e) \tag{4.21d}$$

which is a form of the quasi-neutrality condition and

$$\mathbf{E} = -\nabla \phi \tag{4.21e}$$

In obtaining these equations we have assumed there is no neutral wind, we have neglected the inertial terms and the pressure-driven terms in the equations for conservation of momentum, and we have assumed that the $\mathbf{g} \times \mathbf{B}$ electron velocity term is small due to the small electron-ion mass ratio.

The electrostatic potential ϕ is divided into a zero-order term, ϕ_0, and a perturbation term, $\delta\phi$. If we require the zero-order ion velocity to be zero, then we must have $\nabla\phi_0 = Mg/e$. This zero-order electric field is the order of a microvolt/meter and is somewhat artificial, but it creates an equilibrium about which to perturb the system. In reality, a larger zero-order electric field generally exists. The electron continuity equation evaluated in a frame of reference moving with the $(\mathbf{E}_0 \times \mathbf{B}/B^2)$ velocity is

$$\partial n/\partial t - (1/B)\left(\nabla\delta\phi \times \hat{a}_y\right) \cdot \nabla n = 0 \tag{4.22}$$

where we have dropped the subscript on the n due to the quasi-neutrality condition $n_e \approx n_i \approx n$. Since $(M/e)\mathbf{g} + \mathbf{E}_0 = 0$, the ion velocity in the earth-fixed frame is

$$\mathbf{V}_i = -\frac{\nu_{in} M \nabla \delta\phi}{eB^2} + \frac{-\nabla\delta\phi \times \mathbf{B}}{B^2}$$

The electron velocity in that frame is

$$\mathbf{V}_e = \frac{\mathbf{E}_0 \times \mathbf{B}}{B^2} + \frac{-\nabla\delta\phi \times \mathbf{B}}{B^2}$$

(Recall that the current is the same in the two reference frames.) Now, using $\mathbf{E}_0 = -M\mathbf{g}/e$, (4.21d) becomes

$$\nabla \cdot \left(n\mathbf{g} \times \mathbf{B} - n\nu_{in}\nabla\delta\phi \right) = 0 \tag{4.23}$$

If we linearize the density terms in (4.22) and (4.23) and appropriately convert to the other coordinate system, we recover the dispersion relation (4.14) with $Q = 0$. Equations (4.22) and (4.23) are the two equations that have been solved numerically for n and $\delta\phi$ subject to a variety of geophysical conditions and initial perturbations (see review by Ossakow, 1981).

In typical simulations, such as the ones illustrated in Figs. 4.16a–c, the initial vertical density profile is specified, as is an initial density perturbation. After a time, the order of 1000 s, the initial modulation grows to a sufficient enough amplitude that the low-density region (bubble) pushes through to the topside of the density profile, where the linear theory predicts a stable system (Ossakow, 1981). One of the crucial problems posed by Farley et al. (1970) is thus resolved. Although the system is linearly unstable only on the bottomside, the nonlinear evolution yields structure well above the F peak (Woodman and La Hoz, 1976).

Computer simulations of plumes have been made to determine the ionospheric parameters to which plume development is most sensitive. Not too surprisingly, the higher the layer and the steeper the vertical density gradient, the faster the bubbles grow. This is also clear from the linear GRT growth rate. In addition, the simulations illustrated in Figs. 4.16a and b show that longer horizontal wavelengths result in larger density depletions and larger uplift velocities. For example, at a fixed height the topside density drops by only 10% for a 3 km horizontal wavelength, while it drops by two orders of magnitude for a 150 km initial perturbation. Comparison of the electric potential distribution around the upwelling plasma shows why this is the case. The charge buildup that leads to the upward flow creates fringing electric fields that penetrate deeply into the bottomside region. Low-density plasma thus convects upward to altitudes above the F peak. In fact, since the flow field is incompressible in the two-dimensional approximation used in the simulation, the density at a given point changes only when a low-density region is advected to that height by the $\delta\mathbf{E} \times \mathbf{B}_0$ drift. Each density contour in the set of figures could thus be specified by its initial height as well as by the density. This idea explains the frequent observations of metallic and molecular ions at F-region altitudes in the equatorial zone. In other words, the chemical composition is also to some extent a label of the original altitude. Returning to Fig. 4.1, the downward penetration of electric fields predicted by the simulations can be seen from the upwelling of features as low as 120 km in conjunction with the large-amplitude topside modulations.

Another simulation is shown in Fig. 4.16c. Here, a constant eastward neutral wind is included, as well as the effect of finite conducting "end plates" in the Northern and Southern Hemispheres. Notice that these simulations give the viewpoint of an observer looking south toward the equatorial plane. This is the same viewpoint that the Jicamarca slit camera would record for a fully developed

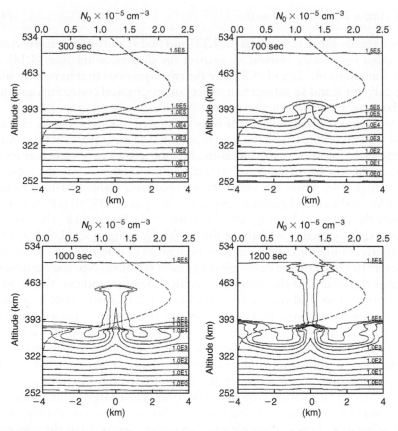

Figure 4.16a   Contour plots showing computer simulations of the Rayleigh-Taylor instability for a 2 km scale perturbation, initially of 5% magnitude. Contours are labeled in units of reciprocal cubic centimeters. [After Zalesak and Ossakow (1980). Reproduced with permission of the American Geophysical Union.]

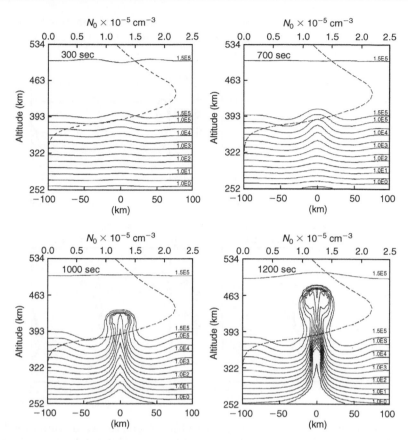

Figure 4.16b Contour plots showing computer simulations of the Rayleigh-Taylor instability for a 100 km scale perturbation, initially of 5% magnitude. Contours are labeled in units of reciprocal cubic centimeters. [After Zalesak and Ossakow (1980). Reproduced with permission of the American Geophysical Union.]

tilted plume drifting eastward across the field of view. The result is a C-shaped structure, as recorded in Figs. 4.1 and 4.2. The scanning radar data plotted in Fig. 4.14 have reversed C-shapes, since the scans are oriented with east to the right and west to the left—the opposite of 4.16c. The data and simulations are thus in excellent agreement. The eastward zonal plasma flow is due to the F-region dynamo, which must be powerful enough to drive the off-equatorial E-region loads. A likely explanation is that the depleted region polarizes and drifts more slowly eastward than the background plasma. For example, consider the region near the top of the plume. The vertical current is given by $\sigma_p(uB + E_z)$, which is upward because $uB > |E_z|$ for an F-region dynamo. Recall that E is negative—that is, the vertical electric field is downward. But inside the plume σ_p is smaller than outside, and thus to maintain the current across the boundary, $|E_z|$ must

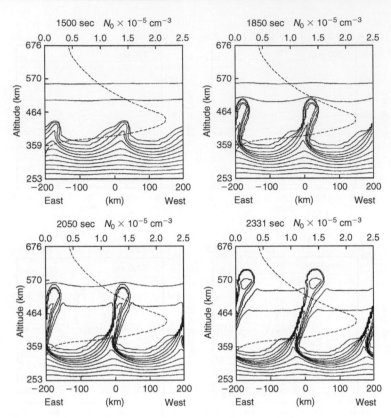

Figure 4.16c Computer simulation including a background eastward neutral wind and finite Pedersen conductivity in the off-equatorial E region. [After Zalesak et al. (1982). Reproduced with permission of the American Geophysical Union.]

decrease. This means that the eastward plasma drift inside the plume is smaller than the background plasma drift, and it lags behind the rest of the ionosphere. The result is a tilt toward the west with altitude.

4.3.2 Simulations Including Seeding and Shear

Huang and Kelley (1996) have simulated the seeding effect of gravity waves on the ionosphere with and without plasma instabilities. To accomplish this, they included a neutral wind of the form

$$U = \left(16\hat{a}_x + 4\hat{a}_z\right) e^{i(\omega t - k_x x + k_z z)} \text{m/s}$$

where **B** remains in the $\hat{a}_y$ direction to the north. They chose $k_x = (2\pi/240)$ (rad/km) and $k_z = (2\pi/60)$ (rad/km), which are reasonable values for a gravity wave perturbation. The values of wind speed and the wave vector are such that

$\nabla \cdot \mathbf{U} = 0$ and also with a downward-phase propagation of the wave (see Chapter 6). The wave frequency is $\omega = (2\pi/1200)$ (rad/s). It is important to note that no initial plasma perturbation was included.

The results are shown in Fig. 4.17, where in Fig. 4.17a, the plasma instability has been turned off by setting $g = 0$, and only a small perturbation occurs. In the upper panel of Fig. 4.17b gravity is included, and two well-developed uplifts with apogee plumes are created within 40 minutes. In the lower panel the waves' eastward phase velocity was exactly matched to the plasma velocity at 100 m/s in the simulation. The structures grow only about 15% faster with the spatial resonance condition met in which these two velocities are equal. As conjectured by Kelley et al. (1981), Fig. 4.17b shows that spatial resonance is not required for gravity wave seeding of CEIS.

These results show that gravity wave winds alone can seed the instability, not just plasma density perturbations. However, the plasma instability clearly amplifies the plasma drift perturbations, yielding much larger $\delta \mathbf{E} \times \mathbf{B}$ drifts than the initial neutral wind. For example, between the times in the upper panels of Fig. 4.17b the average plasma uplift velocity was over 300 m/s compared to $w = 4$ m/s for the initial wind speed. These simulations show that nonlinear processes amplify the initial fluctuations.

Initially it was thought that shear flow would only lead to stabilization of CEIS instabilities (see Section 4.2.4). This was borne out by early nonlinear studies. For example, Sekar and Kelley (1998) studied the effect of a shear in the zonal drift coupled with the reversal of the vertical drift velocity on development of CEIS. Due to the shear, the structures tilted toward the east but still were able to penetrate into the topside. This suggests that the stabilizing effect of the shear is not very effective in suppressing CEIS all by itself. But when they reversed the vertical drift part way into the simulation, there was no penetration of the perturbation above the F peak.

However, these calculations are not self-consistent with respect to the origin of the shear. Hysell and Kudeki (2004) pointed out that strong shear flow in the bottomside equatorial F region implies a region of retrograde plasma motion, where the plasma drifts rapidly westward in strata where the neutral wind is eastward (see Fig. 4.2 and Chapter 3). By extending the analysis of the electrostatic Kelvin-Helmholtz instability (KHI) of Keskinen et al. (1988) into the strongly collisional, inhomogeneous limit, they showed that this configuration is inherently unstable. The growth rate of the instability they found was large enough to compete with that of the ionospheric interchange instability, and the range of preferred wavelengths was a better match to the depletion scale-sizes observed during equatorial spread F. They surmised that this shear-driven instability could precondition the ionosphere for interchange instabilities, explaining the relatively early appearance of fully developed irregularity structures after sunset during spread F events.

Figure 4.18 shows the results of the numerical simulation of the instability. The initial density profile follows a hyperbolic tangent law below the peak and an exponential decay law above it. The initial profile of the electrostatic potential

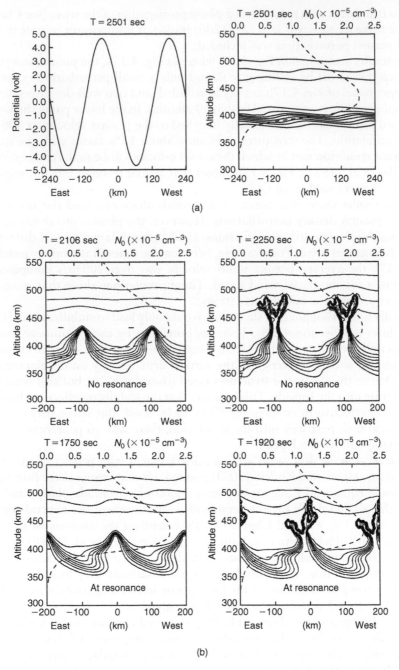

Figure 4.17   (a) Evolution of gravity wave seeking with no amplification by the Rayleigh-Taylor instability. (b) Evolution with gravitation included, both at resonance and without resonance. [After Huang and Kelley (1986). Reproduced with permission of the American Geophysical Union.]

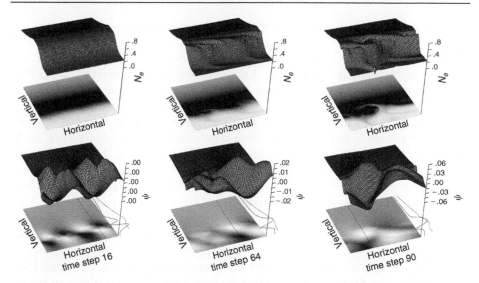

Figure 4.18 Numerical simulation of collisional shear instability in the bottomside F-region ionosphere. The upper and lower row of panels depict number density and perturbed potential for three time steps shown, respectively. Results are shown both in grayscale and relief formats. Time steps represent 100 s of real time and the simulation size is 62.8 km on each side. The solid and dashed lines to the right of the lower panels show the magnitude of the perturbed potential and the zonal plasma drift speed, respectively, taken through the cuts indicated by the dotted lines. Interchange instabilities have been suppressed by setting gravity and the background zonal electric field to zero. [After Hysell and Kudeki (2004). Reproduced with permission of the American Geophysical Union.]

corresponds to the wind-driven dynamo electric field. The perturbed potential plotted in the lower panels is the difference between the current potential and that of the initial time step. By time step 16, evidence of shear instability clearly can be seen in the perturbed potential. Here, we find periodic islands or cells of low and high potential straddling the altitude of the shear node. A dominant horizontal wavelength is also clearly evident by time step 16. Lines of constant potential are streamlines of the flow, and the plasma circulates clockwise and counterclockwise around the cells of low and high potential, respectively.

By time step 64, the instability has grown to the point of being detectable in the plasma density plot. Here, we find elongated regions of depleted and enhanced plasma penetrating above and below the shear node, respectively. A rotational pattern is suggested by the morphology of the enhancements and depletions, and the circulation cells visible at time step 16 have merged into one or two main cells by time step 64.

By time step 90, the unstable flow pattern exhibits a distinct "cat's eye" surrounding the most prominent circulation cell. The primary density irregularity is unstable to secondary, $\mathbf{E} \times \mathbf{B}$ instabilities (note that the electric field is upward in this region), accounting for the intermediate-scale structuring in the cat's eye walls.

The simulation reveals how the collisional shear instability functions. In regions of strong retrograde plasma drift, a vertical Pedersen current flows. The equilibrium configuration is the one in which the current has no divergence. However, if the height of the layer is displaced vertically, appropriate polarization electric fields arise to maintain quasi-neutrality. High potential regions above low potential regions result from downward layer displacements, with the reverse holding for upward displacements. This process is clearly evident in the simulation by time step 16. For instability to occur, the convection driven by the polarization electric fields must deform the plasma such that the initial upward and downward perturbations in layer height are amplified. The growth rate depends strongly on the actual shapes of the number density, collision, and velocity profiles in a complicated way that evolves over time with the profiles.

In KHI simulations, the waveforms that emerged first had a wavelength of about 30 km. Over time, these transient waveforms tended to coalesce to larger scales. A nonlocal analysis predicted the fastest steady-state growth rate for waves with $kL \sim 1/2$ where L is the length scale of the shear. Given values of L of about 20 km in the postsunset bottomside, this implies dominant wavelengths as high as 250 km. The airglow data in Fig. 4.7b show clearly how various scales evolve. The arrow shows the bottomside of the equatorial plasma. At the top left, the two large plumes are separated by 500 km whereas in between, finger-like structures are separated by 100 km. By the top right-hand frame, even the finger-like structures had grown in to major plumes. It is arguable that the largest two plumes at the top left were given a head start by a gravity wave–induced fluctuation whereas the others were seeded by KHI. Alternatively, KHI might explain all the seeding if the two major plumes in the top left image were launched at the peak altitude of the prereversal enhancement (Kudeki et al., 2007) and then drifted overhead.

4.3.3 Summary of Nonlinear Theory Results

The most important result is that plasma depletions push into the linearly stable topside. Other results are as follows:

1. Once an initial seed occurs, subsequent growth depends very little on the scale size of the seed.
2. Plumes can be seeded by plasma density perturbations or by spatially varying electric fields generated by gravity wave wind fields or the collisional KHI instability.
3. Tilting of plumes with altitude is most likely caused by polarization of plumes due to an eastward wind in the plasma reference frame.
4. Plumes with large horizontal scale can create electric fields that penetrate deep into the E layer.

4.4   Linkage of Large and Small Scales in CEIS

4.4.1   Evidence for a Diffusive Subrange

We turn now to the studies of the power spectra of convective ionospheric storms and examine the information they contain concerning the physics of these phenomena. First, the relationship between density perturbation and electric field structure may be derived from (4.13a), which may be written

$$-ikQ\delta n = -ik\delta\phi(Pk/i) \tag{4.24a}$$

or

$$\delta E_x = \left(\frac{Q}{P}\right)\frac{\delta n}{n}$$

which may also be written

$$\delta E_x = -\left(\frac{gB}{\nu_{in}}\right)\frac{\delta n}{n} \tag{4.24b}$$

where δE_x represents the electric field amplitude. In (4.24b) the electric field is proportional to the density perturbation and thus the shape of the power spectrum of the two quantities, δE_x and $\delta n/n_0$, should be the same—that is,

$$\frac{|\delta E|^2}{(\delta n/n)^2} = \left(\frac{gB}{\nu_{in}}\right)^2 \tag{4.24c}$$

Since the relationship between δE_x and $\delta n/n_0$ is different for different plasma phenomena, the ratio may be used as a test for sorting out what processes dominate the physics as a function of altitude and wave number. The ambipolar electric field due to diffusion (see Chapter 10) is given by

$$\delta E = (k_B T/e)(\nabla n/n) = -ik(k_B T/e)(\delta n/n) \tag{4.25}$$

rather than by (4.24b). This expression predicts that the electric field and density fluctuation power spectrum differ by a factor of k^2 due to the presence of the gradient operator.

The ratio of simultaneously measured power spectra is presented in Fig. 4.19 for a rocket experiment. The ratio has been plotted to have the units of electric field squared. For the three likely driving sources (gravity, electric field, wind), the corresponding electric field would be gB/ν_{in}, E, or uB. The flat portion of the plot corresponds to the result in (4.24c); that is, $(\delta E)^2$ and $(\delta n/n)^2$ have the

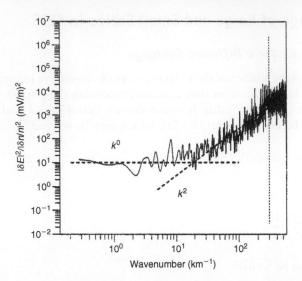

Figure 4.19   Ratio of the spectral density of electric field (total) and density fluctuations as a function of wave number. The ratio is linear at wavelengths longer than 300 meters but adopts k^2 scaling at shorter wavelengths. At very short wavelengths, the electric field measurements are dominated by telemetry noise. [After Hysell et al. (1994). Reproduced with permission of Pergamon Press.]

same spectral form. The electric field indicated by the dashed line is $\sqrt{10}\,\text{mV/m}$, which, in turn, indicates a combination of a gravity and a wind-driven process, since the zonal zero-order electric field is not this large. The physics changes dramatically near the 100 m scale where the ratio begins increasing as k^2. We investigate this transition in the next sections. First, however, we briefly discuss the characteristics of the intermediate wavelength density (or electric field) power spectra. Computer simulations have been used to predict the power spectrum of intermediate wavelength density structures for comparison with experiments (see Fig. 4.5). Keskinen et al. (1980a, b) found isotropic fluctuations at intermediate scales with wave number spectra, which displayed a power law of the form k^{-n} with $n \approx 2.5$. This result is in good agreement with bottomside vertical power spectral measurements on rockets. But horizontal satellite spectra often have spectral indices closer to $n = 5/3$.

An explanation for the anisotropic spectra has been proposed by Zargham and Seyler (1987). Their simulation shows a very strong anisotropy in the development of the irregularities with shock-like structures in the direction of ∇n. A one-dimensional cut was made through the simulation for near-vertical trajectories, and, as shown in Fig. 4.20, the predicted waveforms and their spectra are in excellent agreement with the rocket data in Fig. 4.5a (Kelley et al., 1987). The anisotropy displayed by the Zargham and Seyler simulation is similar to the

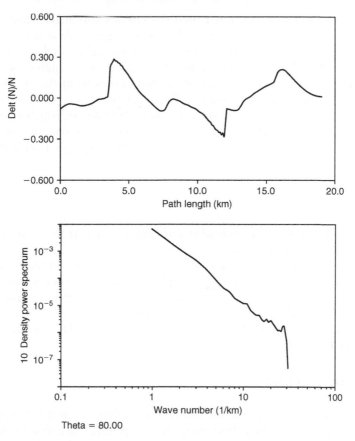

Figure 4.20 One-dimensional cuts through a spread F simulation corresponding to a rocket trajectory. The spectral units are reciprocal kilometers. [After Kelley et al. (1987). Reproduced with permission of the American Geophysical Union.]

early barium cloud simulation work by McDonald et al. (1980), who solved a similar set of equations with a different numerical scheme.

4.4.2 The Diffusive Subrange

The discussion thus far has concentrated on the seeding and generation of long- and intermediate-wavelength flute mode irregularities in the F region. On the other hand, the observational database is determined primarily by scatter from short-wavelength waves—for example, the 3 m waves detected at Jicamarca. Figure 4.5b shows that 3 m is well into the steep portion of the spectrum. By analogy to neutral turbulence as discussed in Section 3.6, it is tempting to consider the steep portion to be in a dissipative wave number range. The steep range in fluid turbulence is called the viscous subrange, where the term $(\eta/\rho)\nabla^2 U$ in the momentum equation damps out small scale structures. The ∇^2 operator becomes k^2 in a Fourier analysis, so large values of k are most effective in dissipation. In our linear long-wavelength stability analysis we ignored a crucial term in the continuity equation that comes from the $(\mathbf{V} \cdot \nabla n)$ term. If we include the diffusion velocity, $D(\nabla n/n)$, we get a term of the form $(D/n)\nabla^2 n$ which in a Fourier analysis becomes $k^2 D(\delta n/n)$, which is analogous to viscous dissipation, and we refer to a diffusive subrange. When included in the linear analysis the total growth rate for the Rayleigh-Taylor instability becomes,

$$\gamma_{RT} = g/\nu_{in}L - k^2 D_\perp \tag{4.26}$$

But what is the proper perpendicular diffusion coefficient to use in a plasma? We take up this problem next.

As noted previously, the lifetime of a horizontal structure in the high-latitude ionosphere depends on how quickly the plasma can diffuse across the magnetic field. Also, whether a particular plasma instability is stable or unstable at a given k value depends on whether some positive contribution to the growth rate exceeds diffusive damping due to terms of the form $- k^2 D_\perp$. In Chapter 5 we study parallel diffusion and show that

$$D_\parallel = (1 + T_e/T_i)D_i \tag{4.27}$$

The factor of T_e/T_i comes from the fact that an ambipolar electric field builds up when electrons attempt to diffuse away quickly parallel to $\mathbf{B}$ due to their small mass. In effect, each electron has to drag a heavy ion with it along $\mathbf{B}$, and the *plasma* diffusion coefficient is then determined by the ions.

In Chapter 10, perpendicular diffusion in the F region is discussed in some detail. In brief, a charged particle on average random walks one gyroradius, r_g, between collisions, so the diffusion coefficient is of the order

$$D_{j\perp} = r_g^2 \nu_j \tag{4.28}$$

This quantity is larger for ions than for electrons and results in an ambipolar perpendicular electric field. This field retards ions and accelerates electrons so that the plasma structure diffuses as a unit and at a rate given by the electron diffusion coefficient. However, if the ambipolar electric field is shorted out by a conducting E region, the structure will decay at a rate given by the ion diffusion coefficient. Thus, the actual $D_{c\perp}$ must fall between the electron and ion rates:

$$D_{e\perp} \leq D_{c\perp} \leq D_{i\perp} \tag{4.29}$$

That the medium is linearly stable at m scales in the direct RT process now can be seen by comparing the growth $g/\nu_{in}L$ to the classical perpendicular diffusive damping rate, as in (4.26). Experiments show that the 3 m structures are highly elongated along the magnetic field (Farley and Hysell, 1996; Hysell and Farley, 1996). Parallel diffusion is so fast that the only structures that can be maintained have $k_\perp \gg k_{||}$ so perpendicular and parallel diffusion is comparable. As indicated in (4.29), $D_{c\perp}$ may range from a minimum of $D_{e\perp}$ to as large as $D_{i\perp}$ depending on the conductivity of the E region. The wavelength at which the RT process is marginally stable, that is, where linear growth equals diffusive damping, is then

$$\lambda_c = 2\pi\left(D_{c\perp}\nu_{in}L/g\right)^{1/2} \tag{4.30}$$

where λ_c is the critical wavelength at marginal stability. This parameter exceeds 10 m for reasonable F-region parameters, even using the smallest value of $D_{c\perp}$ (i.e., $D_{e\perp}$). The RT process is thus linearly stable in the range where most of the radar observations have taken place—that is, for backscatter from waves with $\lambda \leq 3$ m. If no other wave generation process exists, the 3 m waves must receive energy from the longer scales via a cascade process of some sort.

Another possible candidate for meter-scale structure is the collisional drift wave. These waves are driven only by gradients, so they could be a secondary instability due to the primary Rayleigh-Taylor irregularities. However, linear instability studies based on the rocket observations show that such waves are linearly stable (Huba and Ossakow, 1981a, b). This fact, coupled with the featureless k^{-5} spectrum for $\lambda \leq 100$ m, seems to rule out unstable drift waves as a source of 3 m waves. As we shall see next, it appears that classical diffusion may be sufficient to explain the spectrum, provided energy can be coupled from growing large-scale waves to small-scale modes damped by diffusion.

4.4.3 Toward a Unified Theory for the Convective Equatorial Ionospheric Storm Spectrum

Hysell and Kelley (1997) observed the fluctuation spectrum for several sets of plasma bubbles seen on consecutive orbits of the AE-E satellite, one occurring in the 2200–2300 LT period and the next after midnight LT. This is in the period of decay, and the authors estimated the decay of each range of k space using the model

$$\left\langle |\delta n(k, t + \tau)/n|^2 \right\rangle = \left\langle |\delta n(k, t)/n|^2 \right\rangle k^2 D e^{2\gamma(k)\tau} \tag{4.31}$$

Remarkably, $\gamma(k)$ was the order of $-(2 \times 10^{-4})\text{s}^{-1}$ and virtually independent of k. For reference the classical diffusion time constant for $D_{c\perp} = 0.4 \text{ m}^2/\text{s}$ equals this value for $\lambda = 700$ m. So for wavelengths greater than about 500 m, the structures decay faster than the classical rate, and for $\lambda \leq 500$ m, the structures decay more slowly. It seems clear that energy is being transferred out of larger scales and into small scales analogously to neutral turbulence. Hysell and Kelley (1997) discussed this process in terms of a three-wave interaction, one of which is the damped small scale structure. They also created a structure balance model of the form

$$\iint d\mathbf{k}\gamma(\mathbf{k})\left\langle\left|\delta n(\mathbf{k})\right|^2\right\rangle = \iint d\mathbf{k}D_{eff}k^2\left\langle\left|\delta n(\mathbf{k})\right|\right\rangle^2$$

to see if classical diffusion represented by a constant value of D_{eff} can explain the observations. To carry out these integrals they created an anisotropic irregularity model

$$\frac{\left\langle\left|\delta n\right|^2\right\rangle/k_0^2}{1 + \left(k^2/k_0^2\right)\left(1 + k_x^2/k_1^2\right)} \qquad k_0 < k < k_2 \qquad (4.32)$$

where $k^2 = k_x^2 + k_z^2$ with $\mathbf{a}_x$ in the zonal direction and $\mathbf{a}_z$ vertical. This model is meant to represent the spectra from the outer scale, $\lambda_o = (2\pi/k_0) = 60$ km, to some inner scale $\lambda_2 = (2\pi/k_2)$. Projected in the $\mathbf{a}_x$ direction the spectrum has two power law regimes with k^{-1} behavior changing to k^{-3} form at the wave number k_1. Projected in the vertical, this model has a spectral slope that ranges from -1 to -2. If we take the observed value of $\gamma(\mathbf{k})$ to be constant at $-(2 \times 10^{-4})\mathrm{s}^{-1}$ and use observed values for k_1 and k_2, the integrals can be carried out and solved for D_{eff}:

$$D_{eff} = -\gamma\frac{\ln\left(2k_1/k_0\right)}{k_1k_2} \qquad (4.33)$$

Livingston et al. (1981) have shown that $\lambda_1 \approx 1000$ m. Using $k_2 = 2\pi/80$ from Fig. 4.2b yields

$$D_{eff} = 2 \, \mathrm{m}^2/\mathrm{s} \qquad (4.34)$$

which is higher than the classical electron diffusion coefficient for $T_e = 1000$ K and $v_{ei} = 500 \, \mathrm{s}^{-1}$ but less than the typical ion diffusion coefficient.

4.5 Convective Equatorial Ionospheric Storm Summary

Great strides have occurred concerning convective ionospheric storms. Reasonable hypotheses exist for the entire range of observed structures spanning at least six orders of magnitude in spatial scale (0.1–105 m). In order of descending scale, these are the possible contributing processes:

1. Gravity wave seeding and electrodynamic uplift ($\lambda > 200$ km)
2. Shear effects (200 km $> \lambda > 20$ km)

3. The generalized Rayleigh-Taylor instability (0.1 km $< \lambda <$ 20 km)
4. Diffusive damping via wave-wave coupling to damped waves (1 m $< \lambda <$ 100 m)

As an event progresses, the free energy released by the RT process is dissipated via wave-wave coupling, leaving features with $\lambda >$ 1 km intact in the postmidnight period. Classical diffusion and plasma production via sunlight eventually smooth out the ionosphere. Questions such as "Why do storms occur some nights and not on others?" and "Why are certain seasons preferred at certain locations?" are still open. The E-region conductivity, the plasma uplift by electric fields, and the neutral atmospheric density and dynamics all influence the probability of this phenomenon's occurrence and must enter any predictive theory. One of the goals of the National Space Weather Program is just such a predictive capability (Kelley et al., 2006; de La Beaujardiere et al., 2006).

Progress on these topics requires attacking the following issues:

1. Due to the variable declination of the magnetic field with longitude, the terminator aligns with the magnetic meridian at some locations depending on the season. In this case the two E regions conjugate to the equatorial plasma become dark at the same time, reducing the shorting effect (Tsunoda, 1985).
2. The South Atlantic Anomaly in the earth's magnetic field causes energetic particle precipitation in that sector, which affects the E-region conductivity and electric field in that sector. This may explain behavior that is not consistent with item 1 above (Burke et al., 2004).
3. The four-wave number behavior of the diurnal tide (Hagan and Forbes, 2002) modulates the daytime SQ current system as a function of longitude. In turn, this affects the development of the equatorial arcs and possibly the RT growth rate (Immel et al., 2006; Kil et al., 2007).
4. Knowledge of the global electric field and neutral wind is crucial to predicting CEIS and must come from satellite measurements.

4.6   E-Region Plasma Instabilities: The Observational Data Base

As discussed in Chapter 3, the equatorial electrojet is part of the worldwide system of electric fields and currents driven by the dynamo action of the neutral wind. The dynamo currents primarily flow in the E region where the conductivity is greatest, and the current is essentially horizontal except perhaps at high altitudes and high latitudes. The basic reason for the existence of the equatorial electrojet is the large value of the Cowling conductivity close to the dip equator. In the simplest electrojet model the east-west dynamo electric field E_x sets up a vertical polarization field E_z, which completely inhibits the vertical Hall current everywhere. The polarization electric field points upward during the day and has about the same magnitude but points downward at night. Thus, the drift velocity of E-region electrons at night is of the same order of magnitude as the daytime drifts, but the electrical current is much smaller due to the low electron density.

For the case of zero vertical current, the vertical polarization electric field at the magnetic equator is given by

$$E_z = (1 + \sigma_H/\sigma_P)E_x \tag{4.35}$$

which can also be written as

$$E_z \simeq (\nu_i/\Omega_i)[E_x/(1 + \nu_e\nu_i/\Omega_e\Omega_i)] = \left[\kappa_i^{-1}/(1 + \Psi_0)\right]E_x \tag{4.36}$$

where Ω_e, Ω_i, ν_e, and ν_i are the usual electron and ion collision frequencies and gyrofrequencies and $\Psi_0 = \nu_e\nu_i/\Omega_e\Omega_i$ (see Appendix B for plots of Ψ_0). In the equatorial region the east-west electric field is about 0.5 mV/m, and the maximum vertical polarization field (at about 105 km) is estimated from (4.35) to be of the order of 10–15 mV/m (see also Fig. 3.17b). The electrons are magnetized and therefore the $\mathbf{E} \times \mathbf{B}$ drift under the influence of this vertical field yields an east-west electron drift velocity on the order of 400–600 m/s. The drift direction is westward during the day and eastward at night. This flow is sometimes supersonic; that is, the electrons move faster than the acoustic velocity in the medium.

In this section we extend material in the review by Fejer and Kelley (1980) and ask the reader to refer to that publication for numerous references to the early experimental work in this area. The occurrence of an anomalous scattering region in the ionospheric E region close to the dip equator was observed initially from ionosonde records. These echoes were called "equatorial sporadic E" (E_{sq}) echoes because of their apparent similarity to the sporadic E phenomenon (called E_s) occurring at other latitudes. However, the characteristics and generating mechanisms of these two phenomena are now known to be quite different. The intensity of the E_{sq} is well correlated with the electrojet strength, and VHF forward-scattering experiments showed that these echoes are field aligned and are caused by scattering from plasma density irregularities immersed in the electrojet. The most important results concerning the physics of the electrojet scattering region have been obtained from VHF radar measurements performed at the Jicamarca Radar Observatory near Lima, Peru, since 1962. Rocket observations at Thumba, India; Punta Lobos, Peru; Kwajalein Island; and Brazil have also provided valuable information on the electrojet irregularities from electric field and density profiles (see Fig. 3.17b). Multifrequency HF and VHF radar observations have been performed in Central and East Africa, India, and Brazil, and VHF/UHF in Peru.

Equatorial E-region irregularities are present during both day and night. The irregularities are field aligned; that is, the wave number component along the magnetic field ($k_\parallel$) is much less than $k_\perp$. Radar echoes are observed only when the radar wave vector is nearly perpendicular to the earth's magnetic field. The width of the north-south angular spectrum is less than 1° for 3 m irregularities. This property clearly indicates that a plasma process is occurring, since neutral atmospheric turbulence is isotropic at the 3 m scale. Both electron density and electric field but not magnetic field fluctuations have been detected, indicating an electrostatic wave process. Radar spectral studies have shown the existence of two classes of irregularities, called type 1 and type 2, associated with the electrojet. The characteristics of the type 1 or two-stream (the reason for this name will be clear later) irregularities were determined in the early measurements in Peru. The type 2 irregularities have been studied in detail only with the advent of improved sensitivity at Jicamarca. The line of echoes centered near 100 km in Figs. 4.1 and 4.2 is due to the electrojet waves.

The type 1 irregularities have a narrow spectrum with a Doppler shift (120 ± 20 Hz for a 50 MHz radar) that corresponds approximately to the ion acoustic velocity (about 360 m/s) in the electrojet region. These echoes appear nearly simultaneously over a large range of zenith angles when the electron drift velocity is larger than a certain value. Figure 4.21 shows a series of spectra taken during

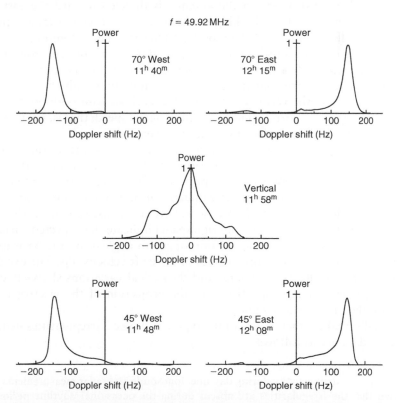

Figure 4.21 Series of Doppler spectra from the equatorial electrojet irregularities at different elevation angles obtained at Jicamarca during a period of relatively strong scattering. The spectra are normalized to the peak value. [After Cohen and Bowles (1967). Reproduced with permission of the American Geophysical Union.]

a period of strong scattering. The mean Doppler shift of the peak in the spectrum was constant when the zenith angle was between 45° and 70°, both to the east and to the west (Bowles et al., 1960). This result remained mysterious until combined rocket and radar data were applied (Kelley et al., 2008). As discussed in Section 4.8.2, they showed that in these scattering volumes there is some region where the total vector drift velocity (zero order plus perturbation drift) in the direction of the radar exceeds the two-stream threshold. Thus, the hypothesis that $V_{ph} = C_s \cos \theta$, where θ is the angle between the *total* drift velocity and the radar beam, explains the observation $V_{ph} \cong C_s$ at all angles.

The data in Fig. 4.21 were normalized to a peak value, since the echo power varies over a large range (up to about 40 dB) throughout the day. Thus, the area below the spectral curves is not proportional to the relative signal strength. In particular, the scattered power from the oblique type 1 echoes is appreciably larger than the power from the overhead echoes coming from vertically traveling "non-two-stream" irregularities. During the daytime the Doppler shift of the dominant peak is positive when the antenna is directed toward the east and negative when the antenna is looking to the west. The reverse is true at night. Therefore the phase velocity of the type 1 irregularities has a component in the direction of the electron flow, since during the day (night) the electron flow is westward (eastward). The fact that the Doppler shift is independent of zenith angle shows that the irregularities are not just advecting with the zero-order electron flow. Often, the vertical antenna detects echoes with both positive and negative spectral peaks corresponding to the acoustic speed; that is, propagating perpendicular to the flow both upward and downward.

The average phase velocity of the type 2 irregularities is smaller than the ion acoustic velocity and is approximately proportional to the cosine of the radar elevation angle (Balsley, 1969a). The spectral width is much broader than that of the type 1 echoes and is often greater than the mean Doppler shift. Figure 4.22 shows the variation of the type 2 spectra and average phase velocity with zenith angle. The solid curves in the bottom right panel indicate the expected variation of the average velocity V_{obs} as a function of zenith angle θ_z on the basis of the relation $V_{obs} = V_D \sin \theta_z$ for three values of the drift velocity V_D. The excellent agreement between the experimental and theoretical variations shows that the type 2 phase velocity is proportional to the projection of the electrojet drift velocity on the radar line-of-sight.

Some additional characteristics of the type 1 and type 2 irregularities deduced from radar data are as follows:

1. *Threshold*: Type 2 irregularities are observed even for very small values of the westward electron drift velocity during daytime. Ionosonde and radar measurements have shown that the irregularities are absent during the occasional daytime periods of westward current flow (counterelectrojet conditions). At nighttime the type 2 irregularities are almost always observed. An exception occurs when the electric field reverses sign. An example of this effect can be seen in Fig. 4.1. When the high-altitude scattering region reached "apogee" at around 2120 LT, the E-region irregularities

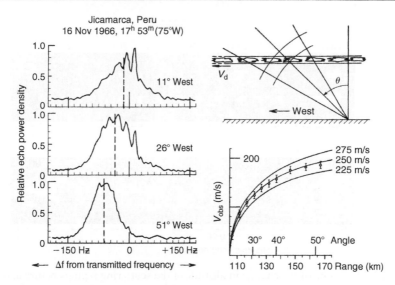

Figure 4.22 Type 2 spectra measured at 50 MHz simultaneously at different antenna zenith angles. The dashed lines indicate the average Doppler shifts. The geometry of the experiment is shown in the top right-hand panel. The results of the experiment, together with three theoretical curves for which a sine dependence of the average phase velocity with zenith angle was assumed, are shown in the bottom right panel. [After Balsley (1969a). Reproduced with permission of the American Geophysical Union.]

ceased, only to reappear when the echoing region above started to fall again. It seems clear that the zonal electric field that causes the F layer to move vertically switched from east to west at this time, going through zero in the process. A dramatic example of the effect of electric field changes is shown in Fig. 4.23. Before 2044, the radar echoes came from two height ranges. For about 10 minutes they disappeared completely, only to return again with one echoing layer exactly midway between the original two echo heights. Thus, a very small threshold electric field seems to be required for type 2 echoes. There is a definite electric field threshold for the excitation of type 1 irregularities, however. These echoes are observed only when the electron drift velocity is somewhat larger than the ion acoustic velocity (about 360 m/s).

2. *Scattering cross section*: The scattering cross section of the type 2 irregularities is approximately proportional to the square of the drift velocity but is independent of zenith angle, while the type 1 scattering cross section increases rapidly with zenith angle, peaking near the horizon. This is where the radar beam is almost parallel to the electron flow in the electrojet.

3. *Altitude dependence of the electrojet echoes*: Fejer et al. (1975) studied the vertical structure of the electrojet scattering region in detail and reported a considerable difference in the echoing region between day and night. Figure 4.24 shows some selected profiles of the electrojet scattered power on February 18–19, 1971, which were observed at Jicamarca using an altitude resolution of 3 km. During daytime, when the electron drift velocity is westward, echoes are generated between 93 and

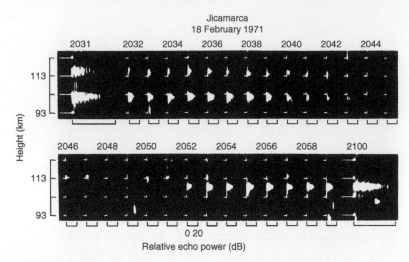

Figure 4.23   Example of the change in altitude of the scattering regions before and after the electrojet reversal, which occurred at about 2048–2050. [After Fejer et al. (1975). Reproduced with permission of the American Geophysical Union.]

   113 km. The daytime power profile as a function of height shows a single peak that remains at a constant altitude (about 103 km). The nighttime scattering is much more structured. Plume-like rising structures in the altitude region of 115 to 125 km with upward Doppler velocities were also observed during nighttime from off-equatorial regions (Patra and Rao, 1999).

4. *Wavelength dependence*: The vast majority of data concerning the electrojet is, of course, at 3 m wavelength. Interferometric techniques have been introduced at Jicamarca, which show that the scattering cells are often organized into kilometer-scale wave structures that propagate horizontally at about one-half the acoustic speed. An example is shown in Fig. 4.25. The most intense backscatter (darkest regions) is related to vertical 3 m waves traveling at the acoustic speed. It seems clear that the large-scale wave is organizing and creating the conditions that produce the vertically propagating acoustic waves. Notice that one wave period requires about 60 s to pass over Jicamarca in this example.

   Occasionally during magnetic activity or special tidal conditions, the daytime electrojet current reverses sign, a counter electrojet. Radar observations in such conditions find that two-stream waves obey the relationship $V_{ph} = C_s \cos \theta$ (Hanuise and Crochet, 1981; Woodman and Chau, 2002). During such conditions the relative directions of the vertical gradient and the electric field are stable. This means that there is no situation for which both two-stream and gradient drift are unstable and large-scale waves do not occur. These results are in agreement with auroral zone data (Bahcivan et al., 2005; see Chapter 10) and a new interpretation of type 1 radar scattering properties as a function of angle (see Section 4.8.2).

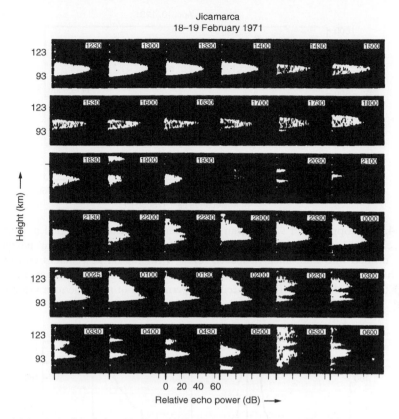

Figure 4.24 Sample of the 50 MHz backscattering power profiles from the electrojet irregularities measured with the large vertically directed incoherent scatter antenna at Jicamarca. Spread F echoes contaminated the data between 0405 and 0550 and perhaps at 1900. [After Fejer et al. (1975). Reproduced with permission of the American Geophysical Union.]

We turn now to complementary data sets obtained from sounding rockets flown through the daytime and nighttime electrojets. Most of the early rocket experiments (prior to 1983) were flown with relatively high apogees and thus passed through the electrojet very quickly. Such high velocities preclude detection of the large-scale waves discussed previously. Nonetheless, some important results were obtained, particularly due to the pioneering efforts of the Indian rocket group (e.g., Prakash et al., 1972). Examples of simultaneous density profiles and density fluctuation measurements taken during daytime and nighttime conditions are shown in Fig. 4.26. During the day, the fluctuations in electron density were located where the gradient is upward, were peaked at 103 km, and were relatively unstructured. The nighttime zero-order density profile is very structured on the negative gradient at about 120 km, with

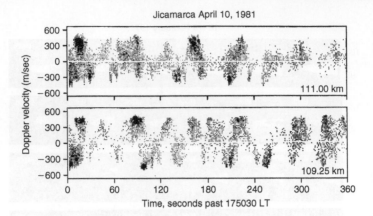

Figure 4.25 Doppler shift spectrogram of the vertical backscatter signal measured at Jicamarca. Each spectrogram is normalized to its own peak power. The power values are divided into nine linearly spaced levels, with the darkest shades corresponding to the largest power values. Negative Doppler velocities indicate downgoing waves. [After Kudeki et al. (1982). Reproduced with permission of the American Geophysical Union.]

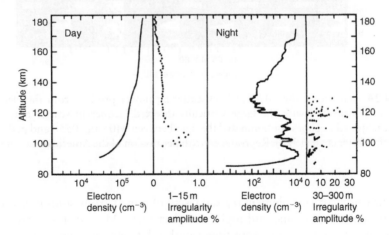

Figure 4.26 Height variations of the electron density and irregularity amplitudes measured at Thumba, India, around noon and midnight. [After Fejer and Kelley (1980). Reproduced with permission of the American Geophysical Union.]

secondary short-wavelength fluctuations also peaking whenever the zero-order gradient shows steep downward-directed gradients. Electric field fluctuation data taken with similar rocket trajectories and similar conditions are shown in Fig. 4.27. Here, the upper panel is daytime data and shows strong electric field signals both on the positive gradient and just above. The nighttime data

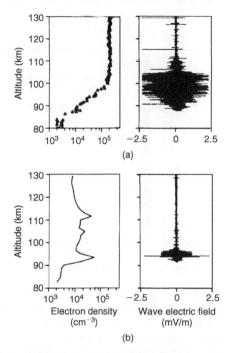

Figure 4.27 (a) Electron density data from a swept Langmuir probe together with high-frequency (34–1000 Hz) electric field data for a daytime equatorial rocket flight. (b) Electron density data from a fixed-bias Langmuir probe on a nighttime flight along with high-frequency (60–1000 Hz) electric field data. The plot in (b) corresponds to an off-equator site (Kwajalein) with a dip angle of 7°. [After Pfaff et al. (1982). Reproduced with permission of the American Geophysical Union.]

show fluctuations on the downward gradient. Curiously, there are no irregularities on two of the three layers, even though a dc electric field would be present at each of them due to the mapping effect. In Chapter 6 we return to this problem and offer an explanation that does not involve the electric field but rather the neutral wind as a source of instability. To summarize, the waves are clearly electrostatic (δE and $\delta n/n$ fluctuations exist), are organized by the direction of the gradient, and have broadband electric field fluctuations of the order of several millivolts per meter, which is comparable to the vertical polarization field.

We turn now to results of the CONDOR daytime rocket launch carried out on March 12, 1983, from Peru (Kudeki et al., 1987; Pfaff et al., 1987a, b). The rocket apogee was kept low to maximize the time and horizontal distance in the electrojet. The zero-order density profile is presented on the left side of Fig. 4.28 for the daytime electrojet flight, during which the magnetic deflection at Huancayo was 130 nT. A sonogram of the signals from a plasma wave receiver

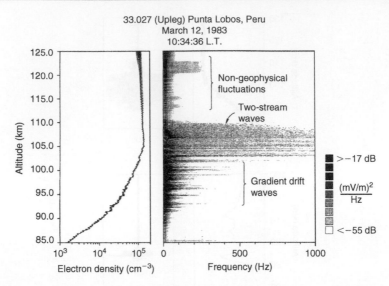

Figure 4.28 Frequency-height sonogram of the horizontal component of the irregularities measured during the upleg of rocket 33.027 during strong electrojet conditions. This instrument had a low-frequency roll-off (3 dB) at 16 Hz. The electron density profile (left) shows the presence of large-scale irregularities. Both panels show nongeophysical "interference" above about 110 km. [After Pfaff et al. (1987a). Reproduced with permission of the American Geophysical Union.]

for electric field fluctuations is plotted on the right side. The data divide clearly into three segments. Above the peak in electron density the profile is smooth, and the wave data extend to high frequencies. Below 102 km and in the region of upward electron density gradients, the plasma density is structured, and the wave data come in strong bursts that do not extend to very high frequencies. Between these regions, an upward gradient exists and the fluctuations are both patchy and extend to high frequencies. The total east-west electric field has been measured as a function of time and altitude during this flight and is shown in the upper panel of Fig. 4.29. The data show very intense alternating electric fields organized in large-scale features very reminiscent of the radar data shown in Fig. 4.25. In fact, large-scale waves of this type were detected by the radar between 103 and 106 km and were a primary launch criterion for the rocket flight. It is of interest to note that as the rocket altitude increased from 92 to 106 km, the signal changed qualitatively from a turbulent sinusoid to a squared-off and seemingly saturated waveform. Similar features were detected during the downleg. In the lower half of Fig. 4.29, a short time interval is blown up to better show the steep edges and saturation-like waveforms. Note that within these waveforms the vertical *perturbation* electron drift velocity, $\delta E \times B/B^2$, exceeds the sound speed. The electron density fluctuation data in this height range display similar, but not identical, waveforms with peak $\delta n/n$ values in

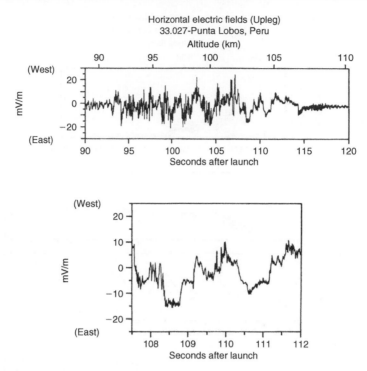

Figure 4.29 Horizontal electric fields observed during the upleg traversal of the electrojet. Note the steepened waveforms and the "flat-topped" nature of the fields in the upper portion of the electrojet (102–107 km), as also seen in the enlargement in the lower panel. [After Pfaff et al. (1987a). Reproduced with permission of the American Geophysical Union.]

the 10–15% range. The electric field and density fluctuations in these data are interpreted as horizontally polarized electrostatic waves with wavelengths of about 2 km. The saturation effect is not instrumental.

At about 106 km altitude, the vertical density gradient changes sign and the large-scale waves abruptly cease. An expanded view of this height range is shown in Fig. 4.30. Notice an almost evanescent behavior of the large-scale waves between 106.5 and 107.3 km. Above this height the total electric field in the second panel shows a nearly sinusoidal waveform with a 2–3 mV/m amplitude. The sonogram above this shows that the frequency of this peak changes with time in a similar fashion, decreasing from about 80 Hz to less than 20 Hz in a 2 km height range as the quasi-dc field from large-scale waves decays with altitude. The lower panels are the raw dc electric field signals, which show that the superimposed wave signal is nearly a pure sinusoid. It appears only when the detector is east-west aligned. The spectrum of this signal is shown in the lower part of Fig. 4.31. The peak in the spectrum occurs near 2.5 m. This means that

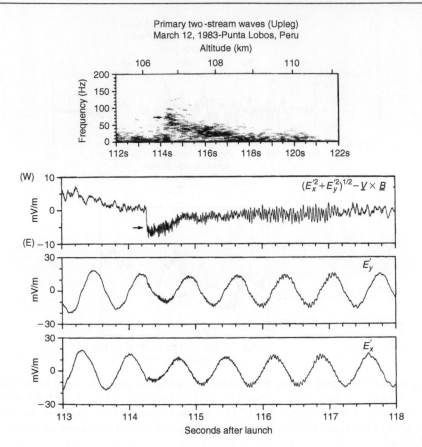

Figure 4.30 Electric field observations of what we believe to be pure two-stream waves for the upleg. The lower panels show the raw dc-coupled data, above which is plotted the square root of the sum of the squares of these waveforms. The upper panel shows a sonogram of these waves. (Note the change in scale of the time axis.) An arrow indicates the onset of the strong burst of primary two-stream waves. [After Pfaff et al. (1987b). Reproduced with permission of the American Geophysical Union.]

the wavelength of the peak in the instability spectrum is shifted by nearly three orders of magnitude from the spectral peak due to the large-scale waves found in the region of the zero-order vertical gradient at a slightly lower altitude.

The waves associated with type 1 irregularities have also been observed at times other than daytime when the electrical conductivity is reduced by an order of magnitude. VHF radar observations over Jicamarca during nighttime (Balsley, 1969b) and over Thumla during evening hours (Ravindran and Krishnamurthy, 1997) reveal the presence of such type 1 irregularities. In situ measurements conducted during early morning hours over Thumla (Gupta, 1986) also reveal the presence of such irregularities.

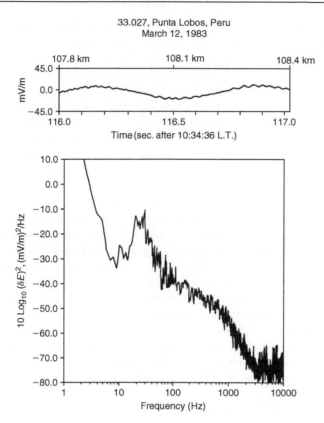

Figure 4.31 Raw data and the spectrum of pure two-stream waves in the daytime electrojet. The peak signal occurs when the detector is parallel to the current and the spectrum maximizes at a wavelength of a few meters. [After Pfaff et al. (1987b). Reproduced with permission of the American Geophysical Union.]

As we shall see following, these data are remarkably well organized by the linear theory into a pure "two-stream" instability at and above the peak in plasma density, a pure "gradient drift" instability well below the peak, and a mixture of the two just below the peak.

4.7 Linear Theories of Electrojet Instabilities

Two mechanisms can efficiently amplify thermal density fluctuations in the equatorial electrojet. These mechanisms result from plasma instabilities known as the two-stream instability and the gradient drift (also known as the cross-field instability).

Many features of the type 1 irregularities are explained by a modified two-stream instability theory developed independently by Farley (1963), using kinetic theory, and by Buneman (1963), using the Navier-Stokes equation. They have shown that the plasma is unstable for waves propagating in a cone of angle ϕ about the plasma drift velocity such that $V_D \cos \phi > C_s$. For smaller drift velocities the plasma still can be unstable, provided there is a plasma density gradient of zero order oriented in the right direction relative to the electric field driving the electrons. This instability was first studied by Simon (1963) and Hoh (1963) for laboratory plasmas and is termed the gradient drift instability. Many features of the type 2 radar echoes are explained by this instability.

We first investigate the linear theory for the two-stream instability. The linear theory is important because two-stream waves have been observed to move with a phase speed of the order of the ion-acoustic speed (which is not necessarily isothermal and which matches the phase speed at the instability linear threshold), no matter what exact nonlinear mechanism has produced them. We will start with the isothermal linear theory. Consider the zero-order condition in which the electrons stream east with a velocity $V_D \hat{a}_x$ driven by a vertically downward zero-order polarization electric field $-E_{z0} \hat{a}_z$ (see Chapter 3). This corresponds to nighttime conditions. The ions are assumed to be at rest to zero order due to the high collision frequency ($\nu_{in} \gg \Omega_i$). For the moment we ignore any zero-order plasma gradient. Perturbations in density, electrostatic potential, and velocity are of the form $\delta n e^{i(\omega t - kx)}$, $\delta \phi e^{i(\omega t - kx)}$, and $\delta V e^{i(\omega t - kx)}$, and we have taken $\mathbf{k}$ to be in the $\hat{a}_x$ direction ($\mathbf{k} = k \hat{a}_x$), which is perpendicular to $\mathbf{B} = B \hat{a}_y$. The linearized electron continuity equation, ignoring production and loss, which are assumed to be in equilibrium in the zero-order equations, is

$$i\omega \delta n + V_D(-ik\delta n) - ikn_0 \delta V_{ex} = 0 \tag{4.37}$$

where the second term comes from the $(\mathbf{V} \cdot \nabla n)$ term and the third from $n(\nabla \cdot \mathbf{V})$. In this expression the common factor $e^{i(\omega t - kx)}$ has been factored out. Notice that since $\mathbf{k}$ is horizontal and there is no vertical gradient, the vertical perturbation drift does not change the electron density. Equation (4.37) can be rewritten as

$$\delta V_{ex} = (\omega/k - V_D)(\delta n/n) \tag{4.38a}$$

where the subscript zero has been dropped from the mean density n_0. Notice that we have already assumed quasi-neutrality by using δn as a variable rather than δn_e. In the electron momentum equation the term $md\mathbf{V}_e/dt$ is dropped because of the small electron mass, leaving

$$0 = (-e/m)(\mathbf{E} + \mathbf{V}_e \times \mathbf{B}) - (k_B T_e/m)(\nabla n/n) - \nu_e \mathbf{V}_e$$

where again we ignore gravity and where ν_e is the electron-neutral collision frequency. We simplify ν_{en} and ν_{in} to ν_e and ν_i in the E region. The linearized version of this in component form is, in the z direction,

$$0 = -\Omega_e \delta V_{ex} - \nu_e \delta V_{ez} \tag{4.38b}$$

In this derivation we have chosen to follow tradition and, unlike in Chapter 2, write all gyrofrequencies as positive numbers (i.e., $\Omega_e = +eB/m$). In the z component of the momentum equation, the zero-order electric field and electron drift terms cancel in the equilibrium state about which we are making our perturbations. For the terms in the x direction we have

$$0 = (-eik/m)\delta\phi + \Omega_e\delta V_{ez} + ik(k_B T_e/m)(\delta n/n) - \nu_e\delta V_{ez}$$

or, equivalently,

$$-\Omega_e\delta V_{ez} + \nu_e\delta V_{ex} = -ik\left[(e\delta\phi/m) - (k_B T_e/m)(\delta n/n)\right] \tag{4.38c}$$

where again the zero-order terms cancel. The corresponding ion equation derived from continuity is

$$i\omega(\delta n/n) - ik\delta V_{ix} = 0 \tag{4.38d}$$

The ion momentum equation yields

$$(i\omega + \nu_i)\,\delta V_{ix} = ik\left[(e\delta\phi/M) + (k_B T_i/M)\,(\delta n/n)\right] \tag{4.38e}$$

where we have used $\Omega_i \ll \nu_i = \nu_{in}$. Since the ions are assumed to be at rest to zero order, the zero-order polarization field is not included in the ion equation of motion. Equations (4.38a)–(4.38e) constitute five equations in the five unknowns δV_{ex}, δV_{ez}, δV_{ix}, $\delta\phi$, and $\delta n/n$. The resulting determinant of coefficients is then set to zero.

$$\begin{vmatrix} 1 & 0 & 0 & 0 & (V_D - \omega/k) \\ -\Omega_e & -\nu_e & 0 & 0 & 0 \\ \nu_e & -\Omega_e & 0 & ike/m & -ik(k_B T_e/m) \\ 0 & 0 & -ik & 0 & i\omega \\ 0 & 0 & (i\omega + \nu_i) & -ike/M & -ik(k_B T_e/M) \end{vmatrix} = 0 \tag{4.39a}$$

This can be evaluated in a straightforward manner to find the relationship between ω and k (e.g., Sudan et al., 1973),

$$(\omega - kV_D) = (-\Psi_0/\nu_i)\left[\omega\,(i\omega + \nu_i) - ik^2 C_s^2\right] \tag{4.39b}$$

where $\Psi_0 = \nu_e\nu_i/\Omega_e\Omega_i$ and $C_s^2 = k_B\,(T_e + T_i)/M$. If we now set $\omega = \omega_r - i\gamma$ and require $\gamma \ll \omega_r$ and ν_i, we find the real and imaginary parts of ω to be

$$\omega_r = kV_D/(1 + \Psi_0) \tag{4.40a}$$

$$\gamma = (\Psi_0/\nu_i)\left(\omega_r^2 - k^2 C_s^2\right)/(1 + \Psi_0) \tag{4.40b}$$

When γ is positive, the waves will grow, and thus the requirement for instability is

$$\omega_r^2 > k^2 C_s^2 \quad \text{or} \quad V_D > (1 + \Psi_0) C_s$$

This is the origin of the term *two-stream* because the electrons must drift through the ions at a speed exceeding the sound speed to generate the waves.

We can now use these mathematical results to gain insight into the physical instability mechanisms involved. First, from the electron continuity equation we have

$$\delta V_{ex} = (\omega/k - V_D)\delta n/n \tag{4.41}$$

But using the two electron momentum equations [(4.38b) and (4.38c)] we also must have

$$\delta V_{ex} = -\left[\nu_e/\left(\Omega_e^2 + \nu_e^2\right)\right]\left[(ike\delta\phi/m) - (ikk_B T_e/m)(\delta n/n)\right] \tag{4.42}$$

If we study the properties of the wave near marginal stability where $\gamma = 0$ (or equivalently assume $\omega_r \gg \gamma$), we can replace $(\omega/k - V_D)$ by $(\omega_r/k - V_D)$, which in turn becomes $-(\Psi_0/[1 + \Psi_0])V_D$, using (4.38a). Setting (4.41) and (4.42) equal and using this result yield

$$\left[\nu_e/\left(\Omega_e^2 + \nu_e^2\right)\right](ike\delta\phi/m) = \left[\nu_e\left(ikk_B T_e/m\right)\left(\Omega_e^2 + \nu_e^2\right)\right.$$
$$\left. + \Psi_0 V_D/(1 + \Psi_0)\right](\delta n/n)$$

Now if we note that $\delta E_x = ik\delta\phi$ and $V_D = E_{z0}/B$, where E_{z0} is the zero-order vertical electric field, this may be written

$$\left[\nu_e e/m\left(\Omega_e^2 + \nu_e^2\right)\right]\delta E_x = \left[\nu_e\left(ikk_B T_e/m\right)\Big/\left(\Omega_e^2 + \nu_e^2\right) + \Psi_0 E_{z0}/B(1 + \Psi_0)\right]\delta n/n$$

and finally,

$$\delta E_x/E_{z0} = \left\{\Psi_0\left(\Omega_e^2 + \nu_e^2\right)\Big/[\nu_e\Omega_e(1 + \Psi_0)] + ikk_B T_e/eE_{z0}\right\}\delta n/n \tag{4.43}$$

This expression shows that the relationship between δE_x and $\delta n/n$ is wavelength dependent. For long wavelengths, the second term in the bracket is negligible and δE_x and δn are either in phase or 180° out of phase, depending on the sign of E_{z0}. Using the good approximation that $\Omega_e \gg \nu_e$ and $\Psi_0 = \nu_e\nu_i/\Omega_e\Omega_i$ we have for this case

$$\frac{\delta E_x}{E_{z0}} = \frac{\nu_i}{\Omega_i(1 + \Psi_0)}\frac{\delta n}{n} \tag{4.44}$$

This is very similar to the result found earlier for the Rayleigh-Taylor instability [i.e., (4.24b)]. The wavelength at which the magnitudes of the first and second terms in (4.43) are equal for $T_i = T_e$ (i.e., $C_s^2 = 2k_B T_e/M$) is given by

$$\lambda_c = \frac{\pi C_s^2 (1 + \Psi_0)}{\nu_i V_D}$$

Setting $\Psi_0 = 0.22$ at an altitude of 105 km, $C_s = V_D = 360 \, \text{m/s}$ and $\nu_i = 2.5 \times 10^3 \, \text{s}^{-1}$ yields $\lambda_c \simeq 0.4 \, \text{m}$. This wavelength is thus quite small and for most wavelengths of interest we may use (4.44) to relate the perturbed electric field to the density.

Using these results, the instability process can now be understood from Fig. 4.32a. The figure is drawn for daytime conditions with the vertical electric field upward. The sinusoidal wave can thus represent both $\delta n/n$ and δE_x for $\lambda \gg \lambda_c$. That is, the eastward perturbation δE_x is positive when $\delta n/n$ is positive. The two quantities are in phase with net positive or negative charges built up as shown, where these charges are associated with the perturbation electric field (i.e., $\rho_c = \varepsilon_0 (\nabla \cdot \mathbf{E}) = -\varepsilon_0 ik\delta E_x$).

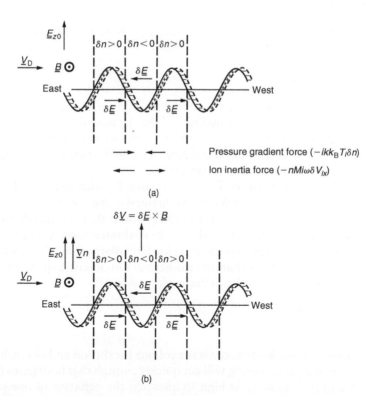

Figure 4.32a Schematic diagrams showing the linear instability mechanism in (a) the two-stream process and (b) the gradient drift process for daytime conditions.

Now, the wave will grow if more plasma moves into a region of high density than leaves that region. Consider the ion motion expressed by (4.38e). The real part of the equation corresponds to the ion velocity in phase with the wave. The $(+ik\delta\phi)$ term is just the electric field, δE_x, which is also in phase and thus is included in the real part of the expression. The imaginary part gives the out-of-phase motion at the spatial positions where $\delta n/n$ and δE_x both vanish and thus the part of the ion motion that causes the wave to grow or decay. The two vectors, pressure gradient force and inertial force, are shown in Fig. 4.32a. For growth, the ion inertial force must be larger than the pressure term so that the plasma moves horizontally from low density to high density. This in turn requires that

$$\omega_r \delta V_{ix} > k(k_B T_i/M)(\delta n/n) \tag{4.45}$$

In other words, the ion inertial force must be greater than the pressure gradient force that is trying to smooth out the density enhancement by diffusion. Using (4.38d) to eliminate δV_{ix}, equation (4.45) can also be written as

$$\omega_r \omega/k^2 > (k_B T_i/M)$$

Near marginal stability, $\gamma \ll \omega_r$, so $\omega \simeq \omega_r$, and using the dispersion relation (4.40a) for ω_r/k we have the requirement that

$$V_D/(1 + \Psi_0) > (k_B T_i/M)^{1/2} \tag{4.46}$$

This is almost but not quite the required threshold condition for growth. The exact result, $V_D > (1 + \Psi_0) C_s$, came out of the detailed determinant analysis, since the electron pressure term adds to the ion pressure via the ambipolar diffusion effect. Ion inertia is thus the destabilizing factor in the two-stream instability, whereas ambipolar diffusion causes damping.

Zero-order temperatures in the E region are usually such that $T_i = T_e = T_n = T$. Then, if the wave process behaves isothermally, the ion acoustic speed is $C_s^2 = 2k_B T/M$. Farley and Providakes (1989) noted that the threshold speed of short-scale E-region irregularities should be evaluated more carefully, based on their observations of high latitude E-region irregularities moving with phase speeds that were clearly faster than the isothermal ion-acoustic speed and much closer to a speed associated with adiabatic electrons:

$$C_s^2 = \frac{\gamma_i T_i + \gamma_e T_e}{M}$$

where $\gamma_{i,e}$ are the specific heats at constant volume for the ion and electron gases. For low-frequency waves, cooling will act quickly enough that both gases behave isotropically and $\gamma_i = \gamma_e = 1$. At high frequencies, the behavior of one or both gases may become adiabatic with a corresponding $\gamma = 5/3$, which will increase the phase velocity. However, since wave phase velocities approach V_D in the upper electrojet, the wave frequency in the drift frame becomes small and the electrons may again be isothermal (Hysell et al., 2007).

St.-Maurice et al. (2003) reported observations of two-step type-I waves in the lower equatorial electrojet moving at speeds higher even than in adiabatic processes and up to 50% higher than the isothermal ion acoustic speed. They named such behavior super-adiabatic and found that only the theory by St.-Maurice and Kissack (2000), which included electron thermal corrections, could explain these waves. The next step, exploring aspect sensitivity (Kagan and St.-Maurice, 2004), showed good correspondence to observations (Kudeki and Farley, 1989) and good qualitative agreement with the observed altitude behavior of Farley-Buneman waves: super-adiabatic at lower altitudes and isothermal at high altitudes. The correspondence of linear phase velocities at marginal stability to observations stimulated further linear theory with nonisothermal electrons, as reported by Kissack et al. (2007a, b). These papers additionally account for nonzero flow angles (essentially the angle between the $\mathbf{E} \times \mathbf{B}$ direction and the center of the radar beam, important for nonzero zenith angle transmissions) and an arbitrary heat source (with possible applications for high latitudes and heating experiments). A step forward by Kissack et al. (2007a, b) was the presentation of thermal corrections, which showed contributions from each physical process. They were also able to recover the results of preceding linear theories.

To test the wavelength dependence predicted by the Kissack et al. theory, Kagan and Kissack (2007) analyzed the multi(3)-frequency experiment of Balsley and Farley (1971) in Jicamarca. Their theoretical predictions showed good correspondence to observations. They also showed that, depending on the frequency of two-stream waves, their altitude behavior changes from super-adiabatic at lower altitudes to isothermal at higher altitudes. The transitional process from super-adiabatic to isothermal is dominated by inelastic electron energy exchange and therefore is much more important at low radar frequencies. In Fig. 4.32b

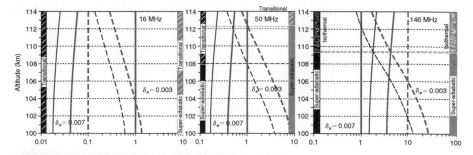

Figure 4.32b Altitude dependences of dimensionless parameters ξ_{AT} (dashed lines) and ξ_T (solid lines) that define dominating physical processes for inelastic electron cooling rates $\delta_e = 0.007$ (black lines) and $\delta_e = 0.003$ (gray lines) for each of three radar frequencies. Calculations are done with code from Kissack et al. (2007b), revised for the time and location of Balsley and Farley (1971). [After Kagan and Kissack (2007). Reproduced with permission of the American Geophysical Union.]

we reproduce their plots of altitude dependences of dimensionless parameters $\xi_{AT} = 3(\omega_r - \mathbf{k} \cdot \mathbf{u}_{0\perp})/2\delta_e \nu_e$ (adiabatic with thermal corrections over inelastic electron energy exchange, shown by dashed lines) and $\xi_T = CD_{e\perp} k^2/\delta_e \nu_e$ (thermal conduction and thermal diffusion over inelastic electron energy exchange, shown by solid lines) that define the dominating physical process for inelastic electron cooling rates of $\delta_e = 0.007$ (black lines) and $\delta_e = 0.003$ (gray lines) for each of three radar frequencies. Here, C is a coefficient and is given following. From Fig. 4.32b, after essentially dominating all altitudes at 16 MHz, this transitional process clearly covers a smaller altitude range and moves to higher altitudes at 50 MHz, then disappears completely at 146 MHz. A direct transition from super-adiabatic to isothermal processes results in a sudden drop in the phase velocity of two-stream waves that could be observed if the radar altitude resolution were less than the height of the transitional region (of about 2 km). This was first reported at 50 MHz by Swartz (1997). Kagan et al. (2008) also observed a sudden change in V_{ph} with the newly employed Prototype Advanced Modular Incoherent Scatter Radar (AMISR-P) at the Jicamarca Radio Observatory. AMISR-P was operated at 430 MHz and had an altitude resolution of 0.6 km at vertical incidence. Range-velocity-intensity plots and individual normalized spectra of these AMISR-P data are shown in the left-hand column of Fig. 4.32c. In the right-hand column of Fig. 4.32c, these observations, plotted together with theoretically predicted phase velocities, show good correspondence between theory and experiment.

For an equatorial electrojet between 100 and 120 km altitudes, Kagan and Kissack (2007) offer a simpler and more explicit expression for a phase velocity of Farley-Buneman waves, that is valid with less than 3% uncertainty. Since observational uncertainty is usually much more than 3%, the Kagan and Kissack (2007) formula is handy for quick estimates and is easy to program. Their Eq. (4.47) following allows easy tracking of wave phase velocity dependence on its wave number and facilitates thermal corrections to the classical expression, contributions from non-zero flow angles, and dominating physical processes such as thermal conduction and inelastic electron cooling.

$$V_{ph} =$$

$$u_{i0} \pm C_{sj} \frac{k_\perp}{k} \sqrt{1 + \frac{T_{e0}}{(T_{e0} + T_{i0})} \frac{\chi^{DS}(2/3)\eta \left[CD_{e\perp} k_\perp^2 + \eta\right] + (1+g)^2 A^2 \left[\Psi_T/(1+\Psi_T)\right]^2 u_0^2 k^2}{\left\{(3/2)\left[\Psi_T/(1+\Psi_T)\right]^2 u_0^2 k^2 + (2/3)\left[CD_{e\perp} k_\perp^2 + \eta\right]^2\right\}}}.$$

$$(4.47)$$

Here, $\mathbf{u}_0 = \mathbf{V}_{i0} - \mathbf{V}_{e0}$ is a current velocity—that is, a relative velocity between ions and electrons; $\delta_e \nu_e$ is the inelastic volume electron-neutral energy exchange rate, where the dimensionless energy exchange factor, δ_e, is essentially constant over the altitudes of interest; $C_{sj} = \sqrt{k_B (T_{e0} + T_{i0})/M}$ is the isothermal ion-acoustic speed where T_{e0} and T_{i0} are background electron and ion temperatures, respectively; k is the magnitude of the wave vector; and $k_\perp$ is the magnitude

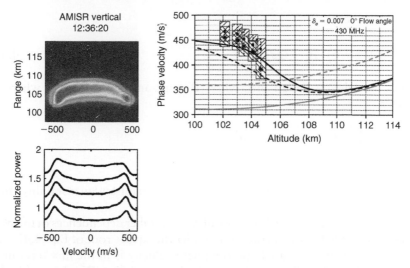

Figure 4.32c Range-Velocity-Intensity plots (top) and individual normalized spectra (bottom) for AMISR-P (left column) and phase velocities at 430 MHz on March 12, 2005, observed at the Jicamarca Observatory and theoretically predicted (right column). Stars and solid squares show data for FB waves moving toward and away from the radar, respectively. Shaded areas around each data point show the altitude (horizontal scale) and velocity (vertical scale) uncertainty. Solid and dashed black lines are theoretically predicted for V_{ph} for 0° and 0.2° aspect angles, respectively. An isothermal ion acoustic speed is shown as a solid gray line. The dashed gray line denotes the adiabatic process. [After Kagan et al. (2008). Reproduced with permission of the American Geophysical Union.] See Color Plate 6.

of the wave vector component perpendicular to the geomagnetic field, **B**. The coefficients are

$$A\left(\frac{k_B^2}{k_\perp^2}\right) = \left[1 + \frac{(\alpha - g)}{(1 + g)}\beta\right] \Big/ [1 + \alpha\beta], \tag{4.47a}$$

$$C\left(\frac{k_B^2}{k_\perp^2}\right) = 5/2 + 2g + g^2 - (g - 5/2)\,\beta - g\,(1 + g)\,(1 - \beta)\,A\left(\frac{k_B^2}{k_\perp^2}\right) \tag{4.47b}$$

$$g = \left[\frac{T_{e0}}{v_e}\frac{\partial v_e}{\partial T_e}\right]_{T_{e0}}, \quad \alpha = 1 + 2g\,(1 + g)/5, \quad \beta = \frac{1}{(1 + 2g/5)}\left(\frac{k_B^2\Omega_e^2}{k_\perp^2 v_{en}^2}\right) \tag{4.47c}$$

$$\Psi_T = \Psi_0\,(1 + \alpha\beta), \quad \Psi_0 = \frac{v_e v_{in}}{\Omega_e\Omega_i}, \quad \eta = \delta_e v_e, \quad D_{e\perp} = \kappa_T T_{e0} v_e \Big/ m\Omega_e^2. \tag{4.47d}$$

The following term,

$$\chi^{DS} = [(1+g)^2 A - 3g\bar{s}/2]\bar{s} \tag{4.47e}$$

describes the Dimant-Sudan instability (Dimant and Sudan, 1997) where the parameter,

$$\bar{s} = \frac{k_\perp^2 u_0^2 \sin\theta_f \cos\theta_f}{D_{e\perp} k^2 \eta} \frac{v_e}{\Omega_e} \frac{\Psi_0}{(1+\Psi_T)} \tag{4.47f}$$

is a function of flow angle θ_f, thereby zeroing the Dimant-Sudan instability effects at flow angles of $0°$ and $90°$.

All of the thermal processes' effects and wave number dependences of the phase velocity are included in the second term under the square root of Eq. (4.47). The greatest effect of the flow angle on the phase velocity is in the first term in the numerator, which describes the Dimant-Sudan instability (Dimant and Sudan, 1997). This term vanishes at 0- and 90-degree flow angles and disappears at higher altitudes. The Dimant-Sudan instability has a cut-off at high wave numbers (or frequencies) so that, for example, at 430 MHz, the change in the flow angle does not affect V_{ph}. This instability occurs for negative flow angles corresponding to the radar looking east for a daytime equatorial electrojet. For positive flow angles corresponding to western transmissions, the Dimant-Sudan thermal instability might still be induced but requires significantly higher threshold velocities than the ion-acoustic speed. Thus, if any radar returns are observed at lower frequencies (≤ 50 MHz), the phase velocity of type 1 irregularities would be higher than for the east and vertical (if the latter are not smeared by gradient-drift processes) transmissions. In an isothermal treatment, Eq. (4.47) reduces to the classical expression in which the phase velocity does not depend on the transmitted frequency.

Turning to the gradient drift instability, we must include the change of density due to advection from the vertical perturbation electron drift in the electron continuity equation. The term comes from $V \cdot \nabla n$, which in linearized form adds a term $1/kL$ to the first row, second column of the determinant where $L = [(1/n)(dn/dz)]^{-1}$ is the zero-order vertical gradient scale length. This added term changes only the imaginary part of ω; the real part is identical. This means that (4.43) also relates the perturbed field δE_x to the density perturbation $\delta n/n$ for the gradient drift mode. To study the stability condition, consider Fig. 4.32b, which shows daytime conditions and includes a zero-order density gradient that is upward. From Fig. 4.32b it is clear that the gradient is a destabilizing factor when it is upward since then the upward perturbation drift $\delta V_{ez} = \delta E_x/B$ occurs in a region where the density is already depleted ($\delta n < 0$). That is, a low-density region is convected upward into a region of higher background density, causing a growth in the relative value of $\delta n/n$. If the gradient is reversed in sign but the

E_{z0} is left the same, the perturbation electric fields will cause high-density plasma to drift into higher-density regions, which is stabilizing. Thus, for instability the zero-order vertical electric field must have a component parallel to the zero-order density gradient. At the equator during normal daytime conditions (see Chapter 3), the unstable gradient direction is upward since E_{z0} is upward, while at night the unstable gradient is downward.

There is no threshold drift requirement for the gradient drift instability in the physics thus far. However, if recombination is considered, a term in the continuity equations of the form $-\alpha n^2$ must be included which, when linearized, becomes of the form $-2\alpha n_0 \delta n$ and a finite threshold thus exists for the gradient drift process. Fejer et al. (1975) found that the electron drift threshold velocity for instability to occur is given by

$$V_D > L\left[2\alpha n_0 \left(1 + \Psi_0\right)\left(\Omega_i/\nu_i\right) + \left(k^2 C_s^2 \nu_e/\nu_i\Omega_e\right)\left(1 + \Psi_0\right)\right]$$

Finally, following Fejer et al. (1975), the complete linear theory must include a finite zero-order ion drift velocity $\mathbf{V}_{Di}$ and the possibility of an arbitrary k vector (but still perpendicular to $\mathbf{B}$), yielding the following set of expressions for the linear theory of both the gradient drift and two-stream modes:

$$\omega_r = \mathbf{k}\cdot(\mathbf{V}_D + \Psi\mathbf{V}_{Di})/(1 + \Psi) \tag{4.48a}$$

$$\gamma = (1 + \Psi)^{-1}\left\{(\Psi/\nu_i)\left[\left(\omega_r - \mathbf{k}\cdot\mathbf{V}_{Di}\right)^2 - k^2 C_s^2\right]\right\}$$
$$+ \left(1/Lk^2\right)\left\{\left(\omega_r - \mathbf{k}\cdot\mathbf{V}_{Di}\right)\left(\nu_i/\Omega_i\right)k_x\right\} - 2\alpha n_0 \tag{4.48b}$$

with

$$\Psi = \Psi_0\left[\left(k_\perp^2/k^2\right) + \left(\Omega_e^2/\nu_e^2\right)\left(k_x^2/k^2\right)\right] \tag{4.49}$$

and under the assumption that $\gamma \ll \omega_r$.

If the latter assumption is relaxed, Kudeki et al. (1982) have shown that waves with $\lambda \simeq L$ propagate more slowly; that is, a more exact result for the linear theory with $\mathbf{k}\cdot\mathbf{B} = \mathbf{V}_{Di} = 0$ is given by

$$\omega_r + i\gamma = \mathbf{k}\cdot\mathbf{V}_D\left(1 + ik_0/k\right)\Big/\left[\left(1 + \Psi_0\right)\left(1 + k_0^2/k^2\right)\right] \tag{4.50}$$

where $k_0 = (\nu_i/L\Omega_i)/(1 + \Psi_0)$. This is the linear theory for large scale waves. Finally, we note that recombination is much more important during daytime conditions than nighttime, since the zero-order density is large and $2\alpha n_0$ is sizable.

The linear theory outlined here has been applied to data from three rocket flights and the results are reproduced in Fig. 4.33 (Pfaff et al., 1985). Panel (a) shows the current density and electron density profiles. Panel (b) shows the

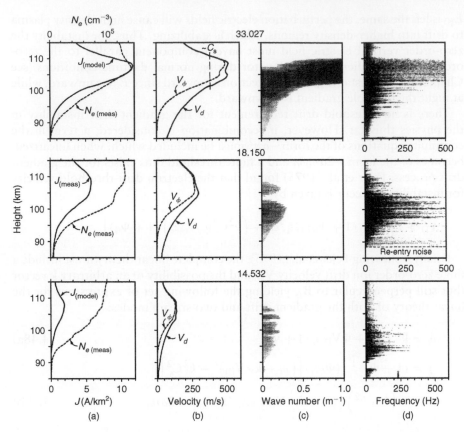

Figure 4.33 (a) Measured or model current and smoothed electron density profiles for each of three experiments; (b) calculated electron drift and phase velocity profiles computed using the values in (a); (c) gray-scale representation of the growth rate calculated from the linear dispersion relation for horizontal waves using the model parameters discussed in the text. Darker shades represent higher values of γ, whereas white indicates that $\gamma < 0$. (d) Frequency-height sonograms showing rocket-measured irregularities. [After Pfaff et al. (1985). Reproduced with permission of Pergamon Press.]

electron drift velocity, the phase velocity for two-stream waves (which must exceed C_s for instability to occur), and the approximate value for C_s. Using these parameters and the gradient of the electron density profile, the growth rate has been calculated as a function of altitude and wave number in panel (c) using a gray scale. Two-stream waves result when the threshold for the two-stream instability is exceeded. This condition occurred only in the CONDOR rocket flight shown on top and is evident in the linear calculation [panel (c)] as well as the wave observations in panel (d). In the calculation, waves are predicted to grow at high wave numbers. Indeed, sonograms of the rocket wave fluctuation data in panel (d) show strong high-frequency (in the rocket frame) waves in the region where the density gradient vanishes or reverses sign, but only for the CONDOR data. Such waves are weak in the middle data set and are not seen at all in the lowest set of data. In the center data set the conditions were close to two-stream instability and the wave activity was stronger than in the case of the lowest data set. However, in both the center and lowest data set the waves were restricted primarily to the region of upward vertical density gradient. Those data are thus in excellent detailed agreement with linear theory.

The fluid theory discussed thus far breaks down at short wavelengths and a kinetic approach is necessary. Such calculations have been carried out by Farley (1963) and more recently by Schmidt and Gary (1973). Pfaff et al. (1987b) compared Schmidt and Gary's calculation of $\gamma(k)$ with the spectrum of electric field fluctuations measured in the CONDOR rocket flight during pure two-stream conditions (above the density gradient) as shown in Fig. 4.34. The two plots are remarkably similar. This strongly suggests that pure two-stream waves grow in a narrow range of wave numbers near the peak in their linear growth rate.

To summarize, the linear theory explains (a) the acoustic speed drift velocity threshold for type 1 irregularities, (b) the low drift velocity threshold for type 2 waves, (c) the association of waves with upward density gradients during the day and downward ones at night, (d) the wavelength and angular dependence of the phase velocity of type 2 waves, (e) the slow phase velocity of large-scale waves, and (f) the peak in the two-stream threshold spectrum.

The linear theory does not explain, among other things, (a) 3 m wave generation by the gradient drift process (since it is linearly stable at that wavelength),

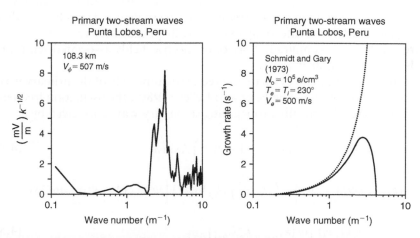

Figure 4.34 Comparisons of the measured electric field wave spectrum and two-stream growth rate calculations for daytime equatorial conditions by Schmidt and Gary (1973). The dotted line in the growth rate calculation is the result of the fluid theory and the solid line represents the kinetic results. [After Pfaff et al. (1987b). Reproduced with permission of the American Geophysical Union.]

(b) the dominance and square-wave nature of kilometer-scale waves, (c) the observation of vertically propagating two-stream waves (perpendicular to the current), (d) the apparent constant phase velocity of type 1 waves at any angle to the current, and (e) details of the observed wave number spectra.

4.8　Nonlinear Theories of Electrojet Instabilities

4.8.1　Two-Step Theories for Secondary Waves

Although strictly speaking not a nonlinear theory, the discussion by Sudan et al. (1973) provides a conceptual framework from which several properties of the fully developed turbulence can be understood. The basic idea is that "primary" waves reach sufficiently large amplitudes that the perturbation electric fields and density gradients are themselves large enough to drive "secondary" waves unstable. For example, referring to the observational data from the CONDOR flights in Fig. 4.29, the primary δE is horizontal, so if it is large enough, $\delta E/B$ will drive two-stream waves in the vertical direction. Indeed, the rocket electric field measurements show that the perturbation electric fields can be of the order of $(1 + \Psi)C_sB$. This is illustrated in Fig. 4.35a, in which the horizontal electric field fluctuation strength has been divided by B and plotted as a function of altitude for the upleg and downleg of the rocket flight. The dotted lines show the threshold velocity for a secondary two-stream instability given by $C_s(1 + \Psi)$. The first thing to note is that between 100 and 105 km the observed perturbation electric field is sufficiently strong to generate a vertical two-stream instability. The region where these waves can be generated agrees remarkably well with the region where vertical up- and downgoing two-stream waves were detected simultaneously by the Jicamarca radar. It is safe to say that vertically propagating two-stream waves are created by large-amplitude horizontal electric fields associated with long-wavelength gradient drift waves.

Quantitatively, one can evaluate the required amplitude of the primary waves as follows. As summarized by Fejer and Kelley (1980), the oscillation frequency and growth rate of vertically propagating secondary waves are given by

$$\omega_r(k_s) = -k_s \left(\Omega_e/v_e\right) \left(\Psi V_D/(1 + \Psi)^2\right) A \sin \delta \qquad (4.51)$$

$$\gamma(k_s) = (\Psi/1 + \Psi) \left\{ - \left(\Omega_e^2/v_e^2\right) (k_p V_D/2) \left[\Psi A^2/(1 + \Psi)^2\right] \sin(2\delta) \right.$$

$$\left. + (1/v_i) \left[\omega(k_s)^2 - k_s^2 C_s^2\right] \right\}. \qquad (4.52)$$

where k_s and k_p are the wave numbers of the secondary vertically propagating and primary horizontally propagating waves, respectively. In this expression, A is the amplitude $(\delta n/n)$ of the primary wave and δ is its phase $(\omega_p t - k_p x)$.

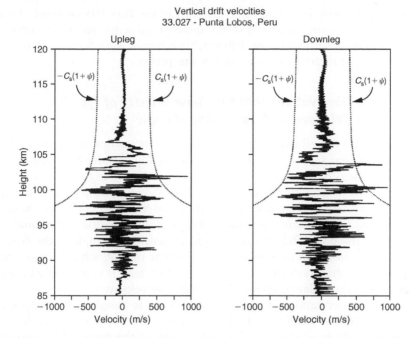

Figure 4.35a Vertical drift velocities associated with the horizontal electric fields measured during the upleg and downleg traversals of the electrojet by rocket 33.027. The boundaries corresponding to $C_s(1 + \Psi)$, which represent the two-stream instability threshold, are also shown for a fixed value of T_e. The figure demonstrates why vertical two-stream waves were only detected in the narrow region of roughly 100–107 km for this day. [After Pfaff et al. (1987b). Reproduced with permission of the American Geophysical Union.]

The necessary condition for the generation of vertically propagating type 1 irregularities is given by the requirement (see Section 4.6) that

$$(\delta E/B) > (1 + \Psi)C_s \tag{4.53}$$

which corresponds to the dotted lines in Fig. 4.35a. Using (4.44) this can also be written

$$A(\nu_i/\Omega_i)\left[V_D/(1 + \Psi)^2\right] > C_s \tag{4.54}$$

where again we have used A to represent the amplitude of the $(\delta n/n)$ variation as defined above. Neglecting ion inertia, secondary gradient drift irregularities are generated if

$$A^2\left(M\nu_i V_D k_p\right)\Big/\left[2m\nu_e(1 + \Psi)^2\right] > k_s^2 C_s^2/\nu_i \tag{4.55}$$

The requirement (4.54) seems to be met only in the 100–105 km height range, but (4.55) is less stringent and it seems quite possible for secondary gradient drift waves to be generated below 100 km. Such waves were detected during the CONDOR rocket flight with radar and in situ probes.

4.8.2 On the Observations That the Phase Velocity of Type I Equatorial Waves Is Independent of Angle

In Chapter 10 we describe the results of auroral zone studies of type 1 electrojet echoes. With a near vertical magnetic field and highly variable flow direction, it is possible to measure the phase velocity as a function of angle. Bahcivan et al. (2005) have conclusively shown that the relation $V_{ph} = C_s \cos \theta$ is the best match to the observations. The implication is that the waves travel at the threshold for instability. This result seems at odds with the classic observation in the equatorial electrojet that the phase velocity is C_s, independent of the angle to the flow. In addition, this result is in disagreement with linear theory, which, for ψ small, predicts a wave phase velocity equal to $V_D \cos \theta$ where θ is the angle between the current and the radar beam and V_D is the flow velocity.

The key to understanding the equatorial results is based on two important factors (Kelley et al., 2008). First, the drift velocity of importance is the total drift velocity, which is equal to the vector sum of the zero-order horizontal drift and the drift induced by intense, large-scale waves. Second, a radar will respond to the most intense wave in the field of view, so it will detect a narrow Doppler shift at the value $C_s \cos \theta$, corresponding to the projection of the largest line-of-sight velocity. The horizontal wavelength of daytime large scale waves is in the range of 2–3 km (Kudeki et al., 1987) and is coherent over an altitude range of 4 km. The vertical drift data presented in Fig. 4.35a can be used along a horizontal drift, based on the rocket data in Fig. 3.17b, to create a vector drift velocity on a two-dimensional grid. Figure 4.35b shows how this region would be interrogated by the JULIA radar at two different zenith angles for the system beam width and range resolution and the same for the AMISR-P at its maximum zenith angle of 20°. The red crosses in the figure correspond to regions in which the total velocity exceeded the factor $(1 + \Psi_0)C_s$ and for which the value of $\cos \theta$ exceeded 0.9, where θ is the angle between the beam and the total flow velocity. For this purpose we used the isothermal value for C_s. The first criterion assures that the total drift velocity exceeds the threshold for instability. Now almost certainly, the most intense waves are those with a phase velocity vector nearest the flow angle, implying that the radar will preferentially detect such waves if they are in the range gate and will thus register their Doppler velocity. By our criterion, this echo will be registered at a phase velocity in excess of $0.9 C_s$. At the equator, clearly the radars will always see a near constant Doppler shift as a function of angle when the horizontal drift velocity is large and intense, large scale waves are present. The long-standing problem thus seems resolved.

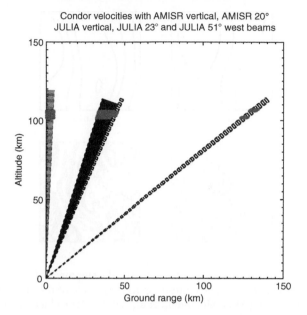

Figure 4.35b Regions in which these radars would record a Doppler shift $\geq .0.9C_s$ in the presence of a zero-order horizontal drift, as in Fig. 3.17b, and vertical perturbation drifts, as in Fig. 4.35a, are indicated in red. [After Kelley et al. (2008). Reproduced with permission of the American Geophysical Union.] See Color Plate 7.

4.8.3 Nonlinear Gradient Drift Theories

Extensive analytical and simulation studies have been performed on the intermediate-scale (≤ 100 m) type 2 or gradient drift irregularities. These studies permit a detailed comparison between theoretical results and the radar and rocket data. McDonald et al. (1974, 1975) studied numerically the nonlinear evolution of small-scale ($\lambda \simeq 10$ m) type 2 irregularities by using a grid of 50×50 points with a mesh spacing of 1.5 m in both horizontal and vertical directions. They observed that intermediate wavelengths ($\lambda < 28$ m) are excited only after the large-scale primary horizontally propagating waves grow to an amplitude of 4%, consistent with the "two-step" theory outlined previously. The quasi-final state is highly turbulent and almost two-dimensionally isotropic, in agreement with the radar observations. Figure 4.36 shows the saturated nonlinear development of the gradient drift irregularities found in these simulations. In the turbulent state the small-scale irregularities have upward and downward motions with speeds comparable to the background horizontal drift. The two-dimensional power spectra of $(\delta n/n)^2$ and $(\delta E)^2$ were found to be proportional to $k^{-3.5}$. This implies a one-dimensional spectrum varying as $k^{-2.5}$.

Additional detailed, quantitative studies of the nonlinear development of the primary gradient drift theory have been reported by Sudan and Keskinen

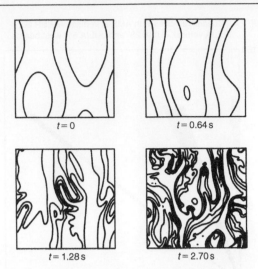

Figure 4.36   Plasma density contours showing the development of the equatorial irregularities at four selected times. The contour spacing is 2.5% of the ambient density, and the grid spacing is 1.5 m. [After McDonald et al. (1975). Reproduced with permission of the American Geophysical Union.]

(1977), Keskinen et al. (1979), and Sudan (1983). The approach is best suited to zero-order conditions such that $V_D < C_s$, and they therefore were able to apply Kadomtsev's (1965) strong turbulence-weak coupling equations to the gradient drift turbulence. These equations are, essentially, the Fourier-transformed version of the direct interaction approximation developed by Kraichnan (1967). One of the crucial features is that the nonlinear damping rate due to wave emission and cascade is much larger than the linear damping rate because these waves interact for a long time. This in turn relies on the fact that the waves are almost dispersionless and therefore all the waves have nearly equal group velocities. Farley (1985) has reviewed this theory, and we follow his approach here. The object is to predict the two-dimensional spectrum of density fluctuation turbulence $I_k(k)$ such that the total density fluctuation strength (dimensionless) is given by

$$\left\langle (\Delta n/n)^2 \right\rangle = \iint I_k(\mathbf{k}) d_\perp \mathbf{k}$$

For isotropic turbulence in the plane perpendicular to $\mathbf{B} = B\hat{a}_y$, $I_k(\mathbf{k}) = I_k(k)$ where

$$k = |\mathbf{k}| = \left(k_x^2 + k_z^2 \right)^{1/2}$$

The goal is to derive a differential equation for $I_k(k)$ that describes the flow of energy in the system as a function of k. There are two main effects: the linear growth rate, which includes both linear growth and damping, and the nonlinear flow of energy from eddy to eddy through the spectrum. In a steady state the input of energy at long wavelengths equals the dissipation at short wavelengths. The eddy process forms a mechanism to transfer energy from the growth portion of the spectrum to the region where damping is strongest. The argument goes as follows. In a range of wave numbers, Δk, the total density fluctuation strength is

$$(\Delta n/n)^2_{\Delta k} = \int_{\Delta k} I_k 2\pi k\, dk \simeq 2\pi I_k k\, \Delta k$$

If we consider the range $\Delta k = k/2\pi$, that is, a bandwidth in k space equal to $k/2\pi$, then

$$(\Delta n/n)^2_{\Delta k} \simeq k^2 I_k$$

Since from (4.44) $\delta V/V_D = [\nu_i/\Omega_i\,(1 + \Psi_0)]\,(\delta n/n)$, we also see that in this range of Δk the velocity fluctuation strength is given by

$$\delta V^2_{\Delta k} \propto V_D^2 k^2 I_k$$

Now the classical turbulence argument is that the eddy decay rate (Γ_k) is given by the inverse of the eddy turnover time τ_k, where

$$\Gamma_k = (\tau_k)^{-1} = k\delta V_{\Delta k} \propto V_D k^2\,(I_k)^{1/2}$$

This is the time it takes for the material in the eddy to move one eddy scale size (k^{-1}). The total energy ε_k in a given eddy is proportional to the $\delta V^2_{\Delta k}$ given above, so the rate of energy loss is given by

$$(\varepsilon_k/\tau_k)_{\Delta k} = \varepsilon_k \Gamma_k \propto V_D^3 k^4 I_k^{3/2} \tag{4.56}$$

The rate of energy gain or loss in the same wavelength band from the linear growth and damping processes is determined by the linear growth/damping rate γ_k averaged over all angles at wavelength k. In a time-stationary steady state it must be the case that the spectrum $\varepsilon_k(k)$ has the property that

$$kd/dk(\varepsilon_k \Gamma_k) = \gamma_k \varepsilon_k \tag{4.57}$$

so the steady-state energy spectral density remains the same in any k interval. From linear theory, for the primary gradient drift process, γ_k is of the form

$$\gamma_k = A - Bk^2 \tag{4.58}$$

where

$$A = \frac{\nu_i V_D}{2(1 + \Psi_0)^2 \Omega_i L} - 2\alpha n_0$$

$$B = \frac{\Psi_0}{\nu_i (1 + \Psi_0)} \left[C_s^2 - V_D^2 (1 + \Psi_0)^{-2} / 2 \right]$$

Substituting (4.55) and (4.58) into (4.57) yields

$$d/dk \left(k^4 I_k^{3/2} \right) \propto V_D^{-1} k I_k \left(A - B k^2 \right) \tag{4.59a}$$

This differential equation may readily be solved to yield

$$I(x) = x^{-8/3} \left[1 - x^{-2/3} - \left(x^{4/3} - 1 \right) / 2S \right]^2 \tag{4.59b}$$

where $x = k/k_{cc}$, k_c is some long wavelength cutoff, $I(x) = (V_0^2 k_c^4 / A^2) I_k$, and $S = A/(Bk_c^2)$. From the definition of γ_k we see that A represents growth and B damping, so S is therefore a "strength parameter" for the process.

A plot of the quantity $xI(x)$ is given in Fig. 4.37 for values of S equal to 10^4 and ∞ and for negative values. The latter correspond to drifts greater than the primary two-stream value. For positive large values of S, the curve rises to a peak

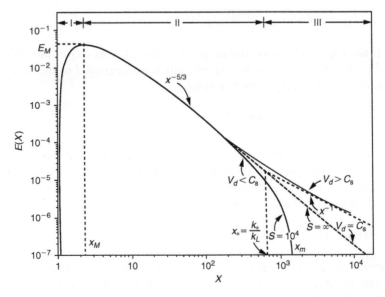

Figure 4.37 Theoretical one-dimensional power spectrum of the electrojet turbulence. [After Sudan (1983). Reproduced with permission of the American Geophysical Union.]

value where $x \approx 3$ and then follows a power law for larger x. The value of the power law index for $E(x) = xI(x)$ is the same as the value that would be measured by a one-dimensional cut through the turbulent plasma using devices that measure either $(\delta n/n)^2$ or $(\delta E)^2$ since the one-dimensional measurement spectrum integrates out one power of k and has the same shape as $xI(x)$ (proportional to kI_k). The power law regime occurs only for large S and occurs in regime II in Fig. 4.37, where the eddy growth rate dominates the linear growth, corresponding to the inertial subrange in neutral turbulence theory (Kolmogorov, 1941). It is interesting that the same one-dimensional spectral form is predicted for this plasma case as for the three-dimensional neutral fluid turbulence in the inertial subrange. For finite S, the spectrum becomes very steep at large k, corresponding to the viscous subrange in neutral turbulence and, in the present case, is due to diffusive damping. For S infinite or negative, the fluid theory breaks down due to the excitation of short-wavelength waves, which require a kinetic description.

There are a number of ways to check this theory. First, the numerical simulations may be tested against the analytic expression (4.59). The numerical results of McDonald et al. (1974, 1975) and Keskinen et al. (1979) seem to disagree with the prediction since they report $I_k \propto k^{-3.5}$ rather than $k^{-11/3}$. This problem can be reconciled as shown in Fig. 4.38. The data points in this figure come from the simulation work reported by Keskinen et al. (1979). The curved line is a fit to the analytic calculation of Sudan (1983) described above for $k_d = 15k_s$, where k_s is the wave number corresponding to the physical size of the grid. The fit is quite good but it must be realized that k_c, the outer scale for the process, must be considerably smaller than k_s to yield such a steep slope in the range of k space plotted. For smaller values of k (not plotted), Sudan's calculations yield a power spectrum of $k^{-2.67}$. Presumably, *if* the simulations and theory are in agreement and *if* the simulation occurred in a larger "box," the calculated power spectra would yield a $k^{-2.67}$ two-dimensional power law at small k. However, such a result has not yet been obtained in any simulation and the comparison shown in Fig. 4.38 must remain somewhat suspect.

4.8.4 Nonlinear Studies of Farley-Buneman (FB) Waves

Linear theory gives conditions for the onset of an instability and characteristics of the initial growing waves but cannot explain saturation or provide the amplitude and spectral characteristics of developed turbulence. A number of theoretical models have been developed to explain saturated FB waves. Hamza and St.-Maurice (1993a, b) proposed a strongly turbulent mode-coupling theory based upon a two-fluid model. Another approach, developed by Albert and Sudan (1991), Sahr and Farley (1995), Otani and Oppenheim (1998), Dimant (2000), and Otani and Oppenheim (2006), uses a truncated three-wave mode-coupling dynamic model to explain instability saturation. None of these theories provides a fully consistent quantitative description of nonlinear saturation of E-region instabilities.

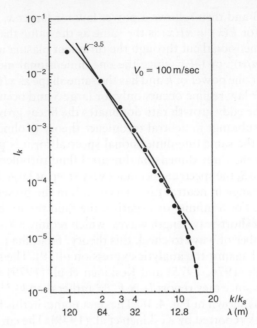

Figure 4.38 Spectra (dots) from the numerical simulation of Keskinen et al. (1979) compared with the theory (solid curve). The curve was normalized at the point marked with the cross and the dissipative cutoff wave number k_d was chosen for the best fit to be $15k_s$, where k_s is determined by the size of the numerical grid and is somewhat analogous to (but in practice considerably larger than) the long-wavelength k_c discussed in the text. [After Sudan (1983). Reproduced with permission of the American Geophysical Union.]

Two-dimensional simulations of equatorial Farley-Buneman waves have enabled researchers to understand a number of key electrojet observations. They show that the dominant nonlinearity arises when the perturbed electric fields interact with the density perturbations and drive energy into modes that are linearly stable or damped. This wave-wave interaction modifies the linear behavior of the waves in a number of observable ways. It leads to mode coupling that saturates the waves when $\langle |\delta E| \rangle \sim E_0$ where $\langle |\delta E| \rangle$ is the average amplitude of the perturbed electric field generated by the waves. This causes a broadening of the turbulent spectrum and also reduces the expected dominant phase velocity below that predicted by linear theory (Oppenheim and Otani, 1996). This results from mode coupling when the perturbed fields of secondary modes, on average, reduce or cancel the driving field, E_0, of the primary modes. This model and the two-dimensional simulations do not predict a strict saturation of the phase velocity at the sound-speed as was inferred from the data by Hanuise and Crochet (1981), Makarevitch et al. (2002), and others. Rather, they predict a saturated phase velocity traveling at roughly the sound speed times the cosine

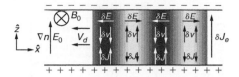

Figure 4.39 Schematic representation of wave-driven currents in the equatorial electrojet. [After Oppenheim (1997). Reproduced with permission of the American Geophysical Union.]

between the wave and the electron drift direction as seen in Woodman and Chau (2002), Shume et al. (2005), and Bahcivan et al. (2005).

On the left of Fig. 4.39, we show the vertical electrojet electric field, E_0, the geomagnetic field, B_0, the plasma density gradient, ∇n, and the electrojet electron drift direction, V_d. If V_d exceeds a threshold then compressional plasma waves develop as shown by the varying shades of grey, darkest where the waves enhance the plasma density and lightest where they reduce it. At the density maxima and minima, we show the direction of the perturbed electric field, δE, the direction in which the electrons drift in response to δE, δv, and the resulting electron current, $\delta J = n\delta v$. This current is larger where the plasma density is enhanced than where it is reduced. On the right, we show the direction of the net, wave-driven, vertical, electron current, δJ_e. An identical mechanism generates wave-driven currents in the auroral electrojet when $\nabla n = 0$, E_0 is horizontal, and B_0 is vertical (Oppenheim, 1997).

The simulations also show that FB waves nonlinearly drive a large-scale (dc) current in the E-region ionosphere, as shown in Fig. 4.39 (Oppenheim, 1997). This current flows parallel to the fundamental Pedersen current and with a comparable magnitude. These currents can restructure the electrojet as shown in Fig. 4.40. Also, by effectively increasing the Pedersen currents, wave-driven currents reduce the electrojet charge and polarization field, E_0, responsible for driving FB waves. This makes the linear phase velocity drop toward the acoustic speed and may play an important role in sound speed saturation of type 1 waves.

A wave-driven current results from two fundamental features of E-region plasma waves. First, electrons travel mostly perpendicular to the electric fields due to the geomagnetic field while ions travel mostly parallel to the fields because ion-neutral collisions make magnetic field effects inconsequential. Second, gradient-drift and two-stream instabilities cause compressional waves where the plasma density enhancements and the perturbed electric fields remain largely in phase. At the plasma density maxima of the propagating wave fronts, electrons move perpendicular to the wave direction and the geomagnetic field. At the density minima, electrons move in the opposite direction with an equal velocity. However, more electrons exist at the maxima than at the minima causing a greater current in one direction than the other, resulting in a net (direct) current.

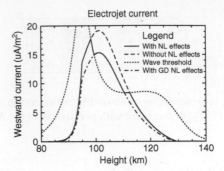

Figure 4.40 Electrojet current density as a function of height. The dashed line shows the current without nonlinear effects, the solid line shows the current with effects of a two-stream, wave-driven nonlinear current; the dashed-dotted line shows the current with effects of a gradient-drift (GD), wave-driven current, and the dotted line shows the minimum current necessary for initiating two-stream waves and the resulting wave-driven current. [After Oppenheim (1997). Reproduced with permission of the American Geophysical Union.]

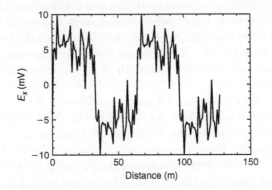

Figure 4.41 The electric field from a one-dimensional, simulated gradient-drift wave system modified by secondary two-stream waves driving a nonlinear current. Note the similarity to Fig. 4.29. [After Oppenheim (1997). Reproduced with permission of the American Geophysical Union.]

The electric fields measured by rockets passing through gradient-drift waves appear as irregular square waves (see Fig. 4.29). Wave-driven electron currents can cause these squared-off electric fields through a two-step process. First, the perturbed electric field of a gradient-drift wave must exceed the threshold necessary to initiate two-stream waves (Sudan et al., 1973). Second, these secondary two-stream waves generate wave-driven electron currents that modify the original gradient-drift waves. This effect has been estimated numerically in Oppenheim (1997) and is shown in Fig. 4.41.

4.9   D-Region Turbulence

Some of the first observations of mesospheric turbulence were made at the Jicamarca Observatory, and we choose to discuss this phenomenon first in this chapter, even though it is a worldwide process. On average, the atmosphere is well mixed below about 100 km, called the turbopause. Above this height the various molecules can separate according to their mass. A detailed study of turbulence per se is beyond the scope here but it does affect the small-scale structure of the D region. Since the atmosphere in the mesosphere is very dense compared to the ionized component, the ions and electrons are pushed around at will by the neutral gas. The charged particles thus act as passive scalars in the vernacular of turbulence theory. As the ionized components are pushed around by turbulent motions, if a gradient exists in their content, as there almost always is, advection will create structure in the ions and electrons. The resulting irregularities have been studied using radars and rockets.

In Fig. 3.35 a typical passive scalar spectrum was shown (Tennekes and Lumley, 1972). In turbulence theory, energy is conjectured to be injected into a fluid at some scale L_i. In a laboratory experiment this might be the spacing of grids placed in the flow. In the atmosphere the injection scale might be related to gravity waves (buoyancy scale) or a Kelvin-Helmholtz instability (KHI). At smaller scales, energy is passed from large eddies to smaller eddies without any energy loss until a small enough scale is reached that viscosity becomes important. These two ranges are called the inertial and viscous subranges, respectively. Viscosity is important at small scales since the term $\nabla^2 u \approx (\eta/\rho)k^2 u$ is large when k^2 is large.

The breakpoint to a very steep spectrum occurs near the Kolmogorov microscale given by

$$\mu = \left(\frac{\nu^3}{\varepsilon}\right)^{1/4}$$

where ν is the molecular kinematic viscosity coefficient (η/ρ) and ε is the energy dissipation rate (Watts/kg). This expression shows that, as ν increases with increasing altitude, μ increases. There is also a weak tendency for μ to decrease with increasing ε, but the fourth-power dependence makes this a less important effect. Since energy is dissipated by viscosity, the dissipation rate (ε) is proportional to the turbulent energy cascading down the spectrum. A plot of various scales for the earth's atmosphere was presented earlier in Fig. 3.36.

In Fig. 3.36, the inner (viscous) scale was calculated following Tatarskii (1971) as

$$l_i \approx 7.4\mu$$

Here, l_i is the wave number associated with the spectral break $(2\pi/l_i)$ to the viscous subrange. At mesospheric heights Hocking finds that $l_i \approx 20$–40 m. This explains why typical radars do not receive echoes from mesospheric altitudes, even though during daytime the region has numerous electrons. The Bragg scattering wavelength at 50 MHz is 3 m and is well into the viscous subrange where turbulent fluctuations are very small. Only the very large Jicamarca radar and the large MU radar receive echoes under such conditions.

A very nice verification of these ideas came during one of the rocket flights from Peru in 1983. The experimenters waited until strong echoes were received at Jicamarca and then launched a high resolution electron density probe (Royrvik and Smith, 1984). The spectrum they measured at 85.5 ± 0.2 km is plotted in the right-hand panel of Fig. 4.42 and shows a very steep spectrum for $\lambda \leq 25$ m. The power spectral density at $k = 2$ m^{-1} was small but was in quantitative agreement with the Jicamarca radar signal strength. The high latitude winter spectrum (central panel) is very similar to the equatorial spectrum. However, the high latitude summer results are very different and are discussed in some detail in Chapter 7 where the importance of charged ice is detailed.

A number of factors lead to atmospheric turbulence. The two most common energy release mechanisms are convective and dynamic instabilities. We return

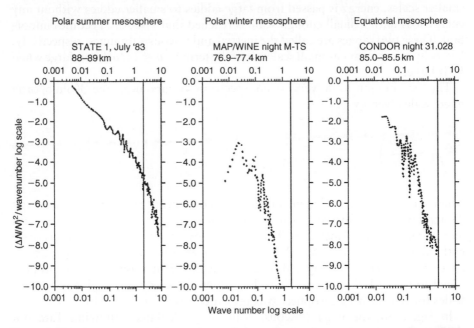

Figure 4.42 Comparison of mesospheric electron density fluctuation spectra from (left) polar summer, (middle) polar winter, and (right) equatorial rocket launches. The vertical line corresponds to the Bragg wave number of a 50 MHz radar. [After Cho and Kelley (1993). Reproduced with permission of the American Geophysical Union.]

to these processes in Chapter 6 after we have discussed the buoyancy properties of the atmosphere. The mesospheric echoes received at Jicamarca are thought to be related to the dynamic instability of a sheared wind field.

4.10   Future Directions

The physics of equatorial plasma instabilities involves a complex interplay between aeronomic, electrodynamic, and plasma physical processes. At the present juncture we have a good understanding of the sources of free energy that lead to the generation of nonthermal fluctuations in the equatorial plasma. The present research thrusts involve (a) understanding the seasonal, local time, and solar cycle control of the onset of instabilities; (b) understanding the nonlinear evolution of the plasma processes themselves; and (c) predicting convective ionospheric storms. In these areas, numerical modeling and data assimilation are becoming very common research tools. Such models allow the researcher to vary different physical parameters and see the effect of the variation. In a geophysical context, such control is not available to the experimenter, so the modeling effort is very important. Major future strides in the space research area will continue to come from direct comparison between experimental results and numerical simulations. Neutral atmospheric effects are becoming more important to aeronomic studies as we begin to study the D-region dynamics more closely.

References

Albert, J. M., and Sudan, R. N. (1991). Three-wave interactions and type II irregularities in the equatorial electrojet. *Phys. Fluids B* 3, 495.

Alex, S., Koparkar, P. V., and Rastogi, R. G. (1989). Spread-F and ionization anomaly belt. *J. Atmos. Terr. Phys.* 51, 371.

Bahcivan, H., Hysell, D. L., Larsen, M. F., and Pfaff, R. F. (2005). The 30 MHz imaging radar observations of auroral irregularities during the JOULE campaign. *J. Geophys. Res.* 110, A05307, doi:10.1029/2004JA010975.

Balsley, B. B. (1969a). Some characteristics of non-two-stream irregularities in the equatorial electrojet. *J. Geophys. Res.* 74, 2333.

———. (1969b). Measurement of electron drift velocities in the nighttime equatorial electrojet. *J. Atmos. Terr. Phys.* 31, 475.

Balsley, B. B., and Farley, D. T. (1971). Radar studies of the equatorial electrojet at three frequencies. *J. Geophys. Res.*, 76(34), 8341–8351.

Basu, Su., Basu, Sa., Aarons, J., McClure, J. P., and Cousins, M. D. (1978). On the coexistence of kilometer- and meter-scale irregularities in the nighttime equatorial F-region. *J. Geophys. Res.* 83, 4219.

Beer, T. (1973). Spatial resonance in the ionosphere. *Planet. Space Sci.* 21, 297.

Berkner, L. V., and Wells, H. W. (1934). F-region ionosphere investigation at low altitudes. *Terr. Magn.* 39, 215.

Bowles, K. L., Cohen, R., Ochs, G. R., and Balsley, B. B. (1960). Radar echoes from field-aligned ionization above the magnetic equator and their resemblance to auroral echoes. *J. Geophys. Res.* **65**, 1853.

Buneman, O. (1963). Excitation of field-aligned sound waves by electron streams. *Phys. Rev. Lett.* **10**, 285.

Burke, W. J., Gentile, L. C., Huang, C. Y., Valladares, C. E., and Su, S.-Y. (2004). Longitudinal variability of equatorial plasma bubbles observed by DMSP and ROCSAT-1. *J. Geophys. Res.* **109**, A12301, doi:10.1029/2004JA010583.

Chau, J. L., and Woodman, R. F. (2001). Interferometric and dual beam observations of daytime spread-F-like irregularities over Jicamarca. *Geophys. Res. Lett.* **28**(18), 3581.

Cho, J. Y. N., and Kelley, M. C. (1993). Polar mesosphere summer radar echoes: Observations and current theories. *Rev. Geophys.* **31**, 243.

Cohen, R., and Bowles, K. L. (1967). Secondary irregularities in the equatorial electrojet. *J. Geophys. Res.* **72**, 885.

de La Beaujardiére, O., Retterer, J. M., Kelley, M. C., Hunton, D., and Jeong, L. (2006). New satellite will forecast ionospheric disturbances. *Space Weather* **4**, S03C05, 10.1029/2005SW000145.

Dimant, Y. S. (2000). Nonlinearly saturated dynamical state of a three-wave mode-coupled dissipative system with linear instability. *Phys. Rev. Lett.* **84**, 622.

Dimant, Y. S., and Sudan, R. N. (1997). Physical nature of new cross-field instability in the lower ionosphere. *J. Geophys. Res.* **102**, 2551–2563.

Dungey, J. W. (1956). Convective diffusion in the equatorial F-region. *J. Atmos. Terr. Phys.* **9**, 304.

Farley, D. T. (1963). A plasma instability resulting in field-aligned irregularities in the ionosphere. *J. Geophys. Res.* **68**, 6083.

———. (1985). Theory of equatorial electrojet plasma waves: New developments and current status. *J. Atmos. Terr. Phys.* **47**, 729.

Farley, D. T., and Hysell, D. L. (1996). Radar measurements of very small aspect angles in the equatorial ionosphere. *J. Geophys. Res.* **101**, 5177.

Farley, D. T., and Providakes, J. F. (1989). The variation with T_e and T_i of the velocity of unstable ionospheric two-stream waves. *J. Geophys. Res.* **94**, 15,415.

Farley, D. T., Balsley, B. B., Woodman, R. F., and McClure, J. P. (1970). Equatorial spread F: Implications of VHF radar observations. *J. Geophys. Res.* **75**, 7199.

Fejer, B. G., and Kelley, M. C. (1980). Ionospheric irregularities. *Rev. Geophys. Space Phys.* **18**, 401.

Fejer, B. G., Scherliess, L., and de Paula, E. R. (1999). Effects of the vertical plasma drift velocity on the generation and evolution of equatorial spread F. *J. Geophys. Res.* **104**, 19,859.

Fejer, B. G., Farley, D. T., Baisley, B. B., and Woodman, R. F. (1975). Vertical structure of the VHF backscattering region in the equatorial electrojet and the gradient drift instability. *J. Geophys. Res.* **80**, 1313.

Gupta, S. P. (1986). Formation of sporadic E layers at low magnetic latitudes. *Planet. Space Sci.* **34**, 1081.

Hagan, M. E., and Forbes, J. M. (2002). Migrating and nonmigrating diurnal tides in the middle and upper atmosphere excited by tropospheric latent heat release. *J. Geophys. Res.* **107**(D24), 4754, doi:10.1029/2001JD001236.

Hamza, A. M., and St.-Maurice, J. P. (1993a). A turbulent theoretical framework for the study of current-driven E-region irregularities at high latitudes: Basic derivation and application to gradient-free situations. *J. Geophys. Res.* **98**, 11,587.

————. (1993b). A self-consistent full turbulent theory of radar auroral backscatter. *J. Geophys. Res.* **98**, 11,601.

Hanuise, C., and Crochet, M. (1981). 5–50-m wavelength plasma instabilities in the equatorial electrojet, 2. Two-stream conditions. *J. Geophys. Res.* **86**, 3567.

Hoh, F. C. (1963). Instability of Penning-type discharge. *Phys. Fluids* **6**, 1184.

Huang, C.-S., and Kelley, M. C. (1996). Nonlinear evolution of equatorial spread F, 1. On the role of plasma instabilities and spatial resonance associated with gravity wave seeding. *J. Geophys. Res.* **101**, 283.

Huba, J. D., and Ossakow, S. L. (1981a). Lower hybrid drift waves in equatorial spread F. *J. Geophys. Res.* **86**, 829.

————. (1981b). Diffusion of small scale irregularities during equatorial spread F. *J. Geophys. Res.* **86**, 9107.

Hysell, D. L., and Burcham, J. D. (1998). JULIA radar studies of equatorial spread F. *J. Geophys. Res.* **103**(A12), 29,155.

Hysell, D. L., and Farley, D. T. (1996). Implications of the small aspect angles in the equatorial ionosphere. *J. Geophys. Res.* **101**, 5165.

Hysell, D. L., and Kelley, M. C. (1997). Decaying equatorial F region plasma depletions. *J. Geophys. Res.* **102**(A9), 20,007.

Hysell, D. L., and Kudeki, E. (2004). Collisional shear instability in the equatorial F-region ionosphere. *J. Geophys. Res.* **109**, A11301, doi:10.1029/2004JA010636.

Hysell, D. L., Kelley, M. C., Farley, D. T., and Seyler, C. E. (1994). Origin and spectral characteristics of violent equatorial ionospheric F-region disturbances. In *Low-Latitude Ionospheric Physics*. F.-S. Kuo (ed.) Proceedings of COSPAR Colloquium on Low-Latitude Ionospheric Physics, Taipei, Taiwan, 1993, vol. 7. Pergamon Press, 97–108.

Hysell, D. L., Kelley, M. C., Swartz, W. E., and Woodman, R. F. (1990). Seeding and layering of equatorial spread F by gravity waves. *J. Geophys. Res.* **95**(A10), 17,253.

Hysell, D. L., Drexler, J., Shume, E. B., Chau, J. L., Scipion, D. E., Vlasov, M., Cuevas, R., and Heinselman, C. (2007). Combined radar observations of equatorial electrojet irregularities at Jicamarca. *Ann. Geophys.* **25**, 457–473.

Immel, T. J., Sagawa, E., England, S. L., Henderson, S. B., Hagan, M. E., Mende, S. B., Frey, H. U., Swension, C. M., and Paxton, L. J. (2006). Control of equatorial ionospheric morphology by atmospheric tides. *Geophys. Res. Lett.* **33**, L15108, doi:10.1029/2006GL026161.

Kadomtsev, B. B. (1965). *Plasma Turbulence*. Academic Press, New York.

Kagan, L. M., and Kissack, R. S. (2007). Energy exchange rate for equatorial electrojet based on a case study of two-stream processes that include thermal corrections. *Geophys. Res. Lett.* **34**, L20806, doi:10.1029/2007GL030903.

Kagan, L. M., and St.-Maurice, J.-P (2004). The impact of electron thermal effects on Farley-Buneman waves at arbitrary aspect angles. *J. Geophys. Res.* **109**(A12), A12302.

Kagan, L. M., Kissack, R. S., Kelley, M. C., and Cuevas, R. (2008). Unexpected rapid decrease in phase velocity of submeter Farley-Buneman waves with altitude. *Geophys. Res. Lett.* **35**, L03106, doi:10.1029/2007GL03245.

Kelley, M. C., and Hysell, D. L. (1991). Equatorial spread-F and neutral atmospheric turbulence: A review. *J. Atmos. Terr. Phys.* 53(8), 695.

Kelley, M. C., Baker, K. D., and Ulwick, J. C. (1979). Late time barium cloud striations and their possible relationship to equatorial spread F. *J. Geophys. Res.* 84, 1898.

Kelley, M. C., Cuevas, R. A., and Hysell, D. L. (2008). Radar scatter from equatorial electrojet waves: An explanation for the constancy of the Type I Doppler shift with zenith angle. *Geophys. Res. Lett.* 35, L04106, doi:10.1029/2007GL032848.

Kelley, M. C., Makela, J. J., and de La Beaujardiére, O. (2006). Convective ionospheric storms: A major space weather problem. *Space Weather* 4(2), S02C04, doi:10.1029/2005SW000144.

Kelley, M. C., Seyler, C. E., and Zargham, S. (1987). Collisional interchange instability 2. A comparison of the numerical simulations with the *in situ* experimental data. *J. Geophys. Res.* 92, 10,073.

Kelley, M. C., Larsen, M. F., LaHoz, C. A., and McClure, J. P. (1981). Gravity wave initiation of equatorial spread F: A case study. *J. Geophys. Res.* 86, 9087.

Kelley, M. C., Makela, J. J., Ledvina, B. M., and Kintner, P. K. (2002). Observations of equatorial spread-F from Haleakala, Hawaii. *Geophys. Res. Lett.* 29, 2003, doi:10.1029/2002GL015509.

Kelley, M. C., LaBelle, J., Kudeki, E., Fejer, B. G., Basu, Sa., Basu, Su., Baker, K. D., Hanuise, C., Argo, P., Woodman, R. F., Swartz, W. E., Farley, D. T., and Meriwether, J. W., Jr. (1986). The Condor equatorial spread F campaign: Overview and results of the large-scale measurements. *J. Geophys. Res.* 91, 5487.

Keskinen, M. J., Mitchell, H. G., Fedder, J. A., Satyanarayana, P., Zalesak, S. T., and Huba, J. D. (1988). Nonlinear evolution of the Kelvin-Helmholtz instability in the high-latitude ionosphere. *J. Geophys. Res.* 93, 137.

Keskinen, M. J., Ossakow, S. L., and Chaturvedi, P. K. (1980a). Preliminary report of numerical simulations of intermediate wavelength collisional Rayleigh-Taylor instability in equatorial spread F. *J. Geophys. Res.* 85, 1775.

———. (1980b). Preliminary report of numerical simulations of intermediate wavelength $E \times B$ gradient drift instability in equatorial spread F. *J. Geophys. Res.* 85, 3485.

Keskinen, M. J., Sudan, R. N., and Ferch, R. L. (1979). Temporal and spatial power spectrum studies of numerical simulations of type 2 gradient drift irregularities in the equatorial electrojet. *J. Geophys. Res.* 84, 1419.

Kil, H., Oh, S.-J., Kelley, M. C., Paxton, L. J., England, S. L., Talaat, E., Min, K.-W., and Su, S.-Y. (2007). Longitudinal structure of the vertical $E \times B$ drift and ion density seen from ROCSAT-1. *Geophys. Res. Lett.* 34, L14110, doi:10.1029/2007GL030018.

Kissack, R. S., Kagan, L. M., and St.-Maurice, J.-P. (2007a). Thermal effects on Farley-Buneman waves for nonzero aspect and flow angles, I. Dispersion relation. *Phys. Plasmas* 15(2), 022901, doi:10.1063/1.2834275.

———. (2007b). Thermal effects on Farley-Buneman waves for nonzero aspect and flow angles, II. Threshold analysis. *Phys. Plasmas* 15(2), 022902, doi:10.1063/1.2834276.

Klostermeyer, J. (1978). Nonlinear investigation of the spatial resonance effect in the nighttime equatorial F-region. *J. Geophys. Res.* 83, 3753.

Kolmogorov, A. N. (1941). The local structure of turbulence in incompressible viscous fluids for very high Reynolds numbers. *Dokl. Akad. Nauk SSSR* 30, 301.

Kraichnan, R. H. (1967). Inertial ranges in two-dimensional turbulence. *Phys. Fluids* 11, 671.

Krishna Murthy, B. V., Ravindran, S., Viswanathan, K. S., Subbarao, K. S. V., Patra, A. K., and Rao, P. B. (1998). Small scale (3 m) irregularities at and off the magnetic equator. *J. Geophys. Res.* **103**, 20,761.

Kudeki, E., and Farley, D. T. (1989). Aspect sensitivity of equatorial electrojet irregularities and theoretical implications. *J. Geophys. Res.* **94**, 426-434.

Kudeki, E., Bhattacharyya, S., and Woodman, R. F. (1999). A new approach in incoherent scatter F region E×B drift measurements at Jicamarca. *J. Geophys. Res.* **104**, 28,145.

Kudeki, E., Farley, D. T., and Fejer, B. G. (1982). Long wavelength irregularities in the equatorial electrojet. *Geophys. Res. Lett.* **9**, 684.

Kudeki, E., Fejer, B. G., Farley, D. T., and Hanuise, C. (1987). The CONDOR Equatorial Electrojet Campaign: Radar results. *J. Geophys. Res.* **92**, 13,561.

Kudeki, E., Akgiray, A., Milla, M., Chau, J. L., and Hysell, D. L. (2007). Equatorial spread-F initiation: Post-sunset vortex, thermospheric winds, gravity waves. *J. Atmos. Solar-Terr. Phys.* **69**(17–18), 2416–2427.

LaBelle, J., Kelley, M. C., and Seyler, C. E. (1986). An analysis of the role of drift waves in equatorial spread F. *J. Geophys. Res.* **91**, 5513.

Livingston, R. C., Rino, C. L., McClure, J. P., and Hanson, W. B. (1981). Spectral characteristics of medium-scale equatorial F-region irregularities. *J. Geophys. Res.* **86**, 2421.

Makarevitch, R. A., Koustov, A. V., Sofko, G. J., André, D., and Ogawa, T. (2002). Multifrequency measurements of HF Doppler velocity in the auroral E region. *J. Geophys. Res.* **107**, 1212, doi:10.1029/2001JA000268.

McDonald, B. E., Coffey, T. P., Ossakow, S., and Sudan, R. N. (1974). Preliminary report of numerical simulation of type 2 irregularities in the equatorial electrojet. *J. Geophys. Res.* **79**, 2551.

———. (1975). Numerical studies of type 2 equatorial electrojet irregularity development. *Radio Sci.* **10**, 247.

McDonald, B. E., Keskinen, M. J., Ossakow, S. L., and Zalesak, S. T. (1980). Computer simulation of gradient drift instability processes in operation Avefria. *J. Geophys. Res.* **85**, 2143.

Mendillo, M., Meriwether, J., and Biondi, M. (2001). Testing the thermospheric neutral wind suppression mechanism for day-to-day variability of equatorial spread F. *J. Geophys. Res.* **106**, 3655.

Narcisi, R. S., and Szuszczewicz, E. P. (1981). Direct measurements of electron density, temperature, and ion composition in an equatorial spread F ionosphere. *J. Atmos. Terr. Phys.* **43**, 463.

Oppenheim, M. M. (1997). Evidence and effects of a wave-driven nonlinear current in the equatorial electrojet. *Ann. Geophys.* **15**, 899.

Oppenheim, M. M., and Otani, N. F. (1996). Special characteristics of the Farley-Buneman instability: Simulations versus observations. *J. Geophys. Res.* **101**, 24,573.

Ossakow, S. L. (1981). Spread F theories—A review. *J. Atmos. Terr. Phys.* **43**, 437.

Otani, N. F., and Oppenheim, M. M. (1998). A saturation mechanism for the Farley-Buneman instability. *Geophys. Res. Lett.* **25**, 1833.

———. (2006). Saturation of the Farley-Buneman instability via three-mode coupling. *J. Geophys. Res.* **111**, 3302.

Patra, A. K., and Rao, P. B. (1999). High-resolution radar measurements of turbulent structure in the low latitude E region. *J. Geophys. Res.* **104**, 24,667.

Perkins, F. W., and Doles, J. H. (1975). Velocity shear and the $E \times B$ instability. *J. Geophys. Res.* **80**, 211.

Pfaff, R. F., Kelley, M. C., Fejer, B. G., Maynard, N. C., and Baker, K. D. (1982). In situ measurements of wave electric fields in the equatorial electrojet. *Geophys. Res. Lett.* **9**, 688.

Pfaff, R. F., Kelley, M. C., Fejer, B. G., Maynard, N. C., Brace, L. M., Ledley, B. G., Smith, L. G., and Woodman, R. F. (1985). Comparative in situ studies of the unstable daytime equatorial E region. *J. Atmos. Terr. Phys.* **47**, 791.

Pfaff, R. F., Kelley, M. C., Kudeki, E., Fejer, B. G., and Baker, K. D. (1987a). Electric field and plasma density measurements in the strongly driven daytime equatorial electrojet. 1. The unstable layer and gradient drift waves. *J. Geophys. Res.* **92**, 13,578.

————. (1987b). Electric field and plasma density measurements in the strongly driven daytime equatorial electrojet. 2. Two-stream waves. *J. Geophys. Res.* **92**, 13,597.

Prakash, S., and Pandy, R. (1980). On the production of large scale irregularities in the equatorial F region. *Int. Symp. Equatorial Aeronomy, 6th*, 3–7.

Prakash, S., Subbaraya, B. H., and Gupta, S. P. (1972). Rocket measurements of ionization irregularities in the equatorial ionosphere at Thumba and identification of plasma irregularities. *Indian J. Radio Space Phys.* **1**, 72.

Raghavarao, R., Gupta, S. P., Sekar, R., Narayanan, R., Desai, J. N., Sridharan, R., Babu, V. V., and Sudhakar, V. (1987). In situ measurements of winds, electric fields, and electron densities at the onset of equatorial spread F. *J. Atmos. Terr. Phys.* **49**, 485.

Raghavarao, R., Nageswararao, M., Sastri, J. H., Vyas, G. D., and Sriramarao, M. (1988). Role of equatorial ionization anomaly in the initiation of equatorial spread F. *J. Geophys. Res.* **93**, 5959.

Ravindran, S., and Krishnamurthy, B. V. (1997). Occurrence of type-I plasma waves in the equatorial electrojet during morning and evening hours. *J. Geophys. Res.* **102**, 9761.

Rino, C. L., Tsunoda, R. T., Petriceks, J., Livingston, R. C., Kelley, M. C., and Baker, K. D. (1981). Simultaneous rocket-borne beacon and *in situ* measurements of equatorial spread F—intermediate wavelength results. *J. Geophys. Res.* **86**, 2411.

Röttger, J. (1973). Wave like structures of large scale equatorial spread F irregularities. *J. Atmos. Terr. Phys.* **35**, 1195.

Royrvik, O., and Smith, L. G. (1984). Comparison of mesospheric VHF radar echoes and rocket probe electron number density measurements. *J. Geophys. Res.* **89**, 9014.

Sahr, J. D., and Farley, D. T. (1995). Three-wave coupling in the auroral E region. *Ann. Geophys.* **13**, 38.

Satyanarayana, P., Guzdar, P. N., Huba, J. D., and Ossakow, S. L. (1984). Rayleigh-Taylor instability in the presence of a stratified shear layer. *J. Geophys. Res.* **89**, 2945.

Schmidt, M. J., and Gary, S. P. (1973). Density gradients and the Farley-Buneman instability. *J. Geophys. Res.* **78**, 8261.

Sekar, R., and Kelley, M. C. (1998). On the combined effects of vertical shear and zonal electric field patterns on nonlinear equatorial spread F evolution. *J. Geophys. Res.* **103**, 20,735.

Sekar, R., and Kherani, E. A. (1999). Effects of molecular ions on the Rayleigh-Taylor instability in the nighttime equatorial ionosphere. *J. Atmos. Solar-Terr. Phys.* **61**, 399.

Sekar, R., and Raghavarao, R. (1987). Role of vertical winds on the Rayleigh-Taylor instabilities of the nighttime equatorial ionosphere. *J. Atmos. Terr. Phys.* **49**, 981.

Shume, E. B., Hysell, D. L., and Chau, J. L. (2005). Zonal wind velocity profiles in the equatorial electrojet derived from phase velocities of type II radar echoes. *J. Geophys. Res.* **110**, 12,308.

Simon, A. (1963). Instability of a partially ionized plasma in crossed electric and magnetic fields. *Phys. Fluids* **6**, 382.

Sridharan, R., Pallam Raju, D., Raghavarao, R., and Ramarao, P. V. S. (1994). Precursor to equatorial spread-F in O I 630.0 nm dayglow. *Geophys. Res. Lett.* **21**, 2797.

Sridharan, R., Chandra, H., Das, S. R., Sekar, R., Sinha, H. S. S., Pallam Raju, D., Narayanan, R., Raizada, S., Misra, R. N., Raghavarao, R., Vyas, G. D., Rao, P. B., Ramarao, P. V. S., Somayajulu, V. V., Babu, V. V., and Danilov, A. D. (1997). Ionization Hole Campaign—A coordinated rocket and ground-based study at the onset of equatorial spread F: First results. *J. Atmos. Terr. Phys.* **59**, 2051.

St.-Maurice, J.-P., and Kissack, R. S. (2000). The role played by thermal feedbacks in heated Farley-Buneman waves at high latitudes. *Ann. Geophys.* **18**, 532–548.

St.-Maurice, J.-P., Choudhary, R. K., Ecklund, W. L., and Tsunoda, R. T. (2003). Fast type-I waves in the equatorial electrojet: Evidence for non-isothermal ion-acoustic speeds in the lower E region. *J. Geophys. Res.* **108**(5), 1170, doi:10.1029/2002JA009648.

Sudan, R. N. (1983). Unified theory of type 1 and type 2 irregularities in the equatorial electrojet. *J. Geophys. Res.* **88**, 4853.

Sudan, R. N., and Keskinen, M. (1977). Theory of strongly turbulent two-dimensional convention of low pressure plasma. *Phys. Rev. Lett.* **38**, 966.

Sudan, R. N., Akinrimisi, J., and Farley, D. T. (1973). Generation of small-scale irregularities in the equatorial electrojet. *J. Geophys. Res.* **78**, 240.

Swartz, W. E. (1997). CUPRI observations of persistence asymmetry reversals in up-down vertical type-I echoes from the equatorial electrojet above Alcântara, Brazil. *Geophys. Res. Lett.* **24**(13), 1675–1678.

Tatarskii, V. I. (1971). *The Effects of the Turbulent Atmosphere on Wave Propagation.* Israel Program for Scientific Translations, Jerusalem.

Tennekes, H., and Lumley, J. L. (1972). *A First Course in Turbulence*, MIT Press, Cambridge.

Tsunoda, R. T. (1981). Time evolution and dynamics of equatorial backscatter plumes. 1. Growth phase. *J. Geophys. Res.* **86**, 139.

———. (1983). On the generation and growth of equatorial backscatter plumes. 2. Structuring of the west walls of upwellings. *J. Geophys. Res.* **88**, 4869.

———. (1985). Control of the seasonal and longitudinal occurrence of equatorial scintillations by the longitudinal gradient in integrated E-region Pedersen conductivity. *J. Geophys. Res.* **90**, 447.

Tsunoda, R. T., Livingston, R. C., McClure, J. P., and Hanson, W. B. (1982). Equatorial plasma bubbles: Vertically elongated wedges from the bottomside F layer. *J. Geophys. Res.* **87**, 9171.

Varney, R. H., Kelley, M. C., and Kudeki, E. (2009). Observations of electric fields associated with internal gravity waves. *J. Geophys. Res.* **114**, A02304, doi:10.1029/2008JA013733.

Whitehead, J. D. (1971). Ionization disturbances caused by gravity waves in the presence of an electrostatic field and background wind. *J. Geophys. Res.* **76**, 238.

Woodman, R. F., and Chau, J. L. (2002). First Jicamarca radar observations of two-stream E-region irregularities under daytime counter equatorial electrojet conditions. *J. Geophys. Res.* **107**(A12), 1482, doi:10.1029/2002JA009362.

Woodman, R. F., and LaHoz, C. (1976). Radar observations of F-region equatorial irregularities. *J. Geophys. Res.* **81**, 5447.

Woodman, R. F., Pingree, J. E., and Swartz, W. E. (1985). Spread-F-like irregularities observed by the Jicamarca radar during the daytime. *J. Atmos. Terr. Phys.* **47**, 867.

Zalesak, S. T., and Ossakow, S. L. (1980). Nonlinear equatorial spread F: Spatially large bubbles resulting from large horizontal scale initial perturbations. *J. Geophys. Res.* **85**, 2131.

Zalesak, S. T., Ossakow, S. L., and Chaturvedi, P. K. (1982). Nonlinear equatorial spread F: The effect of neutral winds and background Pedersen conductivity. *J. Geophys. Res.* **87**, 151.

Zargham, S., and Seyler, C. E. (1987). Collisional interchange instability. 1. Numerical simulations of the intermediate scale irregularities. *J. Geophys. Res.* **92**, 10,089.

5 Hydro- and Electrodynamics of the Midlatitude Ionosphere

The classic aeronomy of the midlatitude ionosphere is discussed in the books by Rishbeth and Garriott (1969) and Schunk and Nagy (2000). Our goal here is to treat what we feel are the most interesting hydrodynamical and electrodynamical processes that arise in midlatitude ionospheric physics. In both a geographic and dynamical sense, the midlatitude zone is a buffer between the low-latitude processes discussed in Chapters 3 and 4 and the high-latitude phenomena presented in later chapters. Both electric fields and perturbed neutral winds penetrate from high-latitude sources, while equatorial plasma streams into the region along the magnetic field lines. Atmospheric tides are quite important, since they grow in intensity with altitude, as do the gravity waves that continually roll in from high latitudes and/or up from stratospheric and tropospheric sources. These are just a few of the dynamical interactions that continue to make the study of midlatitude ionospheric physics challenging and interesting. We explore the dynamics and electrodynamics of the midlatitude ionosphere here and its plasma physics in Chapter 6.

5.1 Introduction to the Tropical and Midlatitude Ionospheres

5.1.1 Background Material

In this context we might define the tropical zone as that region where the magnetic field has a significant dip angle yet cannot be considered to be nearly vertical, as is the case in the auroral zone and polar cap. The latitudes of the Arecibo and, now closed, the St. Santin Radar Observatories are ideally suited for study of this zone (see Appendix A for position data on several sites discussed in the text). The dip angle at Arecibo, for example, is $46.7°$, which allows approximate geometric equality between forces parallel and perpendicular to **B**. Important information also comes from the Millstone Hill Observatory, which is equatorward of St. Santin in geographic coordinates but poleward of St. Santin geomagnetically. Millstone is thus in the transition zone between high and low

The Earth's Ionosphere: Plasma Physics and Electrodynamics

latitudes. We term this region midlatitude as opposed to tropical. Until recently most of our electrodynamical information has come from vector F-region plasma drift measurements made with the incoherent scatter method at these sites. As discussed in Chapter 3, drifts measured perpendicular to B in the F region can be interpreted unambiguously and yield the ambient electric field via the relationship $E = -V \times B$. The parallel drift component is much more complex, however, since gravity, neutral winds, and pressure gradients all contribute. The availability of neutral wind measurements using optical techniques at Arecibo, Fritz Peak, and other midlatitude sites has greatly helped interpretation of the incoherent scatter parallel drift data. Satellite data on neutral and plasma densities and temperatures have been very useful but the dynamics of the midlatitude region are difficult to study at orbital speeds.

The roles of the forces that act on the ionospheric plasma in the direction parallel to B is clear from (2.36a), which we reproduce here:

$$V'_{j\parallel} = \left[b_j E' - D_j \nabla n/n + (D_j/H_j)\hat{g} \right] \cdot \hat{B} \tag{5.1}$$

The primes indicate that the quantities are measured in the frame of reference in which the neutral wind velocity vanishes. Transforming to the earth-fixed frame where the neutral wind has a value U, the parallel component of the velocity V_j is given by

$$V_{j\parallel} = U \cdot \hat{B} + V'_{j\parallel} \tag{5.2}$$

The plasma is thus closely coupled to the neutral gas motion along B, but its velocity is modified by the term $V'_{j\parallel}$. For simplicity, in (5.1) we have continued to ignore temperature gradients, which must be included in a complete treatment of the total pressure gradient. For now, we ignore the neutral wind and external electric fields.

During the day, plasma production or loss by photoionization and recombination dominates the plasma profiles in the E and lower F regions, creating the horizontally stratified, slowly varying plasma content of the ionosphere. Rishbeth and Garriott (1969) and Schunk and Nagy (2000) describe the relevant physics and chemistry in great detail, and we refer the reader to their discussion. The ionization profile increases sharply in the 90–100 km range and then more slowly up to a peak, which is typically near 300 km.

Above the F peak we expect the plasma to be in a state something akin to diffusive equilibrium in the gravitational field. Because of its light mass, the electron gas diffuses much faster than the ions down a pressure gradient, a process that would tend rapidly to destroy any gradient in electron pressure. However, the resultant charge separation is accompanied by an electric field that restrains the electrons and enhances the ion diffusion. Quantitatively, we can argue as follows. In diffusive equilibrium there should be no net flow of the electron gas or of the ion gas. Our calculations are in the neutral gas frame (or, equivalently, we can assume $U = 0$), and we take the applied perpendicular electric field to be zero for now. Parallel to B the velocity for each species is given by (5.1). Taking $T_i = T_e = T$ for algebraic ease, $n_i = n_e = n$, and setting both velocities equal to zero gives for the Northern Hemisphere

$$b_e\left[E_{||} + \frac{k_B T}{ne}\nabla_{||}n - \frac{mg_{||}}{e}\right] = 0 \tag{5.3a}$$

$$b_i\left[E_{||} - \frac{k_B T}{ne}\nabla_{||}n + \frac{Mg_{||}}{e}\right] = 0 \tag{5.3b}$$

Remember that the mobility parameter b_j carries the sign of the charge and that e is a positive number. Dividing each equation by the appropriate mobility, taking $m \ll M$, and adding yields

$$E_{||} = (-M/2e)g_{||}$$

Likewise, setting the perpendicular components of velocity equal to zero in (2.36b) and using the large-κ case, which is suitable for the upper F region (recall that $\kappa_j = \Omega_j/\nu_j$), we have

$$0 = \frac{b_e}{\kappa_e}\left[E + \frac{k_B T}{ne}\nabla n - \frac{mg}{e}\right] \times \hat{B} \tag{5.4a}$$

$$0 = \frac{b_i}{\kappa_i}\left[E + \frac{k_B T}{ne}\nabla n + \frac{Mg}{e}\right] \times \hat{B} \tag{5.4b}$$

We can solve these two equations for the perpendicular electric field to find $E_\perp = -Mg_\perp/2e$, and thus combining both components,

$$E = -(M/2e)g$$

Finally, substituting this result into (5.4b) yields

$$\nabla n/n = (M/2k_B T)g$$

This shows that the equilibrium density gradient is vertically downward, even though the magnetic field is inclined at an arbitrary angle. This particular case is an example of a general theorem from statistical mechanics which states that in thermal equilibrium a magnetic field cannot affect the distribution of any fluid, ionized or not. The plasma scale height is equal to $2k_B T/Mg$ or more generally $k_B(T_i + T_e)/Mg$ if $T_e \neq T_i$. The plasma acts like a neutral gas in a gravitational field with mean mass equal to the average of m and M. The electric field is $Mg/2e$, and thus has the value $0.8\mu V/m$ in an O^+ plasma. This value is quite important when projected parallel to **B**, due to the high conductivity in that direction, which keeps other sources of $E_{||}$ small. The force on the ions associated with this electric field, plus that of gravity, is countered by the plasma pressure gradient. Perpendicular to **B**, this electric field is trivial.

Since the assumed dynamic equilibrium is between electrons and the dominant ion species, light minor ions such as He^+ and H^+ can be accelerated outward from the ionosphere by the parallel electric field. A considerable literature has developed concerning the effect of this $E_{||}$ on minor light ions and a complete theory of the topside ionosphere requires its consideration. We take up this topic in some detail in the polar case treated in Chapter 9, but for midlatitudes the reader is referred to Schunk and Nagy (2000).

Since hydrogen atoms and molecules can escape the earth's gravitational field, the earth is surrounded with a hydrogen gas "geocorona," which interacts with oxygen ions via the charge exchange reaction

$$O^+ + H \underset{\leftarrow}{\overset{\rightarrow}{}} H^+ + O$$

which is a very rapid process. The result is that a transition occurs between an oxygen and a hydrogen plasma between 500 and 1000 km, depending on seasonal and other effects (Vickrey et al., 1979; Gonzalez, 1994). Since light ions such as hydrogen can escape gravity, our simple assumption of diffusive equilibrium breaks down and a net upward flux of plasma is possible during the day at the "top" of the ionosphere. The closed dipole magnetic flux tubes at tropical and midlatitudes act as a reservoir for plasma, called the plasmasphere, which is created during the day by photoionization and carried upward by light ions. At night the plasma can flow back down, tending to maintain the density in the ionosphere. The result is a complex interaction between the ionosphere and a region of hydrogen plasma trapped by the dipole magnetic field.

To summarize, without winds and electric fields, photoionization coupled with diffusion, recombination, and charge exchange with the geocorona would completely determine the properties of the ionosphere. Rishbeth and Garriott (1969) discuss this situation in great detail, deriving the so-called Chapman layer form for the ionosphere and discussing the various ionospheric layers, E, F_1, F, and so on. A diurnal variation in peak plasma density would be expected, and some balance would arise between the low-altitude production and outflow of plasma along magnetic field lines during the day and low-altitude recombination and inflow of plasma at night. We explore simple models for these conditions following.

The situation is much more complicated than this, however, since neutral winds and electric fields also move the plasma. These forces greatly affect the altitude of ionospheric plasma, particularly in the F region. Since the density of molecular ions determines the speed with which O^+ recombination occurs and since these densities are exponentially dependent on altitude, dynamic processes also affect the plasma content. As an example of the variability of the midlatitude ionosphere, data showing the change in altitude of the F-region peak density

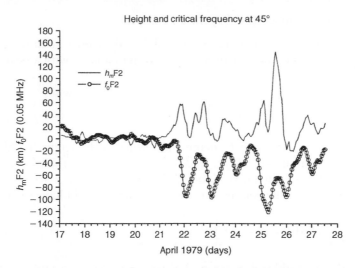

Figure 5.1 A plot of the height of the F peak and the value of the peak plasma frequency at middle latitudes relative to the long-term hourly average for April as a function of time for eleven days in April 1979. A chain of stations near the Japanese longitude was used. [After Forbes et al. (1988). Reproduced with permission of the American Geophysical Union.]

$(h_m\text{F2})$ relative to average hourly values are presented in Fig. 5.1 for an 11-day interval. The first four days were very quiet geomagnetically, but the next six days were very active. Also plotted is the maximum value of the plasma frequency in the F region of the ionosphere $(f_0\text{F2})$ relative to its hourly average. The electron density n is related to the plasma frequency f_p by $n = [f_p/8900]^2$ (see Appendix B). In the plot, both parameters vary drastically after April 21 in a manner called an ionospheric storm. The general trend is for a decrease in plasma density, an effect referred to as a negative ionospheric storm. A negative storm is thought to be due to a composition change and, more specifically, to a decrease in the ratio $[\text{O}]/[\text{N}_2]$, which affects both the production and the recombination rates of the plasma. But superimposed on the density decrease are variations of $f_0\text{F2}$, which correlate with the height variations—that is, a high ionosphere corresponds to high electron density. This correlation is due in part to reduced recombination at high altitudes but is complicated by an exchange of plasma with the plasmasphere.

The plasma content also affects the electrical conductivity of the ionosphere (and thus the electric field) as well as ion drag (and thus the neutral wind). Throwing in the conjugate hemisphere, which in general will have a different neutral wind and conductivity (but presumably the same electric field), one has a very complex coupled electrodynamic/dynamic system (without even mentioning external influences such as high-latitude electric fields and aurorally driven winds!). We consider some of these complexities later in this chapter.

5.1.2   On the Height of the Daytime F2 Layer

For an overhead ionization source with a highly absorbing neutral gas (E and F1 regions), the production function has the form of a layer with a peak at some $z = z_0$:

$$q(O^+) = q_0 \exp\left\{1 - z' - e^{-z'}\right\} \tag{5.5}$$

where $z' = (z - z_0)/H$, with H being the vertical scale height (Rishbeth and Garriott, 1969; Schunk and Nagy, 2000). Equation (5.5) is the Chapman function. In a steady state, the curves labeled E and F1 in Fig. 5.2 are of this form. The predicted electron density in a quasi-steady state would have the form

$$n(z') = \frac{q_0}{\alpha} \exp\left\{1 - z' - e^{-z'}\right\} \tag{5.6}$$

where α is the recombination for molecular ions in a molecular neutral gas, which is dependent on height only through a temperature dependence. For a combination of E and F1 production functions and without atomic ions, the ionosphere would have a single layer with a peak near 120 km.

The situation is quite different in the F2 or main F region. Neglecting transport, during the day the continuity equation for the main F region for O^+ production

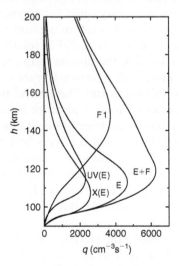

Figure 5.2   Electron production profiles $q(h)$ for the E and F1 regions for vertically incident radiation at sunspot number $R \cong 60$. The curves refer to the following wavelength bands: X(E), 8–140 Å; UV(E) 796–1027 Å; E = UV(E) + X(E); F, 140–796 Å; E + F, total 8–1027 Å, where 1 Å = 0.1 nm. [After Allen (1965). Reproduced with permission of Springer Netherlands.]

at the rate of $q\left(O^+\right)$ is as follows:

$$\frac{dn(O^+)}{dt} = q(O^+) - \gamma_1(N_2)n(N_2)n(O^+) - \gamma_2(O_2)\,n(O_2)\,n(O^+)$$
$$\cong q(O^+) - \beta n(O^+) \tag{5.7}$$

where the reaction rates have been combined to form $\beta = \gamma_1(N_2)n(N_2) + \gamma_2(O_2)\,n(O_2)$. Although the density of N_2 is about ten times that of O_2, the corresponding γ's have the opposite tendency so the two components of β are comparable. In a steady state,

$$n\left(O^+\right) = \frac{q(O^+)}{\beta} \tag{5.8}$$

However, (5.8) cannot be correct as is for the F2 layer since, as β goes to zero with altitude, the electron density would go to infinity. Balance is restored by diffusion, which rapidly increases with altitude. As a rule of thumb, diffusion dominates above the altitude where the chemical lifetime equals the diffusive time scale. The chemical time scale is

$$\tau_{ch} = \beta^{-1} \tag{5.9}$$

and the time needed to diffuse one scale height is

$$\tau_D = H^2/D_a \tag{5.10}$$

where D_a is the ambipolar diffusion coefficient. The F2 peak thus occurs where these two time scales are equal:

$$\tau_{ch} = \tau_D \tag{5.11}$$

For the earth's atmosphere, balance occurs around 300 km. Above this height, ignoring plasma storage and release from high altitudes, diffusive equilibrium occurs (see Section 5.1.1).

5.1.3 Equations Including Vertical Flux Without Winds or Electric Fields

It is instructive first to consider a vertical magnetic field since the equations are less complex. We finally return to a coordinate system with **B** in the $-\hat{a}_z$ direction for the rest of this text. For a stratified ionosphere with $\mathbf{B} = -B_0\hat{a}_z$ and $\mathbf{g} = -g\hat{a}_z$, the z component of the steady state ion and electron momentum equations are

$$0 = -\frac{\partial(n_ik_BT_i)}{\partial z} + n_ieE_z - n_iMg - n_iMv_{in}V_i'$$

$$0 = -\frac{\partial}{\partial z}(n_ek_BT_e) - n_eeE_z - n_emg - n_emv_{en}V_e'$$

where we now have allowed for finite vertical velocities. If there is no field-aligned current, $V_i' = V_e' \equiv W_D$, the vertical *plasma* velocity. Taking $n_i = n_e = n$, $m \ll M$ and adding these two equations yields

$$\frac{\partial}{\partial z} \left[nk_B \left(T_e + T_i \right) \right] = -nMg - nMv_{in} W_D$$

Solving for W_D yields,

$$W_D = -\frac{1}{nMv_{in}} \frac{\partial}{\partial z} \left[nk_B(T_e + T_i) \right] - \frac{g}{v_{in}} \tag{5.12a}$$

W_D is often confusingly called the diffusion velocity, even though it includes a gravitational term. The continuity equation is then

$$\frac{\partial n}{\partial t} + \nabla \cdot (nW_D) = q - l \tag{5.12b}$$

Before proceeding, some simplifications are useful. Let $T_e = T_i = T_n = T = $ constant and let the ion and neutral mass be identical. We also ignore the change of the cross section of a magnetic flux tube with height. Then the neutral scale height, H, is independent of height. Using $z' = (z - z_0/H)$, where z_0 is a reference height, we have

$$\frac{\partial}{\partial z} = \frac{1}{H} \frac{\partial}{\partial z'}$$

Using the ambipolar diffusion coefficient $D_a = \frac{2k_B T}{Mv_{in}}$, we obtain,

$$W_D = -D_a \left\{ \frac{1}{nH} \frac{\partial n}{\partial z'} + \frac{1}{2H} \right\} = -\frac{D_a}{H} \left\{ \frac{1}{n} \frac{\partial n}{\partial z'} + \frac{1}{2} \right\} \tag{5.13}$$

This follows, since $v_{in} \propto n_{neutral} \propto n_0 e^{-z'}$ and so $D_a = D_0 e^{+z'}$ and finally, $\frac{\partial D_a}{\partial z'} = D_a$. Evaluating the spatial derivative in (5.12b) we have

$$\frac{\partial}{\partial z'} (nW_D) = -\frac{D_a}{H^2} \left\{ \frac{\partial^2 n}{\partial z'^2} + \frac{1}{2} \frac{\partial n}{\partial z'} + \frac{\partial n}{\partial z'} + \frac{n}{2} \right\}$$

If we define a diffusion operator, $\mathcal{D} \equiv \frac{1}{H^2} \left[\frac{\partial^2}{\partial z'^2} + \frac{3}{2} \frac{\partial}{\partial z'} + \frac{1}{2} \right]$, then the continuity equation becomes,

$$\frac{\partial n}{\partial t} = -D_a \mathcal{D}n + q(z') - l(z') \tag{5.14}$$

In the next sections we explore some solutions to (5.14).

5.1.4 F-Layer Solutions with Production, Diffusion, and Flux

Consider first the steady-state behavior at high altitudes. Then $\partial/\partial t = 0$, both $l(z')$ and $q(z') = 0$, and (5.14) becomes

$$\mathcal{D}n = \frac{1}{H^2}\left[\frac{d^2 n}{dz'^2} + \frac{3}{2}\frac{dn}{dz'} + \frac{1}{2}n\right] = 0$$

It is straightforward to show that the solutions for $n(z')$ are of the form

$$n(z') = A_1 e^{-\frac{1}{2}z'} + A_2 e^{-z'} \tag{5.15}$$

The corresponding vertical velocity for each solution is found from 5.13. The A_1 solution yields

$$W_D = -\frac{D_a}{H}\left[-\frac{1}{2} + \frac{1}{2}\right] = 0$$

which corresponds to the hydrostatic equilibrium with a scale height of $2H$ as found in section 5.1.1. The A_2 solution has a vertical velocity given by

$$W_D = -\frac{D_a}{H}\left[-1 + \frac{1}{2}\right] = \frac{D_a}{2H} = \frac{D_0 e^{z'}}{2H}$$

Thus, for this solution there is a net vertical flux, which is given by

$$F = n(z')W_D = \frac{A_2 D_0}{2H}$$

which is independent of z' since $n(z') = A_2 e^{-z'}$. With no back pressure from the plasmasphere, W_D is always ≥ 0. Boundary conditions such as a plasmaspheric source and low altitude production determine the relative roles of the A_1 and A_2 solutions. Note that a nonvanishing A_2 term modifies the topside scale height.

Consider now a solution to the time independent equation (5.14) without the recombination term but including production of the form $q(z') = q_0 e^{-z'}$, which is the high altitude term in the Chapman function. The solution to this differential equation is

$$n(z') = n_0 \exp(-z'/2) + \frac{2q_0 H^2}{3D_0}\left[\exp(-z'/2) - \exp(-2z')\right]$$

$$\pm \frac{2FH}{D_0}\left[\exp(-z'/2) - \exp(-z')\right] \tag{5.16}$$

where n_0 is the electron density at the low boundary ($z' = 0$) and F is the plasma flux at the upper boundary ("+" and "−" correspond to inward and outward flux, respectively). At high altitudes the electron density distribution for $F = 0$

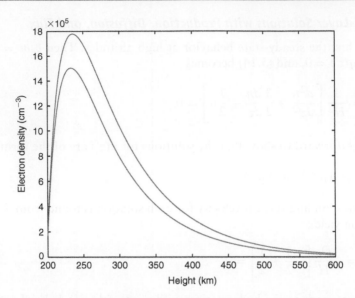

Figure 5.3a Electron density profiles calculated from (5.16) with a plasma flux of $F = 0$ (upper curve) and an upward plasma flux of $F = 3 \times 10^8 \text{cm}^{-2}\,\text{s}^{-1}$ (lower curve). As can be seen from Fig. 5.3a, upward flux results in a scale height that differs from that for diffusive equilibrium at high altitudes. (Figure courtesy of Michael Vlasov.) See Color Plate 8.

corresponds to diffusive equilibrium because $\exp(-z'/2) \gg \exp(-2z')$. Height profiles of the electron density calculated using (5.16) with and without flux are shown in Fig. 5.3a.

5.1.5 More General Nighttime Solutions

After sunset the molecular ions disappear almost immediately, leaving the F peak well defined at around 300 km. Now $q = 0$ but $\partial n/\partial t \neq 0$. We have

$$\frac{\partial n}{\partial t} = -\beta n + D_a \mathcal{D}n \tag{5.17a}$$

Unfortunately, the β term does not have the same altitude dependence as the other terms, since recombination depends upon the molecular neutral density whereas D_a is determined by the atomic oxygen ion mass. That is, $\beta(z') \propto e^{-1.75z'}$, which pertains for $M = \text{O}^+$ and $M_n = \text{N}_2$. However, if we take $\beta(z') = \beta_0 e^{-z'}$ we can find an exact solution that is very interesting (Martyn, 1956). Consider, then, this modified differential equation,

$$\frac{\partial n}{\partial t} = -\beta_0 e^{-z'} n + D_0 e^{z'} \mathcal{D}n \tag{5.17b}$$

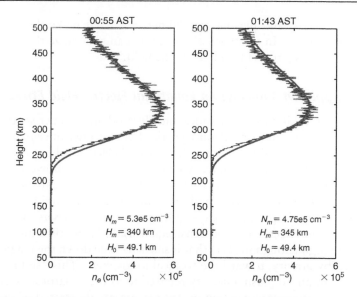

Figure 5.3b Two nighttime plasma density profiles along with a fit to an α-Chapman layer. H_0 is the neutral scale height. (Figure courtesy of Jonathan Makela.)

In the Problem Set, the reader is asked to show that a solution of the form

$$n(z', t) = N_m e^{-\lambda t} \exp\left[\frac{1}{2}\left(1 - z' - e^{-z'}\right)\right] \tag{5.17c}$$

exists if and only if $\lambda = D_0/4H^2 = \beta_0$ where β_0 is the recombination coefficient at $z' = 0$, the altitude where the Chapman function peaks. Note the similarity to (5.10). The profile form, because of the factor of 2, is called an α-Chapman layer, even though the recombination is due to the parameter β. Diffusion and recombination thus balance in such a way that the shape of $n(z')$ is constant while the decay of the whole layer is determined by β_0, the recombination rate at the F peak. Figure 5.3b shows typical layers measured over Arecibo at night, along with a fit to an α-Chapman layer, which gives credence to this simple model.

It can be shown that for this solution W_D is a constant with altitude and is equal to g/ν_{in} evaluated at the F peak. This remarkable result can be explained as follows. At high altitudes the plasma falls through the F peak and recombines below it. This is clearly a nonequilibrium case because the plasma density decreases with time, but the layer shape is independent of time. Without any other forces acting, except for gravity and pressure, the layer would stay at the same height but steadily decrease in density. This is not quite what happens in the real ionosphere, though, since neutral winds and electric fields exist and control the height of the F layer at night. These effects, along with the role of a finite dip angle, neutral waves, and plasma instabilities, are what make the midlatitude ionosphere interesting. In the tropical zone, one can generalize these results by projecting the forces parallel to the magnetic field.

5.1.6   The Appleton Anomaly: An Equatorial Electric Field Effect

Before discussing midlatitude dynamics and electrodynamics per se, there is an interesting and important tropical ionospheric effect arising from *equatorial* electrodynamics. As discussed in Chapter 3, the zonal electric field at the magnetic equator is eastward during the day, which creates a steady upward $E \times B/B^2$ plasma drift. Just after sunset this eastward electric field is enhanced and the F-region plasma can drift to very high attitudes where recombination is slow, while the low-altitude plasma decays quickly once the sun sets. The result is called the fountain effect, since the dense equatorial plasma rises until the pressure forces are high enough in (5.1) that it starts to slide down the magnetic field lines, assisted by gravity, toward the tropical ionosphere. This "sliding" results in a region of enhanced plasma density referred to as the equatorial or Appleton Anomaly. The various forces acting on the plasma are illustrated schematically in Fig. 5.4a.

If the zonal electric field is taken as a given quantity based, for example, on the experimental data shown in Chapter 3, then the actions of production and recombination can be combined with the vertical $E \times B$ motion of the ionosphere

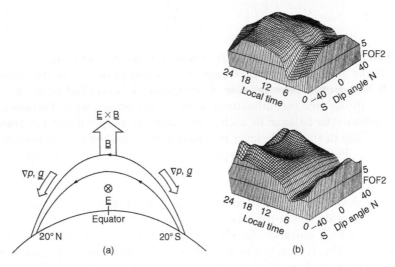

Figure 5.4   (a) Schematic diagram of how plasma uplift via electric fields transports plasma from equatorial to tropical zones. (b) Contour plots of FOF2 in megahertz (the peak F-region electron plasma frequency, which is proportional to the square of the electron density) for zero zonal electric field (upper plot) and for a typical diurnal variation of the zonal (lower right) electric field. (Figure courtesy of D. Anderson.)

to yield a reasonable model for the diurnal and latitude variation of the ionospheric plasma density. The effect has been illustrated in model calculations by
D. Andersen. In the upper-contour plot of Fig. 5.4b, the zonal electric field was
taken to be zero, while in the lower plot it had the typical variation with local
time measured at Jicamarca. A considerable difference in the tropical ionospheric
density is predicted in the two different electrodynamic states. In the latter case
the zonal electric field drove the plasma to heights where recombination was
small. The plasma thus survived longer and had time to flow down the magnetic
field lines to higher-latitude regions, forming two symmetrical enhancements on
each side of the equator through the combined action of electrodynamic uplift,
pressure gradients, and gravity. In either model, a low-density trough occurs
just before sunrise. The calculation, which includes a zonal electric field, fits
observations of the equatorial anomaly quite well.

A visual indication of the equatorial anomaly obtained from an image taken
from the *Dynamics Explorer-1* satellite located nearly 8000 km above the earth
is reproduced here in Fig. 5.5. The image corresponds to an emission line that is
produced by O^+ ions when they recombine with electrons. In the photograph,
the dayside of the earth is very bright, and a ringlike halo surrounds the polar
regions. The latter is the auroral oval. Of interest here are the parallel bands

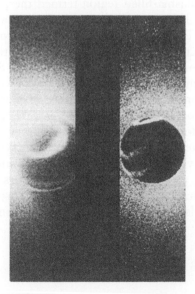

Figure 5.5 Two photographs of the earth's disk with images due primarily to neutral
oxygen emissions made from the imager on the *Dynamics Explorer-1* satellite. The ring
of airglow emissions seen at polar latitudes is due to the aurora, and the outer glow is
scattered light from the geocorona, escaping hydrogen from the earth's atmosphere. Of
interest here are the emission bands at midlatitudes. (Figure courtesy of J. D. Craven,
L. A. Frank, and R. L. Rairden.)

of light just off the magnetic equator, a remarkable visual representation of the equatorial anomaly region. In recent years, this emission line has been detected by instruments on the IMAGE and TIMED satellites.

5.1.7 The Corotation Electric Field and Formation of the Plasmasphere

An important dynamical effect involves the nontrivial fact that the plasma on low-latitude flux tubes to first order corotates with the earth. That this must occur can be proved as follows. If the ionospheric plasma did not corotate, in the nonrotating plasma frame there would be a very large neutral wind U_R and a current $J = \sigma \cdot (U_R \times B)$. Since σ depends on altitude, this current is not divergence-free, and an electric field will build up in the E region of the nonrotating frame until $J = \sigma \cdot (E_R + U_R \times B) = 0$. The current vanishes when the plasma $E \times B$ drifts at exactly the same velocity as the earth rotates. Transforming back to the rotating frame, we find $E = 0$. Now it is a remarkable fact that the electric field in the nonrotating frame is transmitted along the magnetic field and causes the entire inner magnetosphere to corotate with the earth, even though there is virtually no neutral gas at high altitudes on those flux tubes.

Turning this argument around, it is exactly on those flux tubes that corotate with the earth that cold ionospheric plasma can build up to the extent that is observed in the dense plasma-filled region termed the plasmasphere. A crude estimate of the latitude below which corotation dominates can be formed by equating the corotation electric field to the electric field of magnetospheric and solar wind origin. The comparison is properly done in the nonrotating frame. At latitudes where the former dominates, flux tubes and the plasma attached to them make a complete circuit of the earth in one day, allowing the trapping and buildup of a cold hydrogen plasma of ionospheric origin (taking into account the charge exchange process that converts O^+ into H^+, of course). At higher latitudes the flux tubes follow trajectories in which they are at times connected to the solar wind or extend to very great distances down the magnetic tail. In either case, the flux tube volume is so large that plasma almost continuously flows outward, never building up a high density such as occurs in the plasmasphere.

Quantitatively, we can proceed as follows. The corotational electric field at ionospheric heights is in the meridional direction and is given by Mozer (1973) as

$$E_c = 14 \cos\theta \left(1 + 3 \sin^2\theta\right)^{1/2} \text{ mV/m} \qquad (5.18)$$

where θ is the latitude and a centered dipole field has been assumed. The coefficient 14 in this expression is determined by the rotation speed of the earth, its radius, and the magnitude of the magnetic dipole moment. For a planet such as Jupiter, these parameters are all larger and the latitude region of corotation dominance is quite a bit higher than for the earth. Notice that for $\theta = 0$ and

$B_{eq} = 3.1 \times 10^{-5}$ T, E_c/B_{eq} equals the zonal rotation speed of the earth at the equator. To relate this field to the magnitude of the magnetospheric electric field, we treat the magnetic field lines as equipotentials. Following the discussion in Mozer (1970; see also Chapter 2), the meridional electric field component maps from the ionosphere (I) to the magnetosphere (M) (and vice versa) as

$$E_I/E_M = 2L\left(L - \frac{3}{4}\right)^{1/2} \tag{5.19}$$

where L is the equatorial crossing distance of a dipole field line measured in earth radii. In Fig. 5.6 the corotation field from (5.18) is plotted, along with dashed lines giving the ionospheric electric field associated with two different magnitude magnetospheric source fields of 1 and 0.4 mV/m, mapped to the ionosphere using (5.19). In comparing the magnitudes and effects of these two electric field sources (rotation versus solar wind), it is important to note that the relevant reference frame is the one fixed with respect to the sun. In that frame the solar wind blows by and interacts with the magnetosphere, creating the electric field of magnetospheric origin discussed above. The earth rotates in that reference frame, generating the corotation electric field we refer to. The plasma at the equatorial plane $\mathbf{E} \times \mathbf{B}$ then drifts at the whim of whichever source of electric field dominates.

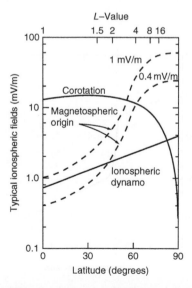

Figure 5.6 Typical ionospheric level electric fields observed in a nonrotating frame of reference that arise from corotation, the interaction of tidal neutral winds with the ionosphere, and the interaction of the solar wind with the terrestrial magnetic field. [After Mozer (1973). Reproduced with permission of the American Geophysical Union.]

The lower-value magnetospheric curve $(0.4\,\text{mV/m})$ crosses the corotation field at an L value near 7. The value corresponding to the higher magnetospheric field matches the corotation field at about $L = 3$. For this range of magnetospheric sources and the properties of the earth's dipole field and rotation rate, we see that planetary rotation dominates the physics at the equator within, say, 3–7 earth radii. The "extraterrestrial" source dominates at higher L values.

The region referred to as the plasmasphere corresponds to latitudes where the flux tubes corotate, since such tubes can fill with plasma on the dayside. Sunlight copiously creates an F-region atomic oxygen plasma, which in turn flows out along the field lines, exchanging charge with neutral hydrogen atoms. At night there is not sufficient time to deplete these flux tubes, and, on average, a dense, cold hydrogen ion plasma fills the upper plasmasphere. The average profile of plasma density in the magnetospheric equatorial plane reproduced in Fig. 5.7 shows how drastic the decrease in electron density can be near what is called the plasmapause (at $L = 4$ in this case). Since the magnetospheric electric field is quite variable, Fig. 5.6 shows that the position of the plasmapause and thus the plasmasphere's effect on the midlatitude ionosphere as a plasma reservoir are strong functions of the applied magnetospheric electric field (see Chapter 8). Magnetic activity and solar wind conditions thus affect the midlatitude ionosphere through electron dynamics as well as chemistry (because of neutral composition changes) and, as we shall see, thermospheric winds driven by high latitude heating and momentum transfer.

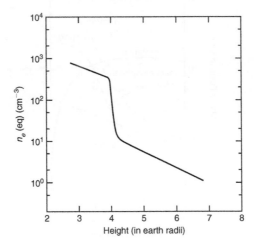

Figure 5.7 Average equatorial profile of electron density. The magnetic condition represented is one of steady, moderate agitation, with K_P in the range 2–4. [After Angerami and Carpenter (1966). Reproduced with permission of the American Geophysical Union.]

5.2 Electric Fields in the Tropical and Midlatitude Zone

We turn now to consideration of the electric field that exists in the earth-fixed reference frame, where, of course, the corotation field vanishes. In some sense this corresponds to the "electrodynamic weather" just as neutral atmospheric weather-related winds are measured in the rotating frame.

5.2.1 Electric Field Measurements

The lowest line in Fig. 5.6 is a plot of the average value of the Sq (solar quiet) dynamo electric field as a function of latitude. This electric field source has been discussed briefly in Chapter 3 and originates from the divergence in the electric currents driven by tidal motions of the neutral atmosphere in the highly conducting E layer (Matsushita, 1967, 1971). The solar heating source dominates the tides, although lunar gravitational effects are also important. Chapman and Lindzen (1970) have discussed tidal theory in great detail, and several applications of this theory involving electrodynamic calculations have been made (e.g., Matsushita, 1967; Richmond et al., 1976). The mathematics in these models is complicated but the physics is relatively straightforward, and we will not discuss these models extensively here. To first order they yield reasonable agreement with quiet-time ionospheric electric fields, particularly in the daytime low-latitude ionosphere. The results are not at all good at night and also deteriorate with increasing latitude. (A comparison of tidal theory with Jicamarca measurements, which show some of these features, was given in Chapter 3.) In this text we emphasize the processes that are less well understood and thus concentrate on the nighttime period at low (but not equatorial) latitudes and the influence of auroral zone effects at the interface between the high-latitude and midlatitude zones.

To organize our study, we use incoherent scatter data from the Northern Hemisphere sites Arecibo (18.5°; 31°), St. Santin (44.1°; 40°), and Millstone Hill (42.6°; 57°). The first number in the parentheses is the geographic latitude and the second the geomagnetic latitude. Other relevant details concerning these sites are included in Appendix A. In the discussion we distinguish between the tropical and the midlatitude ionosphere. The former is characterized by an intermediate dip angle and a sufficient enough distance from the auroral zone that most of the time the ionosphere is unaffected by high-latitude electrodynamics. The Arecibo data set is clearly tropical (31° geomagnetic latitude; 48° dip angle). St. Santin and Millstone Hill have similar midlatitude geographic positions, but the latter is much farther north geomagnetically (57° versus 40°). These two sites will thus be the prototypes for our study of the differences between tropical and midlatitudes and will be closely investigated for evidence of auroral effects.

Our starting point is the empirical study by Richmond et al. (1980), some results of which are presented here in Fig. 5.8a–d. Magnetically quiet periods

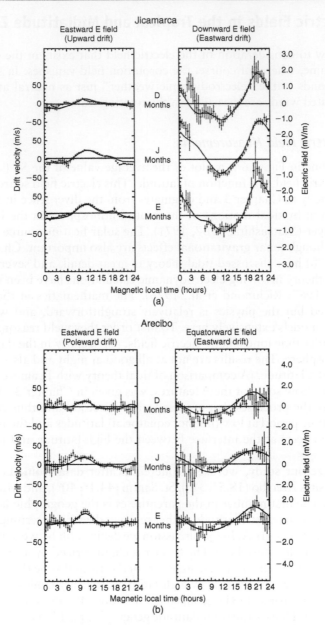

Figure 5.8 Seasonally averaged quiet-day drifts and electric fields (points with error bars) and model drifts (solid lines) perpendicular to the geomagnetic field at 300 km for (a) Jicamarca, (b) Arecibo.

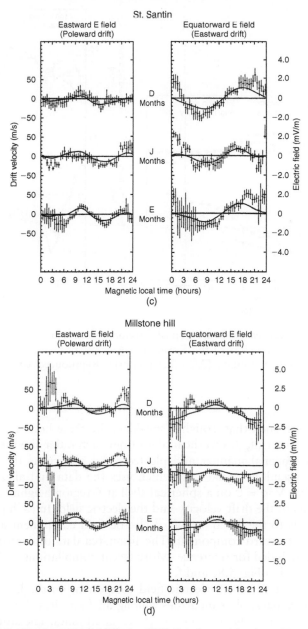

Figure 5.8 (c) St. Santin, and (d) Millstone Hill. D refers to Northern Hemisphere winter, J to Northern Hemisphere summer, and E to equinox. [After Richmond et al. (1980). Reproduced with permission of the American Geophysical Union.]

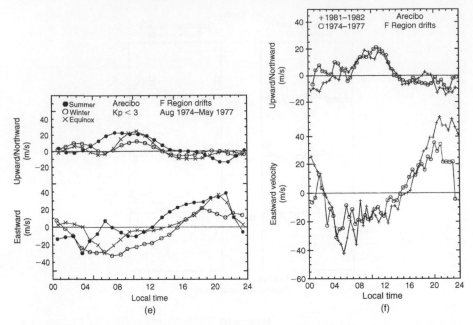

Figure 5.8 (cont.) (e) Seasonal variation of F-region average plasma drifts over Arecibo near solar minimum. (f) Plots of the F-region of Arecibo average plasma drifts during solar maximum and minimum. [After Fejer (1991). Reproduced with permission of Elsevier.]

have been chosen such that, roughly speaking, the three-hour K_P value is less than, or of the order of, 3. Overhead data from each radar site then have been averaged in a seasonal manner and plotted. Also plotted as the solid line is an empirical "pseudopotential" fit, which uses the data and also satisfies the requirement that $\nabla^2 \phi = 0$ on a spherical shell at F-region heights. The data are plotted in terms of the drift velocity and the electric field. In each panel the left side corresponds to the zonal eastward electric field component and the right side to the equatorward component. The Jicamarca data discussed in detail in Chapter 3 are included for reference. More recent compilations of Arecibo drift data are presented in Fig. 5.8e and f.

Considering first the zonal electric field component on the left side of each plot, there are remarkable similarities among all four sites during the period 0300–1400 and among the nonequatorial sites at all times. The latter three data sets seem to exhibit a particularly strong semidiurnal behavior. The Jicamarca zonal component seems to be primarily diurnal. This result is quite reasonable, since it is fairly well established that the semidiurnal atmospheric tide is not very important at equatorial latitudes but that it dominates in the geographic midlatitude region (Chapman and Lindzen, 1970). Thus, as just noted, these results are reasonably consistent with a tidal E-region source, in particular for the daytime zonal electric field component. (The equatorial prereversal enhancement does not show up very well in this early presentation.)

The meridional electric field component in the right-hand panels is generally larger and is dominated by a diurnal variation at all four sites. It seems curious that one component exhibits a semidiurnal modulation and the other a diurnal one! It may be that the two dynamos (E and F regions) conspire to yield a diurnal pattern for the meridional electric field (zonal drift) at all these latitudes. There is an important difference to note between data from the three lowest latitude sites (Jicamarca, Arecibo, and St. Santin) and the Millstone Hill data. In the 1600–2400 local time period even the algebraic sign of the Millstone Hill meridional component is different from the other three sites. That is, Millstone registers a poleward field, while all three other facilities register an equatorward field ("downward" at Jicamarca). We return to this point later when auroral zone effects are discussed.

The electric field vectors in the plane perpendicular to the magnetic fields deduced from barium cloud drift data shown in Fig. 5.9 bear out these last comments as well. Notice that in the evening twilight periods the releases all show equatorward electric fields and that the releases were all at latitudes less than 35° geomagnetically. These data agree with St. Santin and Arecibo quite nicely but disagree with Millstone Hill. Mozer (1973) and Gonzales et al. (1978) have previously pointed out this difference (also see Section 5.2.4).

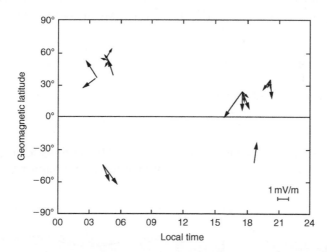

Figure 5.9 Examples of electric field vectors measured by several midlatitude barium releases. [After Mozer (1973). Reproduced with permission of the American Geophysical Union.]

5.2.2 Neutral Wind Effects

The thermospheric wind vector can be measured using the Doppler shift of the redline emission occurring at 630 nm, and the meridional wind component can be found by using the radar technique. Early observations were compiled by Burnside et al. (1980) for Arecibo. The most important feature for the nighttime ionosphere is that prior to midnight, the meridional wind is usually southward. Because of the finite dip angle, this means that the plasma is held up against gravity until about midnight. After this the "midnight collapse" occurs, the F layer falls, recombines, and sometimes nearly disappears (Nelson and Cogger, 1971).

In 1973, Perkins created a remarkably robust concept: the field line–integrated ionosphere. The huge advantage of this idea is that pressure gradients disappear from the equations. The entire ionosphere is a gigantic particle affected by winds, gravitational forces, and electric fields. Figure 5.10 shows how the equatorward wind (southward in the Northern Hemisphere) supports this "particle" against gravity. In the Northern Hemisphere, the steady state, the plasma velocity antiparallel to the magnetic field due to the action of a southward wind and gravity is

$$\mathcal{W}_{\parallel} = u_s \cos I - (g/v_{in}) \sin I$$

where I is the inclination angle, which is equal to the dip angle. If the first term dominates, the ionosphere will rise. But since v_{in} exponentially decreases with height, eventually the second term will cancel the first and the parallel velocity will vanish. In general, we are most interested in the vertical component of ionospheric motion. Projecting the parallel velocity to vertical, an equilibrium holds when $\mathcal{W}_{\parallel} = 0$—that is, when

$$(g/v_{in}) \sin^2 I = u_s \cos I \sin I \tag{5.20}$$

This equality holds if and only if

$$v_{in} = (g/u_s) \tan I \tag{5.21a}$$

This equation represents a subset of Perkins' (1973) solutions for the height of the ionosphere based on wind and gravitational forces but ignoring any electric

Figure 5.10 A ping-pong ball model for the midlatitude ionosphere. [After Kelley (2002). Reproduced with permission of Elsevier.]

field. But what should we choose? Perkins showed that the key value is actually the density-weighted, field line–integrated collision frequency

$$\langle \nu_{in} \rangle = \frac{\int n\nu_{in}ds}{\int n ds} \tag{5.21b}$$

For a given profile, $\langle \nu_{in} \rangle$ is just a number. But if the wind or electric field (see following) changes, the height of the F peak will vary to satisfy (5.21a) with ν_{in} replaced by $\langle \nu_{in} \rangle$. In an exponential neutral atmosphere, the value of $\langle \nu_{in} \rangle$ is a monotonically decreasing function of altitude and thus (5.21b) can be mapped into altitude. Remarkably, for a Chapman layer $\langle \nu_{in} \rangle$ is exactly equal to the value of ν_{in} at the peak of the Chapman profile.

Data illustrating these concepts are presented in Fig. 5.11, in which two and one-half days of electron density data are present. A modest magnetic storm was under way, and on the night of September 15–16, a negative ionospheric storm occurred with lower nighttime electron densities than usual. Such storms are thought to be due to enhanced recombination that, in turn, is due to neutral composition changes. Figure 3.30a shows how a pulse of wind from the north brings temperature changes with it. This causes the $[O]/[N_2]$ ratio to decrease, changing both the production and loss. On the next night the early-evening plasma density returned to normal levels. Both nights show a high and steady altitude for the F peak after sunset followed by the midnight collapse. The night of September 15–16 behaves classically. The southward wind died down, and

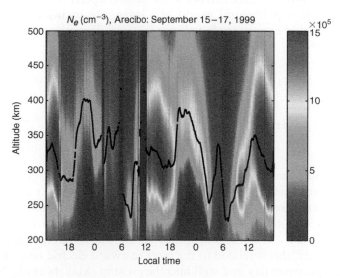

Figure 5.11 Two and a half days in the life of the ionosphere over Arecibo. The lines trace HmF2. [After Vlasov et al. (2003). Reproduced with permission of Elsevier.] See Color Plate 9.

the ionosphere simply fell and recombined. On the next night, even though the layer fell, the density remained high, even at 260 km, where recombination is very fast. On this night it seems that flux from the plasmasphere maintained the high F-region density and furthermore that electric field effects were important (Vlasov et al., 2003). We take up such effects next.

5.2.3 Combined Effects of Electric Fields and Neutral Winds

More generally, to find the equilibrium height at which the vertical velocity vanishes, we have

$$(g/\langle v_{in} \rangle) \sin^2 I = (E_e/B) \cos I + u_s \cos I \sin I$$

where E_e is the eastward component in the electric field. Solving for $\langle v_{in} \rangle$ yields

$$\langle v_{in} \rangle = \left(\frac{g}{E_e/B + u_s \cos I} \right) \sin I \tag{5.22}$$

The wind and electric field can be lumped into an effective southward wind, $u^s_{eff} = (E_e/B)(\cos I)^{-1} + u_s$. Then the height can be found from (5.21b) since $\langle v_{in} \rangle$ is a unique marker for altitude. Turning this around, if we know the height of the layer we can deduce the effective wind. Ionosondes can be used for this application. Equation (5.22) is the Perkins equilibrium condition. We investigate its stability or lack thereof in the next chapter. Note that (5.22) predicts the eventual (equilibrium) height of a layer, not its instantaneous value. A more general set of equations that allows for time–dependent calculations is as follows:

$$V_z(\tau) = -(g/\langle v_{in} \rangle (\tau)) \sin^2 I + u_s(\tau) \cos I \sin I + (E_e(\tau)/B) \cos I \tag{5.23a}$$

$$h(t) = \int_0^t V_z(\tau) d\tau + h(0) \tag{5.23b}$$

These dynamic equations can account for the motion of the layer on September 16–17 but what about the high plasma density? Why doesn't recombination destroy the layer below 300 km? As noted previously, the explanation (Vlasov et al., 2003) seems to be that a high influx of plasmaspheric ions (5×10^8 cm^{-2} s^{-1}) flooded the tropical ionosphere during this relatively small magnetic storm. Most likely, this is related to the large eastward electric field measured at Jicamarca the previous day and well into the evening (Makela et al., 2003), and a well-established fountain effect occurred. Indeed, a large southward gradient in plasma density was observed over Arecibo, indicating that it was located on the poleward edge of the anomaly, which is usually over a thousand kilometers south of the observatory. Note that a poleward displacement of the anomaly accompanies an enhanced fountain effect.

5.2.4 Complexities of the Real Nighttime Tropical Ionosphere

The detailed dynamics of the earth's ionosphere is governed by a complex interplay between motions of the neutral atmosphere, gravity, electromagnetic forces, pressure gradients, and plasma production and loss. In the nocturnal low- to midlatitude F layer, the latter two processes are relatively unimportant and can be ignored for our purposes once the molecular ions, which have a high recombination coefficient compared to O^+, are removed from the system. Since molecular ions are found at low altitudes, the highly conducting daytime E region virtually disappears, which allows the F-layer plasma greater control of the electrodynamics of the region. This can be seen in Fig. 5.12, where the ratios of F region to E-region height-integrated nighttime conductivities over Arecibo are plotted for several months (Burnside, 1984). Each plot corresponds to an average of 5 days. The ratio Σ_{PF}/Σ_{PE} considerably exceeds unity for all but a few data points. The 1981–1982 data (near solar maximum) show evening values as high as 20, whereas in the 1983 data, the largest values approach 10. Earlier, Harper and Walker (1977) found the ratio to be closer to unity at solar minimum. This result most likely reflects the lower F-layer plasma densities that occur at solar minimum.

The Arecibo Observatory is particularly well suited to the study of interrelationships between these forces, and much of our knowledge concerning the tropical off-equatorial ionosphere comes from Arecibo data. The near-45° dip angle allows each of the important forces an "equal vote" in the control of the F-region plasma dynamics. This is not entirely a geometric effect but stems also from the near equality of the following four characteristic "velocities" at altitudes near the F-layer peak:

$$E/B \; ; \; U \; ; \; g/\nu_{in} \; ; \; lV_i^{th}/L \tag{5.24}$$

where V_i^{th} is the ion thermal speed, l is the ion mean free path, and L is the plasma pressure gradient scale length. The last term is a measure of the term $\nabla p_i/\rho_i \nu_{in}$ in the ion equation of motion, and is the velocity at which a pressure gradient would drive the ion flow velocity against friction with the neutral gas (ignoring ambipolar effects). Representative values and their sum are plotted in Fig. 5.13 for a particular (observed) electron density profile over Arecibo. If any of these terms dominated the dynamics, the F-layer as we know it could not be limited to the relatively modest attitude excursions found experimentally. Turning this argument around, we have argued above that the F layer seeks out an attitude where a balance between these factors is reached. But does this balance occur?

Numerous measurements of vector plasma velocities have been conducted at Arecibo. The "natural" coordinate system for display of ionospheric F-region

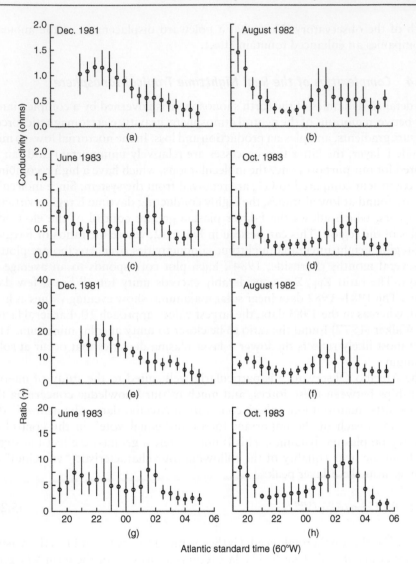

Figure 5.12 Five-day average values for different months of the height-integrated F-region Pedersen conductivity and the ratio between the F- and E-region–integrated Pedersen conductivities just past the peak in the solar cycle. (Reproduced with permission of R. Burnside.)

drift data is, of course, geomagnetic, so the Arecibo data have been displayed in this manner. The seemingly remarkable result first reported by Behnke and Harper (1973), an example of which is shown in Fig. 5.14, is the existence of a strong anticorrelation between the components of the drift velocity in the

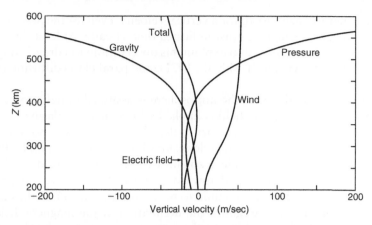

Figure 5.13 Magnitudes of the four contributions to the vertical plasma flow velocity over Arecibo for a typical plasma density profile, neutral wind, and zonal electric field component. (Data supplied by R. Burnside and R. Behnke.)

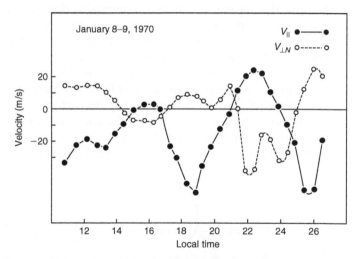

Figure 5.14 Example of the observed anticorrelation between components of plasma drift perpendicular to **B** (positive is geomagnetic northward) and parallel to **B**. (Here and in most Arecibo plots a positive $V_{||}$ is upward along $-\mathbf{B}$.) [After Behnke and Harper (1973). Reproduced with permission of the American Geophysical Union.]

geomagnetic meridian plane—that is, between the components antiparallel and perpendicular to **B**. The resulting vector plasma motion was therefore nearly horizontal, which shows that vertical balance does often occur. The parallel and perpendicular dynamics, which seem so nicely separated by the dominance of the magnetic field effect, are thus not at all decoupled. Just as with the neutral atmosphere, nature seems to abhor a vertical velocity, even for the plasma and an equilibrium height is sought, as just indicated theoretically. At least three explanations have been put forward to explain this observation: ion drag (Dougherty, 1961), the F-region dynamo (Rishbeth, 1971), and parallel ion diffusion (Stubbe and Chandra, 1970).

In the first of these theories, the atmosphere is assumed initially to be at rest and an external zonal electric field is applied which puts the plasma in motion via the $E \times B$ drift. (The reader is warned that this mechanism is not thought to be important at tropical or midlatitude regions but does work at high altitudes, as seen in Chapter 9.) For example, a zonally eastward electric field would cause the plasma to move northward and upward at a near 45° angle over Arecibo with velocity $-(E_e/B)\hat{a}_{y'}$, where $\hat{a}_{y'}$ is a unit vector perpendicular to B in the magnetic meridian (see Fig. 5.15). We ignore the declination of the magnetic field. The moving ions would, after a while, set the neutral gas in motion via the ion drag effect. Scaling arguments (Holton, 1979) show that the neutrals cannot have a very large vertical velocity so they would eventually attain a poleward horizontal velocity $U = v\hat{a}_y$ due to momentum transfer from the ions. Once this horizontal flow begins, it entrains the plasma flow with its component downward, parallel to B, which would oppose the upward velocity component. We estimate the acceleration time below. Note that we are using the standard meteorological notation with components (u, v, w) positive in the directions east, north, and up, and use primes for the geomagnetic coordinate system. The plasma carries a zonal current, $J_\perp = \sigma_P E_\perp$, where we can use a scalar Pedersen conductivity, since we are considering perpendicular motion in the F region where σ is diagonal. The $J \times B$ force due to this zonal current is, in fact, the origin of the poleward ion drag force on the neutrals. Once the neutrals begin to move with some velocity U the current is given by $J = \sigma_P(E + U \times B)$. In the final equilibrium force-free

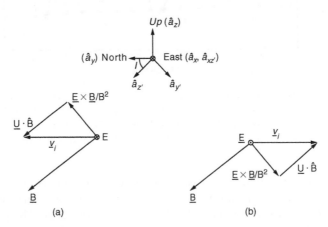

Figure 5.15 Schematic diagram of motor (a) and dynamo (b) electromagnetic phenomena in the tropical F-region ionosphere.

state, however, $\mathbf{J} \times \mathbf{B} = 0$ and no net electric current can flow. In this state, projection of the plasma and neutral flow perpendicular to $\mathbf{B}$ must be equal and, when added to the parallel motion, the two fluids move horizontally together. This is illustrated in Fig. 5.15a and can be seen mathematically as follows. Once the neutrals begin to move with velocity U, they produce field-aligned motion of the ions such that $V_{i\parallel} = U_{\parallel}$. The final vector plasma velocity in geomagnetic (primed) coordinates is given by

$$\mathbf{V}_i = -(E_e/B)\,\hat{a}_{y'} + v\cos I\,\hat{a}_{z'} \tag{5.25}$$

where $\hat{a}_{z'}$ is a unit vector parallel to $\mathbf{B}$. In the final state the neutrals will be driven geographically northward with the velocity $v = E(B\sin I)^{-1}$. Finally, substituting this result into (5.2) yields the plasma *geographic* velocity $V_y = E(B\sin I)^{-1}$, and $V_z = 0$. The plasma thus does not end up with a vertical velocity component at all, but moves horizontally at the same speed as the neutrals. The time constant for horizontal (meridional) acceleration of the neutrals by a zonal electric field can be estimated from the initial $\mathbf{J} \times \mathbf{B}$ force via

$$\partial v/\partial t = (\mathbf{J} \times \mathbf{B}) \cdot \hat{a}_y/n_n M = (\sigma_P E B/n_n M)\sin I$$

In the F-region, $\sigma_P = ne^2 v_{in}/M\Omega_i^2$, which is proportional to the neutral density due to the v_{in} term. Thus, the n_n terms cancel out, and the acceleration depends only on the plasma density. The acceleration time is thus of the order of

$$\delta t = \left(2 \times 10^{15}/v\right)(\delta v B/E)s$$

and for a density of $3 \times 10^{11}\ \mathrm{m}^{-3}$, a time of the order of 1 h is required to accelerate the neutral atmosphere to a velocity v comparable to E/B. Thus, although this mechanism does lead to a net horizontal motion for the plasma, a considerable time is required to establish the effect. In addition to the long time constant, this explanation ignores the problem of the origin of the electric field, which presumably requires an external magnetospheric source or a field applied from the other hemisphere that maps to the local ionosphere along $\mathbf{B}$. This process is thus not likely to explain the common observation of horizontal ion motion. Notice that $\mathbf{J} \cdot \mathbf{E} \geq 0$ initially, as required when electrical energy is converted to the mechanical energy in the neutral atmospheric flow. This process can therefore be termed a "motor."

A dynamo explanation is more promising, since the process self-consistently generates an electric field and operates on a much faster time scale, which is of the order of $(\Omega_i)^{-1}$. For example, if we start with zero electric field but a neutral meridional wind, $v\hat{a}_y$, in the equatorward direction (see Fig. 5.15b), a zonally eastward current given by $\mathbf{J} = \sigma_P \mathbf{U} \times \mathbf{B}$ will flow with magnitude $\sigma_P |v| B \sin I$. If boundary conditions are applied that force the net zonal current to be zero, a westward electric field must build up such that $E = vB \sin I \hat{a}_{x'}$. The net plasma velocity will again be due to the combined effect of the electric field and the neutral velocity. In this case the total plasma velocity due to the $E \times B$ drift and the component of U parallel to B, when projected into geographic coordinates, becomes $V_y = v \sin^2 I + v \cos^2 I = v$ and $V_z = v \sin I \cos I - v \cos I \sin I = 0$. Again, the plasma does not move vertically at all and matches the horizontal neutral speed. Here, $J \cdot E \leq 0$ initially, since electrical energy is created from a mechanical source as in a dynamo. Note that v is negative in Fig. 5.15b (southward winds).

Finally, sedimentation can also act to create horizontal ion motion. Since the downward velocity due to gravity is proportional to $(v_{in})^{-1}$, which increases rapidly with height, there is a natural limitation on the altitude to which either an equatorward wind (which pushes plasma up along B) or an eastward electric field (which $E \times B$ drifts the plasma upward and northward) can drive the plasma. Eventually the material falls along B as fast as it is pushed up and no net vertical motion occurs. This is the equilibrium discussed in the previous section. Note that for the equatorial neutral wind case (and if no F-region dynamo electric field is generated), the net horizontal plasma flow is actually zero, whereas for the applied eastward electric field case, the net horizontal plasma flow is poleward. Curiously, this gravitational effect on the plasma flow only acts for equatorward wind and/or eastward electric field. The opposite signs merely act in consort with gravity to drive the plasma deep into the atmosphere, where it is lost to recombination—the postmidnight collapse effect. In the next chapter we show, after Perkins (1973), that in the former case the equilibrium represented by (5.22) is stable, provided that no meridional electric field and no zonal wind exist.

We now turn to some experimental data aimed at trying to sort out these various processes. All these effects (and more!) seem to compete for control of the midlatitude F-layer plasma. An example in which the local dynamo effect seems to explain everything is illustrated in Fig. 5.16a and b. Two components of the plasma drift, the height of the F peak (h_{max}) and the maximum value of the electron density, are plotted in Fig. 5.16a. The standard coordinate system in which the Arecibo data are presented differs from ours. In Fig. 5.14 and Figs. 5.16–5.18, the Arecibo-measured $V_{||}$ is positive for drifts anti-parallel to $\hat{a}_{z'}$ ($V_{||}$ is positive in the $-B$ direction) and $V_{\perp N}$ is positive along the $\hat{a}_{y'}$ direction. In Fig. 5.16a, the two middle curves are anticorrelated until about 0340 LT, when sunrise occurred in the *conjugate* hemisphere, which is a nonlocal effect. The plasma motion is therefore nearly horizontal all night long. Indeed, h_{max}, only slowly changes from 2300 until 0300, which also shows that the vertical velocity was very small, averaging less than 1 m/s (downward) during this time. Fabry-Perot wind measurements are available for this night, and the measured northward neutral wind (v) and the northward horizontal component of the

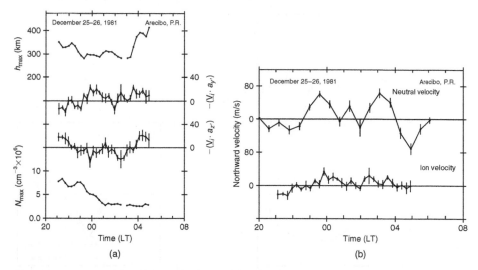

Figure 5.16 (a) Height of the peak electron density (h_{max}), meridional plasma drift velocity components ($V_{\perp N}$ and $V_{||}$), and peak in electron density (n_{max}) measured over Arecibo. [After Behnke et al. (1985). Reproduced with permission of the American Geophysical Union.] (b) Horizontal neutral wind and ion drift velocities (positive northward) for the same night as the data shown in (a). [After Behnke et al. (1985). Wind data supplied by R. Burnside. Reproduced with permission of the American Geophysical Union.]

ion velocity are compared in Fig. 5.16b. The agreement between the meridional neutral wind *direction* and that of the corresponding horizontal ion wind is clear. The ion velocity is much smaller, however, suggesting that there may be a partial shorting out of the electric field in the local E region or in the conjugate hemisphere.

Although consistent with an F-layer dynamo process, even this event raises some interesting questions. The relatively constant large ratio between local neutral wind and ion velocity implies that the electrical loading effect was larger than suggested in Fig. 5.12 for a local E-region load. Also the loading was relatively constant with time. This seems surprising considering the length of time involved. However, we note that the local height-integrated conductivity, which is proportional to nv_{in}, may have been relatively constant since the layer descended (larger v_{in}) at the same time as the peak density was decreasing (2100 to 0100 LT). The density and height of the layer were subsequently relatively constant until 0330. If the conjugate F layer had a smaller neutral wind and a relatively constant height-integrated conductivity that was slightly higher than the local F-region conductivity, then it would act as a load and the data could be explained.

Turning to the effects of an applied electric field, the magnetosphere supplied such an event on October 10–11, 1980, as illustrated in Fig. 5.17a. The

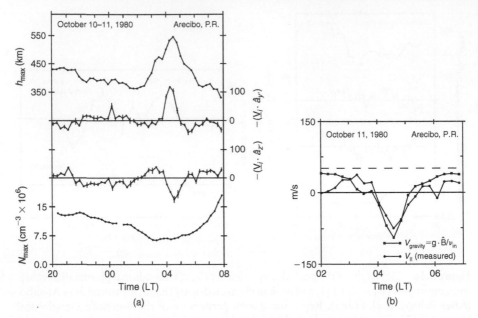

Figure 5.17 (a) Height of the peak electron density (h_{max}), meridional plasma drift velocity components ($V_{\perp N}$ and $V_{||}$), and peak in electron density (n_{max}) measured over Arecibo. (b) Calculated diffusion velocity and observed ion motion for the event illustrated in (a). [After Behnke et al. (1985). Reproduced with permission of the American Geophysical Union.]

large eastward perturbation in **E** seen at Arecibo was also detected at Jicamarca and was a clear magnetospheric effect (see the earlier discussion in Chapter 3). Unfortunately, no neutral wind data were taken on this night. The anticorrelation between $V_{\perp N}$ and $V_{||}$ and slow change of h_{max} were occurring as usual until about 0300, when the tropical and low-latitude ionospheric dynamo was interrupted by the auroral event. The F-layer height increased dramatically due to the large eastward electric field, which was only partially compensated by the anticorrelation effect. To test this, the large observed downward $V_{||}$ has been superimposed on a calculated value of $(g/\nu_{in}) \sin I$ in Fig. 5.17b. The two curves match very well, illustrating the tendency of gravity to counter extreme vertical electrodynamic forcing. The peak height of the F layer surged upward when the applied field existed (trying to reach equilibrium) and then began to fall when the penetration electric field decreased.

Although complex, these two days at least can be more or less explained in a straightforward manner. To avoid leaving the reader with the impression that this is always the case, we discuss an event recorded on June 15–16, 1980. Data from this night are presented in Fig. 5.18a and b in the same format as Fig. 5.16a and b. Comparison of Fig. 5.18a and b shows that the field-aligned

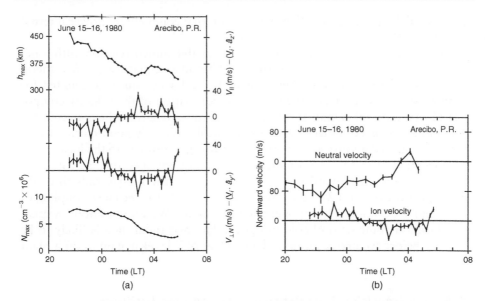

Figure 5.18 (a) Height of the peak electron density (h_{max}), meridional plasma drift velocity components ($V_{\perp N}$ and $V_{||}$), and peak in electron density (n_{max}) measured over Arecibo for June 15–16, 1980. (b) Horizontal neutral winds and ion drift velocities (positive northward) for the same night as the data shown in (a). [After Behnke et al. (1985). Reproduced with permission of the American Geophysical Union.]

ion motion, $V_{||}$, was directed *oppositely* to the parallel component of the local neutral wind until 0100. Although counterintuitive, this situation can occur if a strong downward gravity-driven velocity is simultaneously present due to the high altitude of the layer. In fact, Burnside et al. (1983) have calculated the gravity term for this data set and have found that indeed the downward flow is sufficient to overpower the oppositely directed neutral wind during this time. In this same period (from 2130 LT until 0100), the perpendicular ion flow shown in Fig. 5.18a was upward, which corresponds to an eastward electric field. But the neutral wind was equatorward at this time, which would tend to generate a westward electric field component by the dynamo process. This rules out a local F-region dynamo process as a source of the electric field. Magnetic activity was low, so there is no reason to expect a magnetospheric origin for the eastward zonal electric field. A more likely explanation is an electric field that mapped from the other hemisphere. In fact, if the neutral wind extended into the Southern Hemisphere with the same direction (toward the south pole) and generated an F-layer dynamo field at the conjugate point, the zonal component in the Northern Hemisphere would be eastward as observed. This postulated south-poleward wind in the Southern Hemisphere would also tend to keep the F layer low, which in turn would enhance the Southern Hemisphere Pedersen conductivity. If two ends of a magnetic field line with differing dynamo wind fields compete for the electric field, the region with the higher internal conductivity (lower internal resistance) will determine the electric field.

Comparison of $V_\parallel$ and $V_{\perp N}$ shows a remarkable anticorrelation with a net poleward horizontal ion flow prior to 0030 LT and an equatorward flow thereafter. Although the individual behavior of $V_\parallel$ and $V_{\perp N}$ can be accounted for by invoking nonlocal effects, the strong anticorrelation of the velocity components observed on this night cannot be explained using any of the conventional mechanisms discussed above. Certainly the local F-layer dynamo mechanism is not operating, as demonstrated by the ions moving oppositely to the local wind. The external electric field/gravity mechanism is also not operating, as witnessed by the lack of a net upward layer motion. (In fact, the F layer falls throughout the period 2100 to 0230 LT.) What is responsible for the close coupling of the $V_\parallel$ and $V_{\perp N}$ velocity components on this night? We do not know. A great deal of careful comparison between optical data and radar data is likely to be necessary before we arrive at an understanding of F-layer dynamics on a night like this.

5.2.5 The Transition Zone Between Mid- and High Latitudes

The magnetic dip angle increases very rapidly with increasing latitude. One effect is to decouple parallel and perpendicular dynamics. To some extent this simplifies the physics, since gravity and pressure gradients are most important in the vertical direction, while electric fields and neutral winds are most associated with horizontal motions of the plasma and neutral constituents. Atmospheric motions driven by solar heating still create E- and F-region dynamo electric fields, of course, but these sources are in competition with other processes. We have already pointed out the marked difference between the St. Santin/Arecibo nighttime observations and those at Millstone Hill, even during relatively quiet times. The strong implication is that some dominating high-latitude factor is present at $L = 3.2$, even during very modestly active times.

This result is further emphasized in Fig. 5.19, where hourly averages of the electric field/zonal drift obtained on days with $\Sigma K_p < 14$ are plotted for the evening period (filled circles) along with the "low" K_p Millstone Hill values published by Richmond et al. (1980), which were presented earlier in Fig. 5.8. The "superlow" K_p electric field is found to be equatorward in the evening period, as is the normal evening case for both St. Santin and Arecibo. It seems that even in what one might consider magnetically quiet times, electrodynamics in the Millstone Hill area are strongly affected by high-latitude process.

Two possible explanations were advanced by Gonzales et al. (1978). One is the direct penetration of the high-latitude electric field into the subauroral region. As we shall see in Chapter 8, this field has the proper sign, since the auroral zone electric field is poleward in the evening-to-midnight period, in agreement with the line plot in Fig. 5.19. This effect was studied at midlatitudes by Blanc (1983).

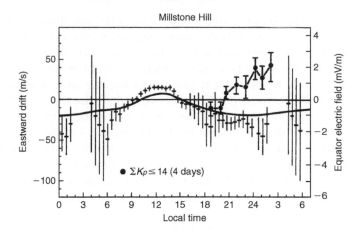

Figure 5.19 Average Millstone Hill equatorward electric field component measurements at equinox reproduced from Fig. 5.8 along with 1-hour averages of the same component during times of very low activity [Figure courtesy of C. Gonzales.]

However, the situation is not so simple since the high-latitude field changes sign near midnight, switching from poleward to equatorward, which is not observed at Millstone Hill. Also, for Kp around 2 it is unlikely that high latitude electric fields can penetrate to such low L values.

A second possibility is that the electric field is, after all, created by an F-region dynamo but that the local solar-driven winds are modified by high-latitude energy and momentum sources. The heat sources stem both from Joule heating and from particle precipitation into the atmosphere. The momentum source arises from the $\mathbf{J} \times \mathbf{B}$ forcing term where $\mathbf{J} = \boldsymbol{\sigma} \cdot \mathbf{E}$. In our previous discussion of this term, we showed that at low latitudes it yields the ion drag effect. In the neutral wind dynamo region $\mathbf{J} \cdot \mathbf{E} < 0$ and the electric field is generated by the winds. At high latitudes the electric field is imposed on the ionosphere and $\mathbf{J} \cdot \mathbf{E} > 0$. This shows that electrical energy is available for Joule heating as mentioned previously and that the $\mathbf{J} \times \mathbf{B}$ force is not so much a frictional drag on the neutrals as it is a mechanism for accelerating the thermospheric neutrals. At high latitudes the applied electric fields to first order form a two-celled plasma flow pattern which is virtually always present. It is not surprising then that the high-latitude neutral atmospheric motions are greatly affected by electrodynamic forcing. Suppose, as discussed in Chapter 9, that winds in the thermosphere are indeed driven across the polar cap by the plasma flow. The plasma turns to follow the auroral oval, but the wind has no such constraint. A flywheel-like effect may then occur with disturbance winds blowing out of the auroral oval even during relatively quiet times. Such winds would be reinforced by the equatorward pressure gradient due to Joule and particle heating in the auroral oval. Once equatorward of the oval, the Coriolis force will deflect the wind toward the west. This wind, coupled with a negative meridional conductivity gradient, would then create a disturbance dynamo electric variation of field in the poleward direction. Just such a disturbance dynamo electric field may be responsible for the Millstone Hill measurements at very low Kp values.

During very active times there is good evidence for such a wind pattern. Quiet-time neutral wind measurements at Fritz Peak, Colorado (39.9°N, 105.5° $W(L = 3)$, for six nights are gathered in Fig. 5.20a, along with the predictions of a thermospheric global circulation model (Hernandez and Roble, 1984). The data and model both show eastward winds in the evening sector. On the other hand, measurements made on an active day and shown in Fig. 5.20b display strong westward winds until 0200 LT and an equally strong equatorward wind from 2300 until 0400 LT. The unusual wind pattern was detected only north of the station.

5.3   Midlatitude Lower Thermosphere Dynamics

5.3.1   Tidal Effects

When the E-region conductivity is high, the electrodynamics are driven by tidal modes in the E-region. (These have already been discussed in Chapter 3.) As noted, the semidiurnal tidal mode becomes important at midlatitudes, and we expect that the daytime electric field will be dominated by semidiurnal tides at Millstone Hill and St. Santin. Both diurnal and semidiurnal tides should contribute at Arecibo. However, since the F-region dynamo and the high-latitude electric field sources are both primarily diurnal in form, we might expect the composite picture to be quite complex. This is borne out by the electric field data in Fig. 5.8. Arecibo, St. Santin, and Millstone Hill all display semidiurnal variations in the zonal electric field component and diurnal behavior in the meridional field.

Some of the richness of lower thermospheric dynamics may be visualized using motion of the layers that form in the ionosphere as natural tracers. Plasma density profiles over Arecibo for the period including sunset, sunrise, and the nighttime hours are presented in Figs. 5.21 and 5.22. The F layer slowly undulates with a several-hour period, whereas the lower-altitude layers exhibit both lower- and higher-period variations. E-region layers are very common over Arecibo, and the tidal modes can be visualized quite well. In the next chapter we discuss how the layers are formed, but for now we take their existence as an experimental fact and study their behavior. These lower-thermospheric oscillations are more obvious at night than during the day, since plasma production by sunlight tends to wash them out during the daytime. However, long-period oscillations such as diurnal and semidiurnal tides can be traced both day and night if some care is taken. Notice that the valley between the E and F layers has more plasma on the day with high Kp (Fig. 5.22) than on the previous night. This is due to energetic particle precipitation (Voss and Smith, 1979, 1980).

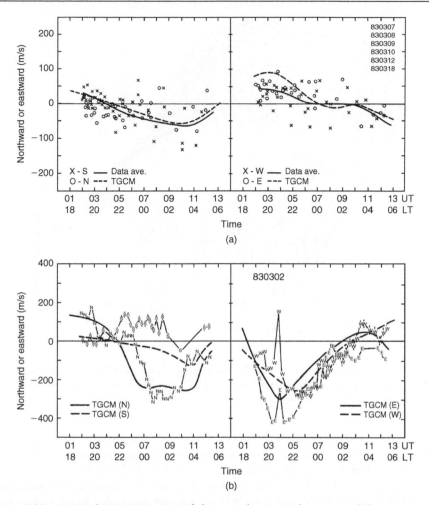

Figure 5.20 (a) Nighttime variation of thermospheric winds measured during six geomagnetic quiet days in early March 1983 with the year, month, and day given in the upper right corner. The meridional and zonal wind measurements, positive northward and eastward, are given at the left and right, respectively, with the solid line being an average of the data points and the dashed line representing TGCM predictions for geomagnetic quiet conditions. (b) Nighttime thermospheric winds measured on March 2, 1983: (left) meridional winds (positive northward) measured to the north (N) and south (S); (right) zonal winds (positive eastward) measured to the east (E) and west (W) of Fritz Peak Observatory. The solid and dashed curves represent TGCM predictions for constant-pressure surface near 300 km and for grid points north and south (left) and east and west (right) from Fritz Peak. [After Hernandez and Roble (1984). Reproduced with permission of the American Geophysical Union.]

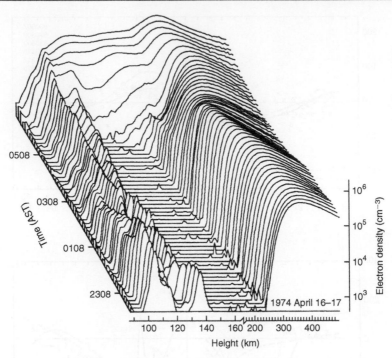

Figure 5.21 Electron density profiles for the night of April 16–17, 1974. This night was extremely quiet magnetically. [After Shen et al. (1976). Reproduced with permission of the American Geophysical Union.]

The altitude of the peak in the plasma density for a number of low-altitude layers has been determined in a study that includes daytime hours. The results are plotted in Fig. 5.23 for 78 consecutive hours. A well-defined motion is seen in the central part of the time period (January 4) in which a layer starts descending from 150 km at 0600 LT, reaching 90 km at about midnight. Parts of the same pattern can be discerned on the previous day as well as on the following day, which suggests that a solar diurnal mode (S1) exists. However a similar pattern occurs 12 hours before and 12 hours later so the tide could be semidiurnal (S2) or an in-phase combination of S1 and S2. Furthermore, on January 5 there is a prominent six-hour tide (S4). A study of tides near the ground using surface pressure and GPS data shows that S1 through S6 are all energized (Humphreys et al., 2005). Higher-frequency motions are also clearly indicated in the figure. When discussing atmospheric waves, as the frequency increases above the tidal range, it is no longer necessary to discuss the atmospheric winds in terms of tidal modes because the horizontal wavelengths become much smaller than the radius of the earth. In this case the appropriate modes are termed internal inertiogravity waves, a topic we take up in some detail in the next chapter. The high-frequency features of the data in Fig. 5.23 are thus geophysical in origin and are probably

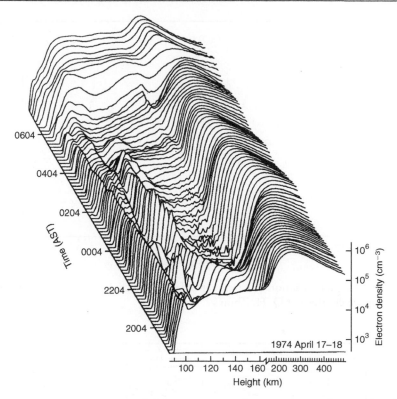

Figure 5.22 Electron density profiles for the night of April 17–18, 1974. This night was somewhat disturbed, and an intermediate layer can be seen. Note also the undulations of the F layer. The index on this night ranged from 3 to 6. [After Shen et al. (1976). Reproduced with permission of the American Geophysical Union.]

due to short-period gravity waves. The predawn noiselike behavior is most likely related to increased meteor activity near dawn.

A number of ionospheric parameters have recently been found to exhibit a four-wave longitudinal variation at a fixed local time when averaged over several days. These parameters include the intensity of Appleton Anomaly emissions (Immel et al., 2006), the vertical drift/zonal electric field at the magnetic equator (Kil et al., 2007), and the magnitude of the equatorial electrojet (England et al., 2006). The vertical drift case is shown in Fig. 5.24a, along with a study of the semidiurnal (S2) tide (Hagan and Forbes, 2003), which indicates both the migrating S2 tide and a mode 3 nonmigrating tide. The agreement is excellent and can be explained using Fig. 5.24b. The complex tide creates corresponding currents whose divergence leads to the electric field patterns. This longitudinal variability of the zonal equatorial electric field may explain the longitudinal dependence of CEIS/ESF.

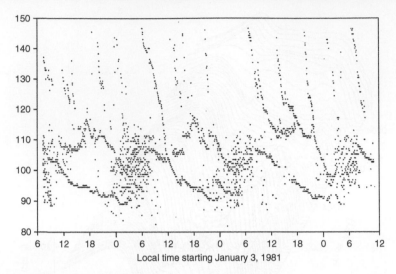

Figure 5.23 Altitude variation of a number of layers detected during a 78 h period over Arecibo. (Figure courtesy of Q. H. Zhou.)

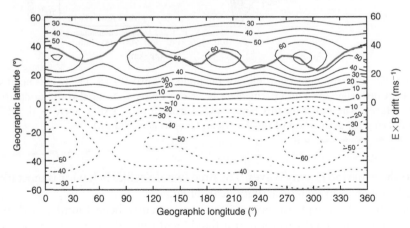

Figure 5.24a Contours of temperature perturbation for the diurnal tide on which the equatorial vertical drift is superposed. [After Kil et al. (2007). Reproduced with permission of the American Geophysical Union.] See Color Plate 10.

In summary, the electrodynamics of the midlatitude ionosphere are controlled by tidal and higher-frequency atmospheric wave modes in the E-region during the day. At night a very complex combination of the thermospheric wind dynamo, high-latitude electric field penetration, conjugate hemisphere effects, and gravity waves all play a role. These effects are modulated by a strong diurnal variation in the E-region conductivity which allows the F-layer dynamo to gain control at night.

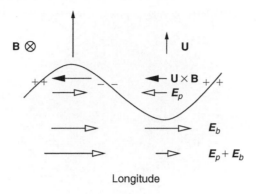

Figure 5.24b   Cartoon showing how the tidal winds create zonal electric fields.[After Kil et al. (2007). Reproduced with permission of the American Geophysical Union.]

5.3.2   Wind Profiles

Over the last few decades, nearly a hundred trimethyl aluminum (TMA) trails have been laid behind sounding rockets at all latitudes. Many of winds deduced from these trails are superimposed in Fig. 5.25a. The peak observed winds are much larger than predicted by tidal theory. On the other hand, they do have tidal properties. For example, the wind vector rotates with altitude as predicted for the tides. Another example is presented in Fig. 5.25b. Here a chemiluminescent meteor trail is shown 82 seconds after a Leonid meteor hit the atmosphere. A nearly perfect circle (corresponding to a helix-like structure) is seen on the image (Drummond et al., 2001), suggesting that a large amplitude inertiogravity wave is the source.

   Lasers can also be used to study winds and thus have become a powerful diagnostic. If tuned to the 589 nm sodium line, a resonance fluorescent interaction increases the cross section enormously, making the sodium layer visible. The straight line from the middle of Fig. 5.25b to the right-hand side was generated by a sodium lidar. Created by meteor ablation, the sodium layer density can exceed $10^4 \mathrm{cm}^{-3}$. One event in which TMA and lidar methods were used simultaneously is shown in Fig. 5.26a and b. Here, lidar sodium echoes were obtained on a night when several rockets were fired. We have used the wind profile to generate the neutral atmospheric Richardson number, R_i, for the event where

$$R_i = \frac{\omega_b^2}{\left(d|\mathbf{U}|/dz\right)^2} \tag{5.26}$$

and ω_b is the Brunt-Vaisala frequency. If $R_i < 1/4$ the medium is unstable to the Kelvin-Helmholtz instability of the atmosphere and the wind shears are strong enough to create unstable neutral layers. The KHI is most likely responsible for

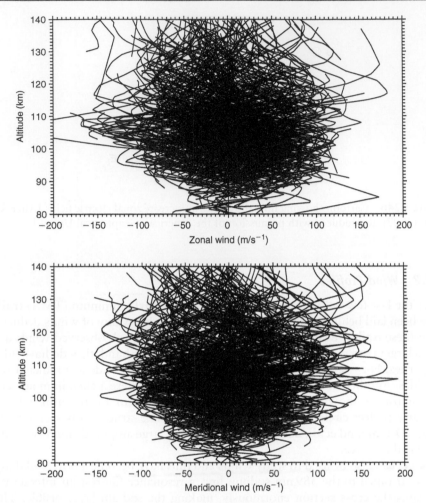

Figure 5.25a Superposition of numerous wind measurements using TMA trails. [After Larsen (2002). Reproduced with permission of the American Geophysical Union.]

the billows seen in the sodium near 23 LT. These neutral instabilities in turn can create structure in the collision-dominated ions (see Chapter 6). The downward streaks in the sodium are very related to the progression of the tidal ion layers operating in conjunction with sodium and sodium ion chemistry (Collins et al., 2002).

The double trail at low altitudes is a fascinating phenomenon that may be due to a vortex train left by rising hot ablation gases (Zinn and Drummond, 2005) or a combination of a buoyant gas trail and a falling dust trail (Kelley et al., 2003).

Figure 5.25b   Distortion of a long-lived meteor trail due to mesospheric winds. The size of the photo is about 1.5 km. Note that the trail exhibits both laminar and turbulent behavior. Since altitude increases clockwise in this spiral trail, we see that the upper portion of the trail (above 95 km) is turbulent and the lower part is laminar. [After Kelley et al. (2000). Reproduced with permission of the American Geophysical Union.]

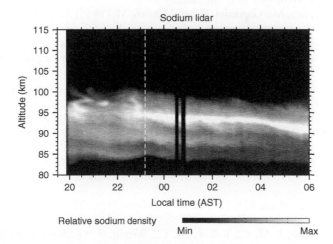

Figure 5.26a   Sodium density measured during the night in which two rockets were launched. (Figure courtesy of S. Collins.)

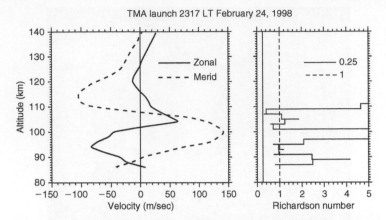

Figure 5.26b   On the left-hand side the zonal and meridional wind components are plotted versus altitude. On the right-hand side the Richardson number is plotted using a standard model for the temperature profile. (Figure courtesy of S. Collins.)

References

Allen, C. W. (1965). The interpretation of the XUV solar spectrum. *Space Sci. Rev.* **4**, 91–122.

Angerami, J. J., and Carpenter, D. L. (1966). Whistler studies of the plasmapause in the magnetosphere. 2. Electron density and total tube electron content near the knee in magnetospheric ionization. *J. Geophys. Res.* **71**, 711.

Behnke, R. A., and Harper, R. M. (1973). Vector measurements of F region ion transport at Arecibo. *J. Geophys. Res.* **78**, 8222.

Behnke, R. A., Kelley, M., Gonzales, G., and Larsen, M. (1985). Dynamics of the Arecibo ionosphere: A case study approach. *J. Geophys. Res.* **90**, 4448.

Blanc, M. (1983). Magnetospheric convection effects at mid-latitudes. 3. Theoretical derivation of the disturbance convection pattern in the plasmasphere. *J. Geophys. Res.* **88**, 235.

Burnside, R. G. (1984). *Dynamics of the Low-Latitude Thermosphere and Ionosphere.* Ph.D. thesis, University of Michigan, Ann Arbor.

Burnside, R. G., Behnke, R. A., and Walker, J. C. (1983). Meridional neutral winds in the thermosphere at Arecibo: Simultaneous incoherent scatter and airglow observations. *J. Geophys. Res.* **88**, 3181.

Burnside, R. G., Meriwether, J. W., Jr., and Walker, J. C. G. (1980). Airglow observations of the OI 7774 Å multiplet at Arecibo during a magnetic storm. *J. Geophys. Res.* **85**, 767.

Chapman, S., and Lindzen, R. S. (1970). *Atmospheric Tides: Thermal and Gravitational.* Gordon & Breach, New York.

Collins, S. C., Plane, J. M. C., Kelley, M. C., Wright, T. G., Soldán, P., Kane, T. J., Gerrard, A. J., Grime, B. W., Rollason, R. J., Friedman, J. S., González, S. A., Zhou, Q., Sulzer, M. P., and Tepley, C. A. (2002). A study of the role of ion-molecule chemistry in the formation of sporadic sodium layers. *J. Atmos. Solar-Terr. Phys.* **64**, 845.

Dougherty, J. P. (1961). On the influence of horizontal motion of the neutral air on the diffusion equation of the F-region. *J. Atmos. Terr. Phys.* **20**, 167.

Drummond, J. D., Grime, B. W., Gardner, C. S., Liu, A. Z., Chu, X., and Kane, T. J. (2001). Observations of persistent Leonid meteor trails, 1, Advection of the "Diamond Ring." *J. Geophys. Res.* **106**, 21,517.

England, S. L., Maus, S., Immel, T. J., and Mende, S. B. (2006). Longitudinal variation of the E-region electric fields caused by atmoshpheric tides. *Geophys. Res. Lett.* **33**, L21105, doi:10.1029/2006GL027465.

Fejer, B. G. (1991). Low latitude electrodynamic plasma drifts: A review. *J. Atmos. Terr. Phys.* **53**, 677.

Forbes, J. M., Codrescu, M., and Hall, T. J. (1988). On the utilization of ionosonde data to analyze the latitudinal penetration of ionospheric storm effects. *Geophys. Res. Lett.* **15**, 249.

Gonzales, C. A., Kelley, M. C., Carpenter, L. A., and Holzworth, R. H. (1978). Evidence for a magnetospheric effect on mid-latitude electric fields. *J. Geophys. Res.* **83**, 4397.

Gonzalez, S. (1994). *Radar, Satellite and Modeling Studies of the Low Latitude Protonosphere*. Ph.D. thesis, Utah State University.

Hagan, M. E., and Forbes, J. M. (2003). Migrating and nonmigrating semidiurnal tides in the upper atmosphere excited by tropospheric latent heat release. *J. Geophys. Res.*, **108**(A2), 1062, doi:10.1029/2002JA009466.

Harper, R. M., and Walker, J. C. G. (1977). Comparison of electrical conductivities in the E and F regions of the nocturnal ionosphere. *Planet. Space Sci.* **25**, 197.

Hernandez, G., and Roble, R. G. (1984). Nighttime variation of thermospheric winds and temperatures over Fritz Peak Observatory during the geomagnetic storm of March 2, 1983. *J. Geophys. Res.* **89**, 9049.

Holton, J. R. (1979). *An Introduction to Dynamic Meteorology*. 2nd ed., Academic Press, New York.

Humphreys, T. E., Kelley, M. C., Huber, N., and Kintner, Jr., P. M. (2005). The semidiurnal variation in GPS-derived zenith neutral delay. *Geophys. Res. Lett.* **32**, L24801, doi:10.1029/2005GL024207.

Immel, T. J., Sagawa, E., England, S. L., Henderson, S. B., Hagan, M. E., Mende, S. B., Frey, H. U., Swenson, C. M., and Paxton, L. J. (2006). Control of equatorial ionospheric morphology by atmospheric tides. *Geophys. Res. Lett.* **33**, L15108, doi:10.1029/2006GL026161.

Kelley, M. C. (2002). Ionosphere. *Encyclopedia of Atmospheric Science*. James R. Holton, John A. Pyle, and Judith A. Curry, eds. Academic Press: Elsevier Science, London, 1022.

Kelley, M. C., Gardner, C., Drummond, J., Armstrong, W. T., Liu, A., Chu, X., Papen, G., Kruschwitz, C., Loughmiller, P., and Engleman, J. (2000). First observations of long-lived meteor trains with resonance lidar and other optical instruments. *Geophys. Res. Lett.* **27**, 1811.

Kelley, M. C., Kruschwitz, C., Drummond, J., Gardner, C., Gelinas, L., Hecht, J., Murad, E., and Collins, S. (2003). A new explanation of persistent double meteor trains. *Geophys. Res. Lett.* **30**, 2202, doi:10.1029/2003GL018312.

Kil, H, S.-J. Oh, M. C. Kelley, L. J. Paxton, S. L. England, E. Talaat, K.-W. Min, and S.-Y. Su (2007). Longitudinal structure of the vertical $E \times B$ drift and ion density seen from ROCSAT-1. *Geophys. Res. Lett.* **34**, L14110, doi:10.1029/2007GL030018.

Larsen, M. F. (2002). Winds and shears in the mesosphere and lower thermosphere: Results from four decades of chemical release wind measurements. *J. Geophys. Res.* 107, 1215, doi:10.1029/2001JA000218.

Makela, J. J., Kelley, M. C., González, S. A., Aponte, N., and Sojka, J. J. (2003). Mid-latitude plasma and electric field measurements during space weather month, September 1999. *J. Atmos. Solar-Terr. Phys.* 65, 1077, doi:10.1016/j.jastp.2003.07.002.

Martyn, D. F. (1956). The physics of the ionosphere. *Aust. J. Phys.* 9, 161.

Matsushita, S. (1967). Solar quiet and lunar daily variation fields. *Physics of Geomagnetic Phenomena* (S. Matsushita and W. H. Campbell, eds.), 1, 301. Academic Press, New York.

———. (1971). Interactions between the ionosphere and the magnetosphere for Sq and L variations. *Radio Sci.* 6, 279.

Mozer, F. S. (1970). Electric field mapping from the ionosphere to the equatorial plane. *Planet. Space Sci.* 18, 259.

———. (1973). Electric fields and plasma convection in the plasmasphere. *Rev. Geophys. Space Phys.* 11, 755.

Nelson, G. J., and Cogger, L. L. (1971). Dynamical behavior of the nighttime ionosphere at Arecibo. *J. Atmos. Terr. Phys.* 33, 1711.

Perkins, F. W. (1973). Spread F and ionospheric currents. *J. Geophys. Res.* 78, 218.

Richmond, A. D., Matsushita, S., and Tarpley, J. D. (1976). On the production mechanism of electric currents and fields in the ionosphere. *J. Geophys. Res.* 81, 547.

Richmond, A. D., Blanc, M., Emery, B. A., Wand, R. H., Fejer, B. G., Woodman, R. F., Ganguly, S., Amayenc, P., Behnke, R. A., Calderon, C., and Evans, J. V. (1980). An empirical model of quiet-day ionospheric electric fields at middle and low latitudes. *J. Geophys. Res.* 85, 4658.

Rishbeth, H. (1971). The F-layer dynamo. *Planet. Space Sci.* 19, 263.

Rishbeth, H., and Garriott, O. K. (1969). *Introduction to Ionospheric Physics*. Int. Geophys. Ser. Vol. 14. Academic Press, New York.

Schunk, R. W., and Nagy, A. F. (2000). *Ionospheres: Physics, Plasma Physics, and Chemistry*. Atmospheric and Space Science Series, Cambridge Univ. Press, Cambridge.

Shen, J. S., Swartz, W. E., Farley, D. T., and Harper, R. M. (1976). Ionization layers in the nighttime E region valley above Arecibo. *J. Geophys. Res.* 81, 5517.

Stubbe, P., and Chandra, S. (1970). The effect of electric fields on the F-region behavior as compared with neutral wind effects. *J. Atmos. Terr. Phys.* 32, 1909.

Vickrey, J. F., Swartz, W. E., and Farley, D. T. (1979). Postsunset observations of ionospheric-protonospheric coupling at Arecibo. *J. Geophys. Res.* 84, 1310.

Vlasov, M. N., Kelley, M. C., Makela, J. J., and Nicolls, M. J. (2003). Intense nighttime flux from the plasmasphere during a modest magnetic storm. *J. Atmos. Solar-Terr. Phys.* 65, 1099.

Voss, H. D., and Smith, L. G. (1979). Nighttime ionization by energetic particles at Wallops Island in the altitude region 120 to 200 km. *Geophys. Res. Lett.* 6, 93.

———. (1980). Rocket observations of energetic ions in the nighttime equatorial precipitation zone. In *Low Latitude Aeronomic Processes* (A. P. Mitra, ed.). Pergamon Press, Oxford, 131.

Zinn, J., and Drummond, J. (2005). Observations of persistent Leonid meteor trails: 4. Buoyant rise/vortex formation as mechanism for creation of parallel meteor train pairs. *J. Geophys. Res.* 110, A04306, doi:10.1029/2004JA010575.

6 Waves and Instabilities at Midlatitudes

In this chapter we discuss analogues to the plasma processes presented in Chapter 4. At midlatitudes the length of magnetic field lines containing high-density plasma is much shorter, gravity acts parallel to the magnetic field, and wind effects are at least as important as electric fields. We finally discuss internal gravity waves in some detail and explore their importance as well as that of energetic particle precipitation. Familiar plasma instabilities are extended to include finite dip angle and wind effects and some new processes, including the Perkins (F-layer) instability, the sporadic E-layer instability, neutral Kelvin-Helmholtz instabilities, and thermal effects, are addressed.

6.1 Mesoscale Vertical Organization of Ionospheric Plasma: General Considerations

Dynamical considerations, of course, do affect the dayside midlatitude ionospheric content and altitude. However, production and recombination are sufficiently rapid to mask much of the structure in the medium other than large-scale vertical and horizontal gradients. At night, however, dynamics rules, and many interesting dynamic and electrodynamic processes contribute to the formation of mesoscale structures (50–1000 km horizontally and 0.5–50 km vertically).

Two graphic examples of consecutive plasma density profiles detected in the nighttime Arecibo ionosphere were presented in Figs. 5.21 and 5.22. The April 16–17, 1974, night shown in Fig. 5.21 was very quiet magnetically, while the other night was moderately disturbed. The solar influence can be clearly seen at sunrise, when the deeply depressed plasma in the F-layer "valley" between 160 and 240 km altitude fills in and causes even the E-region structure to merge into the fairly featureless daytime ionospheric profiles at the top of each figure. The magnetically active night had a much higher valley plasma density. Some features are common to these and most other nights over Arecibo. The high-density F layer displays undulations with a typical period of 2 h in which the layer rose and fell by many tens of kilometers. In the E region between 90 and

The Earth's Ionosphere: Plasma Physics and Electrodynamics

120 km, very intense layers developed on each night and lasted from sunset to sunrise.

Whenever there are electrons to scatter from, the radar can detect motion and organization of the plasma. For example, often a piece of the F region seems to "peel off" the bottomside of the layer at sunset and to propagate downward into the F-layer valley region. Such a structure has been termed an intermediate layer by Shen et al. (1976), since it occurs between the F layer and the more classical sporadic E layers below. Since ionosonde signals are often reflected by the intense lower-altitude layers, the intermediate structures are not visible with an ionosonde and can be studied only via rockets or incoherent scatter radars such as the one at the Arecibo Observatory. The sporadic nature of the various lower layers, which gives them the name sporadic E, is evident in these profiles. The strongest intermediate layer lasted almost all night on April 17–18, while it died out at 0230 LT on April 5–6. Other more sporadic and weaker intermediate layers came and went on April 17–18.

In the next two sections we discuss the formation and dynamics of these layers, which are primarily due to oscillatory behavior of the neutral atmosphere. Such motions are classified as either tides or gravity waves depending on their frequency. Although, as we shall see, these neutral atmospheric motions are capable of organizing the plasma into layered structures, this is not the entire story. The intermediate-layer ionization in particular cannot entirely be explained by photoionization processes, and we briefly discuss the observation and effect of ionizing energetic particle fluxes in this regard. F-region midlatitude plasma instabilities can amplify (or even create?) mesoscale structures as well.

Finally, once the layers are formed, they may be subject to primary or secondary plasma instability processes similar to those discussed for the equatorial zone, which lead to small scale structures, and we end the chapter with a brief discussion of such processes.

6.2 Oscillations of the Neutral Atmosphere

From the discussion in Section 5.1, it is clear that when $\kappa_i = \Omega_i/\nu_i \gg 1$, it is difficult to move plasma across magnetic field lines with a neutral wind. But if the neutral atmosphere has a velocity component parallel to the magnetic field, the plasma will be carried along $\mathbf{B}$ with the same velocity as the parallel component of the neutral atmosphere. Any horizontal neutral velocity with a component in the magnetic meridian will therefore create an F-region plasma velocity projected in the direction of the magnetic field. We previously discussed the fact that mean meridional neutral wind patterns cause F-region plasma motion upward or downward along $\mathbf{B}$. Here we expand this discussion to include tides and gravity waves. Traditionally, *upper* thermospheric dynamics are discussed in terms of wind patterns rather than tidal modes, even though they display clear diurnal variations. On the other hand, long-period *lower* thermospheric dynamics are discussed in terms of propagating diurnal and semidiurnal atmospheric tides. This distinction is partly historical but is also related to the origin of the atmospheric forcing. The upper thermospheric winds are driven in situ by solar heating, Joule heating, and momentum transfer with the plasma, whereas the lower E-region winds are usually ascribed to upward-propagating tides generated at tropospheric and/or stratospheric heights. The oscillation periods of the semidiurnal and diurnal propagating tides are, of course, 12 and 24 h, respectively. Higher-order tides also occur, but as the oscillation period nears several hours, the motions are usually referred to as gravity waves. Upper thermospheric forcing via solar UV heating has a strong diurnal component and drives an in situ diurnal tide.

The upper atmosphere is continuously bombarded with gravity waves from a number of sources. These include tropospheric weather fronts, tornadoes and thunderstorms (Kelley, 1997; Hung et al., 1978), impulsive auroral zone momentum injection and heating events (Richmond and Matsushita, 1975; Nicolls et al., 2004), and even earthquakes (Kelley et al., 1985). The famous monograph entitled *The Upper Atmosphere in Motion* by Hines (1974) is an excellent annotated collection of gravity wave studies published by Hines and coworkers over about a 10-year period. The reader is referred to that work for details about gravity waves and tidal oscillations, as well as to the excellent review of tidal theory by Chapman and Lindzen (1970) mentioned earlier. Here our approach is much more modest in scope, aiming at physical intuition rather than detailed analysis.

We study gravity waves first, in effect finding the normal modes of a flat nonrotating inviscid atmosphere. These results will be valid as long as the periods do not approach the tidal range and the wavelengths are not long enough that the curvature of the earth matters. We assume an isothermal, inviscid atmosphere initially in hydrostatic equilibrium, so that if ρ_0 and p_0 are the zero-order mass density and pressure, the relation

$$\rho_0 \mathbf{g} = -\nabla p_0$$

applies. In addition, it can be shown that ρ_0 and p_0, which vary only in the vertical direction in this model, are of the form

$$\rho_0, p_0 \propto e^{-z/H}$$

where H is the scale height of the atmosphere—that is, $1/H = -(1/\rho_0)(d\rho_0/dz)$. Here we again choose our coordinates using the meteorological convention and take x eastward, y northward, and z vertically upward. We assume there are no neutral winds in the unperturbed atmosphere. The equations governing the behavior of the atmosphere are the mass continuity equation (2.2), the equation of motion (2.20), and the adiabatic condition (see Yeh and Liu, 1974). In the equation of motion, only terms due to gravity, pressure gradients, and inertia are retained.

Now consider atmospheric oscillations in the presence of gravity. We assume there are small perturbations in the mass density, pressure, and wind velocity denoted by $\delta\rho$, δp, and $\mathbf{U} = (u, v, w)$. Without the Coriolis or viscous forces there is no coupling between oscillations in the y-z plane and those in the x direction, so we can ignore the x component of velocity, making the problem two-dimensional. For meridional propagation we define a column vector $\mathbf{F}$ by

$$\mathbf{F} = \begin{vmatrix} \delta\rho/\rho_0 \\ \delta p/p_0 \\ v \\ w \end{vmatrix}$$

and assume that atmospheric perturbations can be described by plane waves of the form

$$F \propto e^{i(\omega t - k_y y - k_z z)} \tag{6.1}$$

Substituting $\rho = \rho_0 + \delta\rho$, $p = p_0 + \delta p$, $\mathbf{U} = (0, v, w)$ into the equations describing the atmosphere (see Chapter 2) and retaining terms up to first order in $\delta\rho$, δp, and $\mathbf{U}$ gives the linearized forms of the mass continuity, motion, and adiabatic state equations—that is,

$$\partial(\delta\rho)/\partial t + \mathbf{U} \cdot \nabla\rho_0 + \rho_0 \nabla \cdot \mathbf{U} = 0 \tag{6.2a}$$

$$\rho_0 \partial v/\partial t + \partial(\delta p)/\partial y = 0 \tag{6.2b}$$

$$\rho_0 \partial w/\partial t + \partial(\delta p)/\partial z + \delta\rho g = 0 \tag{6.2c}$$

$$\partial(\delta p)/\partial t + \mathbf{U} \cdot \nabla p_0 - C_0^2 \partial(\delta\rho)/\partial t - C_0^2 \mathbf{U} \cdot \nabla\rho_0 = 0 \tag{6.2d}$$

We have taken the atmosphere to be isothermal with temperature T. In (6.2d), C_0 is the speed of sound, given by

$$C_0^2 = \gamma p_0/\rho_0 = \gamma g H$$

where γ is the ratio of specific heats at constant pressure and constant volume and $H = k_B T/Mg$ is the scale height. Viscosity has been ignored (inviscid fluid). Using (6.1) and the condition for hydrostatic equilibrium, (6.2) can be rewritten as a matrix equation:

$$\begin{vmatrix} i\omega & 0 & -ik_y & -1/H - ik_z \\ 0 & -ik_y C_0^2/\gamma & i\omega & 0 \\ g & -C_0^2(1/H + ik_z)/\gamma & 0 & i\omega \\ -i\omega C_0^2 & i\omega C_0^2/\gamma & 0 & (\gamma - 1)g \end{vmatrix} \cdot \mathbf{F} = 0 \tag{6.2e}$$

In deriving (6.2e) we used the fact from (6.1) that $\delta p \propto p_0 \exp i(\omega t - k_y y - k_z z)$ and that the zero-order pressure varies only vertically. This leads to $\partial(\delta p)/\partial z = \delta p[(1/p_0)(dp_0/dz) - ik_z]$ and eventually to the corresponding entry in row 3. Setting the determinant of the 4×4 matrix equal to zero yields the dispersion relation for linear modes of a nonrotating neutral atmosphere on a flat earth,

$$\omega^4 - \omega^2 C_0^2 \left(k_y^2 + k_z^2\right) + (\gamma - 1) g^2 k_y^2 + i\gamma g \omega^2 k_z = 0 \tag{6.3}$$

A variety of possible wave modes are buried in this dispersion relation. Suppose we take the limit that $g = 0$. Then (6.3) reduces to

$$\omega^2 = C_0^2 \left(k_y^2 + k_z^2\right)$$

which is the dispersion relation for sound waves propagating without attenuation, growth (pure real ω and $\mathbf{k}$), or dispersion ($\omega/k = $ constant).

We now turn to the gravity wave case. If there are no sources of energy or dissipation (viscosity was ignored), waves will not grow or decay in time at a fixed point in space, so we can assume ω is real. If we are including gravity, however, it can be shown that there are no solutions of (6.3) with both k_y and k_z purely real. Anticipating the final result, let us assume k_y is purely real and investigate k_z. This corresponds to a wave propagating in an unattenuated fashion with a component in the horizontal direction. Then we can write (6.3) as

$$\omega^4 - \omega^2 C_0^2 k_y^2 + (\gamma - 1)g^2 k_y^2 = -i\gamma g \omega^2 k_z + \omega^2 C_0^2 k_z^2 \tag{6.4}$$

where the left-hand side is purely real. Now if we let k_z be a complex number,

$$k_z = k_z' + ik_z''$$

it is straightforward to show that the right-hand side of (6.4) is purely real if and only if

$$k_z'' = (1/2H)$$

Dropping the superscript (prime) notation, we can now see that the solutions for the quantities in the column vector $\mathbf{F}$ are of the form

$$e^{i(\omega t - k_y y - k_z z)} e^{z/2H} \tag{6.5}$$

In (6.5) ω, k_y, and k_z are real.

Atmospheric waves that propagate in the manner described by (6.5) are termed internal gravity waves (IGW). Some of the complexity of the wind patterns that arise in the 90–120 km height range due to such waves can be gauged from the photograph of a trimethyl aluminum (TMA) vapor trail deployed by a sounding rocket, which is shown in Fig. 6.1. This photograph yields only one perspective

TMA Trail
June 11, 1978
0634 UT
(Wallops Island, VA)

Figure 6.1   A TMA trail deployed from Wallops Island, Virginia on June 11, 1978, at 0634 UT. The trail was photographed from the NASA C54 airplane. (Courtesy of I. S. Mikkelsen.)

on the distortion of the trail by the ambient winds, but numerous reversals and shears are evident. From (6.5) the theoretical prediction is that the wave amplitude should grow as it propagates upward (positive z). The physical explanation for this somewhat bizarre result is that to conserve the wave perturbation energy (e.g., terms of the form $\rho_0 v^2$) as ρ_0 decreases with z, v^2 must increase. The factor of 2 in the exponential form occurs, since, in order to keep $\rho_0 v^2$ constant, v need only e-fold over a height interval $2H$ when ρ_0 decreases by a factor of e in the height interval H. The classical observation supporting this result is shown in Fig. 6.2a. The dashed curve shows the mean wind. The actual wind fluctuates about this mean with an amplitude that increases with height. In Fig. 6.2b a detrended version of the same data is given after subtracting the mean and multiplying by $\exp[(112-z)/2H]$ where z and H are measured in kilometers. This procedure "undoes" the effect of the exponential atmosphere, and the wave component appears constant with altitude. A schematic representation

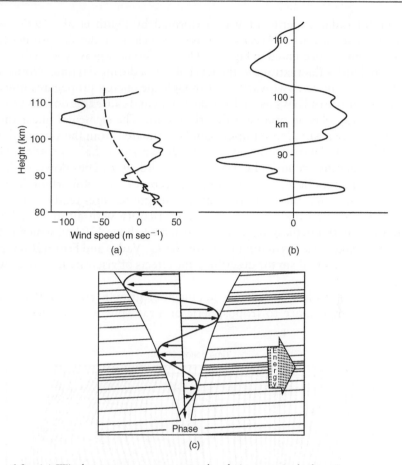

Figure 6.2 (a) Wind components at meteor levels in a vertical plane in one representative case, derived by Liller and Whipple (1954) from the distortion of a long-enduring meteor trail. (b) Normalized wind profile at meteor heights, measured to the right and to the left from the "0" position, deduced from (a) by removal of the general shear and amplification of the residual by a factor proportional to $\rho_0^{1/2}$. (c) Pictorial representation of internal atmospheric gravity waves. Instantaneous velocity vectors are shown, together with their instantaneous and overall envelopes. Density variations are depicted by a background of parallel lines lying in surfaces of constant phase. Phase progression is essentially downward in this case, and energy propagation obliquely upward; gravity is directed vertically downward. [After Hines (1974). Reproduced with permission of the American Geophysical Union.)

of the effect is shown in Fig. 6.2c. Many data sets show that the fluctuating component of the neutral wind velocity increases with increasing height. Furthermore, the perturbations about the average profile are often comparable to the mean wind.

A beautiful radar experiment was performed by Djuth et al. (1997) using plasma line observations at Arecibo. Consecutive electron density profiles from that experiment are presented in Fig. 6.3. Here we see how gravity waves can create electron density fluctuations at the level of $\pm 1\%$ during daytime. Notice that the phase fronts move downward even though the energy propagates upward. This is a key feature of IGWs as well as tidal waves and causes the downward progression of many plasma processes described here. The elemental linear theory cannot include viscosity, but damping can be estimated from the viscous damping rate $k_z^2(v/\rho_0)$. Two conclusions can be drawn from this expression: short vertical wavelengths are more heavily damped and, as ρ_0 goes down with altitude while v changes little, viscous damping becomes more important. Turning to Fig. 6.3, notice that the short vertical wavelength features tend to disappear above 110 km due to the rapidly increasing viscosity with altitude. A longer vertical wavelength fluctuation still can be detected above 130 km, which is consistent with weaker viscous damping for smaller k_z. Vadas and Fritts (2005) have developed a new formalism for describing the effects of viscosity in gravity wave propagation.

An important implication of this exponential growth effect is that waves of little or no importance to tropospheric or stratospheric dynamics grow to

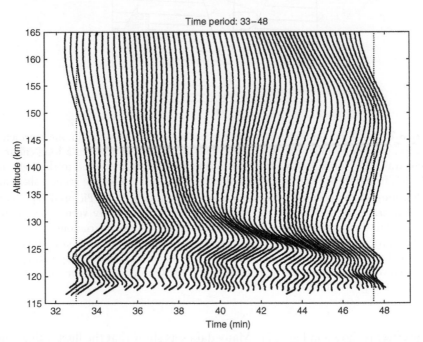

Figure 6.3 Electron density profiles measured using the Arecibo radar's plasma line. The amplitude of the perturbation at 130 km is about 4%. [After Djuth et al. (1997). Reproduced with permission of the American Geophysical Union.]

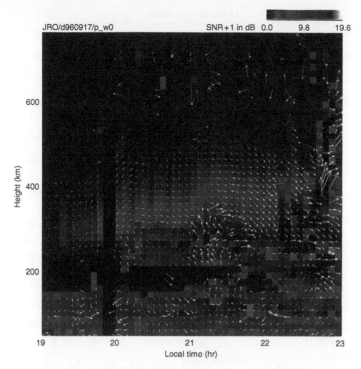

Color Plate 1 (Figure 3.13) A combination of irregularity and $\mathbf{E} \times \mathbf{B}$ drifts and a backscattered power map for September 17, 1996 (sunset time 18:03 LT, Kp = 3−). A data gap exists at ∼20:00 LT. [After Kudeki and Bhattacharyya (1999). Reproduced with permission of the American Geophysical Union.]

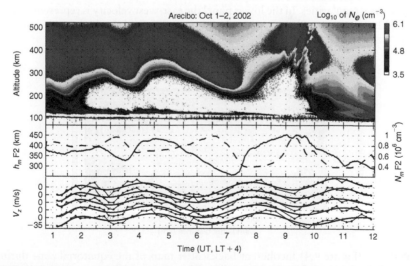

Color Plate 2 (Figure 3.30b) Periodic oscillations of the ionosphere over Arecibo with the downward phase progression typical of gravity waves are shown in the top panel. One of these oscillations triggered a turbulent upwelling over Jicamarca near dawn. [After Nicolls et al. (2004). Reproduced with permission of the American Geophysical Union.]

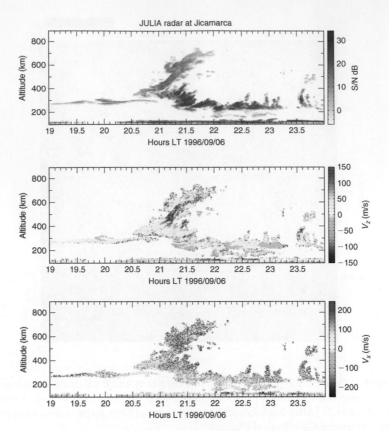

Color Plate 3 (Figure 4.2) An RTI map for September 6, 1996, is reproduced in the top panel. In the central panel the map is color scaled to represent the vertical line-of-site velocity of the irregularities. In the lower panel the east-west velocity is represented. [After Hysell and Burcham (1998). Reproduced with permission of the American Geophysical Union.]

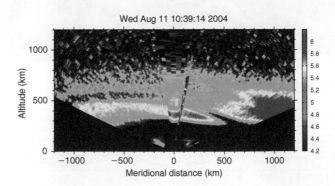

Color Plate 4 (Figure 4.4) Incoherent backscatter map of the equatorial zone during a CEIS event. The depleted plasma region follows the contour of the magnetic field lines. The strong signals just off vertical were detected when the radar was pointed perpendicular to the magnetic field lines. The magnetic equator is located near 5 degrees south of the radar. The equatorial anomaly can be seen at the right. The color scale corresponds to $\log_{10} n$ (from 4.2 to 6.2). (Figure courtesy of D. Hysell.)

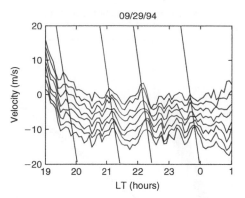

Color Plate 5 (Figure 4.13b) Stack plot of the vertical drift velocities at individual heights. The lowest height is 225 km and each subsequent line is 15 km higher and has been shifted by 2 m/s upward. The red portions of the plot are points inferred by interpolation. The slanted black lines are identical and have been aligned with peaks in the data to illustrate the downward-phase velocity.

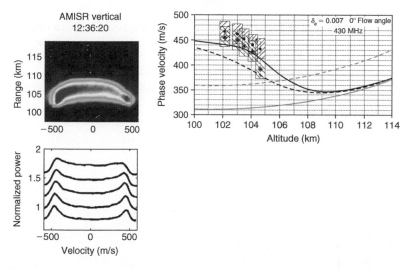

Color Plate 6 (Figure 4.32c) Range-Velocity-Intensity plots (top) and individual normalized spectra (bottom) for AMISR-P (left column) and phase velocities at 430 MHz on March 12, 2005, observed at the Jicamarca Observatory and theoretically predicted (right column). Stars and solid squares show data for FB waves moving toward and away from the radar, respectively. Shaded areas around each data point show the altitude (horizontal scale) and velocity (vertical scale) uncertainty. Solid and dashed black lines are theoretically predicted for V_{ph} for 0° and 0.2° aspect angles, respectively. An isothermal ion acoustic speed is shown as a solid gray line. The dashed gray line denotes the adiabatic process. [After Kagan et al. (2008). Reproduced with permission of the American Geophysical Union.]

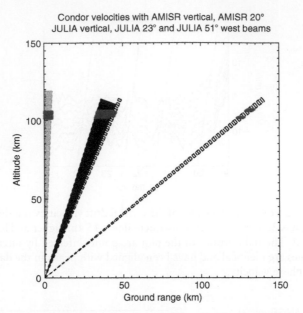

Condor velocities with AMISR vertical, AMISR 20°
JULIA vertical, JULIA 23° and JULIA 51° west beams

Color Plate 7 (Figure 4.35b) Regions in which these radars would record a Doppler shift $\geq .0.9C_s$ in the presence of a zero-order horizontal drift, as in Fig. 3.17b, and vertical perturbation drifts, as in Fig. 4.35a, are indicated in red. [After Kelley et al. (2008). Reproduced with permission of the American Geophysical Union.]

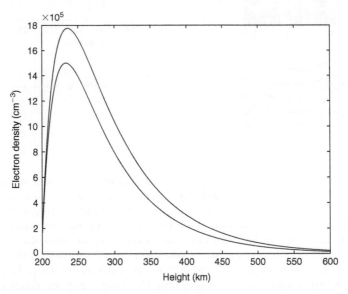

Color Plate 8 (Figure 5.3a) Electron density profiles calculated from (5.16) with a plasma flux of $F = 0$ (upper curve) and an upward plasma flux of $F = 3 \times 10^8 \mathrm{cm}^{-2}\mathrm{s}^{-1}$ (lower curve). As can be seen from Fig. 5.3a, upward flux results in a scale height that differs from that for diffusive equilibrium at high altitudes. (Figure courtesy of Michael Vlasov.)

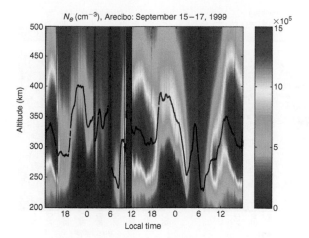

Color Plate 9 (Figure 5.11) Two and a half days in the life of the ionosphere over Arecibo. The lines trace HmF2. [After Vlasov et al. (2003). Reproduced with permission of Elsevier.]

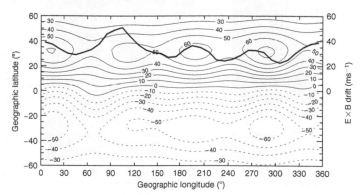

Color Plate 10 (Figure 5.24a) Contours of temperature perturbation for the diurnal tide on which the equatorial vertical drift is superposed. [After Kil et al. (2007). Reproduced with permission of the American Geophysical Union.]

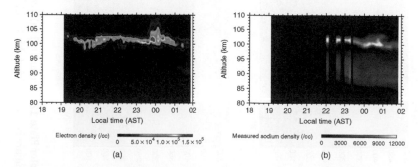

Color Plate 11 (Figure 6.7) Intense sporadic E event (a) measured by the Arecibo radar, accompanied by release of sodium atoms detected by a sodium lidar (b). [After Swartz et al. (2002). Reproduced with permission of the American Geophysical Union.]

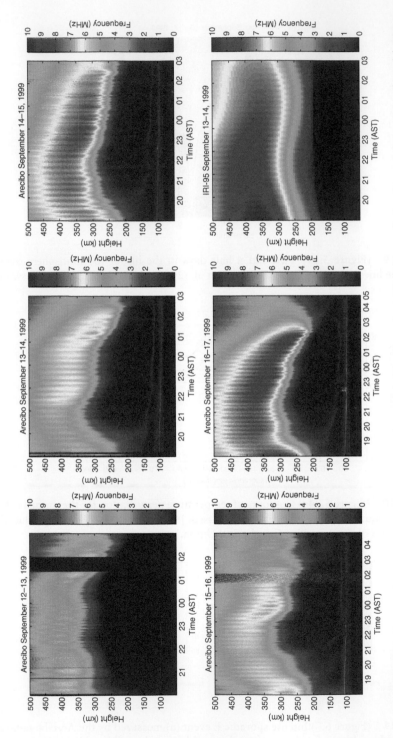

Color Plate 12 (Figure 6.14a) Five consecutive nights of Arecibo radar data along with the prediction from a climatological model (IRI). (Figure courtesy of J. J. Makela.)

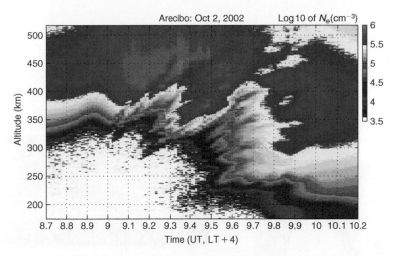

Color Plate 13 (Figure 6.14b) Summary of Arecibo ISR data for the night of October 1–2, 2002. The plasma went unstable just before 9 UT (near dawn). [After Nicolls and Kelley (2005). Reproduced with permission of the American Geophysical Union.]

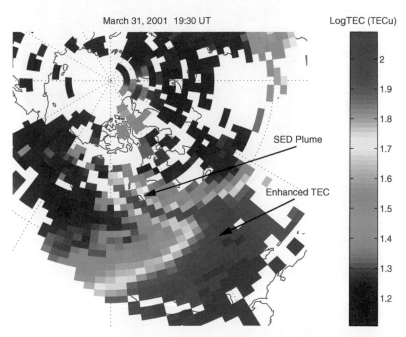

Color Plate 14 (Figure 6.14c) Poleward extension of the anomaly zone and its capture of a plume by antisunward flow into the polar cap. (Figure courtesy of John Foster.)

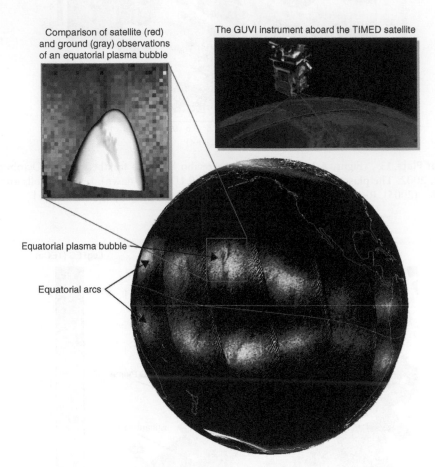

Comparison of satellite (red) and ground (gray) observations of an equatorial plasma bubble

The GUVI instrument aboard the TIMED satellite

Equatorial plasma bubble

Equatorial arcs

Color Plate 15 (Figure 6.15b) GUVI observations of the 135.6-nm emission on September 22, 2002, along with simultaneous ground-based observations. [After Kelley et al. (2003b). Reproduced with permission of the American Geophysical Union.]

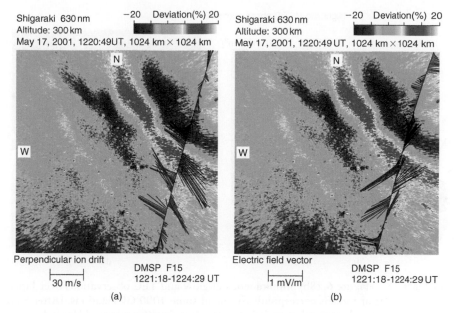

Color Plate 16 (Figure 6.26) Polarized electric field in nighttime MSTID. 630 nm airglow for May 17, 2001, along with perturbation drifts and electric fields from a DMSP satellite pass. [After Shiokawa et al. (2003). Reproduced with permission of the American Geophysical Union.]

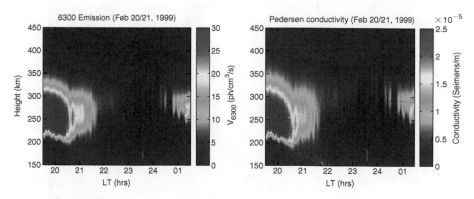

Color Plate 17 (Figure 6.28a) A comparison of the modeled 630.0 nm airglow emission line (left) and the calculated Pedersen conductivity (right) for February 20–21, 1999. [After Kelley et al. (2000c). Reproduced with permission of the American Geophysical Union.]

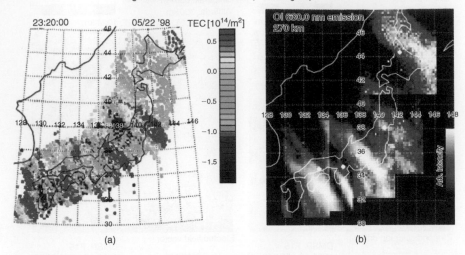

Nighttime MSTID observations (TEC, Airglow)

(a)   (b)

Color Plate 18   (Figure 6.28b) Simultaneous airglow and TEC observations over Japan. Each data point at the left corresponds to one of some 1000 GPS stations. [After Saito et al. (2001). Reproduced with permission of the American Geophysical Union.]

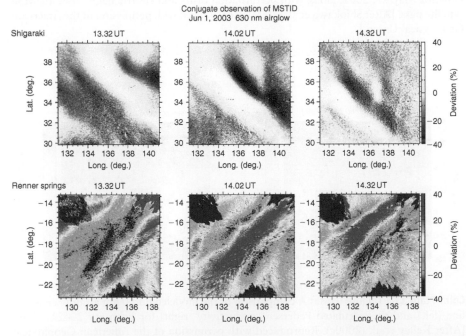

Color Plate 19   (Figure 6.29) Simultaneous images of MSTIDs in both hemispheres from June 1, 2003. Note that when projected along the magnetic field, a feature in the top left-hand side of the Shigaraki image (upper panel) maps to a feature on the bottom left-hand side of the Renner Springs image (bottom panel). [After Otsuka et al. (2004). Reproduced with permission of the American Geophysical Union.]

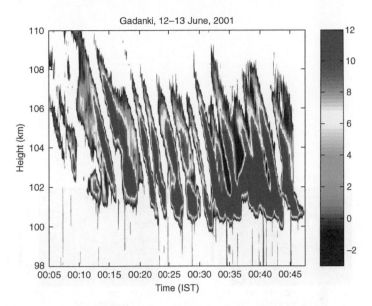

Color Plate 20 (Figure 6.32) Q-P radar echoes observed with the Gadanki radar. (Figure courtesy of R. K. Choudhary.)

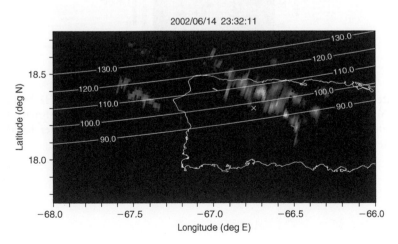

Color Plate 21 (Figure 6.34) Radar image of coherent echoes received at 23:32 UT. Note that UT = LT + 4. Blue corresponds to echo velocity toward the southeast and yellow-green toward the northwest. [After Hysell et al. (2004). Reproduced with permission of the European Geosciences Union.]

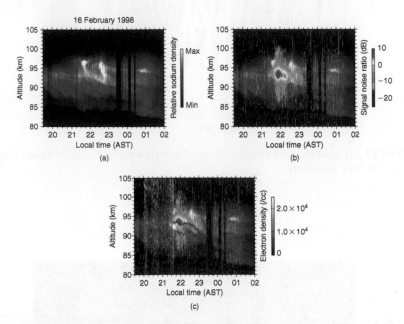

Color Plate 22 (Figure 6.35) Sodium atom density contours (a), with VHF backscatter overlay (b), and with ISR sporadic E density overlay (c) for the nights of February 16–17, 1998. Unfortunately, the radar data are noisy due to interference. [After Kane et al. (2001). Reproduced with permission of the American Geophysical Union.]

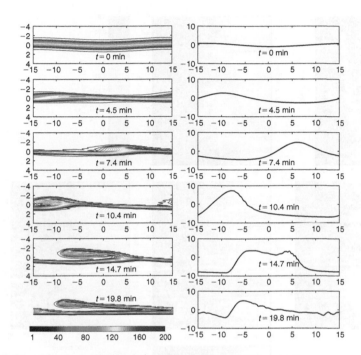

Color Plate 23 (Figure 6.40) The left panels show isodensity contours on cross sections through the E_sL, from numerical simulations, for six successive times. At time $t = 0$ the layer is perturbed by a $\pm 1/4$ km sinusoidal altitude modulation, and the subsequent panels show the growth phase of the E_s layer instability. (The cross-sectional plane is rotated about the horizontal southwest-northeast line—(the horizontal axis)—so it contains the magnetic field.) The right-hand panels show the corresponding electric fields at the original layer altitude of 105 km, plotted versus position on the horizontal axis. All scales are kilometers, except for the electric field ordinate, which is mV/m. The integrated Pedersen conductivities of the E_s and F layers (Σ_{PE} and Σ_{PF}) and the integrated Hall conductivity of the E_s layer (Σ_H) satisfy $\Sigma_H = 2.5(\Sigma_{PE} + \Sigma_{PF})$. The line color indicates the percentage of initial peak density.

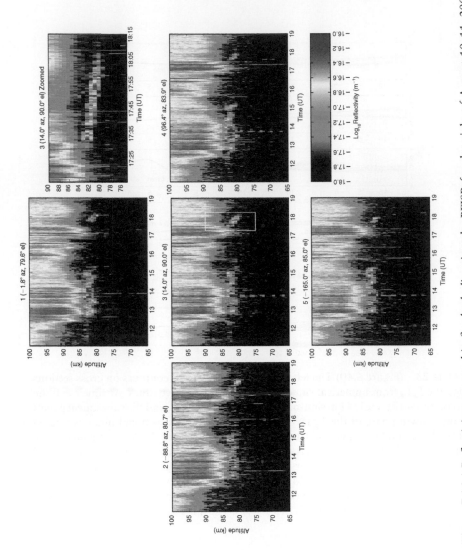

Color Plate 24 (Figure 7.14a) Reflectivity measured in five look directions by PFISR for the night of August 10–11, 2007. The boxed area is expanded at the top right. Lidar echoes were obtained at that same height and time period.

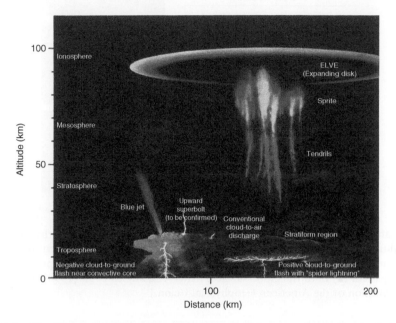

Color Plate 25   (Figure 7.18) Phenomena associated with upward coupling by lightning. (Figure courtesy of W. Lyons and C. Miralles.)

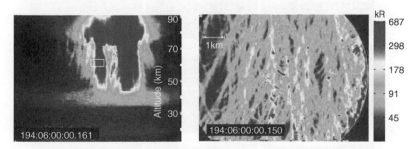

Color Plate 26   (Figure 7.20) Telescopic imaging of sprites. Wide (L) and narrow (R) field-of-view images of a bright sprite event. [After Gerken et al. (2001). Reproduced with permission of the American Geophysical Union.]

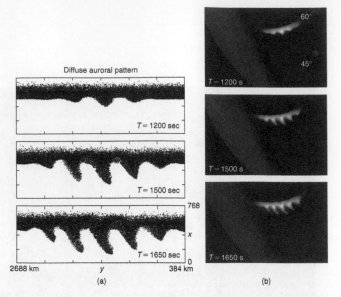

Color Plate 27 (Figure 10.4b) (a) Simulation of the temporal evolution of the distribution of the diffuse aurora on the poleward edge of the trough. (b) A three-dimensional picture of giant undulations with imitative color, obtained by projecting the computer-produced auroral pattern (a) onto the globe surface. [After Yamamoto et al. (1994). Reproduced with permission of the American Geophysical Union.]

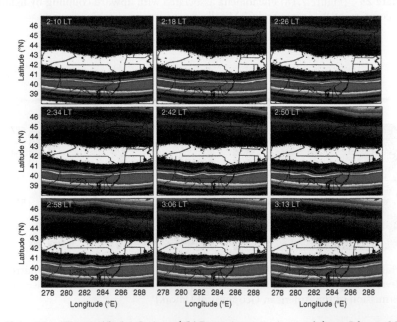

Color Plate 28 (Figure 10.6a) Series of SAR arc images observed from Ithaca, NY, on October 28–29, 2000. The arc is centered at 40° geographical latitude, while the edge of the diffuse aurora is near 43°N. [After Nicolls et al. (2005). Reprinted with permission of the Institute of Electrical and Electronics Engineers, Inc. (© 2005, IEEE).]

monumental proportions in the E and F regions. Eventually, either these waves break down nonlinearly due to their large amplitude or due to the fact that the kinematic viscosity (ν/ρ_0) gets so large (as ρ_0 decreases) that viscous dissipation balances growth and eventually destroys the wave. Wave breaking is discussed in Chapter 7.

Equation (6.4) can also be written in the form:

$$\omega^4 - \omega^2 C_0^2\left(k_y^2 + k_z^2\right) + \omega_b^2 C_0^2 k_y^2 - \omega_a^2 \omega^2 = 0 \tag{6.6}$$

where $\omega_b^2 = (\gamma - 1)g^2/C_0^2$ is the square of the Brunt-Väisälä frequency, $\omega_a^2 = C_0^2/4H^2$ is the acoustic frequency, and where k_y and k_z are real. At a given value of k_y, if ω is large enough (the high-frequency branch with $\omega > \omega_a$) the first and second terms dominate and we recover the sound wave dispersion relation found previously for $g = 0$. The low-frequency branch corresponds to gravity waves that propagate only for $\omega < \omega_b$. Physically, the Brunt-Väisälä frequency is the frequency at which a parcel of air oscillates about its equilibrium position when it is initially displaced from that position. For a nonisothermal atmosphere, one in which C_0^2 varies with height, it can be shown (Holton, 1979) that

$$\omega_b^2 = (\gamma - 1)g^2/C_0^2 + (g/C_0^2)(dC_0^2/dz) \tag{6.7a}$$

and also

$$\omega_b^2 = \left(\frac{g}{\theta}\right)\frac{d\theta}{dz} \tag{6.7b}$$

where θ is the potential temperature (the temperature that a parcel of air would attain if adiabatically brought to the ground).

Both branches of this dispersion relation (and one more!) are shown in Fig. 6.4. For $\omega > \omega_a$ we have normal sound waves. There is a forbidden band where $m(k_z)$ is imaginary (dotted region) and then for $\omega < \omega_b$ we have inertio-internal gravity waves, which become modified near the inertial period, f, where $f = 2\Omega \sin\theta$ (Ω being the radian frequency corresponding to a day and θ being the latitude). Inertial waves are discussed following. Notice that IGWs travel much slower than the sound speed (for the same k, ω is much less), which allows them to interact with winds. There is somewhat of a similarity here between the role of the plasma frequency in electrodynamics and the Brunt-Väisälä frequency in neutral dynamics. For a plasma to have propagation, you must have $\omega > \omega_p$, the plasma frequency, while for IGWs $\omega < \omega_b$ is required for propagation. Both ω_p and ω_b represent natural oscillations of the medium. Representative values for the buoyancy period $T_b = (2\pi/\omega_b)$ derived from (6.7a,b) are plotted in Fig. 6.5 as a function of height. Clearly, the 1–3 h oscillations in the ionospheric parameters just discussed fall in the gravity wave branch $\tau = (2\pi/\omega) > T_b$.

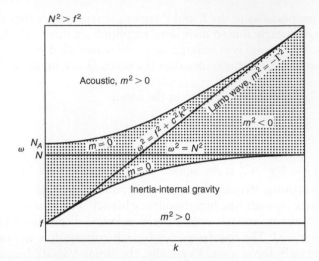

Figure 6.4 Dispersion relation for waves on a rotating planet but ignoring earth curvature (no tides). Three modes are shown: acoustic, Lamb (surface waves that attenuate as a function of z), and inertio-gravity waves. Here, as in many publications, $N(\omega_b)$ is used for the Brunt-Väisälä frequency and $N_A(\omega_a)$ for the acoustic frequency, k is the horizontal wavenumber, and m is the vertical wavenumber. [After Cornish (1987). Reproduced with permission of C. R. Cornish.]

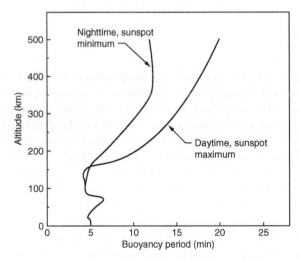

Figure 6.5 Vertical structure of the buoyancy period computed for a standard (non-isothermal) atmosphere. [After Yeh and Liu (1974). Reproduced with permission of the American Geophysical Union.]

A simple way to describe the dispersion relation for $\omega \ll \omega_a$ is

$$\omega^2 = \frac{\omega_b^2 k_y^2}{k_y^2 + k_z^2 + (1/2H)^2} \tag{6.8}$$

(Note that many authors use k for k_y and m for k_z.) The reader is asked to show that the product of the vertical phase and group velocities, $(\omega/k_z)(d\omega/dk_z)$, is negative. Thus, in the important case of upward energy propagation, when $d\omega/dk_z$ is positive, the vertical phase velocity is downward. This downward phase propagation is evident in Fig. 6.3 and, as we shall see, explains why the plasma layers in Fig. 5.23a and c all move downward, at least, on average.

It is crucial to note that in (6.8) ω is the intrinsic frequency, the wave frequency in the neutral wind reference frame. The Doppler formula relates ω to ω',

$$\omega' = \omega + \overline{k} \cdot \overline{u}$$

where ω' is the frequency in the frame in which $\overline{u}$ is measured (usually the earth frame). When the horizontal wave phase velocity in the earth frame matches the wind velocity, $\omega = 0$ and the wave simply vanishes. This is called a critical layer. Such a layer cannot happen for the sound wave branch, since the wind would have to be supersonic. But IGW waves are much slower, and critical layers happen all the time. For example, consider an isotropic wave source in the lower atmosphere propagating up into an atmospheric jet. As long as $\omega/k < u_{peak}$, a critical layer must be found for waves nearly parallel to the jet. Waves in the other direction are simply upshifted and go right through. (This is illustrated in Fig. 7.5 in the next chapter.) Waves propagating into the critical layer add their momentum to the layer when they are absorbed. This accelerates the jet even more. A classic example is the quasi-biennial oscillation in which a critical layer moves downward as gravity waves are absorbed in the stratosphere. The process requires almost two years to complete a cycle of zonal wind reversal at a given height (Holton, 1979).

If we ignore the curvature of the earth but allow the wave frequency to approach $f = 2\Omega \sin\theta$, then the rotation of the earth cannot be ignored. The frequency f is called the inertial frequency and is equal to $2\pi/\tau_F$, where τ_F is the time it takes for the plane of oscillation of a Foucault pendulum to return to its initial plane at $t = 0$. At the latitude of Arecibo this period is 52 hours. Such long-period waves are not easy to study, but they have been observed in the stratosphere (Cornish and Larsen, 1989; Cho, 1995). Another observation that seems to be due to such a wave at 35° latitude in this frequency range is shown in Fig. 6.6. Here a long-duration Leonid meteor trail was photographed 82 seconds after the meteor struck the atmosphere (Kelley et al., 2003c). The trail formed into a helix, and the mean wind, $\overline{u}$, the perturbation wind, $\delta\overline{u}$, and the vertical wave number (k_z) could be measured (Drummond et al., 2001). The ambient temperature profile was determined using a lidar located at the same site

(Starfire Optical Range in New Mexico). The characteristics of inertial period waves are an extension of the gravity wave dispersion relation and can be written in the form

$$\omega^2 = \frac{\omega_b^2 k_y^2 + k_z^2 f^2}{k_z^2 + k_y^2} \tag{6.9}$$

where we have taken $k_z^2 \gg (1/2H)^2$ for an isothermal atmosphere (Eckart, 1960). This is the lowest frequency wave in Fig. 6.4. Note that if $f = 0$ and $k_z^2 \gg 1/4H^2$, (6.8) becomes

$$\omega^2 = \frac{\omega_b^2 k_y^2}{k_z^2 + k_y^2} \tag{6.10}$$

which agrees with (6.8) and (6.9) for $f = 0$. By using the fact that the wind vector rotated with altitude, Drummond et al. (2001) found that the wave responsible for the trail shape in Fig. 6.6 had the properties of an inertial wave. In general, an elliptical pattern arises for waves with $\omega_i = 2\pi f < \omega < \omega_b$ with the ratio of

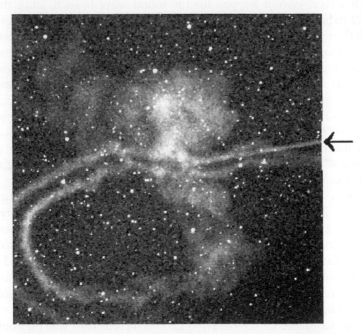

Figure 6.6 Photo of a long-duration meteor trail 82 seconds after impact. The circular feature shows that the wind rotates with altitude. Puffy areas are turbulent regions. The arrow shows a laser beam. [After Kelley et al. (2003c). Reproduced with permission of the American Geophysical Union.]

major axis to minor axis being equal to $|\omega_i/\omega|$, where it must be emphasized that ω is the intrinsic frequency. Due to the lidar data, this event is important because it is the most complete data set ever obtained. Other examples of rotating winds exist from TMA rocket releases but simultaneous lidar data did not exist. This event shows that an 8-hour wave in the earth frame, which might be considered a gravity wave or a higher-order tide, is actually an inertial gravity wave.

Turning to the tidal case, complications arise in the analysis because of the requirement that the solutions satisfy boundary conditions on the spherical earth. The solutions must exhibit certain altitudinal, latitudinal, and longitudinal forms referred to as Hough functions. Which propagating tidal normal modes actually are generated depends on how well the forcing function—for instance, solar or lunar forcing—matches the radial form of the mode structure. As alluded to earlier in the Chapter 3 discussion of electrodynamics, the diurnal tide due to atmospheric heating is important at low latitudes, but the response is small above 30° latitude. The semidiurnal forcing is smaller, but the altitude profiles of ozone and water vapor content fit the so-called (2,2) semidiurnal tidal mode quite well. The local heating due to these minor constituents thus couples well to the semidiurnal tide, explaining its importance at the higher latitudes. The solar heating function is similar to a half-wave rectifier, since it is on for only half of a day. This means higher-order Fourier components exist, and thus higher-order tides are expected. In the layer motions of Fig. 5.23c, some are separated regularly by about 6 hours. This may be a higher-order tide, but the analysis of Drummond et al. (2001) shows that an 8-hour period in the earth frame was actually a 20-hour inertio-gravity wave, not a third-order tide. Thus, great care must be taken in interpreting long-period waves. Recently a number of studies have revealed that a wave number three, nonmigrating diurnal tide (Hagan et al., 2003) creates a four-model perturbation in many ionospheric parameters, including the equatorial electric (Kil et al., 2007) and magnetic (England et al., 2006) fields and the intensity of the anomaly plasma density (Immel et al., 2006).

6.3 Role of Gravity Waves and Tides in Creating Vertical Ionospheric Structure

Given that gravity waves, inertio-gravity waves, and tides can be generated by a variety of sources and that they grow to respectable amplitudes by the time they reach ionospheric heights, we can now investigate their effect on the ionization.

For ω much less than ω_b and λ_z, the vertical wavelength, much less than the scale height, the first row in (6.2e) corresponds to

$$k_y v + k_z w = 0 \tag{6.11}$$

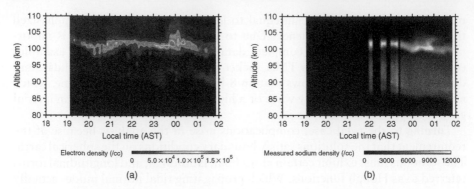

Figure 6.7 Intense sporadic E event (a) measured by the Arecibo radar, accompanied by release of sodium atoms detected by a sodium lidar (b). [After Swartz et al. (2002). Reproduced with permission of the American Geophysical Union.] See Color Plate 11.

This equation is equivalent to the assumption of incompressible flow $\nabla \cdot \mathbf{U} = 0$ when expressed in Fourier form and holds for all the low frequency waves considered here. This is the Boussinesq approximation, which uses $\nabla \cdot \mathbf{U} = 0$ for wave perturbations when $\omega < \omega_b$ but allows for compressibility in the zero-order equations. Equation (6.11) can also be written

$$|v/w| = |\lambda_y/\lambda_z|$$

Typically, $\lambda_y \gg \lambda_z$ for the periods of interest here, so $v \gg w$. That is, the horizontal component of the perturbation wind is much stronger than the vertical, and we explore its effect on the plasma next.

An example of the plasma layering that occurs is presented in Fig. 6.7. Here the Arecibo radar has been used in a high-resolution mode to best characterize these thin layers. Interesting structures occur for many hours. Remarkably, the neutral sodium atoms revealed by a sodium lidar also seem to be layered. At first this is surprising, since the plasma layers only occur due to magnetic forces, which do not affect the neutrals. However, recent advances in our understanding of ion and neutral sodium chemistry seem capable of explaining these atom layers (Cox and Plane, 1998; Swartz et al., 2002). We return to this topic later in this section.

As a first study, consider the F-layer valley where κ_i is large. For a wave having a small vertical wavelength, we have sketched the horizontal component of wind fluctuations for a typical gravity wave as a function of height in Fig. 6.8. This figure illustrates the case in which the perturbation wind lies in the magnetic meridian and, as also shown, the magnetic field has a finite dip angle. Because the plasma is constrained to move with the component of the neutral wind motion parallel to $\mathbf{B}$ when κ_i is large, in regions where the perturbation wind changes

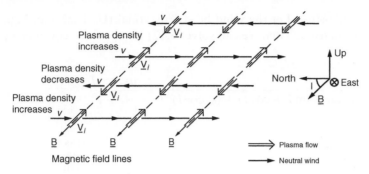

Figure 6.8 Meridional and neutral wind (horizontal vector) and ion flow vectors (large arrows) due to a gravity wave with a large wavelength in the y direction and a smaller wavelength in the z direction.

sign, the plasma either converges to or diverges from the altitude where the shear occurs. A layer of ionization will arise in the convergence zone. This wind shear theory for layer formation was first suggested by Dungey (1959), was extended by Whitehead (1961), and seems to explain many of the observations quite well for intermediate layers. The situation illustrated in Fig. 6.8 is most effective in the lower F region. At higher altitudes, diffusion parallel to **B** keeps sharp layers from forming. This can be seen by comparing the time constant at which plasma converges, which is of the order of $(k_z v \cos I)^{-1}$, with the time constant at which it diffuses, which is of the order of $(k_z^2 D_A)^{-1}$. For

$$D_A \geq v \cos I / k_z$$

diffusion is too strong for layers to form. For $\lambda_z \leq 2\pi H$ and $v \cos I \simeq 10$ m/s, we can estimate the critical value for D_A to be roughly 10^4 m²/s at 200 km. For molecular ions at 500° K, $D_A = 64 \times 10^4 / v_n$ m²/s and thus we find that gravity waves are not very efficient in producing layers above about 200 km because diffusion is too fast.

At heights below 130 km, the high collision frequency with neutrals merely pushes the ions horizontally, and no layers are formed due to meridional winds. However, shears in the zonal wind component can take over the production of layers below about 130 km. The mechanism stems from the Lorentz force felt by the ions when subject to a wind field. For example, referring to (2.22b), which gives a general relation for the ion velocity, we are interested in the equilibrium case where $d\mathbf{V}_i/dt = 0$. In addition, we ignore the effects of **E**, ∇n, and **g** and consider collisions between ions and neutrals only. Then, expressing (2.22b) for ions in the earth-fixed frame, we have

$$\kappa_i(\mathbf{V}_i \times \hat{B}) = \mathbf{V}_i - \mathbf{U} \tag{6.12}$$

where $\kappa_i = eB/Mv_{in}$. Let the wind be strictly zonal ($\mathbf{U} = u\hat{a}_x$) and ignore the magnetic declination. Then we can solve (6.12) for the three components of the ion velocity

$$V_{iz} = \frac{u}{\kappa_i(\tan I \sin I + \cos I) + (\kappa_i \cos I)^{-1}} = \frac{u\kappa_i \cos I}{\kappa_i + 1} \tag{6.13a}$$

$$V_{ix} = \frac{V_{iz}}{\kappa_i \cos I} \tag{6.13b}$$

$$V_{iy} = V_{iz} \tan I \tag{6.13c}$$

For small I and at a height where $\kappa_i = 1$, which occurs at about 130 km, (6.13a) and (6.13b) yield:

$$V_{ix} = V_{iz} = u/2, \ \ V_{iy} = 0$$

The ions therefore have a component parallel to $\mathbf{U}$ but are also deflected in the direction of the Lorentz force $q(\mathbf{U} \times \mathbf{B})$, giving a net motion at a 45° angle to the neutral wind $\mathbf{U}$, the deflection being upward for an eastward wind. The net velocity is illustrated in Fig. 6.9a. For larger κ_i the deflection angle is larger and for smaller κ_i the ion motion is nearly parallel to $\mathbf{U}$. The mechanism that creates the plasma layers is illustrated in Fig. 6.9b for the case of a vertical shear in the zonal wind. As shown by (6.13a), ions above the shear point drift downward, while those below drift upward. Plasma accumulates where the zonal wind is zero.

Note that in the last sentence we shifted from a discussion of *ion motion* to a statement about *plasma accumulation*. In the geometry of Figs. 6.9a, with the magnetic field perfectly horizontal, the highly magnetized (high κ_e) electrons could not move perpendicular to $\mathbf{B}$ to join the converging ions. Furthermore, the field lines bend into a very low plasma region, and there is no source of electrons to move along the magnetic field lines and neutralize the ions. A huge space charge electric field would build up, and the whole process would grind to a halt.

However, if there is even a slight dip angle, electrons can move along the magnetic field from a region of ion divergence to one of ion convergence in response

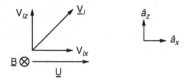

Figure 6.9a Ion velocity vector ($\mathbf{V}_i$) and its components subject to an eastward neutral wind when $\kappa_i = 1$.

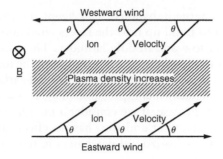

Figure 6.9b Illustration of the wind shear mechanism that operates at E-region and lower F-region heights. The angle θ depends upon κ_i and I. [After Hines (1974). Reproduced with permission of the American Geophysical Union.]

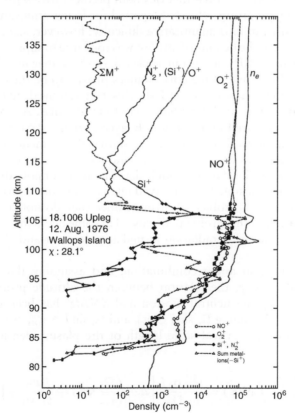

Figure 6.9c Electron and ion density profiles from a rocket flight. Below 108 km the individual ion measurements are shown. Above this altitude the curves represent running averages from four consecutive mass spectrometer sweeps. Localized peaks in the electron density profile are associated with metallic ions and are responsible for the daytime tidal ion layers plotted in Fig. 5.23c. (Courtesy of L. Smith.)

to a very slight initial charge imbalance. This keeps the plasma charge neutral and allows the plasma to build up into the layers cited above. This mechanism is thus most efficient at low and middle latitudes. The wind shear process can be considered as an example of the compressibility of the ionospheric plasma in the direction perpendicular to $\mathbf{B}$ when $\kappa_i \approx 1$. As noted earlier, F-region perpendicular plasma flow is virtually incompressible ($\nabla_\perp \cdot \mathbf{V}_\perp$) = 0 but in the lower F and E regions this is not the case. If we consider a meridional wind, the Lorentz force only deflects ions east-west and thus has no effect on producing vertical layering, but it could create horizontal patchiness in the layers if the winds are periodic, as in gravity waves.

The wind shear theories outlined here yield a very nice dynamical explanation for ionospheric layering. The downward progression of the layers is accounted for, since gravity wave and tidal theories both predict downward phase propagation for an upward group velocity (expected for lower atmospheric sources). The theory initially ran into quantitative difficulty, however, since the standard recombination rate at the heights of interest was too large to support the observed layer densities (up to $10^6 \, \text{cm}^{-3}$). But rocket data show that metallic ions such as Mg^+, Si^+, and Fe^+ due to meteoric sources proved to be the dominant ions in these layers (Herrmann et al., 1978). This fact removed objections to the wind shear theory, since such ions have very long lifetimes. An example of the ion composition measured in a sporadic E layer during a daytime rocket flight is presented in Fig. 6.9c. The $\Sigma \, M^+$ curve shows all the metal ions that track the peaks in the electron density quite well and constitute more than half of all the ions, even during the daytime. Such data strongly support the notion that metallic ions are responsible for long-lived intense sporadic E layers. The intermediate layers remain a problem, however, since they do not contain metallic ions. One resolution of this problem, discussed in some detail in the next section, involves additional ionizing radiation at midlatitudes over and above the usual photoionization sources.

Neglecting ionization and recombination, and assuming that all variables are functions of z' only, the relation between the altitude profile of plasma density produced by the neutral motion and dN/dz' has been calculated by Gershman (1974). Assuming $\Omega_e \sin I \gg v_{en}$ and $\Omega_i \sin I \ll v_{in}$, as in Gershman's derivation, the inverse vertical scale length of the plasma density gradient is given by

$$L^{-1} = \frac{dN(z')}{dz'} \bigg/ N(z') = \frac{\Omega_i \cos I}{v_{in} D_A} \left[u - \left(\frac{\Omega_i}{v_{in}} \sin I + \frac{v_{en}}{\Omega_e \sin I} \right) v \right] \quad (6.14)$$

where u_e and v_n correspond to the maximum zonal (positive eastward) and meridional (positive northward) winds about the shear point. D_A is the ambipolar diffusion coefficient. Substituting reasonable values for these parameters, we find $L < 100 \, \text{m}$, which is smaller than observed. Kagan and Kelley (1998) concluded that plasma instabilities must create an anomalously large D_A^*, which then limits the sharpness of the gradients and the thickness of the layer.

Finally, we note that a wind shear is not necessary for all layer formation. A westward Northern Hemispheric wind will drive ions downward but into an ever increasing atmosphere. Collisions (low κ_i) cause the vertical velocity to decrease with decreasing altitude and a layer forms. A southward electric field has the same effect. As seen in the various examples of sporadic E layers, as their altitude decreases, they seem to come nearly to rest below 95 km. In this height range, ion chemical reactions by metal ions become much more frequent and eventually they are destroyed chemically.

These metallic ions have their origin in the ablation of meteors. This source also leads to copious metallic atoms in the 90–100 km altitude range. The existence of sodium, iron, and potassium atoms allows the use of resonant lidar scatter to study this important height range. In the case of sodium, Cox and Plane (1998) first seriously studied sodium ion chemistry and attempted to explain the atom layers. A slightly modified version of their chemical model is illustrated in Fig. 6.10. A crucial reaction rate was incorrect in the earlier study, and the initial results did not fit the data very well. In conjunction with new data taken at Arecibo and a revised reaction rate for the $NaO^+ + N_2 \rightarrow Na^+N_2 + O$ and $NaO^+ + O_2 \rightarrow Na^+ + O_3$ branches, Collins et al. (2002) have produced much better agreement. Through such interactions, sodium ions can maintain sodium atom layers (see Fig. 6.7).

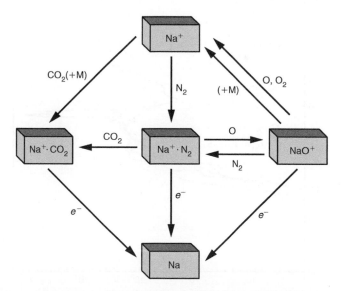

Figure 6.10 Ion chemical reactions involving sodium. [After Collins et al. (2002). Figure reproduced with permission of Elsevier Science Ltd.]

6.4   Effects of Particle Precipitation at Midlatitudes

Shen et al. (1976) concluded that both the intense intermediate layers in Fig. 5.22 and the general enhancement of F-layer valley ionization were associated with energetic particle precipitation. To test this idea, extensive studies involving a number of rocket flights at midlatitudes have been conducted by Voss and Smith (1979, 1980a, b). Summary plots are reproduced in Fig. 6.11a–c. Figure 6.11a shows plasma density profiles for four different values of the magnetic activity index K_p. The highest K_p profile shows F-layer valley densities an order of magnitude higher than the lowest K_p profile. A layer with peak density at 135 km is also evident in the high-K_p profile. Figure 6.11b shows both the particle flux and the ionization rate as functions of a specialized magnetic

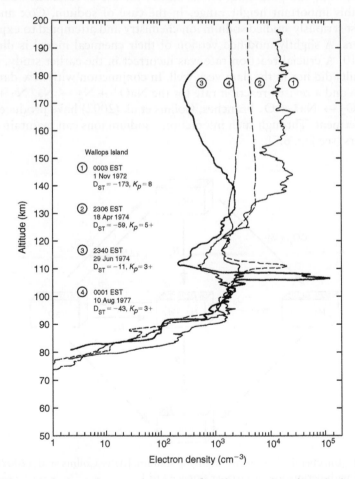

Figure 6.11a   Nighttime electron density profiles from rockets launched at Wallops Island. [After Voss and Smith (1979). Reproduced with permission of the American Geophysical Union.]

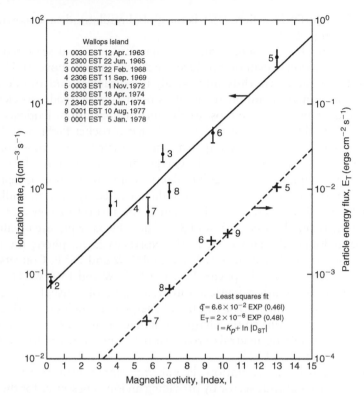

Figure 6.11b Variation with magnetic activity of the nighttime ionization rate in the upper E region and of the particle energy flux obtained from rocket experiments. [After Voss and Smith (1979). Reproduced with permission of the American Geophysical Union.]

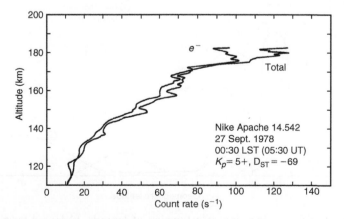

Figure 6.11c Total energetic particle count rate and count rate of electrons only as a function of altitude. (Courtesy of L. Smith.)

index, $I = K_p + \ln(\text{DST})$, where DST is the global midlatitude average magnetic deflection in units of nT.

It is clear that photon sources such as galactic x-rays and scattered Lyman α (from the geocorona) cannot produce much ionization above 105 km. Particle ionization sources are thus surprisingly important at midlatitudes, and it seems that the primary component involves positively charged particles rather than electrons, which produce most of the auroral ionization. Evidence in support of this hypothesis is given in Fig. 6.11c for a rocket flight over Wallops Island, Virginia ($L = 2.6$). In comparing the two curves, electrons contribute only about 10% of the count rate in the energy range surveyed. A schematic latitude profile of precipitating particles based on satellite observations is reproduced in Fig. 6.12. Positive particles dominate in three bands: equatorial, midlatitude, and the subauroral zone. Electrons dominate in the auroral zone proper and in a "low" latitude region between 20° and 30°. The schematic picture is valid when averaged over longitude. However, in the Northern Hemisphere, the midlatitude zone disappears almost entirely between 45° W and 75° E. Conversely, the low-latitude electron zone disappears between 135° W and 150° W. These latter effects are due to the South Atlantic anomaly, a region of unusual magnetic field strength that modifies the mirror heights of energetic particles quite drastically. The equatorial zone is due to the charge exchange of ring current ions with the geocorona. The energetic neutrals cross the field lines and are focused in the equatorial zone (Pröllss, 1973).

In summary, particle precipitation plays an important role in creating structure in the mid- and low-latitude sector by providing an ionizing source for the F-layer

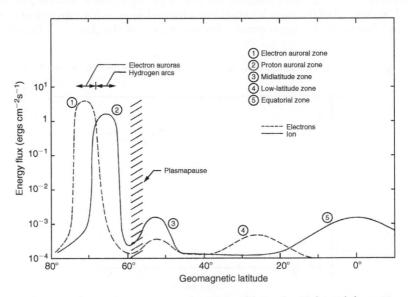

Figure 6.12 Particle precipitation as a function of latitude. [Adapted from Voss and Smith (1980b). Reproduced with permission of Pergamon Press.]

valley. The production rate is very low, but neutral atmospheric waves gather the ions and produce easily observable features in the plasma profile. To this are added meteoric sources of metallic ions at lower altitudes. There, long-lived ions are also gathered together by wind and wave patterns to form the ionization layers observed.

6.5 Horizontal Structure in the Midlatitude Ionosphere

Studies by Bowman (1981, 1985) seem to show clearly that much of midlatitude spread F as registered on ionosondes is due to altitude modulation of the electron density in the F layer, most likely due to gravity waves. Some of the more violent disturbances that include small-scale structure most likely also involve the plasma instabilities discussed following. Since these features propagate, they are referred to as traveling ionospheric disturbances or TIDs.

We have concentrated thus far on vertical ionospheric structure, since it is far easier to observe. Ionosondes usually are directed vertically but can detect horizontal variations due to their broad antenna patterns. Bowman (1981, 1985) has described and reviewed the results of such observations. Figure 6.13a shows one interpretation of ionosonde observations of a medium-scale traveling ionospheric

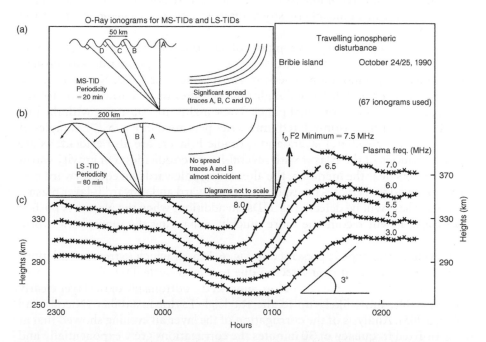

Figure 6.13 Cartoons (a, b) showing how a corrugated ionosphere leads to ionosonde spreading. Also, actual ionosonde data (c) showing downward phase progressions for a large-scale TID. [After Bowman (1981). Reproduced with permission of Elsevier Science.]

disturbance. A single vertically pointed ionosonde would obtain echoes from several ranges due to these tilts—hence the name range spread F. Such plasma observations of TIDs were part of the data set that led Hines to develop the neutral wave theory described in Section 6.2. One must remember, though, that the neutral gas is not measured directly in these and in other TID observations. Figure 6.13b shows that longer wavelength features do not create the same sort of spread in the trace. However, Fig. 6.13c shows that the associated long-period features can be detected by ionosondes using time variation of plasma contours. Here, 1.5-hour periodicity is found, which has the characteristic downward phase progression associated with gravity waves. Bowman (1981, 1985) used the term "large" to include the 200 km scale feature in Fig. 6.13c. Here we reserve that term for structures greater than 500 km and use the term medium scale or mesoscale for features in the range of 50 to 500 km.

Large-scale TIDs are usually very fast and come equatorward out of the auroral zone. A computer simulation of a neutral atmospheric disturbance launched by a high-latitude heating event was presented in Section 3.5. It is clear that such a pulse would drive the midlatitude ionosphere upward due to the southward wind and that this uplift would seem to travel equatorward. Less obvious is the fact that a temperature enhancement also changes the composition of the lower thermosphere, decreasing the O/N_2 ratio. Such a composition effect changes the production and recombination rates in the F region in such a way that a negative midlatitude ionospheric storm often accompanies a geomagnetic storm (see Fig. 5.1).

An example of the negative phase of an ionospheric storm is shown in Fig. 6.14a, in which five consecutive nights of Arecibo plasma density profiles are shown during a modest magnetic storm in September 1999. Nights 1, 2, and 4 in this sequence had depressed nighttime densities relative to the climatological prediction shown in the final panel. These nights are consistent with the low values of f_0F_2 in the storm-related time series presented in Fig. 5.1. Examples of descending (intermediate) layers and sporadic E layers are also evident, as are height variations of the F layer with several-hour periodicities. For gravity waves with $\mathbf{k}$ vectors near the meridian, the alternating poleward and equatorward perturbation wind will create corresponding downward and upward F-layer plasma motions. These will be detected as TIDs. In addition, when the plasma is pushed downward, it will recombine more quickly, enhancing the ionospheric storm effect.

An event of this type is presented in Fig. 6.14b, in which extreme oscillations of the F layer occurred over Arecibo due to a large-scale TID. When the layer reached its highest altitude just before sunrise, the bottomside of the layer clearly went unstable, most likely up to the Rayleigh-Taylor instability (Nicolls and Kelley, 2005). Analysis of the corrugation of the layer all evening showed that at an earth-fixed frequency of 30 minutes the corrugations grew exponentially and broke into the striations seen in Fig. 6.14b. This is the only example we know of

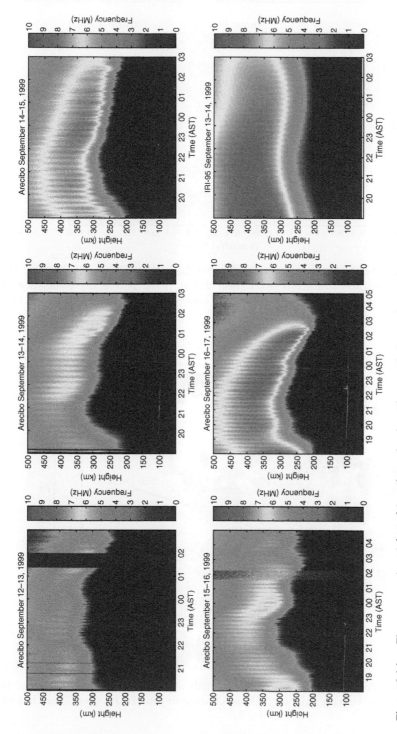

Figure 6.14a Five consecutive nights of Arecibo radar data along with the prediction from a climatological model (IRI). (Figure courtesy of J. J. Makela.) See Color Plate 12.

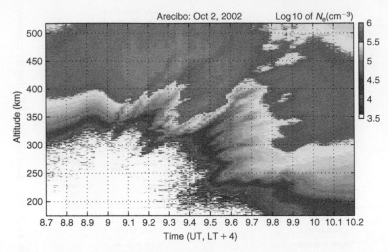

Figure 6.14b Summary of Arecibo ISR data for the night of October 1–2, 2002. The plasma went unstable just before 9 UT (near dawn). [After Nicolls and Kelley (2005). Reproduced with permission of the American Geophysical Union.] See Color Plate 13.

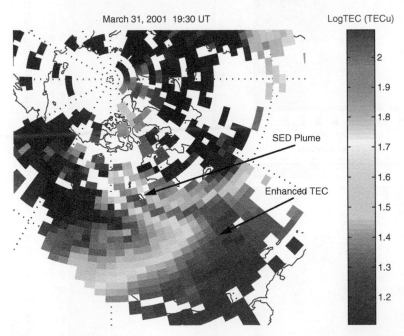

Figure 6.14c Poleward extension of the anomaly zone and its capture of a plume by antisunward flow into the polar cap. (Figure courtesy of John Foster.) See Color Plate 14.

in which an off-equatorial plasma is unstable due to gravitational currents. We take up other midlatitude instabilities in the next section.

With the advent of GPS observations a remarkable structure has been found to develop during magnetic storms. These storm-enhanced density (SED) features, one of which is presented in Fig. 6.14c, exhibit total electron content values of nearly $200 \times 10^{16}\,\mathrm{cm}^{-2}$. They typically develop in the late afternoon and form a channel of high density at midlatitudes, which becomes entrained in the convection and moves into the auroral oval and polar cap (Foster et al., 2002). The origin seems to be the dayside equator where high eastward electric fields continually pump solar-produced plasma up and into the equatorial fountain (Vlasov et al., 2003). We have already discussed prompt penetrating electric fields (PPE) and their effect at the equator in Chapter 4. SED events are very likely a midlatitude effect of both zonal and meridional electric fields penetrating to low latitudes. Meridional PPE fields are more difficult to study using the ISR technique, although Fig. 3.23 shows how a shift to B_z north decreases an auroral zonelike penetrating meridional field over Millstone Hill ($L = 3.2$). Rowland and Wygant (1998) have used double probe electric field instruments (see Appendix A) in the CRESS satellite for this purpose. They showed that as Kp increases, the radial component of the electric field penetrates deeper and deeper into the mid- to low latitudes. Note that the radial electric field in the equatorial plane maps to a meridional component in the ionosphere. For a Kp of 5 or more, this component is affected for L values as low as 2.

6.6 Midlatitude F-Region Plasma Instabilities

In this section the plasma physics of the midlatitude F-region ionosphere is discussed. Many of the processes are similar to those discussed already in Chapter 4 and thus need only be briefly reintroduced. E-region processes are presented in Section 6.7.

6.6.1 F-Region Plasma Instabilities in the Equatorial Anomaly (Equatorial Arc) Region

The plasma bubbles discussed in Chapter 4 are electrodynamically produced, so they involve uplift of the entire flux tube. Indeed, the radar map in Fig. 4.4 and the all-sky image from Christmas Island in Fig. 4.7a show that the depletions are field aligned. These flux tubes reach sufficiently high altitudes that they map into the equatorial anomaly. Just such an effect is shown in Fig. 6.15a, an all-sky camera airglow photograph taken in the anomaly region over Ascension Island (geographic position 8° S, 14° W). The dark bands correspond to plasma depletions near 300 km altitude, where the airglow originates. They extend from horizon to horizon in the north-south direction and are 50–100 km in east-west horizontal size. When mapped to the equatorial plane along magnetic field lines,

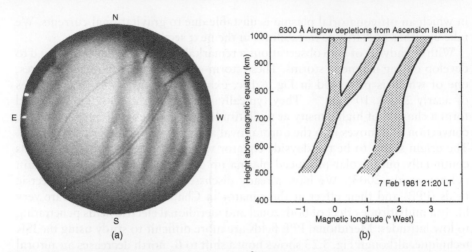

Figure 6.15a   (a) All-sky photograph at 6300 Å taken from Ascension Island at 2120 LT on February 7, 1981. (b) Linear and bifurcated airglow features recorded in the all-sky camera photograph after mapping to the equatorial geomagnetic plane. [After Mendillo and Tyler (1983). Reproduced with permission of the American Geophysical Union.]

these structures take on the vertical wedge-like shapes shown in Fig. 4.7b. These wedges tilt to the west as noted in Chapter 4 and seem to give definitive proof of the wedge versus bubble geometry for the equatorial depletions. Observations of the equatorial anomaly from the TIMED satellite often display dark bands. Comparison with an airglow camera on the ground as in Fig. 6.15b shows that they are of CEIS origin.

These depletion regions owe their origin to an equatorial phenomenon but in turn create the seed for very strong local off-equatorial F-region instabilities. This stems from their associated east-west horizontal density gradients. Such gradients are unstable to the generalized Rayleigh-Taylor instability discussed in Chapter 4. To first order, gravity does not play a role, since $\mathbf{g}$ is perpendicular to ∇n for a horizontal gradient. For $\mathbf{g} = 0$, the criterion for instability discussed in Chapter 4 is

$$(\mathbf{E}' \times \mathbf{B}) \cdot \nabla n > 0 \tag{6.15a}$$

where

$$\mathbf{E}' = \mathbf{E} + \mathbf{U} \times \mathbf{B} \tag{6.15b}$$

is the electric field that would be measured in the neutral frame of reference, while $\mathbf{E}$ and $\mathbf{U}$ are the electric field and neutral wind vectors in the earth-fixed frame, respectively. As noted earlier, $\mathbf{B}$ is unchanged in such a transformation if $\mathbf{U} \ll c$, the speed of light. At midlatitudes it is usually the case that $|\mathbf{U} \times \mathbf{B}| > |\mathbf{E}|$. For $\mathbf{E} = 0$, (6.15a) implies that an eastward wind is destabilizing to a westward

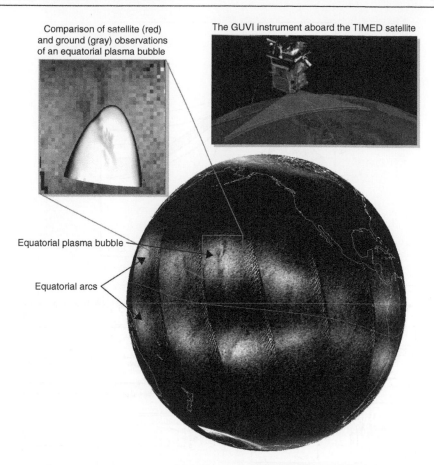

Figure 6.15b GUVI observations of the 135.6-nm emission on September 22, 2002, along with simultaneous ground-based observations. [After Kelley et al. (2003b). Reproduced with permission of the American Geophysical Union.] See Color Plate 15.

density gradient and vice versa. The depletions should thus structure on one side or the other, depending on the zonal neutral wind direction.

Since the equatorial anomaly region is one of very high electron density—the highest average electron density anywhere in the earth's ionosphere—the effect of such structuring on transionospheric propagation is very pronounced. The scintillation technique discussed in Appendix A has been used extensively in the anomaly region to characterize the structure there and to better understand the scintillation process itself. Even at a transmission frequency of 6 GHz, up to 6 dB of scintillation-induced amplitude variation has been detected in this region. An example of data from the anomaly is shown in Fig. 6.16a and b. In Fig. 6.16a, a schematic diagram shows the relative motion of an airplane conducting scintillation measurements using satellite transmission through an

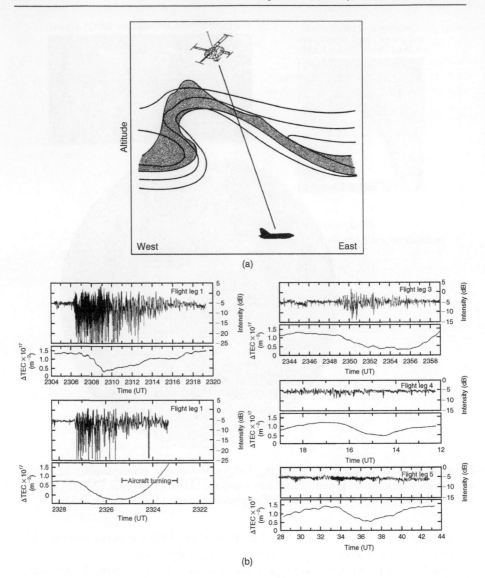

Figure 6.16 (a) Schematic diagram of an airplane flight path near Ascension Island and a satellite transmission link through a plasma uplift. (b) Data from five consecutive passes similar to the one illustrated in (a). For each leg the scintillation intensity and total electron content are plotted. The structure is clearly enhanced on one side of the bubble and decays with time. (Courtesy of R. Livingston.)

uplifted plasma wedge near Ascension Island in the Appleton Anomaly. Data from five consecutive traversals are shown in Fig. 6.16b. The line-of-sight total electron content (TEC) and the scintillation level at 249 MHz are depicted for each pass. The westward edge is clearly more structured than the eastward edge, which is consistent with the preceding discussion and the fact that the wind is usually eastward at this local time and latitude. Structures on the top and eastward regions are probably due to instabilities caused by large-scale secondary perturbation electric fields δE, which can also create $\delta E \times B$ instabilities on the gradients.

Many "active experiments" have been conducted at midlatitudes to model this process by creating artificial plasma density gradients perpendicular to B and observing the resulting structure. The active technique is to inject large amounts of barium gas into the ionosphere from a rocket, where the barium is vaporized from the metal state by the intensely energetic thermite chemical reaction. The barium is ionized by sunlight and, if released at sunset or sunrise, results in a visible long-lived plasma made usable by resonant scattering of sunlight (see Appendix A for more details). A photograph of such a release made over the Gulf of Mexico is given in Fig. 6.17. Notice that striated regions form in the plasma and that they occur in only a portion of the cloud.

The ultimate active experiment was conducted in the mid-1960s when a nuclear explosion was carried out in the ionosphere. The right-hand side of Fig. 6.18 shows the result, which has a similar appearance to Fig. 6.17 and to another barium cloud photograph to the left. Remarkably, even the spacing of the striations are comparable, on the order of 500–1000 m. The physics here is that of the deceleration of the blast as it expands outward, carrying a plasma with it.

Figure 6.17   A photograph of a striated barium cloud taken at right angles to the magnetic field. (Courtesy of W. Boquist.)

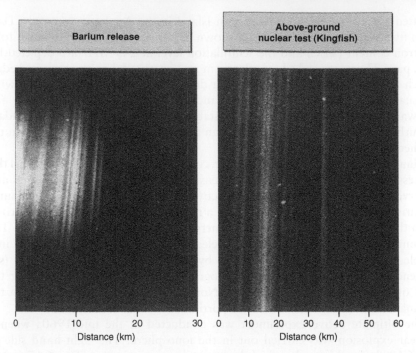

Figure 6.18 Comparison of a striated barium cloud and a late time nuclear explosion on the same distance scale. Both events occurred in the 150–200 km height range. [After Kelley and Livingston (2003). Reproduced with permission of the American Geophysical Union.]

The effective gravity is then $v_{in}(\mathbf{U} - V_i)$, which points outward. This condition is then linearly unstable to the generalized Rayleigh-Taylor instability. After the nuclear test ban treaty, barium cloud releases fueled further understanding of the nuclear case as well as CEIS/ESF and led to the computer simulations, which eventually led to the equatorial simulations presented in Chapter 4.

Whether a cloud is large or small depends on its electrical properties. A small cloud is required whenever the barium technique is used for tracer experiments, such as those discussed in relation to Fig. 3.12. The circuit diagram analogy in Fig. 6.19a applies if we associate R_2 with the reciprocal of the field line-integrated Pedersen conductivity in the cloud [$R_2 = (\Sigma_{Pc})^{-1}$], while R_1, is the inverse of the field line-integrated conductivity in the background ionosphere [$R_1 = (\Sigma_{Pb})^{-1}$]. if $R_1 \ll R_2$, the full ambient potential appears across the cloud and it drifts with the full velocity $\mathbf{E} \times \mathbf{B}/B^2$. If the cloud is large ($\Sigma_{Pc} > \Sigma_{Pb}$), the ambient electric field is shorted out and the cloud moves more slowly than the background. This has in fact been observed during a rocket flight, which traversed a dense barium plasma cloud. The data in Fig. 6.19b show that the electric field was indeed lower inside the barium cloud than outside (Schutz et al., 1973).

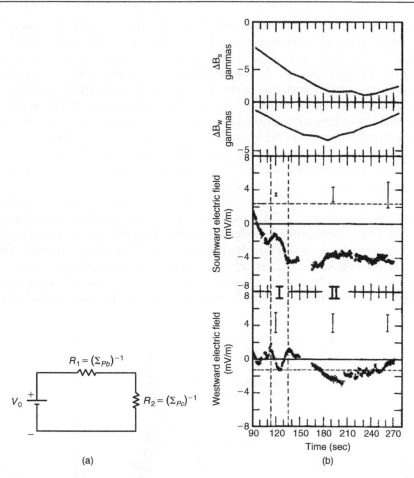

Figure 6.19 (a) Circuit diagram analog for the effect of a barium cloud on the electro-dynamics of the ionosphere. Here, V_0 is the applied voltage due to the ambient electric field, $R_1 = (\Sigma_{Pb})^{-1}$ is the resistivity of the ionosphere and $R_2 = (\Sigma_{Pc})^{-1}$ is the resistivity of the cloud. (b) Electric field signatures during an ionospheric rocket flight through a midlatitude barium cloud. The measurements were made in situ and show that **E** is con-siderably smaller inside the barium cloud (between the dashed lines) than outside. The magnetic field instrument was located on the ground. [Part (b) after Schutz et al. (1973). Reproduced with permission of the American Geophysical Union.]

We have already discussed in Chapter 4 why an electric field produces an instability in the plasma. Another way to understand the instability of a barium cloud and formation of the striations that are apparent in the photograph in Figs. 6.17 and 6.18 involves the shorting effect discussed previously. Consider alternating regions of enhanced and depleted plasma density. That is, consider a small initial sinusoidal perturbation that occurs on the particular side of a cylindrically symmetric plasma cloud where $(\mathbf{E'} \times \mathbf{B}) \cdot \nabla n > 0$. If we work in the neutral frame of reference where $\mathbf{U} = 0$, then $\mathbf{E'}$ is the electric field and $\mathbf{E'} \times \mathbf{B}/B^2$ is the background plasma flow velocity in that reference frame. A region of enhanced plasma density will have an elevated field line-integrated Pedersen conductivity and, due to the shorting effect, will drift slower than the adjacent depleted region for which Σ_P is lower. Now since $(\mathbf{E'} \times \mathbf{B})$ is parallel to ∇n, the lower-density region will move up the gradient to regions of higher surrounding density, while the high-density region will move down the gradient. Thus, relative to the surroundings, both enhancements and depletions seem to grow in intensity, and we say the configuration is unstable. This argument is equivalent to that associated with Fig. 4.9.

Midlatitude barium cloud striations have been penetrated by instrument-laden sounding-rocket payloads. Data from one such experiment are reproduced in Fig. 6.20. The smooth, enhanced-density profile due to the plasma cloud is interrupted on one edge by fingers of alternating high and low plasma density. These are the unstable electrostatic perturbations just discussed. Such a process is compared to the naturally occurring phenomenon of bottomside equatorial spread F in Fig. 6.21 by comparing the power spectrum of the barium fluctuations

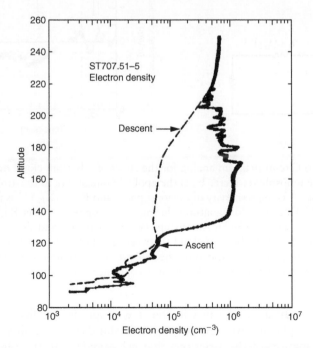

Figure 6.20 Electron density profiles from a probe rocket flight through a barium cloud. The dashed curves are the measurements of the undisturbed F-region profile on rocket descent. [After Kelley et al. (1979). Reproduced with permission of the American Geophysical Union.]

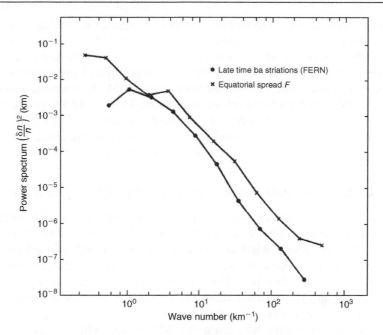

Figure 6.21 Filled circles indicate the wave number power spectrum of the data shown in Fig. 6.20. Also shown (crosses) is a spectrum from a rocket flown into equatorial spread F conditions. [After Kelley et al. (1979). Reproduced with permission of the American Geophysical Union.]

with spread-F data, which is driven by the gravitational term in the generalized Rayleigh-Taylor instability. The similarity is remarkable and gives further evidence that the active barium cloud experiment indeed mirrors natural phenomena quite well. This barium cloud spectrum is included in the set presented in Fig. 4.5a, which dramatically shows the similarity to CEIS/ESF spectra.

The growth rate for the E×B instability for a "local" calculation, ignoring, among other things, the plasma effects due to electrical coupling along the magnetic field lines is (from Chapter 4),

$$\gamma = E'/BL \tag{6.16a}$$

where L is the inverse gradient scale length of the cloud. For typical midlatitude conditions $E'/B = 50\,\text{m/s}$ while typical barium clouds have $L \sim 6000\,\text{m}$, and thus $\gamma \cong (2\,\text{min})^{-1}$. For the natural case the growth rates are lower since the gradients are weaker but they are still significant.

A first-order correction to (6.16a; see Francis and Perkins, 1975) is given by

$$\gamma(k) = \left[\Sigma_{Pb}/(\Sigma_{Pb} + \Sigma_{Pc})\right] (E'/BL) \left[k^2/\left(k^2 + k_0^2\right)\right] \tag{6.16b}$$

which takes into account the shorting effect of a conducting background E or F region and the finite size of the cloud, since $k_0 = 2\pi/L$. Considerable effort has gone into the barium cloud striation problem, and a vast literature exists. Much of this research is applicable to naturally occurring spread F phenomena as well. Some subtleties that arise involve the effect of velocity shear on the instability, which is a stabilizing factor (Perkins and Doles, 1975), and the production of image striations in the background ionosphere (Goldman et al., 1976). Both of these processes introduce additional scale size factors into the linear growth rate expression. These processes are discussed further in Chapter 10.

6.6.2 Local Midlatitude F-Region Plasma Instabilities: A New Process

A few early experimental results indicated that something new was happening at midlatitudes. Most notable were data reported by Behnke (1979) using the Arecibo radar. He found what he called height-layer bands aligned from NW to SE and traveling in the southwest direction. Adjacent bands sometimes were displaced from each other by 80 km in altitude. One structure was found to have an internal electric field five times the average background electric field and corresponding to drifts exceeding 400 m/s.

The next breakthrough was from the MU radar, which reported regular patches of 3 m irregularities moving across the various beams (Fukao et al., 1990; Kelley and Fukao, 1991). An example of such data is presented in Fig. 6.22. Three patches of high-echo strength are shown in the top panels, and the corresponding line-of-sight Doppler shifts are plotted in the lower panels. Large away drifts accompany the strong echoes, and at the edges, as the signals disappear, they weaken and even downward irregularity motions occur. The patches moved westward from beam to beam.

Airglow observations at 630 nm have greatly clarified the situation (Mendillo et al., 1997; Garcia et al., 2000a; Saito et al., 2001). Figure 6.23 shows five midlatitude examples over Puerto Rico and Hawaii and contrasts them with an image taken from Christmas Island during equatorial CEIS conditions. The latter features are very elongated and aligned with the magnetic meridian. The midlatitude structures are more localized and make a large angle with the magnetic meridian. Another key difference is the motion of these structures; their phase fronts almost invariably move in the southwest direction, as illustrated in Fig. 6.24. Since the process is not yet understood for mesoscale structures, next we will consider the only existing theoretical development before presenting more observations to provide a context for interpretation.

Figure 6.25 shows the plasma density and plasma drift/electric field observations associated with two uplift events over Arecibo. From 2215 to 2345 LT and again from 0100 to 0215 the ionosphere was elevated by 100 km altitude relative to adjacent time periods, and the associated electric field was highly unusual. A perpendicular southward perturbation drift was accompanied by a perpendicular east drift (e.g., at 2340 AST). The former was related to the decrease in

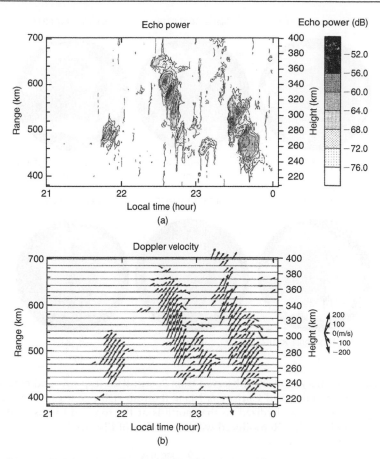

Figure 6.22 (a) Range-time-intensity radar plot obtained looking in the direction of due magnetic north at a zenith angle of 57.8° for June 8, 1987. (b) Line-of-sight Doppler velocity for the same period. The velocity corresponds to the angle of the arrow from the horizontal line as shown on the right-hand side. [From Fukao et al. (1991). Reproduced with permission of the American Geophysical Union.]

altitude of the layer that reached its low point at 0015 AST. Prior to this, a cycle of opposite phase can be seen with perturbation north and western drifts occurring before the uplift occurred. The uplift/downdraft cycle is due to the perturbation electric fields in an electrostatic wave for which $\nabla \times E = -ik \cdot E = 0$. This was even more dramatically shown by Shiokawa et al. (2003) and is reproduced in Fig. 6.26. Here, plasma flow (left panel) and electric field vectors (right panel) measured on a DMSP satellite are superposed on an airglow image. These vectors are consistent with the Arecibo data and the electrostatic wave hypothesis. Kelley and Miller (1997) even suggested a new wave mode—an electrobuoyancy wave—might be involved. Saito et al. (1995) investigated midlatitude electric

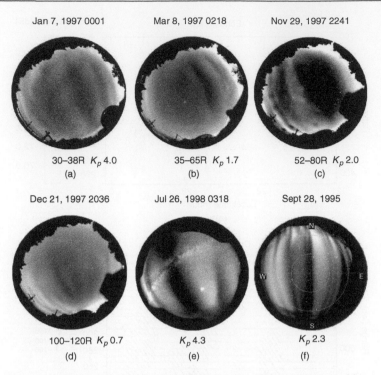

Jan 7, 1997 0001	Mar 8, 1997 0218	Nov 29, 1997 2241
30–38R K_p 4.0	35–65R K_p 1.7	52–80R K_p 2.0
(a)	(b)	(c)

Dec 21, 1997 2036	Jul 26, 1998 0318	Sept 28, 1995
100–120R K_p 0.7	K_p 4.3	K_p 2.3
(d)	(e)	(f)

Figure 6.23 Summary of events observed from Puerto Rico (top row and bottom left), Hawaii (bottom middle), and Christmas Island (bottom right, courtesy of M. Taylor). Each image has been standardized so that north is at the top and east is to the right. [After Kelley et al. (2002). Reproduced with permission of Elsevier.]

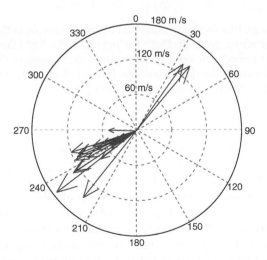

Figure 6.24 A compass plot showing the direction and velocity of structures observed using the all-sky imagers. [After Garcia et al. (2000a). Reproduced with permission of the American Geophysical Union.]

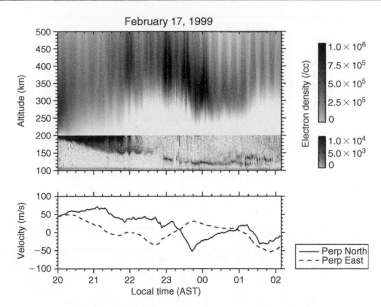

Figure 6.25 The top panel shows the electron density measured by the Arecibo incoherent scatter radar during the night of February 17–18, 1999. Note the scale change at 200 km altitude. Below, the two components of the drift perpendicular to **B** are given. [After Kelley et al. (2000b). Reproduced with permission of the American Geophysical Union.]

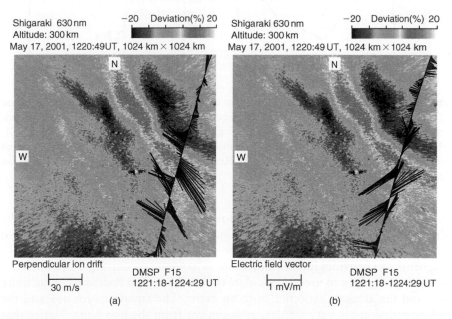

Figure 6.26 Polarized electric field in nighttime MSTID. 630 nm airglow for May 17, 2001, along with perturbation drifts and electric fields from a DMSP satellite pass. [After Shiokawa et al. (2003). Reproduced with permission of the American Geophysical Union.] See Color Plate 16.

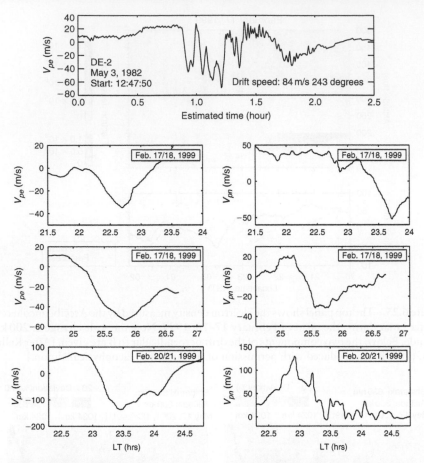

Figure 6.27a Eastward plasma drift velocity estimated with the electric field observation by DE-2 (top panel) and the average drift velocity of airglow features (three left-hand side panels) measured by the Arecibo radar during the nights of February 17–18 and 20–21, 1999. V_{pe} is shown in the left-hand panels and V_{pn} is on the right. V_{pe} is to be compared to the DE data. [After Kelley et al. (2000c). Reproduced with permission of the American Geophysical Union.]

fields measured on the DE satellite and found a band of large electric fields at the poleward edge of the anomaly region (Fig. 6.27a, b). These fields were compared to Arecibo observations and found to be similar (Fig. 6.26a).

Figure 6.28a provides further indication as to what is occurring. Here the radar observations have been used to calculate the local F-region Pedersen conductivity (σ_{PF}) and the airglow expected from an event. The equations for σ_{PF} and the airglow emission are very similar, as is evident from the two plots. Notice that the uplifted region has a very low σ_{PF} and that it is also a low-airglow zone. We wish to point out that, although low airglow can be produced simply by

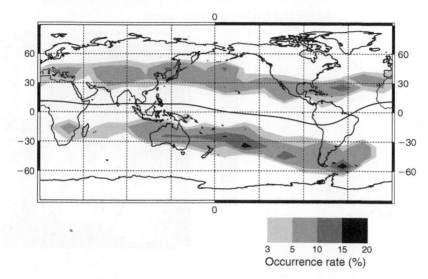

Figure 6.27b Global distribution of the occurrence rate of the MEFs observed by the DE-2 satellite between 250 and 900 km altitude. [After Kelley et al. (2002). Reproduced with permission of Elsevier Press.]

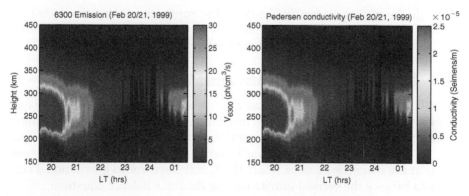

Figure 6.28a A comparison of the modeled 630.0 nm airglow emission line (left) and the calculated Pedersen conductivity (right) for February 20–21, 1999. [After Kelley et al. (2000c). Reproduced with permission of the American Geophysical Union.] See Color Plate 17.

raising the layer, these regions are also observed to be depleted of plasma (Kelley et al., 2002). Once formed, the low-TEC regions propagate toward the southwest and can last many hours. Figure 6.28b shows simultaneous TEC and airglow observations (Saito et al., 2001).

Airglow observations have been made in both hemispheres during a mesoscale TID event (Otsuka et al., 2004). Figure 6.29 shows the structures in geographic coordinates. But when the southern image is mapped along the magnetic field to

Nighttime MSTID observations (TEC, Airglow)

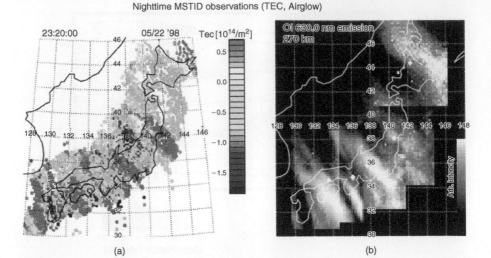

(a)   (b)

Figure 6.28b Simultaneous airglow and TEC observations over Japan. Each data point at the left corresponds to one of some 1000 GPS stations. [After Saito et al. (2001). Reproduced with permission of the American Geophysical Union.] See Color Plate 18.

the north, the images are nearly identical. The implication is that the electrification of the structures is dominant, since it is very unlikely that identical neutral gravity waves would occur in both hemispheres.

6.6.3 Linear Theory for the Perkins Instability

By the late 1970s, Behnke (1979) had already suggested that these height-layer bands were due to the Perkins (1973) instability. Before studying any instability, it is important to first describe the equilibrium for which stability is addressed. First, we note that without an eastward electric field, equilibrium can only exist exactly at the magnetic equator where the $\mathbf{J}_g \times \mathbf{B}$ force due to gravitational currents can balance the vertical gravity term in the force balance equation:

$$\mathbf{J}_g \times \mathbf{B} + \rho\mathbf{g} = 0$$

At the equator, $\mathbf{J}_g$ has magnitude $\rho g/B$ and is directed toward the east. The Rayleigh-Taylor theory given in Chapter 4 shows that this equilibrium is not stable. In general, using (2.36b) for $m \ll M$, $\mathbf{J}_g = (\rho/B^2)\mathbf{g} \times \mathbf{B}$, and if the magnetic field has a dip angle I, then the $\mathbf{J}_g \times \mathbf{B}$ force is

$$\mathbf{J}_g \times \mathbf{B} = +\rho g \cos I \hat{a}_z \tag{6.17}$$

where the coordinates are defined in Fig. 5.15. This force cannot balance the gravitational force $\rho\mathbf{g} = -\rho g \hat{a}_z$ except at the equator, and no stability exists

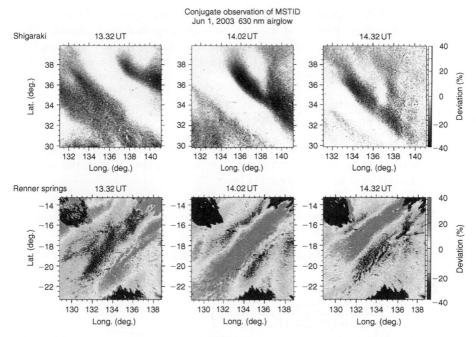

Figure 6.29 Simultaneous images of MSTIDs in both hemispheres from June 1, 2003. Note that when projected along the magnetic field, a feature in the top left-hand side of the Shigaraki image (upper panel) maps to a feature on the bottom left-hand side of the Renner Springs image (bottom panel). [After Otsuka et al. (2004). Reproduced with permission of the American Geophysical Union.] See Color Plate 19.

outside the equatorial region. The plasma simply falls down the magnetic field lines. The situation in Fig. 6.14b is unique in that it seems to describe a situation that is not in equilibrium but remains at a high enough altitude for a long enough time that gravitational current can drive the Rayleigh-Taylor instability.

Perkins noted, however, that an equilibrium was possible if the plasma was supported against gravity by a southward neutral wind, U_s. In this case, the condition for no vertical ion motion is:

$$U_s \cos I \sin I = (g/<v_{in}>) \sin^2 I \qquad (6.18a)$$

The situation is illustrated in Fig. 6.14a. On Sept. 16–17, 1999, the postsunset ionosphere was held aloft stably for several hours and then fell quickly into the atmosphere and almost disappeared. In the stable equilibrium state, if the ionosphere is displaced upward, $\langle v_{in} \rangle$ decreases and gravity causes it to fall back. If it is displaced downward, the wind (or electric field) causes it to move back up to the equilibrium height. On other nights it seems that the effective wind was variable, causing modulations of the equilibrium, but near midnight a general collapse occurred.

Perkins' breakthrough idea was to note that the entire layer could equally well be considered a single particle if the content was integrated over altitude. This is the field line-integrated ionospheric model, also known as the Ping-Pong ball model, which is illustrated in Fig. 5.10. In (6.18a), $\langle v_{in} \rangle$ is the plasma density-weighted collision frequency

$$\langle v_{in} \rangle = \frac{\int (n(z)v_{in}(z)dz}{\int n(z)dz} = \frac{\int n(z)v_{in}(z)dz}{N}$$

The field line-integrated Pedersen conductivity is thus $\Sigma_P = Ne^2 \langle v_{in} \rangle / M\Omega_i^2$.

Equivalently, an eastward electric field can support the layer in the vertical direction. In this case the condition becomes

$$\frac{E_e}{B} \cos I = (g/\langle v_{in} \rangle) \sin^2 I \tag{6.18b}$$

If u_s and E_e are given, 6.18a and b can be combined and solved for $\langle v_{in} \rangle$, which can be expressed as

$$\langle v_{in} \rangle(h) = \left(\frac{g \sin^2 I}{u_s \cos I \sin I + (E_e/B) \cos I} \right) \tag{6.19}$$

Since $\langle v_{in} \rangle (h)$ is a monotonically decreasing function of height, this equation determines the height of the F layer. In the electric field case there is a net poleward horizontal motion as the ionosphere moves up perpendicular to **B** and falls down parallel to **B**. For a pure southward wind there is no net equatorward displacement of the layer. For a Chapman alpha-layer it can be shown that $\langle v_{in} \rangle = v_{in}(h_{max})$. Perkins used the height-integrated Pedersen conductivity as one of his variables

$$\Sigma_P = \frac{e^2}{M_i \Omega_i^2} \int n(z)v_{in}(z)dz = \frac{e^2 N}{M_i \Omega_i^2} \langle v_{in} \rangle \tag{6.20}$$

Thus, Σ_P is also a surrogate for height. Since the electric field does not vary along **B**, the potential is also a "layer" variable. Perkins derived two time derivatives of importance:

$$\frac{\partial N}{\partial t} = 0 \tag{6.21a}$$

$$\frac{\partial \Sigma_P}{\partial t} = \frac{eg \sin^2 I}{\Omega BH}N - \frac{\Sigma_P E_e \cos I}{BH} \tag{6.21b}$$

where H is the neutral scale height. The first equation is simply a statement that the TEC does not change in this model. Perkins thus studied an ionosphere high enough that recombination could be ignored. The second equation can be written as:

$$\frac{\partial \Sigma_P}{\partial t} = K_1 N - K_2 \Sigma_P \tag{6.22}$$

Consider a small perturbation, $\Sigma_P = \Sigma_{P0} + \Sigma'$. Then,

$$\frac{\partial \Sigma'}{\partial t} = K_1 N_0 - K_2 (\Sigma_{P0} + \Sigma') = -K_2 \Sigma'$$

since, in equilibrium, $\partial \Sigma_{P0}/\partial t = 0$, the first two terms cancel, leaving

$$\left(\frac{1}{\Sigma'}\right) \frac{\partial \Sigma'}{\partial t} = \gamma = -K_2 \tag{6.23}$$

The growth rate is negative and the perturbation decays.

Perkins (1973) pointed this out but went on to show that if, in addition to the equilibrium wind/electric field requirement, an eastward wind and/or northward electric field is present, the system is unstable. The full growth rate is (Hamza, 1999; Garcia et al., 2000b):

$$\gamma = \frac{-E'_e \cos I}{BH} + \frac{k_e}{k_\perp^2} \frac{\cos I}{BH} \mathbf{E}'_{0\perp} \cdot \mathbf{k}_\perp \tag{6.24}$$

where $\mathbf{E}'_{0\perp} = \mathbf{E} + \mathbf{U} \times \mathbf{B}$ and E'_e is the eastward component. The real part of the frequency is

$$\omega_r = \mathbf{k}_\perp \cdot (\mathbf{E}_0 \times \mathbf{B})/B^2 \tag{6.25}$$

Note that the real part of ω only depends on $\mathbf{E}_0$, the electric field in the earth-fixed frame, not the wind vector. The phase velocity is simply the $\mathbf{E} \times \mathbf{B}$ velocity component parallel to $\mathbf{k}_\perp$. Expressions can be derived for pure electric field or pure wind-driven processes; following Garcia et al. (2000b), these are

$$\gamma_E = \left[\frac{k_e k_n}{k_\perp^2} E_e + \left(\frac{k_n^2}{k_\perp^2} - 1\right) E_n\right] \frac{\cos I}{BH} \tag{6.26a}$$

$$\gamma_W = \left[\frac{k_e k_n U_e}{k_\perp^2} + \left(\frac{k_n^2}{k_\perp^2} - 1\right) U_s \sin I\right] \frac{\cos I}{H} \tag{6.26b}$$

where subscripts e and n stand for east and north, respectively. The most unstable k vector for case (6.26a) lies halfway between $\mathbf{E}_\perp$ and the eastward direction (Perkins, 1973). For $E_n = E_e$ this would be 22.5° north of east. For $U_s \sin I = U_e$ the same condition applies. For the latter case, since $\omega_r = 0$ these structures would grow in place and be aligned from NW to SE as observed. This directionality is a major triumph of the Perkins theory, since the observed wave fronts do make an angle with the magnetic meridian. However, the growth rate is very slow and is unlikely to provide significant amplification of thermal noise.

For the pure electric field case, the real part of ω is negative for the most unstable wave vector, since $k_y E_e > k_x E_N$ and the wave would seem to travel toward the southwest as observed (e.g., Fig. 6.23f). The problem with this seemingly positive result is that under typical conditions, both the background stability and instability are established by the winds, not the electric field. In fact, since the average electric field is southeastward after sunset, ω_r is positive and wind-driven instability structures should drift toward the east, not the west. This is a major problem for the linear theory.

Figure 6.28b shows that the TEC does vary in these events, so unlike, in the Perkins model, $\partial N/\partial t \neq 0$. This is likely a three-dimensional effect. The first results of a three-dimensional simulation of this phenomenon show promise for new theoretical insights (Yokoyama et al., 2008). In the next section, a similar layer instability is discussed for the E region, and we take a new look at coupling between the layers.

6.7   Midlatitude E-Region Instabilities

For many years extrapolation of equatorial instabilities to midlatitudes was thought to be trivial. Today we know this is far from the truth. One key difference is that electric fields share importance with neutral winds. Another is that the plasma gradients are sharper at midlatitudes than elsewhere due to the layering process discussed above. Even pure neutral atmospheric instabilities, such as the Kelvin-Helmholtz process, seem to play a role in creating plasma structures.

6.7.1   Radiowave Observations of Nighttime Midlatitude E-Region Instabilities

Figure 6.30a illustrates a well-documented sporadic E event over Arecibo. As an intermediate layer descends from around 180 km, it is most likely supported against diffusion by a meridional wind shear in a tidal wave mode. As NO^+ and O_2^+ decay away, a core of metallics maintains the layer and it steepens even more if subjected to a zonal wind shear. This mode change may have occurred at 2148 when the layer peak briefly began to rise. Quickly thereafter, as shown in Fig. 6.30b, the layer goes unstable to long-period variations in the Doppler shift, similar to equatorial large-scale waves, waves with λ much greater than the layer thickness. The mean Doppler shift of 30 m/s toward the radar could be due to an

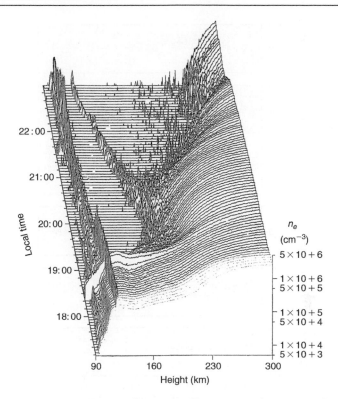

Figure 6.30a Consecutive electron density profiles measured over Arecibo on May 7, 1983. Notice the intense sporadic E layer, which existed even prior to sunset and the intermediate layer which descended through the F-layer valley. [After Riggin et al. (1986). Reproduced with permission of the American Geophysical Union.]

$E \times B$ drift or a mean wind toward the radar. The Doppler oscillations, however, must be due to meridional perturbation electric fields since the radar was located in St. Croix, which is east of Puerto Rico. As we shall see below, equatorial-like geometries may indeed occur at midlatitudes and lead to two-stream conditions.

Figure 6.30c shows a more spectacular event detected in the same geometry in which the spectra seem to saturate at close to the speed of sound. Schlegel and Haldoupis (1994) presented several examples of midlatitude Type 1 waves just like this case, one of which is reproduced in Fig. 6.31. They concluded that a modified two-stream condition existed. By "modified" they mean that the narrow two-stream spectra saturate at a relatively low velocity compared to the equatorial case. They concluded that this is most likely due to metallic ions and to the correspondingly lower acoustic speed. It is not at all obvious how such large drifts could occur, since the equatorial case has a unique geometry

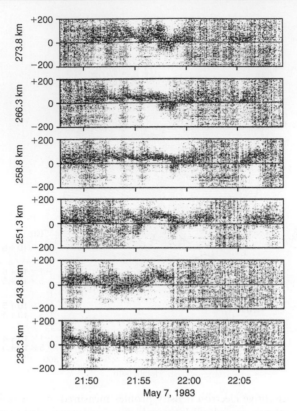

Figure 6.30b   Doppler spectrogram (measured from St. Croix with the CUPRI 50 MHz radar) for the event associated with Fig. 6.30a. The gray scale denotes intensity normalized to the peak intensity in a given spectrum and the vertical axis gives the Doppler shift in meters per second. The range is noted in kilometers. [After Riggin et al. (1986). Reproduced with permission of the American Geophysical Union.]

and the typical electric fields are small. As we shall see following, equatorial-like geometries may indeed occur at midlatitudes and lead to two-stream conditions.

Many midlatitude events have been characterized as quasi-periodic or Q-P echoes. An example from the Gadanki radar in India is illustrated in Fig. 6.32. Typically, the echo slants downward in range as time goes by, as shown. Q-P echoes have been reported from many sites (Yamamoto et al., 1991, 1992; Huang et al., 1995; Chau and Woodman, 1999; Tsunoda et al., 1999). Due to the high aspect sensitivity of plasma scattering, the characteristic Q-P slants, which seem to be altitude dependent, may in fact be related to a change in range as such a blob moves into and out of the beam at a fixed height. For example, a structure slanted from northwest to southeast and moving toward the west (which as we

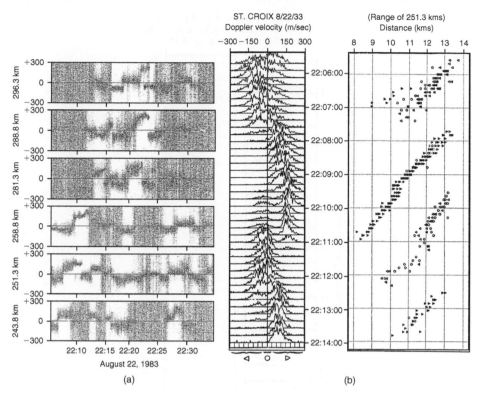

(a)   (b)

Figure 6.30c Examples of square waves in the Doppler velocity using gray scale (to the left) and spectra plots (middle). The plots on the right are interferometric velocities across the beam. A range of 250 km corresponding to 105 km altitude. [After Riggin et al. (1986). reproduced with permission of the American Geophysical Union.]

show next is typical) would enter the field of view at a range that decreases with time, even if it was unstable at a fixed height.

Various observations have implied that E_s layers tend to organize into frontal structures with phase fronts aligned northwest to southeast (northeast to south-west) in the Northern (Southern) Hemisphere. Sinno et al. (1965) analyzed time-delay measurements of Loran transmissions in the Northern Hemisphere and concluded that they could be explained by an organization of E_s layer plasma into frontal structures with the orientation described. Goodwin and Summers (1970) came to a similar conclusion by analyzing data from a spaced ionosonde network in the Southern Hemisphere. Goodwin (1966) suggested that the fronts could be 1000 km long. Bowman (1989) modeled scintillations from E_s layers in the Southern Hemisphere as opaque high-density strips arranged in a frontal structure, also with the alignment described. The fact that the frontal alignment mirrors about the equator suggests an electrodynamic cause.

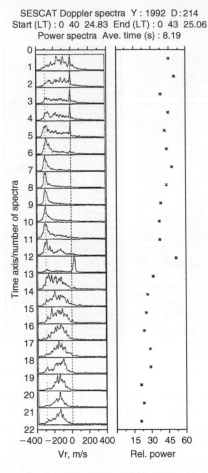

SESCAT Doppler spectra Y : 1992 D : 214
Start (LT) : 0 40 24.83 End (LT) : 0 43 25.06
Power spectra Ave. time (s) : 8.19

Figure 6.31 A detailed sequence of self-normalized Doppler power spectra prior, during, and after the occurrence of typical type 1 echoes (numbers 7–11), characterized by an invariant peak at velocities about −300 m/s. The 22 boxes correspond to 22 ranges. These are generated by the Farley-Buneman instability during a brief period of an impulsive electric field that exceeds the instability threshold near the ion acoustic speed. Also seen are typical type 2 echoes (numbers 1–4 and 15–22), characterized by a broad and variable spectrum centered at low Doppler shifts. Type 2 irregularities are believed to be secondaries generated from longer wavelength waves during conditions of strong plasma turbulence. [After Schlegel and Haldoupis (1994). Reproduced with permission of the American Geophysical Union.]

Most convincing are the results obtained by Goodwin and Summers (1970), who used a network (near Brisbane, Australia) of two ionosondes, and four receivers that were geographically separated. With this network, 10 reflection points in the ionosphere could be monitored. They categorized E_s into two types:

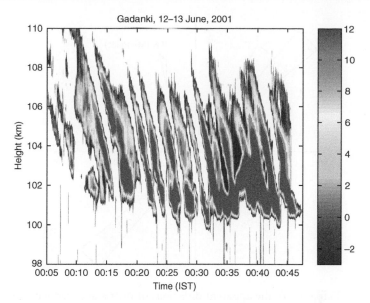

Figure 6.32 Q-P radar echoes observed with the Gadanki radar. (Figure courtesy of R. K. Choudhary.) See Color Plate 20.

"steady" and "changing." At night, 77% of the observed E_s were of the changing type, with a much lower occurrence probability during the day. Most of the changing types were found to be elongated structures that appeared to move horizontally without change of form in any 15-minute interval. The distribution of drift directions associated with "ridges" in 110 E_s frontal structures are presented in Fig. 6.33. The finding that drift azimuths were tightly clustered about the northwest and southeast directions is consistent with the passage of frontal structures. Similar results were obtained earlier by Clark (1965), whose results have been reproduced in Bowman (1985).

More recent VHF observations from St. Croix have been made in conjunction with high-resolution incoherent scatter observations of the plasma density profiles using the dual beam system (Hysell et al., 2004). These also found large-scale waves with occasional type 1 characteristics. Interferometric data showed that the Q-P-like echoes came from near 105 km event, although the range versus time presentation could have been misinterpreted as a height change. The echoing periods were collocated with patches of high density plasma at 105 km moving across the dual beam system. The horizontal interferometry (see Fig. 6.34) showed that the echoing regions were aligned from northwest to southeast.

Finally, Urbina et al. (2000) reported on a weak echoing region over Puerto Rico located below 100 km, which seems to be almost always present at night. The signal is relatively featureless with low Doppler shifts toward the radar and narrow spectra.

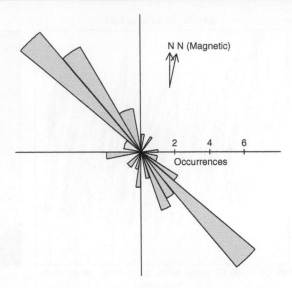

Figure 6.33 Histogram of the apparent directions of drift associated with 110 cases of ridges in E_s structures. [Adapted from Goodwin and Summers (1970). Reproduced with permission of Elsevier Science Ltd.]

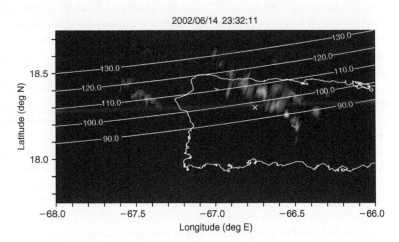

Figure 6.34 Radar image of coherent echoes received at 23:32 UT. Note that UT = LT + 4. Blue corresponds to echo velocity toward the southeast and yellow-green toward the northwest. [After Hysell et al. (2004). Reproduced with permission of the European Geosciences Union.] See Color Plate 21.

6.7.2   Multiexperimental Observations of Midlatitude Structures

Surprisingly, neutral atmosphere perturbations, some of them periodic, have also been seen in conjunction with VHF radar echoes (Kane et al., 2001). Figure 6.35 shows a spectacular set of sodium atom pillars over Arecibo. Panel b shows that these were accompanied by strong VHF echoes related to the plasma layers, which are added to the plot in panel c. We are just beginning to understand this type of coupling between ion and neutral plasma instabilities.

A number of rocket/radar experiments have been dedicated to the study of midlatitude E-region plasma processes. An early one (Kelley et al., 1995) found that the plasma layer was organized horizontally by periodic structures with a scale of 15 km and an amplitude of the order of 50% ($\delta n/n$). One of these features had additional fluctuations that extended from a peak at 30 m to well into the m-scale. The rocket only caught the end of the event in time and space, but the VHF radar data clearly showed Q-P-type organization. The spectrum of density fluctuations was almost identical to that of the localized electric field structures presented in Section 6.7.7.

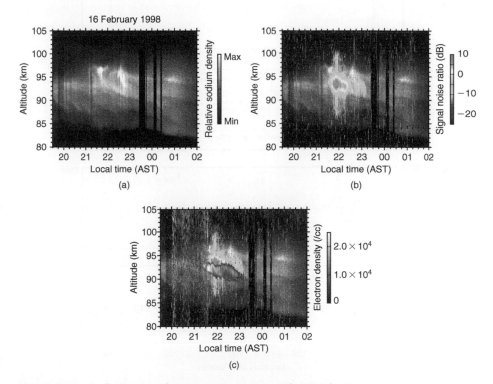

Figure 6.35   Sodium atom density contours (a), with VHF backscatter overlay (b), and with ISR sporadic E density overlay (c) for the nights of February 16–17, 1998. Unfortunately, the radar data are noisy due to interference. [After Kane et al. (2001). Reproduced with permission of the American Geophysical Union.] See Color Plate 22.

Two major rocket/radar campaigns called SEEK I and SEEK II (Sporadic-E Experiment over Kyushu) were carried out in Japan (Fukao et al., 1998; Yamamoto et al., 2005) during Q-P conditions. Plasma density and electric field observations are presented in Fig. 6.36. As reported earlier by Kelley et al. (1995), the fluctuations are organized in time of flight, not altitude. For an electrostatic process, $\mathbf{k} \cdot \mathbf{E} = 0$ and thus, considering the vectors in the lower plot, $\mathbf{k}$ is either aligned from northeast to southwest or 180° opposite. Since the radar line of sight was about 20° east or north, a slanted structure at a 45° angle to the north and a slanted structure with the most intense radar signals at a fixed height would yield radar echoes that come from the largest range first, as observed. These data

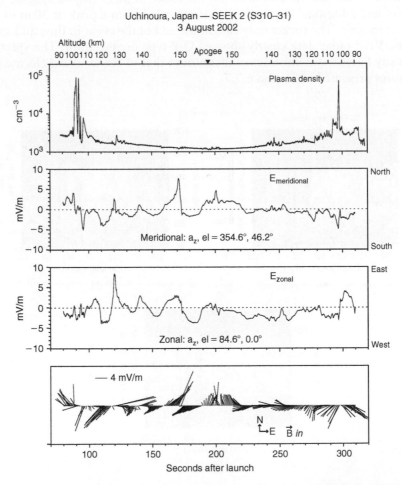

Figure 6.36 The DC electric field solution for the SEEK II experiment in magnetic coordinates with the plasma density shown in the uppermost panel. [After Pfaff et al. (2005). Reproduced with permission of the European Geosciences Union.]

are thus consistent with echoing regions and large-scale electrostatic waves organized horizontally with about a 15 km separation traveling toward the southwest. It appears clear then based on the Hysell et al. (2004) interferometry that Q-P echoes should not be considered as altitude changes but rather as horizontal echoing regions making an angle to the radar beam and moving through it.

6.7.3 Midlatitude E-Region Instabilities: Difficulties with Simple Explanations

Consider first the instabilities generated by global scale dynamo electric fields and global scale neutral winds (tides) interacting with vertically stratified plasma layers. Both the two-stream and gradient drift instabilities have difficulties in this case. As discussed in Chapter 4, the condition under which primary two-stream waves occur is given by

$$V_D \simeq E_0/B > (1 + \Psi_0)C_s \qquad (6.27)$$

where $\Psi_0 = \nu_i\nu_e/\Omega_i\Omega_e$, C_s is the acoustic speed, and E_0 is the dc electric field. The magnetic field increases quickly with latitude, while the midlatitude electric field is actually smaller than the field at the equator. Thus, (6.27) is difficult to satisfy and seems to preclude primary two-stream waves at midlatitudes. However, in the next section it is argued that if the layers are patchy horizontally, a two-stream instability can be driven by polarization of the patches.

The problem with the classic electric field-driven gradient drift instability source is that the layers are not nicely perpendicular to the magnetic field. As illustrated in Fig. 6.37a, when a vertically stratified layer is subject to a perpendicular

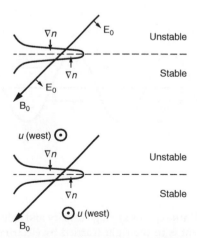

Figure 6.37a Geometry for stability of a vertically stratified layer in the presence of a southward electric field (which must map along **B**) and a uniform westward wind (which does not, however, have to be uniform with height).

electric field, the layer is stable on the side where **E** is parallel to ∇n but stable on the other side. Likewise, as shown in the bottom panel of Fig. 6.37b, a uniform zonal neutral wind is unstable where $U \times B$ is parallel to ∇n. But in both cases, a layer is unstable on one side but stable on the other. At first glance this seems to preclude any instability at all when the growth rate is integrated over the layer, a fact that makes midlatitudes as interesting as they are. In the next sections we explore how additional complexities in the midlatitude ionosphere can be used to help explain midlatitude E-region radar echoes. For reference, and due to the potential importance of neutral winds, we show the conditions for the

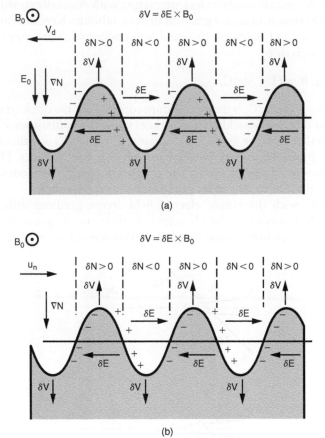

Figure 6.37b A schematic diagram for gradient drift instability driven by (a) an electric field or (b) a neutral wind. Dark regions are high density and light regions are low density. In (a) the classic Hall current is to the right (carried by electrons $E_0 \times B$ drifting to the left). A small; perturbation then grows due to the $\delta E \times B$ drift. In (b) the neutral wind (u_n) simply drags the ions with it to the right with the same effect. (Figure courtesy of L. Kagan.)

gradient drift instability in Fig. 6.37b for the topside of a plasma layer similar to that shown in Fig. 6.37a. The perturbation charges and electric fields that characterize the electrostatic wave are also shown. The electric field-driven case is shown in (a) and the neutral wind-driven case in (b).

In summary, the simple extension of equatorial instability theory to mid-latitude is not very satisfying. Applying global scale electric fields and horizontal winds to horizontally stratified layers cannot produce plasma structures with scale sizes larger than a few tens of meters (see Section 6.7.7) due to the mapping of the large scale wave electric fields along **B** to a stable region.

6.7.4 The Effect of a Wind Shear: The Kelvin-Helmholtz Instability as a Source of Q-P Echoes

The wind shears discussed in Section 5.3.2 are often high enough to create a neutral atmospheric Kelvin-Helmholtz instability (KHI). This would result in periodic uplifts of the neutrals such as seen in Fig. 6.35a and in the plasma as well, since the ions are tightly controlled by the neutral gas (Larsen, 2000). One explanation for Q-P echoes involves just such a KHI process. The two rocket campaigns carried out in Japan to study these instabilities included TMA releases. Large shears were indeed seen during the events (Larsen et al., 1998, 2005). Bernhardt (2002) has simulated the KHI in the lower thermosphere and its effects on the plasma layers. Figure 6.38 shows KHI billows made visible by a TMA trail during the SEEK-2 project. When present the KHI must affect the plasma but it is not yet clear if they are the primary cause of the Q-P echoes.

Figures 6.35a–c are a graphic example of the coupling of neutral and plasma fluids. The sodium atom tracers of the neutral gas shown in Fig. 6.35a reveal multiple billowing structures typical of Kelvin-Helmholtz shear-driven instability

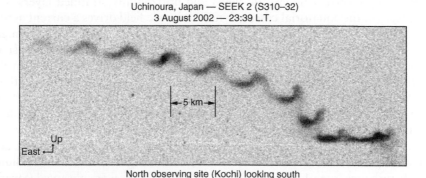

Uchinoura, Japan — SEEK 2 (S310–32)
3 August 2002 — 23:39 L.T.

North observing site (Kochi) looking south

Figure 6.38 A photographic negative of the TMA chemical release for the Kochi observing site showing the billowing structure in the neutral wind. [After Pfaff et al. (2005). Reproduced with permission of the European Geosciences Union.]

(KHI). This causes a violent mixing of the fluid, and, as seen in the 50 MHz echoes presented in Fig. 6.35b, the plasma responds, even at small scales. Evidence for unstable shears—that is, shears with a Richardson number less than 1/4—were presented in Section 5.3.2 and seem to be very common in this height range. In such a process it is also common for periodic billows to form with separation scales several times the shear scale. In the case of the E region, the result would be separations in the range of 5–15 km of the radar echoes, which is typical of Q-P spacings (Larsen, 2000).

Since the Kolmogorov microscale at 105 km is many tens of meters, it is not at all obvious how 3 m scales are created, which is needed to explain the radar scatter. A simple passive scalar-mixing argument will not produce structures at such small scales, nor will it produce structures with k strictly perpendicular to B. Some plasma process is required to allow coupling across the scales from tens of kilometers to meters. Such a process in polar mesospheric clouds involving charged ice particles is described in the next chapter. However, that mechanism is not likely to occur at middle latitudes, and, in any case, the scattering is much more isotropic in the polar summer mesosphere than was observed in sporadic E echoes. Several examples from Kane et al. (2001), including Fig. 6.35c, revealed 3 m echoes only on the top of the plasma layer. If the wind was strictly westward across the layer, only the topside would be unstable (see Section 6.7.7). Since the KHI seems very common in the midaltitude E region, neutral atom billows have been directly associated with VHF backscatter (see Fig. 6.35) and billow spacings are the right spacing; it seems the KHI plays a role in at least some Q-P echoes.

6.7.5 The Role of Horizontal Structure: Amplification by the Cowling Effect

The occasional observation of Type 1 or two-stream instabilities is difficult to explain using global scale electric fields and horizontally stratified layers. For example, as in the equatorial case, a dynamo electric field drives a current across the magnetic field lines. Since the layers are very sharply bounded, one might think that they would polarize and the Cowling effect would occur. But since the magnetic field lines are oriented at an angle, any polarization field in the meridian plane will be shorted out along the magnetic field lines. Thus, there is no Cowling effect and no horizontal type 1 waves. Haldoupis et al. (1996, 1997) proposed that sharp plasma boundaries in the zonal direction could lead to large amplification of an applied meridional electric field. These horizontal structures have been observed over Arecibo by Smith and Miller (1980) and clearly are present in Fig. 6.7. In fact, during that E-region plasma and sodium-atom disturbance, type 1 echoes and large F-region drifts were observed (Swartz et al., 2002). The geometry for this process is presented in Fig. 6.39 for a southward electric field, the usual polarity for the evening to early morning hours. This direction electric field drives a westward Hall current through the sporadic

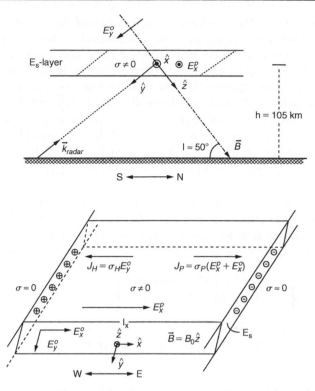

Figure 6.39 Geometry for generating large polarization electric fields. [After Haldoupis et al. (1996). Reproduced with permission of the American Geophysical Union.]

E layer. But if the layer is localized zonally by sharp boundaries and if no charge leaks off to the local F region and/or the conjugate E and F regions, a large eastward polarization field will build up to keep $\mathbf{V} \cdot \mathbf{J} = 0$. Using their geometry and notation, the eastward polarization field is denoted by E_x^P and we have,

$$E_x^P = \frac{\sigma_H}{\sigma_P} E_y^0 - E_x^0 \tag{6.28}$$

where E_x^0 is the eastward component and E_y^0 is the southward component of the dynamo electric field. Since σ_H/σ_P can easily exceed a factor of 10, the polarization field can drive the two-stream instability. Similarly, a zonal wind (u_e) will polarize a zonally localized patch with an electric field of value (σ_H/σ_P)$u_e B$, which then drives a meridional current that can be quite large. But in either case, one might wonder, why doesn't the thin sporadic E layer polarize and stop the current? In fact, it does polarize but only in such a way as to stop the vertical component of the current. This is easy to do since the conductivity along $\mathbf{B}$ is very large.

The existence of metallic ions helps the generation of the two-stream waves in several ways: (1) they reduce the acoustic speed and thus the threshold drift; (2) their meteoric origin most likely leads to the patchy horizontal structures that polarize; and (3) they yield a higher σ_H/σ_P ratio than molecular ions (Haldoupis et al., 1997). Operating against all these positive features is the load the polarization electric field must maintain. Typically, $\Sigma_P^F > \Sigma_P^E$ and the source must drive two F regions and the conjugate E region. This reduces E_x^P and most likely explains why the midlatitude two-stream instability is relatively rare. On the other hand, $\Sigma_H^E \cong neh/B$, where h is the layer thickness, may be the appropriate internal conductance. For $h = 5$ km, $n_{max} = 10^5$ cm^{-2}, $\Sigma_H^E = 2$ mho, which is sizeable.

Before leaving this subject, we explore the general case including neutral winds and the possibility of polarization fields in *both* directions. In the plane perpendicular to **B**, the two-dimensional current vector in the coordinate system used by Haldoupis et al. (1997) is given by:

$$J = \begin{pmatrix} \sigma_P & -\sigma_H \\ +\sigma_H & \sigma_P \end{pmatrix} \begin{pmatrix} E_x^0 + E_x^P + u_s B \cos I \\ E_y^0 + E_y^P - u_e B \end{pmatrix} \tag{6.29}$$

where u_s is the southward wind, u_e is the eastward wind, and I is the dip angle. We have included the polarization electric fields in both directions. Solving for J_x and J_y, we get

$$J_x = \sigma_P \left(E_x^0 + E_x^P + u_s B \cos I \right) - \sigma_H \left(E_y^0 + E_y^P - u_e B \right)$$

$$J_y = \sigma_P \left(E_y^0 + E_y^P - u_e B \right) + \sigma_H \left(E_x^0 + E_x^P + u_s B \cos I \right)$$

For the ideal case in which the local F region and conjugate E and F regions are ignored, we set $J_x = 0$ and $J_y = 0$ and solve for the polarization fields

$$E_x^P = +\frac{\sigma_H}{\sigma_P} \left(E_y^0 + E_y^P - u_e B \right) - E_x^0 - u_s B \cos I \tag{6.30a}$$

$$E_y^P = -\frac{\sigma_H}{\sigma_P} \left(E_x^0 + E_x^P + u_s B \cos I \right) - E_y^0 + u_e B \tag{6.30b}$$

Simplifying the two equations, we get

$$E_x^P = -E_x^0 - u_s B \cos I \quad ; \quad E_y^P = -E_y^0 + u_e B \tag{6.30c}$$

and, finally,

$$E_x = -u_s B \cos I \quad ; \quad E_y = u_e B \tag{6.30d}$$

In this case, (6.30c) shows that large polarization fields do not build up. Furthermore, (6.30d) shows that since the electrons **E**×**B** drift in the total electric field and move parallel to $\mathbf{B}$ at the velocity $-u_s \cos I$ (as do the ions), the plasma just moves with the neutral wind. Thus, to obtain the large electric fields, the patches must have one dimension much larger than the other.

To summarize, the very existence of two-stream instabilities at midlatitudes seems to require patchy but elongated sporadic E layers, and these indeed have been observed. The driving mechanism is most likely to be large zonal neutral winds and wind shears. But when the large scale F-region structures occur, as discussed earlier in this chapter, large meridional electric fields exist, and the F-region contribution to the load is greatly reduced where the plasma is at high altitude (low airglow region). Both effects contribute to enhanced E-region instabilities. Furthermore, Swartz et al. (2002) and Kelley et al. (2003a) show examples in which F-region structures are highly correlated with strong E-region irregularities. Since electric fields map both directions it is not obvious which region is dominant when, as Swartz et al. (2002) found, the height-integrated conductivities are comparable in the two zones.

Coupling between the E and F regions has been discussed for a long time, going back to Farley (1959). At midlatitudes, the nighttime F-region conductivity exceeds that of the E region so it usually wins the electrodynamic battle. But Tsunoda and Cosgrove (2001), Haldoupis et al. (2003), Cosgrove and Tsunoda (2004a, b), and Shalimov and Haldoupis (2005) have shown that E-region sources can affect the F region.

At first, there was debate about the configuration of Fig. 6.39: Because the polarization electric field would drive a strong secondary Hall current down the length of the structure (southward), there appeared to be no way for this current to close. Hence, it appeared that a secondary polarization effect would quickly shut down the process, as indicated by (6.30c,d). However, Shalimov et al. (1998) showed that if the structures were narrow in the east-west direction and extended in the north-south direction, then closure of the secondary Hall current could occur through the F region. This closure mechanism relied on the assumption that the zonal polarization electric field extended over a short enough spatial scale that it would not map effectively to the F region. Hysell and Burcham (2000) and Hysell et al. (2002) verified the Haldoupis et al. (1996) theory with the Shalimov et al. (1998) current closure path using simulations.

Tsunoda (1998) showed that the geometry of Fig. 6.39 can be generalized and that the long direction of the plasma structure can have any alignment in the horizontal plane, as long as there is a component of the Hall current (from the background electric field or neutral wind) perpendicular to the long direction. Also, Cosgrove and Tsunoda (2001) showed that if the mechanism was driven by a sheared neutral wind, then closure of the secondary Hall current could occur through other E_s patches, eliminating the need for long narrow structures and electrical contact with the F region. In fact, Cosgrove and Tsunoda (2002a) showed that, simply given the configuration of an E_s layer located in a vertical wind shear (such as the one that supposedly formed the layer), generic altitude or field-line-integrated (FLI) density distortions create polarization electric fields through essentially the mechanism of Haldoupis et al. (1996), with current closure occurring entirely within the E_s layer. Given the large midlatitude wind shears found by Larsen (2002), relatively smooth altitude modulations of an E_s layer can result in substantial polarization electric fields through this mechanism with more or less arbitrary orientations.

Finally, although we have concentrated on the two-stream case in this section, the sizable polarization electric field and the existence of the zonally structured sporadic E patches suggest that the gradient drift process will occur even more often. Indeed, in Fig. 6.31, even during times when two-stream is not occurring, type-II echoes are evident. The spectra are a bit narrower than the equatorial case, which suggests that Sudan-like strong turbulence may be less common at midlatitudes (Haldoupis et al., 2003).

6.7.6   Spontaneous Structuring by the E_s-Layer Instability

As previously described, Perkins (1973) studied altitude modulations of the F layer and thereby discovered the Perkins instability. Armed with the theory of polarization of E_s layers in a wind shear brought about by altitude (or field line-integrated (FLI) conductivity) modulation, Cosgrove and Tsunoda (2002b) undertook a similar study of the stability of E_s layers. They found that the equilibrium configuration of an E_s layer at a zonal wind shear node is unstable at night to altitude and FLI density modulations. (During the day, polarization fields are heavily loaded by the highly conducting E region.) The unstable modulations are horizontally distributed as plane waves, and the growth rate maximizes when the plane wave phase fronts are aligned northwest to southeast (southwest to northeast) in the Northern (Southern) Hemisphere. Hence, this E_s-layer instability (E_sLI) is in the same class as the Perkins instability, since it shares a similar geometry. However, it differs from the Perkins instability in that it involves Hall currents and derives its free energy from a wind shear instead of gravitation. In addition, the compressibility of the E region leads to the coupling of FLI density modulations to altitude modulations, which does not occur for the Perkins instability.

Tsunoda et al. (2004) have summarized the experimental evidence for the E_sLI, which can explain wavelengths from about 500 meters up to hundreds of kilometers. Hence, it applies to the longer wavelengths for which the gradient drift instability does not apply (Kelley and Gelinas, 2000; see Section 6.7.6). It also provides an alternative to the Kelvin-Helmholtz instability (see Section 6.7.2) as a way to create horizontal structuring of E_s layers, which applies when the wind shear is hydrodynamically stable. Therefore, through the Haldoupis et al. (1996) polarization mechanism, it can also lead to large polarization electric fields, which are needed to explain the observations of type 1 and large-spread type 2 Doppler spectra. Simulations by Cosgrove and Tsunoda (2003) have shown the nonlinear evolution of the E_sLI and demonstrated the generation of large polarization electric fields, even when loaded by the F region. An example simulation is shown in Fig. 6.40, with the integrated F-region conductivity (Σ_{PF}) is set so that $\Sigma_H = 2.5(\Sigma_{PE} + \Sigma_{PF})$, where Σ_{PE} and Σ_H are the integrated Pedersen and Hall conductivities, respectively, of the E_s layer. This conductivity ratio is critical for determining the activity level of the instability, and a detailed investigation of its morphology is needed in the future. Fig. 6.40 shows isodensity contours for cross sections of the E_s layer at successive times on the left and the

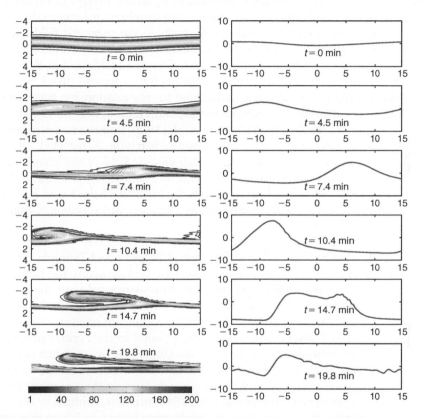

Figure 6.40 The left panels show isodensity contours on cross sections through the E_sL, from numerical simulations, for six successive times. At time $t = 0$ the layer is perturbed by a $\pm 1/4$ km sinusoidal altitude modulation, and the subsequent panels show the growth phase of the E_s layer instability. (The cross-sectional plane is rotated about the horizontal southwest-northeast line—(the horizontal axis)—so it contains the magnetic field.) The right-hand panels show the corresponding electric fields at the original layer altitude of 105 km, plotted versus position on the horizontal axis. All scales are kilometers, except for the electric field ordinate, which is mV/m. The integrated Pedersen conductivities of the E_s and F layers (Σ_{PE} and Σ_{PF}) and the integrated Hall conductivity of the E_s layer (Σ_H) satisfy $\Sigma_H = 2.5(\Sigma_{PE} + \Sigma_{PF})$. The line color indicates the percentage of initial peak density. See Color Plate 23.

corresponding electric field at the equilibrium layer altitude on the right. In this example, the layer evolution becomes quickly nonlinear, and there is an 8 mV/m electric field pulse associated with the instability growth phase.

6.7.7 Coupling of E_s Layers and the F Layer

As far back as 1960 (Bowman, 1960), researchers have noted the apparent relationship between structures in the F layer and in E_s layers. Farley (1959) demonstrated that electric fields of sufficient spatial scale should map between the E and F regions due to the extremely high conductivity in the magnetic field-aligned direction, thus giving the beginning of a theoretical foundation for the observations. Tsunoda and Cosgrove (2001) accounted for the Haldoupis et al. (1996) polarization mechanism and described a scenario for positive feedback between an E_s layer and the F layer as a way to account for the relationship between E_s-layer and F-layer structure. In companion papers, Haldoupis et al. (2003) and Kelley et al. (2003a) invoked the Haldoupis et al. (1996) polarization mechanism to explain new observations of E_s-F-layer coupled behavior. Kelley et al. (2003a) noted that an eastward E-region perturbation electric field of sufficiently large scale would cause the F region to rise, reducing its conductivity and thus its load on the E region. All of this led up to a unified treatment of the F and E_s layers and of the Perkins and E_s-layer instabilities as a single coupled system (Cosgrove and Tsunoda, 2004a,b).

Using the assumption that electric fields of sufficient spatial scale map unattenuated along **B** between the E and F regions, Cosgrove and Tsunoda (2004a) solved the equations of motion associated with the Perkins instability (see equation 6.21b) together with the equations of motion associated with the E_sLI to obtain the growth rate for the coupled E_s-F-layer system. In general, they found two unstable modes. Under conditions in which a significant relative drift existed between the E_s and F layers, one mode corresponded to the Perkins instability and the other corresponded to the E_s-layer instability. However, when the relative drift between the E_s and F layers was small, there was only a single unstable mode, indicating a true coupling of the two instabilities. In this case they found that the overall system growth rate was normally more than doubled by the coupled electrodynamic effect. Of course, the implication is that the E-region neutral wind and the F-region $E \times B$ drift are equal, which would be rare.

To treat the coupled condition when the wavelength is not long enough to assume 100% efficiency of the electric field mapping between the E and F regions and to obtain an approximate analytic expression for the coupled system growth rate, Cosgrove and Tsunoda (2004b) employ a circuit model approach similar to that given for the barium cloud in Fig. 6.19. Figure 6.41 shows the development of this circuit model, which is generally useful for the two-layer coupled problem, under the Farley (1960) mapping criteria. The resistors R_0, R_F, and R_{Es} are defined in terms of the field-aligned conductivity (σ_0), the FLI Pedersen

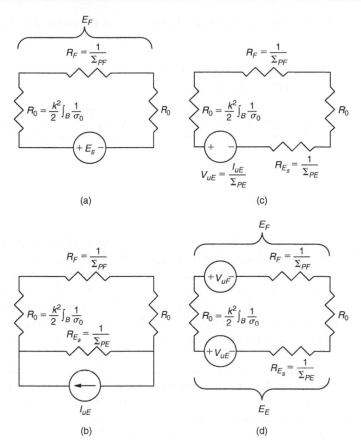

Figure 6.41 Circuit models for E-F coupling. (a) Mapping of an electric field from an E_s layer to the F layer. (b) Wind driven E_s layer source. (c) Thévenin equivalent for (b). (d) Wind-driven sources in both E_s and F layers.

conductivity of the F region (Σ_{PF}), and the FLI Pedersen conductivity of the E_s layer (Σ_{PE}) as shown. It is assumed that the source has a sinusoidal form with wavenumber k. Figure 6.41a shows how the electric field in the F layer due to an electric field applied in the E_s layer is given by a voltage divider between the field-aligned resistor R_0 and the F-layer resistor R_F. Figure 6.41b shows the voltage divider applicable to determining the F- and E_s-layer electric fields due to a wind-driven Hall current, such as in the wind-driven Haldoupis et al. (1996) polarization mechanism. Figure 6.41c shows the Thevenin equivalent of Fig. 6.41b, and Fig. 6.41d shows the generalization to sources in both the E_s and F layers.

Using Fig. 6.41d, Cosgrove and Tsunoda (2004b) derive the following approximate expression for the coupled system growth rate in term of the isolated

Perkins instability growth rate λ_F (Equation 6.24), and the isolated E_s LI growth rate λ_{E_s}:

$$\lambda_{EFCL} = \frac{\lambda_{E_s}(R_F + 2R_0) + \lambda_F(R_{E_s} + 2R_0)}{2[R_{E_s} + R_F + 2R_0]}$$

$$\pm \sqrt{\left(\frac{\lambda_{E_s}(R_F + 2R_0) + \lambda_F(R_{E_s} + 2R_0)}{2[R_{E_s} + R_F + 2R_0]}\right)^2 - \frac{[\lambda_{E_s}\lambda_F 2R_0]}{[R_{E_s} + R_F + 2R_0]}}$$

(6.31)

$$\lambda_{E_s} = k_s u_e \sin\theta' \cos\theta' \cos I \frac{\Sigma_{HE}}{\rho_i \Sigma_{PE}}$$

(6.32)

where I is the magnetic field dip angle, θ' is the azimuthal angle of the phase fronts with the north-south direction in magnetic field-aligned coordinates (45° is northwest to southeast phase fronts; see Cosgrove and Tsunoda 2003), and $u_e k_s$ is the zonal wind shear at the $E_s L$ altitude. Equations (6.32) and (6.31) are valid whenever $(\vec{E} + \vec{u}_{E_s} \times \vec{B}) \cdot \hat{e} = 0$, where $\vec{u}_{E_s}$ is the wind velocity at the $E_s L$ altitude, and $\hat{e}$ is a unit vector toward the east. When this relation is not satisfied, (6.30) may still provide approximations to the growth rates. The reader should refer to Cosgrove and Tsunoda (2004a).

Equations (6.31), (6.32), and (6.24), together with the definitions in Fig. 6.41, allow study of the growth rates for structure in the E_s and F layers for various amounts of electrical coupling (values of R_0). For the highly coupled case ($R_0 = 0$), which occurs for long wavelengths under conditions such that the relative velocity of the E_s and F layers is zero, equation (6.31) gives the two growth rates:

$$\lambda_{EFCL}|_{R_0=0} = \begin{cases} = \dfrac{\lambda_{E_s} R_F + \lambda_F R_{E_s}}{R_F + R_{E_s}} \\ 0 \end{cases}$$

(6.33)

Equation (6.33) shows only a single unstable mode for which the isolated growth rates combine according to the internal impedances of the E_s- and F-layer sources, in the same way as the mapped polarization electric field. For the uncoupled case ($R_0 = $ infinity), which occurs for short wavelengths, equation (6.31) gives the growth rates

$$\lambda_{EFCL}|_{R_0 \to \infty} = \begin{cases} \lambda_{E_s} \\ \lambda_F \end{cases}$$

(6.34)

In this case there are two unstable modes: the growth rate of the Perkins instability, and the growth rate of the E_s layer instability.

6.7.8 The Wavelength Limiting Effect and Small-Scale Instabilities

If a meridional electric field causes the instability or if the layer is subject to a zonal wind of only one sign, then only one side of a layer is unstable. When this occurs a very interesting phenomenon occurs. The outer scale of the instability is limited to waves which do not map to the other side of the layer where damping occurs. This is due to the Farley mapping effect. That is, if $\lambda_\perp (\sigma_0/\sigma_\rho)^{1/2}$ is greater than the layer thickness, the wave cannot exist.

A very nice example of this effect is presented in Fig. 6.42. The wind was westward during the event and waves were found only on the top of the layer. Furthermore, the outer scale was 30 m, exactly as predicted by the mapping effect. This type of release of free energy in scales of 1 to 30 m may be the primary way that m-scales are created from plasma patches and billows. Urbina et al.

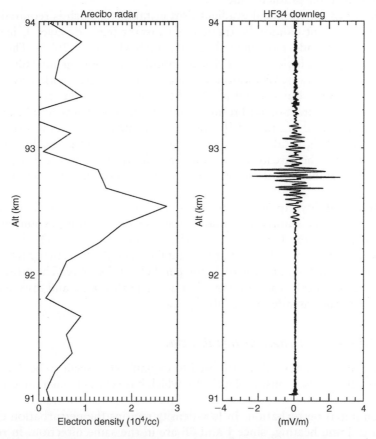

Figure 6.42 Sporadic E layer (left-hand panel) along with fluctuating electric fields. [After Kelley and Gelinas (2000). Reproduced with permission of the American Geophysical Union.]

(2000) have reported the common occurrence of low altitude echoes (≤ 95 km) that have low Doppler shifts and low Doppler widths. It seems quite likely that fluctuations in electron density associated with the electric fields presented in Fig. 6.42 are responsible for these VHF echoes and that a wind-driven instability could be the source. More importantly, perhaps, is the fact that the free energy release associated with the wavelength-limited process is manifested in fluctuations with an outer scale only an order of magnitude larger than the typical 3 m observing scale.

Given the primary perturbation structures associated with any of the processes described above (e.g., KHI, E_s layer, and drift instabilities), it is highly likely that patches with kilometer scales are present multiple places in a typical radar range gate/beam width area. Then fluctuation spectra similar to that in Fig. 6.31 will develop as secondary instabilities due to the mean wind or zero-order electric field on the unstable gradient side.

The theory of the nonlocal gradient-drift instability, which is required when the density gradient cannot be treated as constant (e.g., E_s layers), has been worked out by Rosado-Roman et al. (2004) and Seyler et al. (2004). The theory involves using a realistic vertical density profile and looking for solutions of the linearized equations of motion as sums of plane waves with wavevectors ranging over a two-dimensional plane. The resulting solutions are localized in the magnetic field direction and reside mainly on the side that would be unstable under ordinary local gradient-drift theory. This theory provides a formulation of the gradient drift instability that can be applied to E_s layers. Such a theory is needed to explain the cascade of irregularity scales from the km-scale E_s patches down to the few meter-scale irregularities that are detected by coherent scatter radars.

Another instability, applicable to the midlatitude region, was discovered by Hysell (2002) while simulating polarization and the resulting evolution of E_s patches coupled to the F region in a background electric field. The instability, which he refers to as a collisional drift instability, derives free energy from field-aligned currents through gradients in the parallel conductivity. The growth rate for the instability peaks for about one km wavelength waves, and thus may also be involved in the cascade of irregularity scales.

6.7.9 Wind-Driven Thermal Instabilities

Finally we turn to a new idea proposed to explain what seem to be 3 m waves where no layer exists. Consider Fig. 6.43, which has echoes from a huge altitude range. Figure 6.44 illustrates the proposed thermal effect. A wind is assumed as well as an initial perturbation. In low-density regions the perturbation electric field causes Joule heating, since **J** and δ**E** are in the same direction. In regions of plasma enhancement the opposite effect occurs and no heating occurs. If the heating is great enough, the pressure force drives out more plasma and the depletion deepens—a plasma instability with no gradient needed (Kagan and

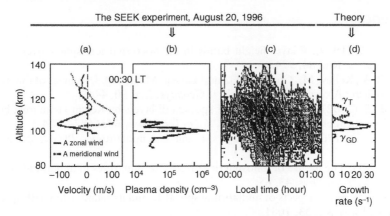

Figure 6.43 Comparison between experiment and theory for the data from the SEEK campaign: altitude profiles of winds (left-hand panel) measured by M. Larsen (Clemson University); an altitude profile of electron density (left-center panel) measured by T. Ono (Tohoku University); altitude-time distribution of 6.1-m backscatter intensity (right-center panel) measured by R. Tsunoda (SRI International) and altitude dependencies of theoretical positive growth rates (right-hand panel) for the neutral wind-driven gradient drift (GD) and thermal instabilities (TI) calculated for the neutral wind distribution on the left-hand panel.

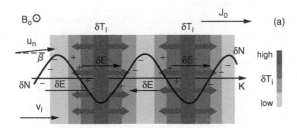

Figure 6.44 A schematic diagram of the thermal instability.

Kelley, 2000). This theory is compared with the experiment in Fig. 6.43. The thermal growth rate is plotted in the right-hand panel along with the wind-driven gradient drift growth rate. The thermal instability greatly increases the altitude range of the instabilities.

Dimant and Sudan (1995) have proposed a similar thermal instability driven by strong auroral zone electric fields and Dimant and Oppenheim (2004) have revisited the Kagan and Kelley (2000) idea. One caveat is in order, however. At this time there is some debate over whether the altitudes listed in figures such as Fig. 6.43 are correct or whether, in fact, the signals are coming from the antenna sidelobes or are just due to the decreasing range of Q-P structures (see Section 6.7.1).

References

Behnke, R. A. (1979). F layer height bands in the nocturnal ionosphere over Arecibo. *J. Geophys. Res.* **84**, 974.

Behnke, R. A., Kelley, M., Gonzales, G., and Larsen, M. (1985). Dynamics of the Arecibo ionosphere: A case study approach. *J. Geophys. Res.* **90**, 4448.

Bernhardt, P. (2002). The modulation of sporadic-E layers by Kelvin-Helmholtz billows in the neutral atmosphere. *J. Atmos. Solar-Terr. Phys.* **64**, 1487–1504.

Bowman, G. G. (1960). Some aspects of sporadic E at midlatitudes. *Planet. Space Sci.* **2**, 195.

———. (1981). The nature of ionospheric spread-F irregularities in midlatitude regions. *J. Atmos. Terr. Phys.* **43**, 65.

———. (1985). Some aspects of midlatitude spread-E, and its relationship with spread-F. *Planet. Space Sci.* **33**, 1081.

———. (1989). Quasi-periodic scintillations at midlatitudes and their possible association with ionospheric sporadic E structures. *Ann. Geophys.* **7**, 259.

Chapman, S., and Lindzen, R. S. (1970). *Atmospheric Tides.* D. Reidel Publishing Co., Norwell, MA.

Chau, J. L. and Woodman, R. F. (1999). Low-latitude quasi-periodic echoes observed with the Piura VHF radar in the E region. *Geophys. Res. Lett.* **26**, 2167–2170.

Cho, J. Y. N. (1995). Inertio-gravity wave parameter estimation from cross-spectral analysis. *J. Geophys. Res.* **100**, 18, 727–18, 737, 1995.

Clark, R. H. (1965). *Off-Vertical Ionospheric Reflections.* M.Sc. Thesis. University of Queensland, Brisbane, Australia.

Collins, S. C., Plane, J. M. C., Kelley, M. C., Wright, T. G., Soldán, P., Rollason, R. J., Kane, T. J., Gerrard, A. J., Grime, B. W., Friedman, J. D., González, S. A., Zhou, Q., and Sulzer, M. P. (2002). A study of the role of ion-molecule chemistry in the formation of sporadic sodium layers. *J. Atmos. Solar-Terr. Phys.* **64**, 845–860.

Cornish, C. R. (1987). *Observations of Inertial Period Waves and Vertical Velocities in the Tropical Middle Atmosphere Using the MST Radar Technique.* Ph.D. Thesis, Cornell University, Ithaca, NY.

Cornish, C. R., and Larsen, M. F. (1989). Observations of low-frequency inertia-gravity waves in the lower stratosphere over Arecibo. *J. Atmos. Sci.* **46**, 2428–2439.

Cosgrove, R. B., and Tsunoda, R. T. (2001). Polarization electric fields sustained by closed-current dynamo structures in midlatitude sporadic E. *Geophys. Res. Lett.* **28**, 1455.

———. (2002a). Wind-shear-driven, closed-current dynamos in midlatitude sporadic E. *Geophys. Res. Lett.* **29**, 7–1.

———. (2002b). A direction-dependent instability of sporadic-E layers in the nighttime midlatitude ionosphere. *Geophys. Res. Lett.* **29**(18), 1864, doi:10.1029/2002GL014669, 11–1.

———. (2003). Simulation of the nonlinear evolution of the sporadic-E layer instability in the nighttime midlatitude ionosphere. *J. Geophys. Res.* **108**(A7), 1283.

———. (2004a). Instability of the E-F coupled nighttime midlatitude ionosphere. *J. Geophys. Res.* **109**(A4), A04305, doi:10.1029/2003JA010243.

————. (2004b). Coupling of the Perkins instability and the sporadic E layer instability derived from physical arguments. *J. Geophys. Res.* **109**, A06301, doi:10.1029/2003JA010295.

Cox, R. M. and Plane, J. M. C. (1998). An ion-molecule mechanism for the formation of neutral sporadic Na layers. *J. Geophys. Res.* **103**, 6349–6360.

Dimant, Y. S., and Oppenheim, M. M. (2004). Ion thermal effects on E-region instabilities: Linear Theory. *J. Atmos. Solar-Terr. Phys.* **66**(17), 1639–1654, doi:10.1016/j.jastp.2004.07.006.

Dimant, Y. S., and Sudan, R. N. (1995). Kinetic theory of low-frequency cross-field instability in a weakly ionized plasma, 1. *Phys. Plasma* **2**(4), 1157–1168.

Djuth, F., Sulzer, M., Elder, J., and Wickwar, V. (1997). High-resolution studies of atmosphere-ionosphere coupling at Arecibo Observatory, Puerto Rico. *Radio Sci.* **32**(6), 2321–2344.

Drummond, J. D., Grime, B. W., Gardner, C. S., Liu, A., Chu, X., Kelley, M. C., and Kane, T. J. (2001). Observations of persistent Leonid meteor trails: 1. Advection of the "Diamond Ring." *J. Geophys. Res.* **106**(A10), 21, 517–21, 524.

Dungey, J. W. (1959). Effect of a neutral field turbulence in an ionized gas. *J. Geophys. Res.* **64**, 2188.

Eckart, C. (1960). *Hydrodynamics of Oceans and Atmospheres.* Pergamon Press, NY.

England, S. L., Maus, S., Immel, T. J., and Mende, S. B. (2006). Longitudinal variation of the E-region electric fields caused by atmospheric tides. *Geophys. Res. Lett.* **33**, L21105, doi:10.1029/2006GL027465.

Farley, D. T. (1959). A theory of electrostatic fields in a horizontally stratified ionosphere subject to a vertical magnetic field. *J. Geophys. Res.* **64**, 1225.

————. (1960). A theory of electrostatic fields in the ionosphere at nonpolar geomagnetic latitudes. *J. Geophys. Res.* **65**, 869–877.

Foster, J. C., Erickson, P. J., Coster, A. J., Goldstein, J., and Rich, F. J. (2002). Ionospheric signatures of plasmaspheric tails. *Geophys. Res. Lett.* **29**(13), doi:10.1029/2002GL015067.

Francis, S. H., and Perkins, F. W. (1975). Determination of striation scale sizes for plasma clouds in the ionosphere. *J. Geophys. Res.* **80**, 3111.

Fukao, S., Kelley, M. C., Shirakawa, T., Takami, T., Yamamoto, M., Tsuda, T., and Kato, S. (1991). Turbulence upwelling of the midlatitude ionosphere, 1. Observational results by the MU radar. *J. Geophys. Res.* **96**, 3725–3746.

Fukao, S., Sato, T., Tsuda, T., Yamamoto, M., Yamanaka, M. D., and Kato, S. (1990). The MU radar: New capabilities and system calibrations. *Radio Sci.* **25**, 477.

Fukao, S., Yamamoto, M., Tsunoda, R., Hayakawa, H., and Mukai, T. (1998). The SEEK (Sporadic-E Experiment over Kyushu) campaign. *Geophys. Res. Lett.* **25**(11), 1761–1764.

Garcia, F. J., Kelley, M. C., Makela, J. J., and Huang, C.-S. (2000a). Airglow observations of mesoscale low-velocity traveling ionospheric disturbances at midlatitudes. *J. Geophys. Res.* **105**(A8), 18, 407–18, 415.

Garcia, F. J., Kelley, M. C., Makela, J. J., Sultan, P. J., Pi, X., and Musman, S. (2000b). Mesoscale structure of the midlatitude ionosphere during high geomagnetic activity: Airglow and GPS observations. *J. Geophys. Res.* **105**(A8), 18, 417–18, 427.

Gershman, B. N. (1974). *Dynamics of Ionospheric Plasma.* Nauka, Moscow.

Goldman, S. R., Baker, L., Ossakow, S. L., and Scannapieco, A. J. (1976). Striation formation associated with barium clouds in an inhomogeneous ionosphere. *J. Geophys. Res.* **81**, 5097.

Goodwin, G. L. (1966). The dimensions of some horizontally moving E_s-region irregularities. *Planet. Space Sci.* **14**(8), 759–771, doi:10.1016/0032-0633(66)90105-X.

Goodwin, G. L., and Summers, R. N. (1970). E_s-layer characteristics determined from spaced ionosondes. *Planet. Space Sci.* **18**(10), 1417–1432, doi:10.1016/0032-0633(70)90116-9.

Hagan, M. E., and Forbes, J. M. (2003). Migrating and nonmigrating semidiurnal tides in the upper atmosphere excited by tropospheric latent heat release. *J. Geophys. Res.* **108**(A2), 1062, doi: 10.1029/2002JA009466.

Haldoupis, C., and Schlegel, K. (1996). Characteristics of midlatitude coherent backscatter from the ionospheric E region obtained with SESCAT. *J. Geophys. Res.* **101**, 3387.

Haldoupis, C., Farley, D. T., and Schlegel, K. (1997). Type-1 echoes from the midlatitude E-region ionosphere. *Ann. Geophys.* **15**, 908–917.

Haldoupis, C., Schlegel, K., and Farley, D. T. (1996). An explanation for type-1 echoes from the midlatitude E-region ionosphere. *Geophys. Res. Lett.* **23**, 97.

Haldoupis, C., Kelley, M. C., Hussey, G. C., and Shalimov, S. (2003). Role of unstable sporadic-E layers in the generation of midlatitude spread F. *J. Geophys. Res.* **108**(A12), 1446, doi:10.1029/2003JA009956.

Hamza, A. (1999). Perkins instability revisited. *J. Geophys. Res.* **104**(A10), 22, 567–22, 576.

Herrmann, U., Eberhardt, P., Hidalgo, M. A., Kopp, E., and Smith, L. G. (1978). Metal ions and isotropes in sporadic E layers during the Perseid meteor shower. *Space Res.* **18**, 249.

Hines, C. O. (1974). *The Upper Atmosphere in Motion: A Selection of Papers with Annotation. Geophys. Monogr.* **18**, American Geophysical Union, Washington, D.C.

Holton, J. R. (1979). *An Introduction to Dynamic Meteorology.* 2nd ed., Academic Press, New York, 314–317.

Huang, C. M., Kudeki, E., Frank, S. J., Liu, C. H., and Röttger, J. (1995). Brightness distribution of midlatitude E-region echoes detected at Chung-LI VHF radar. *J. Geophys. Res.* **100**, 14, 703–14, 715.

Hung, R. J., Phan, T., and Smith, R. E. (1978). Observation of gravity waves during the extreme tornado outbreak of 3 April 1974. *J. Atmos. Terr. Phys.* **40**(7), 831–843.

Hysell, D. L., and Burcham, J. D. (2000). The 30-MHz radar interferometer studies of midlatitude E-region irregularities. *J. Geophys. Res.* **105**, 12, 797.

Hysell, D. L., Larsen, M. F., and Zhou, Q. H. (2004). Common volume coherent and incoherent scatter radar observations of midlatitude sporadic E-layers and QP echoes. *Ann. Geophys.* **22**, 3277–3290.

Hysell, D. L., Yamamoto, M., and Fukao, S. (2002). Simulations of plasma clouds in the midlatitude E-region ionosphere with implications for Type I and Type II quasiperiodic echoes. *J. Geophys. Res.* **107**(A10), 1313, doi:10.1029/2002JA009291, 17–1.

Immel, T. J., Sagawa, E., England, S. L., Henderson, S. B., Hagan, M. E., Mende, S. B., Frey, H. U., Swenson, C. M., and Paxton, L. J. (2006). Control of equatorial ionospheric morphology by atmospheric tides. *Geophys. Res. Lett.* **33**, L15108, doi:10.1029/2006GL026161.

Kagan, L. M., and Kelley, M. C., (1998). A wind-driven gradient drift mechanism for midlatitude E-region ionospheric irregularities. *Geophys. Res. Lett.* 25, 4141.

———. (2000). A thermal mechanism for generation of type-2 small-scale irregularities in the ionospheric E region. *J. Geophys. Res.* 105, 5291.

Kane, T., Grime, B., Franke, S., Kudeki, E., Urbina, J., Kelley, M., and Collins, S. (2001). Joint observations of sodium enhancements and field-aligned irregularities. *Geophys. Res. Lett.* 28(7), 1375–1378.

Kelley, M. C. (1997). In situ ionospheric observations of severe weather-related gravity waves and associated small-scale plasma structure. *J. Geophys. Res.* 102(A1), 329–335.

Kelley, M. C., and Fukao, S. (1991). Turbulent upwelling of the mid-latitude ionosphere: 2. Theoretical framework. *J. Geophys. Res.* 96, 3747–3753.

Kelley, M. C., and Gelinas, L. J. (2000). Gradient drift instability in midlatitude sporadic E layers: Localization of physical and wave number space. *Geophys. Res. Lett.* 27, 457.

Kelley, M. C., and Livingston, R. (2003). Barium cloud striations revisited. *J. Geophys. Res.* 108(A1), 1044, doi:10.1029/2002JA009412.

Kelley, M. C., and Makela, J. J. (2001). Resolution of the discrepancy between experiment and theory of midlatitude F-region structures. *Geophys. Res. Lett.* 28, 2589–2592.

Kelley, M. C., and Miller, C. A. (1997). Electrodynamics of midlatitude spread F, 3. Electrohydrodynamic waves? A new look at the role of electric fields in thermospheric wave dynamics. *J. Geophys. Res.* 102(A6), 11, 539–11, 547.

Kelley, M. C., Baker, K., and Ulwick, J. (1979). Late time barium cloud striations and their possible relationship to equatorial spread F. *J. Geophys. Res.* 84, 1898–1904.

Kelley, M. C., Livingston, R., and McCready, M. (1985). Large amplitude thermospheric oscillations induced by an earthquake. *Geophys. Res. Lett.* 12, 577–580.

Kelley, M. C., Makela, J. J., and Saito, A. (2002). The midlatitude F region at the mesoscale: Some progress at last. *J. Atmos. Solar-Terr. Phys.* 64, 1525–1529.

Kelley, M. C., Makela, J. J., Saito, A., Aponte, N., Sulzer, M., and González, S. A. (2000c). On the electrical structure of airglow depletion/height layer bands over Arecibo. *Geophys. Res. Lett.* 27, 2837–2840.

Kelley, M. C., Haldoupis, C., Nicolls, M. J., Makela, J. J., Belehaki, A., Shalimov, S., and Wong, V. K. (2003a). Case studies of coupling between the E and F regions during unstable sporadic-E conditions. *J. Geophys. Res.* 108(A12), 1447, doi:10.1029/2003JA009955.

Kelley, M. C., Makela, J. J., Paxton, L. J., Kamaladabi, F., Comberiate, J. M., and Kil, H. (2003b). The first coordinated ground- and space-based optical observations of equatorial plasma bubbles. *Geophys. Res. Lett.* 30(14), 1766,10.1029/2003GL017301.

Kelley, M. C., Riggin, D., Pfaff, R. F., Swartz, W. E., Providakes, J. F., and Huang, C.-S. (1995). Large amplitude quasi-periodic fluctuations associated with a mid-latitude sporadic E layer. *J. Atmos. Terr. Phys.* 57(10), 1165–1178.

Kelley, M. C., Kruschwitz, C., Drummond, J., Gardner, C., Gelinas, L., Hecht, J., Murad, E., and Collins, S. (2003c). A new explanation of persistent double meteor trains. *Geophys. Res. Lett.* 30(23), 2202, doi:10.1029/2003GL018312.

Kil, H., Oh, S.-J., Kelley, M. C., Paxton, L. J., England, S. L., Talaat, E., Min, K.-W., and Su, S.-Y. (2007). Longitudinal structure of the vertical $E \times B$ drift and ion density seen from ROCSAT-1. *Geophys. Res. Lett.* 34, L14110, doi:10.1029/2007GL030018.

Larsen, M. F. (2000). A shear instability seeding mechanism for quasiperiodic radar echoes. *J. Geophys. Res.* **105**, 24, 931–24, 940.

———. (2002). Winds and shears in the mesosphere and lower thermosphere: Results from four decades of chemical release wind measurements. *J. Geophys. Res.* **107**(A8), doi:10.1029/2001JA000218.

Larsen, M., Yamamoto, M., Fukao, S., Tsunoda, R., and Saito, A. (2005). SEEK 2– Observations of neutral winds, wind shears, and wave structure during a sporadic E/QP event. *Ann. Geophys.* **23**, 2369–2375.

Larsen, M. F., Fukao, S., Yamamoto, M., Tsunoda, R., Igarashi, K., and Ono, T. (1998). The SEEK chemical release experiment: Observed neutral wind profile in a region of sporadic E. *Geophys. Res. Lett.* **25**, 1789–1792.

Liller, W., and Whipple, F. L. (1954). High-altitude winds by meteor train photography, in *Rocket Exploration of the Upper Atmosphere*. Spec. Suppl., *J. Atmos. Terr. Phys.* **1**, 112–130.

Mendillo, M. and Tyler, A. (1983). Geometry of depleted plasma regions in the equatorial ionosphere. *J. Geophys. Res.* **88**, 5778.

Mendillo, M., Baumgardner, J., Nottingham, D., Aarons, J., Reinisch, B., Scali, J., and Kelley, M. (1997). Investigations of thermospheric-ionospheric dynamics with 6300-Å images from the Arecibo Observatory. *J. Geophys. Res.* **102**, 7331.

Nicolls, M. J., and Kelley, M. C. (2005). Strong evidence for gravity wave seeding of an ionospheric plasma instability. *Geophys. Res. Lett.* **32**, L05108, doi:10.1029/2004GL020737.

Nicolls, M., Kelley, M. C., Coster, A., Gonzalez, S., and Makela, J. (2004). Imaging the structure of a large-scale TID using ISR and TEC data. *Geophys. Res. Lett.* **31**(9), L09812, doi:10.1029/2004GL019797.

Otsuka Y., Shiokawa, K., Ogawa, T., and Wilkinson, P. (2004). Geomagnetic conjugate observations of medium-scale traveling ionospheric disturbances at midlatitude using all-sky airglow imagers. *Geophys. Res. Lett.* **31**, L15803, 10.1029/2004GL020262.

Perkins, F. W. (1973). Spread F and ionospheric currents. *J. Geophys. Res.* **78**, 218.

Perkins, F. W., and Doles, J. H., III. (1975). Velocity shear and the E×B instability. *J. Geophys. Res.* **80**, 211.

Pfaff, R., Freudenreich, H., Yokoyama, T., Yamamoto, M., Fukao, S., Mori, H., Ohtsuki, S., and Iwagami, N. (2005). Electric field measurements of DC and long wavelength structures associated with sporadic-E layers and QP radar echoes. *Ann. Geophys.* **23**, 2319–2334.

Prölss, G. W. (1973). Radiation production and energy deposition by ring current protons dissipated by the charge-exchange mechanism. *Ann. Geophys.* **29**, 503–508.

Richmond, A. D., and Matsushita, S. (1975). Thermospheric response to a magnetic substorm. *J. Geophys. Res.* **80**, 2839.

Riggin, D., Swartz, W. E., Providakes, J., and Farley, D. T. (1986). Radar studies of long wavelength waves associated with midlatitude sporadic-E layers. *J. Geophys. Res.* **91**, 8011.

Rosado-Roman, J. M., Swartz, W. E., and Farley, D. T. (2004). Plasma instabilities observed in the E region over Arecibo and a proposed nonlocal theory. *J. Atmos. Solar Terr. Phys.* **66**, 1593–1602.

Rowland, D. E., and Wygant, J. R. (1998). Dependence of the large-scale, inner magnetospheric electric field on geomagnetic activity. *J. Geophys. Res.* **103**, 14, 959.

Saito, A., Iyemori, T., Sugiura, M., Maynard, N. C., Aggson, T., Brace, L. H., Takeda, M., and Yamamoto, M. (1995). Conjugate occurrence of the electric field fluctuations in the nighttime midlatitude Ionosphere. *J. Geophys. Res.* 100, 21, 439–21, 451.

Saito, A., Nishimura, M., Yamamoto, M., Fukao, S., Kubota, M., Shiokawa, K., Otsuka, Y., Tsugawa, T., Ogawa, T., Ishii, M., Sakanoi, T., and Miyazaki, S. (2001). Traveling ionospheric disturbances detected in the FRONT campaign. *Geophys. Res. Lett.* 28(4), 689–692, 10.1029/2000GL011884.

Schlegel, K., and Haldoupis, C. (1994). Observation of the modified two-stream plasma instability in the midlatitude E region ionosphere. *J. Geophys. Res.* 99, 6219.

Schutz, S., Adams, G. J., and Mozer, F. S. (1973). Probe electric field measurements near a midlatitude ionospheric barium release. *J. Geophys. Res.* 78, 6634.

Seyler, C. E., Rosado-Roman, J. M., and Farley, D. T. (2004). A nonlocal theory of the gradient-drift instability in the ionospheric E region plasma at midlatitudes. *J. Atmos. Solar Terr. Phys.* 66, 1627–1637.

Shalimov, S., Haldoupis, C., and Schlegel, K. (1998). Large polarization electric fields associated with midlatitude sporadic E. *J. Geophys. Res.* 103, 11, 617.

Shalimov, S., and Haldoupis, C. (2005). E region wind-driven electrical coupling of patchy sporadic E and spread F at midlatitude. *Ann. Geophys.* 23, 2095–2105.

Shen, J. S., Swartz, W. E., Farley, D. T., and Harper, R. M. (1976). Ionization layers in the nighttime E region valley above Arecibo. *J. Geophys. Res.* 81, 5517.

Shiokawa, K., Otsuka, Y., Ihara, C., Ogawa, T., and Rich, F. J. (2003). Ground and satellite observations of nighttime medium-scale traveling ionospheric disturbance at midlatitude. *J. Geophys. Res.* 108(A4), 1145, doi:10.1029/2002JA009, 639.

Sinno, K., Ouchi, C., Nemoto, C., and Futagawa, H. (1965). Structure and movement of E_s detected by LORAN observations (supplemental report). *J. Radio Res. Lab.* 12, 59.

Smith, L. G., and Miller, K. L. (1980). Sporadic layers and unstable wind shears. *J. Atmos. Terr. Phys.* 42, 45.

Swartz, W. E., Collins, S. C., Kelley, M. C., Makela, J. J., Kudeki, E., Franke, S., Urbina, J., Aponte, N., González, S., Sulzer, M. P., and Friedman, J. S. (2002). First observations of an F region turbulent upwelling coincident with severe E region plasma and neutral atmosphere perturbations. *J. Atmos. Solar-Terr. Phys.* 64, 1545–1556.

Tsunoda, R. T. (1998). On polarized frontal structures, type-1 and quasi-periodic echoes in midlatitude sporadic E. *Geophys. Res. Lett.* 25, 2641.

Tsunoda, R. T., Yamamoto, M., Igarashi, K., Hocke, K., and Fukao, S. (1998). Quasi-periodic radar echoes from midlatitude sporadic E and role of the 5-day planetary wave. *Geophys. Res. Lett.* 7, 951.

Tsunoda, R. T., Buonocore, J. J., Saito, A., Kishimoto, T., Fukao, S., and Yamamoto, M. (1999). First observations of quasi-periodic radar echoes from Stanford, California. *Geophys. Res. Lett.* 26(7), 995.

Urbina, J., Kudeki, E., Franke, S. J., González, S., Zhou, Q., and Collins, S. C. (2000). 50 MHz radar observations of midlatitude E region irregularities at Camp Santiago, Puerto Rico. *Geophys. Res. Lett.* 27, 2, 853–2, 856.

Vadas, S. L., and Fritts, D. C. (2005). Thermospheric responses to gravity waves: Influences of increasing viscosity and thermal diffusivity. *J. Geophys. Res.* 110, D15103, doi: 10.1029/2004JD005574.

Vlasov, M. N., Kelley, M. C., Makela, J. J., and Nicolls, M. J. (2003). Intense nighttime flux from the plasmasphere during a modest magnetic storm. *J. Atmos. Solar-Terr. Phys.* **65**, 1099–1105, doi:10.1016/j.jastp.2003.07.003.

Voss, H. D., and Smith, L. G. (1979). Nighttime ionization by energetic particles at Wallops Island in the altitude region 120 to 200 km. *Geophys. Res. Lett.* **6**, 93.

———. (1980a). Rocket observations of energetic ions in the nighttime equatorial precipitation zone. *Low Latitude Aeronomic Processes* (A. P. Mitra, ed.), Pergamon, Oxford, 131.

———. (1980b). Global zones of energetic particle precipitation. *J. Atmos. Terr. Phys.* **42**, 227–239.

Whitehead, J. D. (1961). The formation of the sporadic-E layer in the temperate zones. *J. Atmos. Terr. Phys.* **20**, 49.

Woodman, R. F., Yamamoto, M., and Fukao, S. (1991). Gravity wave modulation of gradient drift instabilities in midlatitude sporadic E irregularities. *Geophys. Res. Lett.* **18**, 1197.

Yamamoto, M., Fukao, S., Ogawa, T., Tsuda, T., and Kato, S. (1992). A morphological study on midlatitude E region field-aligned irregularities observed with the MU radar. *J. Atmos. Terr. Phys.* **54**, 769–777.

Yamamoto, M., Fukao, S., Tsunoda, R., Pfaff, R., and Hayakawa, H. (2005). SEEK-2 (Sporadic-E experiment over Kyushu 2) – Project outline and significance. *Ann. Geophys.* **23**, 2295–2305.

Yamamoto, M., Fukao, S., Woodman, R., Ogawa, T., Tsuda, T., and Kato, S. (1991). Midlatitude E region field-aligned irregularities observed with the MU radar. *J. Geophys. Res.* **96**(A9), 15, 943–15, 949.

Yeh, K. C., and Liu, C. H. (1974). Acoustic gravity waves in the upper atmosphere. *Rev. Geophys. Space Phys.* **12**, 193–216.

Yokoyama, T., Otsuka, T., Yamamoto, M., and Hysell, D. L. (2008). First three-dimensional simulation of the Perkins instability in the nighttime midlatitude ionosphere. *Geophys. Res. Lett.* **35**, L03101, doi:10.1029/2007GL032496.

7 Dynamics and Electrodynamics of the Mesosphere

By definition, the mesosphere is the region of the atmosphere located between the stratopause and the mesopause, between roughly 50 and 90 km, in which the temperature decreases with height. The transition between the mesosphere and the thermosphere is called the mesopause and is the altitude at which the temperature reaches a minimum before increasing with height in the thermosphere. The mesopause also delineates the boundary between the earth's neutral atmosphere and the ionosphere. The latter is collocated with the thermosphere, which has temperatures ten times higher than those at the mesopause. The primary subject in this chapter is the polar summer mesosphere which is by far the most interesting mesospheric region. The highest clouds on earth are found here, and the temperature often reaches values as low as 110 K with one measurement as low as 90 K. This is clearly the coldest place on earth. Most of this chapter is devoted to this region since it exhibits the very interesting electrodynamics and plasma physics associated with dusty or icy plasmas. We present a brief discussion of atmospheric electricity and upwardly propagating lightning and close with observations and theory of mesospheric bore waves, both of which are of great scientific interest.

7.1 Noctilucent Clouds (NLC) and the Solstice Temperature Anomaly

Interest in the polar summer mesosphere began on June 18, 1885, two years after Krakatoa's eruption in 1883. A set of silvery-blue clouds was observed in the Northern Hemisphere under twilight conditions when the sun was below the horizon, but it was still illuminating the mesosphere—that is, just before dawn or just after sunset (Gadsden and Schröder, 1989). Using photographic triangulation, it was found that the height of these clouds was about 82 km—the highest clouds ever seen on earth. Due to their unusual nighttime brightness, they were named "noctilucent clouds" (NLC). It is believed that this major volcanic eruption introduced a considerable amount of water vapor and possibly dust

The Earth's Ionosphere: Plasma Physics and Electrodynamics

into the troposphere, which took two years to be transported to the mesosphere, eventually becoming seeds for the ice particles that formed these clouds. The ice particles were large enough to scatter sunlight and be seen by the naked eye. Excellent scientific and historical references are provided by Gadsden and Schröder (1989) and Schröder (2001). A less-well-known NLC event followed the great Tungiska event of 1908 when a comet or meteor struck Siberia. During the next two nights, the postsunset sky over England was bright enough to read a newspaper (Whipple, 1930). Recent studies of the aftermath of solstice launches of the space shuttle have been accompanied by brilliant NLC displays within a few days (Stevens et al., 2003, 2005). The associated injection of 300 meter tons of water vapor in the height range 100–115 km is clearly a significant perturbation but its transport to polar regions was remarkably rapid and must be related to the huge winds reported by Larsen (2002) described in Chapter 5. Equally curious is the rapid dispersal of the clouds, implying an anomalous horizontal diffusion, perhaps due to two-dimensional turbulence (Kelley et al., 2009). Kelley et al. (2009) connected these results with the Great Siberian Meteor Event of 1908 (Whipple, 1930) and concluded that the impacting object was a comet.

These clouds are only observed during the summer months and usually present a wavy pattern, an effect that is attributed to their interaction with passing gravity waves. The ideal zone for viewing the scattered sunlight due to the largest ice particles is between 53° and 57° latitude because of the long twilight and the polar location of the clouds themselves. The photographic example presented in Fig. 7.1 displays some of the characteristics of noctilucent clouds. As will be seen, the charged component of the polar summer mesosphere allows radar

Figure 7.1 Noctilucent clouds as observed at 2255 UT on July 19, 1997, from Glengarnork, Ayshire, Scotland. (Photo courtesy of Tom McEwan.)

detection in the polar daylight and sometimes at night when auroral precipitation creates a plasma at mesopause heights (Nicolls et al., 2008). Satellite instruments can detect the scattered photons in any lighting conditions, and thus the term *polar mesospheric clouds* (PMC) is a more general term for these clouds.

An important related discovery was made during the International Geophysical Year (IGY) of 1957–1958. Rocket grenades launched from many locations and detonated at high altitudes revealed that, against all expectations, the temperature in the polar summer mesosphere is colder than the temperature in the winter polar zone (Nordberg et al., 1965). In fact, a temperature difference of about 100 K is found between summer and winter. More recent data are presented in Fig. 7.2, which not only reveal these temperature differences quite well but also show that the level of temperature fluctuation is higher in winter (Lübken and von Zahn, 1991). This result supports speculations that the NLC are composed of ice, which forms around dust particles (or large ions) at extremely low temperatures, even at the low water vapor pressure of the mesopause zone (1–2 parts per million).

Gravity waves, which create the interesting visible structures in Fig. 7.1 and the wintertime variability in Fig. 7.2, are of more than passing importance to understanding the low temperature of the polar summer mesosphere. Typical

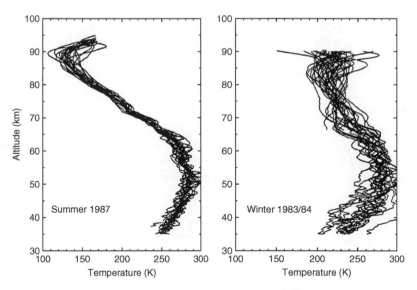

Figure 7.2 Temperature profiles measurements taken using radar tracking of falling spheres during the summer 1987 (left) and winter 1983–1984 (right) over Andøya, Norway. [After Lübken and von Zahn (1991). Reproduced with permission of the American Geophysical Union.]

wave sources include surface wind flow over orographic features, frontal systems, and severe storms; even earthquakes, nuclear explosions, and tsunamis have created substantial waves in the atmosphere (see references in Chapter 6). In brief, if energy is to be conserved as an internal wave propagates upward, every time the background density decreases by a factor of 2, the wave velocity must increase by $\sqrt{2}$ or about 40%. In previous chapters we were most interested in those waves that survived breaking and/or viscosity to create observable ionospheric structure. We now investigate the wave breaking process, which can deposit energy and momentum in the mesosphere.

7.2   Gravity Wave Breaking

Eventually most gravity waves will reach a height where their amplitude is so great that they break like water waves on a beach. A good rule of thumb is that a wave will break when its internal wave-induced velocity perturbation exceeds its phase propagation speed. Then the wave particles overtake the wave phase velocity and it steepens and breaks, as shown in Fig. 7.3 for a surface water wave approaching a beach. An example of the breaking of an internal wave is shown in Fig. 7.4. Here the velocity of a wave detected with the MU radar is plotted along with the predicted change of amplitude with height without breaking. At around 80 km the wave amplitude abruptly decays to values well below the linear prediction ($e^{z/2H}$) curve. Wave breaking clearly has occurred.

For a gravity wave, the breaking criterion that the perturbation velocity exceeds the wave phase velocity in the frame of the mean wind can be written as (Orlanski and Bryan, 1969),

$$\delta u/(c - \overline{u}) > 1 \tag{7.1}$$

where δu is the perturbation amplitude, c is the horizontal wave phase velocity component, and $\overline{u}$ is the component of the mean wind parallel to the wave vector. From gravity wave theory, in which it is assumed that the flow is incompressible

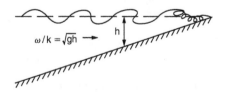

Figure 7.3   Simulation of water waves breaking on a beach.

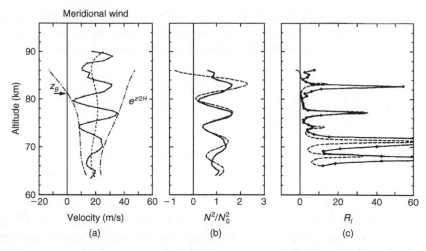

Figure 7.4 A comparison between height profiles of (a) meridional wind velocity, (b) normalized total static stability (N^2/N_0^2) and (c) Richardson number (R_i) in the mesosphere (see Eq. (5.27)). The solid, dashed, and dot-dashed lines in the wind profile indicate the measured meridional wind, the mean flow, and a hypothetical exponential growth of the wind perturbation, respectively. [After Muraoka et al. (1988). Reproduced with permission of the American Geophysical Union.]

and the potential temperature is simply advected, δu can be expressed in terms of the wave-induced potential temperature perturbation ($\delta\theta$) by

$$\delta\theta = \left[\frac{i\,(d\theta/dz)}{m(c - \overline{u})}\right]\delta u \tag{7.2}$$

where $i = \sqrt{-1}$ indicates that $\delta\theta$ is 90° out of phase with δu, $\theta(z)$ represents the background potential temperature profile, and m is the vertical wave number.

The Brunt-Väisälä frequency, N, can be related to θ by

$$N^2 = (g/\theta)(d\theta/dz) \tag{7.3}$$

Following atmospheric science tradition, we use N for the Brunt-Väisälä frequency in this chapter rather than ω_b. Recall that if N^2 is positive, a parcel of air will oscillate at the radian frequency N if it is displaced vertically by a distance of δz from its equilibrium altitude, that is, the subsequent motion is described by the real part of

$$z(t) = \delta z e^{-iNt} \tag{7.4}$$

But if N^2 is negative the parcel does not return to its equilibrium position and the atmosphere is said to be convectively unstable. Taking the spatial derivative of (7.2) and ignoring the second derivative of the background potential temperature profile gives

$$N^2(z) = N_0^2 \left[1 - \frac{\delta u}{c - \overline{u}} \right] \tag{7.5}$$

where N_0^2 is the mean background Brunt-Väisälä frequency. Thus, condition is identical to the convective instability condition (7.1) for the region since for $\delta u > (c - \overline{u})$, $N^2(z)$ is negative.

Panel b in Fig. 7.4 shows the wave-induced $N^2(z)$ for the event using (7.5) for both the actual data set and the extrapolated wave field using the $e^{z/2H}$ factor. Clearly, N^2 was going to become negative at 80 km, the height where the wave broke. The fact that the measured N^2 did not actually become negative may be due to the fact that observations almost always occur in the fully nonlinear regime. But another possibility is shown in panel c. Here the extrapolated Richardson number is plotted. It becomes less than 0.25 at 79 km, suggesting that a shear instability may have set in at a lower height than the convective instability. In any case, the wave clearly broke. The fact that its amplitude dipped to less than one-half of its value where breaking should occur is surprising. Theorists have speculated that the wave amplitude would stay at the amplitude where breaking occurs. But recent simulations (D. Fritts, personal communication, 2001) indicate that wave amplitudes drop well below this value due to rapid energy dissipation and then start to grow again with altitude before breaking again.

7.3 The Polar Summer Mesosphere: A Wave-Driven Refrigerator

But what does this have to with the cold summer mesopause? When waves break on a beach or in the clear air they deposit their energy and momentum back into the local medium (swimmers in a rip current know this very well). The mesosphere is so tenuous that the input of momentum from waves generated in the dense lower atmosphere is very significant. Current theories of the mesosphere argue that the waves that reach these heights come from preferred directions that are different in the two hemispheres. In the summer hemisphere the waves preferentially come from the west depositing a net eastward momentum into the medium. This spins the atmosphere up somewhat and it moves away from the pole. To conserve mass there is a net upflow at high latitudes resulting in adiabatic

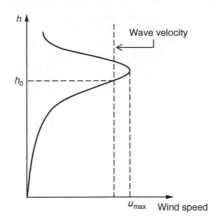

Figure 7.5   Conditions for wave absorption by the jet stream.

cooling, a wave-driven refrigeration process. The opposite effect occurs at the winter pole and the temperature rises.

This preferential direction of momentum transfer could arise in a variety of ways and we only discuss one here, the so-called critical layer effect. Gravity waves are created by weather processes and it is no surprise that their phase velocities are small enough that jet stream winds are often larger than the wave propagation speeds. Suppose, then, that a wave propagates upward to a height where its velocity equals that of the background wind at, say, height h_0 in Fig. 7.5. In general, the Doppler-shifted frequency in the wind frame (ω') is given by

$$\omega' = \omega - \mathbf{k} \cdot \mathbf{u} \tag{7.6}$$

where ω, $\mathbf{k}$, and $\mathbf{u}$ are measured in the earth frame. At $h = h_0$, then, $\omega' = 0$ and the wave is Doppler-shifted to zero. In this case the wave is not a wave at all, just some eddy in the flow, and it ceases to exist. In fact, only waves with horizontal velocities greater than u_{max} get through the jet stream at all. Waves propagating in the other direction, against the flow, are Doppler-shifted upward and pass through easily. Eventually these waves will break at altitudes above the jet and deposit their momentum. Since the jet stream is to the west in the summer and to the east in the winter, gravity wave filtering thus can lead to the refrigeration mechanism just described above and to the observed temperature asymmetry.

Modern global circulation models can only include such effects by parameterizing momentum fluxes. One such model (Roble and Ridley, 1994) is the TIME-GCM (Thermosphere-Ionosphere-Mesosphere-Electrodynamics General Circulation Model). Their calculated yearly variation of temperature at 85 km is

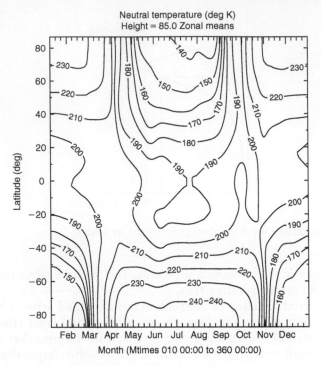

Neutral temperature (deg K)
Height = 85.0 Zonal means

Figure 7.6 TIME-GCM variation of zonal average neutral gas temperature in Kelvin, over a year at 85 km. (Figure courtesy of R. G. Roble.)

shown in Fig. 7.6. A huge difference exists between the hemispheres for solstice conditions at high latitudes. The polar summer hemisphere is indeed colder than the polar winter hemisphere by 100 K, as is observed.

7.4   New Observations of NLC and Related Phenomena

New ways to study NLC have been developed in the latter part of the twentieth century. For example, in the early 1980s satellite measurements detected NLC, but due to the different method of detection, they were called Polar Mesospheric Clouds, or PMC (Thomas, 1991). They are believed to be the same as NLC with the only difference being that, from orbit, they could be observed for 24 hours and at a much less restrictive latitude range. Rocket observations of the electron density have been available as early as those of Pedersen et al. (1970), although such flights are fairly rare. Another observation method that is not hampered by lighting conditions takes advantages of the unexpectedly high radar cross section for VHF radars in the polar summer mesosphere. This radar scattering

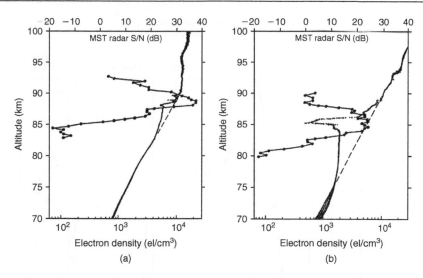

Figure 7.7 Height profiles of the rocket-borne electron density measurements and MST radar echo signal-to-noise ratio (S/N) (solid circles) for the (a) STATE 1 and (b) STATE 3 rocket flights. The dashed lines illustrate a more typical D-region profile. [After Ulwick et al. (1988). Reproduced with permission of the American Geophysical Union.]

process is of considerable interest in its own right and is discussed at length in the next section. Another way to monitor NLC involves lidar, a method analogous to radar but using light waves instead of radio waves (also see below). Finally, rocket instrumentation has been developed in the 1990s to detect dust and ice particles (Havnes et al., 1996; Gelinas et al., 1998).

Figure 7.7 shows examples of simultaneous rocket and radar data taken in Alaska. The radar echoes in the left-hand panel were collocated with a burst of turbulent electron density structure while in the example on the right the echoes are collocated with a bite-out with electron density. It is important to point out that the sun is shining nearly 24 hours a day in this height range and that the mesosphere is a weakly ionized plasma. Thus, the deep plasma density depletion is unexpected since the solar ionization is a smooth source of plasma. The radar echoes are very strong in both cases for this height range—many orders of magnitude higher at VHF than detected elsewhere at this altitude. These early results suggested that more than one mechanism was responsible for the huge radar echoes (Kelley and Ulwick, 1988). STATE 1 is consistent with a turbulent-like process for the scatter (Fig. 7.7a) whereas STATE 3 is quite different (Fig. 7.7b) and may be due to very sharp edge effects called partial reflection.

Figure 7.8 shows a more recent rocket flight. Again, a bite-out in the electron density occurred (panel b) in the center of a strong VHF radar event (panel e). At

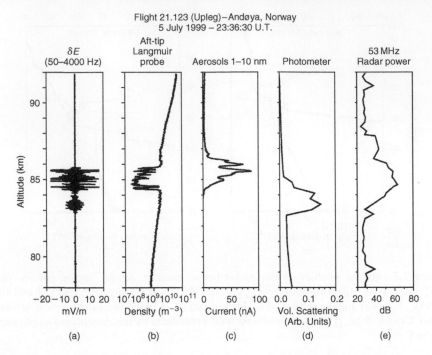

Flight 21.123 (Upleg)–Andøya, Norway
5 July 1999 – 23:36:30 U.T.

Figure 7.8 Composite measurements gathered on the upleg in the PMSE region. The in situ data show raw electric fields (a), aft Langmuir probe density data (b), aerosol measurements (c), and photometer data (d), which indicate the presence of a weak NLC near 84 km. Panel (e) displays vertical backscatter power from the ALWIN radar. [After Pfaff et al. (2001). Reproduced with permission of the American Geophysical Union.]

the upper edge of the bite-out, small aerosols (1–10 nm) were detected while at the lower edge the rocket-borne photometer revealed larger particles, presumably an NLC layer. These data argue for a model in which initially small particles, possibly nucleated by meteoric dust, grow by attaching water molecules, become charged (leading to the bite-outs), and eventually become snow-like aerosols that began to fall. They continue to fall and grow to be detected by the photometer until just below the mesopause where they sublime and disappear. (The electric field fluctuations in panel a are not yet understood but may indicate that either ambipolar fields or a dusty plasma instability is operating.)

NLC are usually visible only at high latitudes, between 50° and 60°, but in June 22, 1999, they were observed in Boulder, Colorado (40°N). This is an indication that NLC are moving south, which could be due either to upper atmospheric cooling or increased water vapor caused by rising levels of CH_4. Either case is related to human activity, a topic we take up later in this chapter.

7.5 Polar Mesosphere Summer Echoes (PMSE)

Early studies of VHF radar echoes from the high-latitude summer mesosphere and lower thermosphere using the Poker Flat MST radar in Alaska (65°N) showed a relatively narrow and surprisingly intense echoing layer centered at about 86 km (Ecklund and Balsley, 1981). The echoes were characterized by their strong VHF (50 MHz) radar backscattering cross section, with backscattered powers 3–5 orders of magnitude greater than typical values observed at low or middle latitudes (in any season) or at high latitudes (in nonsummer periods). The echoes are now referred to as Polar Mesosphere Summer Echoes, or PMSE. PMSE are both intriguing and surprising, because the 3-m irregularities responsible for Bragg backscatter at 6 m radar wavelengths (i.e., the irregularities responsible for VHF backscattering) should lie within the viscous subrange of turbulence at 86 km, and as a consequence, should be strongly damped.

As observed at VHF ($\sim$50 MHz), northern-hemispheric PMSE exhibit the following characteristics:

- They comprise a thin but intense echoing region near and above the summer mesopause ($\sim$85 km), with a typical thickness of 5 km.
- Although some relatively strong, albeit sporadic, echoes have been reported at latitudes as low as 52°, the strongest, most continuous echoes are observed at latitudes poleward of about 65°.
- The echoes appear around mid-May, last until mid-August, and are relatively continuous.
- Both the height range and seasonal variations of PMSE correlate reasonably well with those of the cold mesosphere.
- Joint observations using VHF radar and sounding rockets show that intense PMSE can be often associated with sharp "bite-outs" in the ambient electron density.
- The scatter is aspect sensitive, with the largest signals from the zenith.

Subsequent to their discovery in 1981, many observations related to PMSE and NLC have been made using radar, lidar, and rockets. These observations have helped formulate a number of theories proposed to explain the generation of the intense radar echoes and the remarkable physical conditions associated with them. Subfields in research as disparate as dusty (icy) plasma physics, interplanetary dust cloud studies, meteor ablation, and recoagulation science all have something in common with the polar summer mesopause region.

Many mechanisms have been proposed as being responsible for, or at least partly responsible for, PMSE generation. An example showing 6 hours of radar echo data from this region is presented in Fig. 7.9. The downward echo progression is similar to that of the higher altitude ionospheric layers discussed

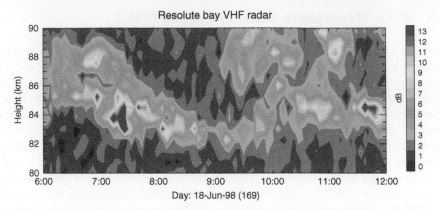

Figure 7.9 Contour plot of the signal-to-noise ratio at Resolute Bay, Canada (75°N, 95°W). The time axis is in UT. [After Huaman and Kelley (2002). Reproduced with permission of Elsevier.]

in Chapter 6 and most likely is related to tidal and/or gravity wave perturbations. (The radar frequency is 51.5 MHz corresponding to backscatter from structures of 2.91 m wavelength, which is smaller than the Kolmogorov microscale in the mesosphere.) Radiowaves scatter most easily from electrons and, when they are present, usually dominate the radiowave physics. Indeed, since the summer polar zone is in full sunlight, the ionosphere actually extends downward to the altitudes of these echoes. But this is true at other latitudes in daytime, so simply having electrons present in a turbulent atmosphere (the mesopause is below the turbopause) is not sufficient to yield these huge echoes. Rayleigh scatter is not a viable explanation.

The rocket data in Fig. 7.7 that show the electron density structure in the medium provided the first clear indication that the echoes were related to the coupling of electrons and small particles (Kelley et al., 1987; Kelley and Ulwick, 1988). Figure 7.10 shows simultaneous VHF scatter echo profile and the electron density during a strong radar event. A severe, sharp, bite-out in the latter is coincident with the echoing region. It seems curious that removing electrons can increase the signal, but if the resulting coupled electron-ice gas has very sharp edges or small eddies, the crucial requirement that structure at half the radar wavelength exists may be accomplished. Analysis of the rocket data indeed shows that for this VHF (53.5 MHz) study the electron gas was indeed structured at this small scale. Wavelets have been used to probe the sharp edges in the data set to see if they are steep enough to partially reflect VHF radar signals. Indeed, as shown in Fig. 7.11, the Canny edge detector wavelets are strong deep into the meter range responsible for edge scatter (Alcala and Kelley, 2001; Alcala et al., 2001). With edges removed, these authors found some regions that appeared turbulent

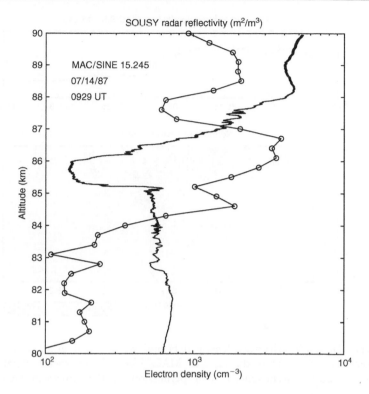

Figure 7.10   Comparison of rocket measurements of the electron density profile (solid line) with the simultaneous 53.5 MHz SOUSY radar observations of radar reflectivity (o) during the MAC/SINE campaign. [After Alcala et al. (2001). Reproduced with permission of the American Geophysical Union.]

and others that did not, but the VHF echoes fill the whole height range. Both turbulent and edge scatter thus seem to occur, with the former dominating.

The complementary roles of active turbulence and sharp edges in PMSE are upheld by Figs. 7.10 and 7.11. Figure 7.12 indicates that a third mechanism plays a role. These data are the first simultaneous measurements of radar echoes, electron density, and neutral density fluctuations (Lübken et al., 1993). The three phenomena clearly are not collocated, suggesting that turbulence in the passive scalar electrons may outlive the active neutral turbulence that created it. This fossil turbulence idea has been quantitatively verified by Rapp and Lübken (2004a, b).

Both passive scalar mixing and edge scatter can be related to one dimensionless parameter: the Schmidt number (S_c), defined by

$$S_c = \nu/D_A \tag{7.7}$$

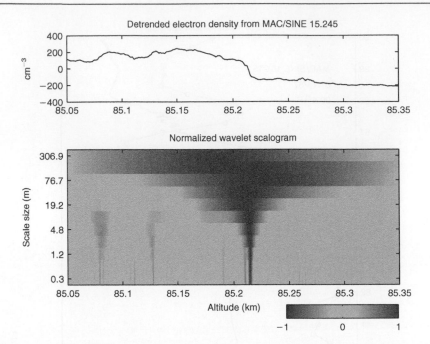

Figure 7.11 Detrended electron density data and wavelet scalogram from a 300 m section centered at 85.2 km during the MAC/SINE 15.245 rocket flight. [After Alcala et al. (2001). Reproduced with permission of the American Geophysical Union.]

where ν is the kinematic viscosity of the neutrals and D_A is the ambipolar plasma diffusion coefficient. Usually $S_c \approx 1$ and in the turbulence case the passive scalar electrons have the same spectrum as the neutral gas with an inner scale of tens of m in the mesosphere. Batchelor (1959) showed that a passive scalar (plasma in this case) can have structure at scales much smaller than the neutral gas if S_c is large. Figure 7.13 shows a comparison of the electron fluctuation spectra measured using rocket probes at three locations. The equatorial and nonpolar summer cases show the usual neutral turbulence-like inner scale (tens of meters) but the polar summer spectrum extends well into the submeter scale sizes. Some 5–7 orders of magnitude of higher spectral density are found at VHF Bragg scales, which directly translate into 50–70 db of increased radar echo strength.

A high Schmidt number is caused by a low diffusion coefficient, which in turn is caused by heavy charged aerosols. An analogy by J. Cho likens electrons to flies buzzing around individual cows slowly lumbering around a field. Once the heavy aerosols tie up around half the charge, D_A drops dramatically (Cho et al., 1992a). The simple model by these authors has been extended by Hill et al. (1999) and Rapp and Lübken (2004a, b), which show that the evolution of D_A

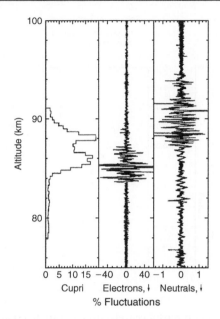

Figure 7.12 S/N ratio from a radar and relative electron and neutral density fluctuations (in %) simultaneously measured in situ in flight NBT5 of the NLC-91 campaign. [After Lübken et al. (1993). Reproduced with permission of the American Geophysical Union.]

is fairly complicated but the basic idea is valid. Schmidt numbers well over 1000 have been indicated (Nicolls et al., 2008).

Turning to edge scatter, it is clear that a high S_c also allows sharp gradients to exist but, unlike the turbulence theory by Batchelor (1959), we as yet have no explanation of the sharp edge phenomenon. The wavelet analysis of rocket data taken during PMSE (shown in Fig. 7.11) illustrates just how sharp these edges can be. The scalogram clearly shows that submeter scale components exist on the edges, and a quantitative analysis shows that VHF scattering can be explained (Alcala et al., 2001). This type of scatter clearly leads to aspect sensitivity, since, for a radar beam off vertical, the specular-like scatter would not yield a return signal. Aspect sensitivity can also occur if the turbulence is itself anisotropic (Bolgiano, 1968). Since the sharp edges on the rocket data are somewhat rare, it is likely that most aspect sensitivity is due to anisotropy in the turbulent structures.

Recently, a UHF radar was installed near Fairbanks, Alaska (Poker Flat Incoherent Scatter Radar). Figure 7.14a shows an example of ice-related echoes at a Bragg wavelength of 33 cm near 82 km. These echoes were obtained at night, implying that auroral particle precipitation produced the plasma necessary to charge the particles. Indeed, the dispersed echoes up to 90 km (and above but

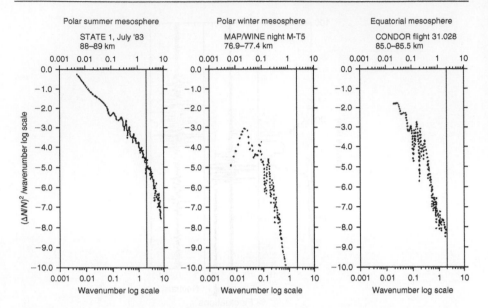

Figure 7.13   Comparison of mesospheric electron density fluctuation spectra from (left) polar summer (Ulwick et al., 1988), (middle) polar winter (Blix, 1988), and (right) equatorial (Røyrvik and Smith, 1984) rocket launches. The vertical line corresponds to the Bragg wave number of a 50-MHz radar. [After Cho and Kelley (1993). Reproduced with permission of the American Geophysical Union.]

not shown) were due to incoherent scatter from this plasma (Nicolls et al., 2008).

Finally, another idea has arisen (Havnes et al., 1990; Hagfors, 1992; La Hoz, 1992) to explain scatter at even higher frequencies that has been observed in the polar mesosphere—for example, at 933 MHz (Röttger et al., 1990) and at 1290 MHz (Cho et al., 1992b). The idea is that an ice particle with charge state $\pm Z$ will attract a cloud of equal and opposite sign within a few Debye lengths of the particle. If the radar wavelength is larger than the cloud, the signal scattered by it is proportional to Z^2, thus enhancing the incoherent scatter component of the return echo (La Hoz et al., 2006).

Radar echoes are thought to be associated with small particles, while the visible NLC are due to those larger ones ($r \approx 30\,\text{nm}$) capable of forward scattering of visible light ($\lambda \approx 500\,\text{nm}$). The simultaneous radar and lidar data in Fig. 7.14b support this idea. Notice that the lidar (which detects the largest ice particles) signal is at the lower edge of the radar signal. This seems a clear indication that the radar detects small particles as they fall and grow larger to be detected by the lidar just before they begin to sublimate as the temperature rises.

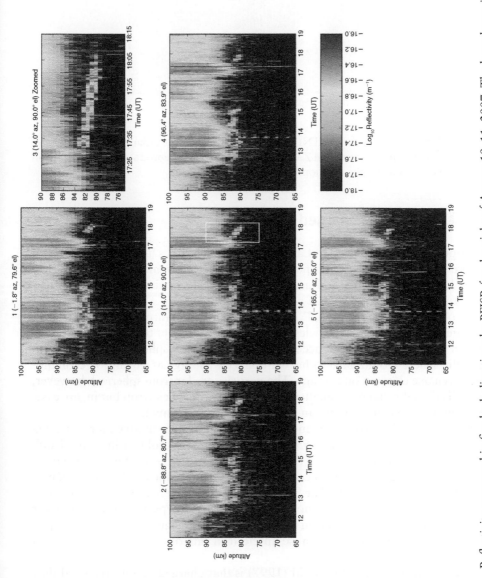

Figure 7.14a Reflectivity measured in five look directions by PFISR for the night of August 10–11, 2007. The boxed area is expanded at the top right. Lidar echoes were obtained at that same height and time period. See Color Plate 24.

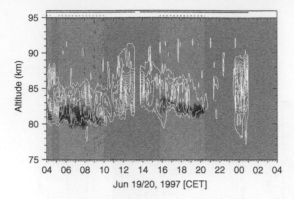

Figure 7.14b   The lidar-observed NLC are dark regions and the radar observed PMSE are the white contour lines. [After von Zahn and Bremer (1999). Reproduced with permission of the American Geophysical Union.]

7.6   The Role of Charged Ice

In Appendix A, a discussion of Langmuir probes describes how an isolated metal sphere charges negatively in a plasma. The higher thermal velocity of the electrons leads to a high flux and negative voltage results, which repel most of the electrons until the positive and negative currents match. The absolute value of this voltage is a few times $k_B T_e/e$, the electron temperature in electron volts. Charged mesospheric ice particles behave similarly except they are so small that only a small charge can be accommodated. For mesospheric temperatures, a metallic sphere would have to have a diameter as large as 30 nm to accommodate even one electron on average. The particles are far from spherical, however, and it is thought that even smaller particles can hold an electron but in any case, the number of electrons that can be accommodated is small.

This simple idea provides some insight but not a quantitative model dealing with how ice charges. Reid (1990) has explored this problem in some detail. His results are summarized in Figs. 7.15a and b. In the former, a variety of species number densities are accounted for during daytime conditions for a particle radius of 1 and 10 nm. The balance equations have been solved for various ionization rates. Reid (1990) quotes $10\,\text{cm}^{-3}\,\text{s}^{-1}$ as a typical ionization rate for daytime mesospheric heights. In both cases he finds significant electron depletions. Charged dust thus can cause an electron bite-out as observed (Pedersen et al., 1970; Ulwick et al., 1988).

The scenario suggested by Reid (1997) is that charged nanometer-sized dust or molecular ions nucleate ice formation in the 90 km height range. As the dust/ice particles grow they begin to fall, growing ever larger, becoming the 10 nm particles in Fig. 7.15b. As the temperature rises below 85 km, the particles begin to sublimate and all but the dust disappears. Since the radar echoes

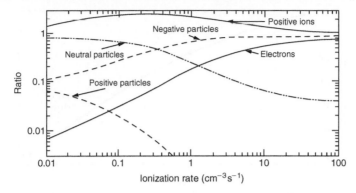

Figure 7.15a Charge distribution for 3000 particles cm^{-3} at 85 km. Particle radius = 1 nm.

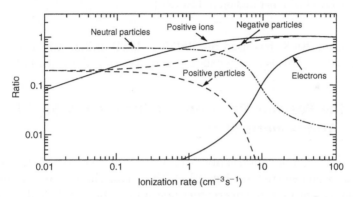

Figure 7.15b Charge distribution for 3000 particles cm^{-3} at 85 km. Particle radius = 10 nm. [After Reid (1990). Reproduced with permission of the American Geophysical Union.]

are thought to be associated with smaller ice particles, the simultaneous lidar and radar data in Fig. 7.14 are in excellent agreement with this scenario (Lübken et al., 1996). The charged particles enhance radar scattering, leading to PMSE.

But what is the origin of these small particles? Reid (1990) used the calculations of meteoric dust or smoke particles supplied by Hunten et al. (1980), which is reproduced in Fig. 7.16. Here the dust concentration is plotted versus particle radius at 90 km. This was determined using meteor impact and recoagulation theory as originally suggested by Rosinski and Snow (1961). Dust detectors have now been flown on sounding rockets, and we are just beginning to measure the earth's dust and dusty plasma layers (Havnes et al., 1996; Gelinas et al., 1998). Recently, based on an idea by Cho et al. (1998), Strelnikova et al. (2007) used the Arecibo incoherent scatter radar to detect dust. R. Varney (personal

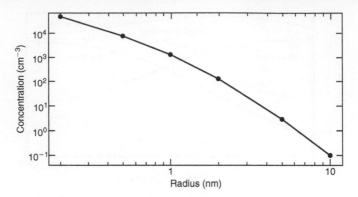

Figure 7.16 Concentration of meteoric smoke and dust particles at 90 km for the assumed initial particle radii indicated on the bottom axis. [From data in Hunten et al. (1980). Reproduced with permission of Elsevier.]

communication, 2008) has used the PFISR radar to utilize this method for ice particle detection. This is a very exciting new area of radar research!

7.7 On the Possible Relationship Between PMSE, NLC, and Atmospheric Change

As time has passed, more observations of NLC have been reported and an increasing trend has been detected (see Fig. 7.17). This change in NLC may be due to the observed increment in atmospheric methane and carbon dioxide due to human activity. A doubling increment in any of these components will produce a cooling of the thermosphere and mesosphere by about 50 K and 10 K, respectively (Roble and Dickinson, 1989). Remember that cold temperatures are necessary conditions for NLC generation, but not the only one: water is also needed. About half of the mesospheric water vapor is believed to come from the photodissociation and oxidation of upwardly transported methane with the chemically active hydroxal radical (OH). Colder temperatures and more water vapor can produce more NLC events, so they can be used as indicators of changes in the atmosphere.

The necessary conditions for the occurrence of PMSE and NLC (or PMC) appear to be similar: they require low temperatures and are apparently related to water vapor (Gadsden and Schröder, 1989). The seasonal PMSE occurrence corresponds well with the high-latitude seasonal occurrence of NLC. Recent studies have shown a correlation of PMC with PMSE in the Northern Hemisphere where the mean long-term PMC occurrence ratio curve fits symmetrically inside the PMSE occurrence ratio. A close correlation between NLC and PMSE has also been observed using lidar and radar data (von Zahn and Bremer, 1999), respectively. These studies were made using a common volume and they

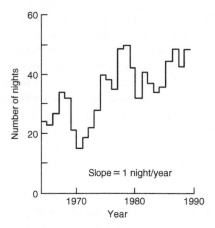

Figure 7.17 Observations of NLC over nearly three decades. [After Gadsden (1990). Reproduced with permission of Pergamon Press.]

agree most of the time. Thus, PMSE are of particular interest in view of their frequent coincident occurrence with NLC and the possible association of recent increased detection of NLC with global warming trends (Gadsden, 1990). If we could monitor PMSE for long periods of time and observe increases/decreases in these events (and their relative strength) over time, we could use such information as a possible indicator of atmospheric change. In fact, these clouds have been called the miner's canary of Global Change (Thomas, 1996a, b). Unfortunately, the injection of water vapor by the space shuttle in the last twenty years may distort this development.

7.8 Upward-Propagating Lightning

The earth is a good electrical conductor compared to the atmosphere and effectively forms a "ground" potential. Thunderstorms charge the earth negatively, which then discharges through the weakly conducting atmosphere. Thus, the system can be modeled as a leaky spherical capacitor. About 250,000 V exist between the earth and the lower ionosphere. The average current density over the earth is about one pA/m^2 and the total is about 1500 A.

Ninety-nine percent of the time, lightning striking the ground brings negative charge, which is how the earth becomes charged. Every such flash consists of several strikes following the same channel. Each such current pulse creates an electromagnetic wave similar to that of a dipole antenna. The fields have electrostatic, inductive, and radiation characteristics. Although the electrostatic field is largest in the near field and right over the strike, all three field types are important at the base of the ionosphere.

Figure 7.18 shows the variety of mesospheric phenomena that can accompany cloud-to-ground lightning (CG). The most well studied of these phenomena are the so-called "sprites" and "elves." "Blue jets" and "gigantic jets" have recently come to the forefront, since gigantic jets appear to provide an electrical connection between cloud tops and the ionosphere. Few observations of gigantic jets have been made thus far, however, and all known gigantic jets have occurred above oceanic storms (Pasko et al., 1997, 2002; Su et al., 2003). Elves are expanding rings of luminosity that are directly linked to a causative electromagnetic pulse (EMP) emanating in a dipole pattern from the cloud-to-ground lightning. When the EMP encounters the base of the ionosphere (at around 80 km altitude), it heats the ionospheric electrons, and optical emissions are produced. Elves have been observed with large lightning discharges of either positive or negative polarity (Inan et al., 1991; Fukunishi et al., 1996; Barrington-Leigh et al., 2001).

Sprites occur between thundercloud tops and the base of the ionosphere and are the optical manifestation of upper atmospheric electrical breakdown caused by cloud-to-ground lightning. They are almost always associated with the more rare strong positive polarity lightning strikes and generally are thought to be due to a quasistatic electric field set up above a thundercloud after charge is deposited to ground by a positive CG strike (Pasko et al., 1997, 2002). Figure 7.19 shows a schematic diagram of the charge configuration in a sprite-producing cloud as

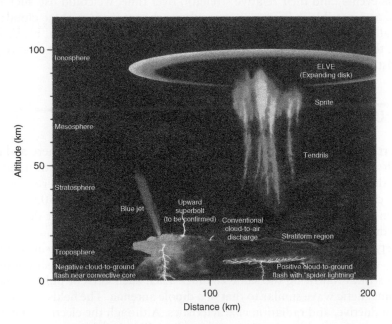

Figure 7.18 Phenomena associated with upward coupling by lightning. (Figure courtesy of W. Lyons and C. Miralles.) See Color Plate 25.

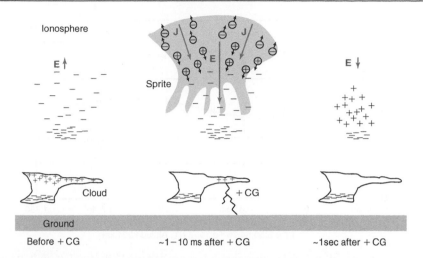

Figure 7.19 Schematic diagram of charge configurations in sprite-producing thundercloud (Pasko et al., 1997). (Figure courtesy of V. Pasko and E. Gerken.)

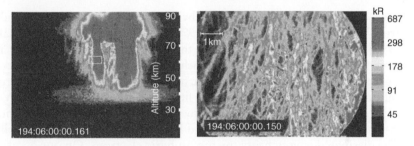

Figure 7.20 Telescopic imaging of sprites. Wide (L) and narrow (R) field-of-view images of a bright event. [After Gerken et al. (2001). Reproduced with permission of the American Geophysical Union.] See Color Plate 26.

based on the work of Pasko et al. (1997). Initially the thundercloud is effectively a vertical dipole with positive charge on top and negative charge on the bottom. The ions and electrons in the overlying mesosphere and lower ionosphere are distributed so as to cancel the electric field created by the thundercloud dipole. When a +CG occurs and positive charge is removed from the cloud, a subsequent breakdown, or sprite, can occur above the thundercloud. Following the sprite, the electrons and ions rearrange to once again reduce the electric field existing above the cloud. Observations using telescopic imagery have shown that the lower portions of sprites consist of filamentary structures that are thought to be streamer discharges (Fig. 7.20; Gerken et al., 2000).

Most of the energy from the lightning EMP is absorbed at the lower boundary of the ionosphere (D region), but some leaks into the ionosphere as a whistler

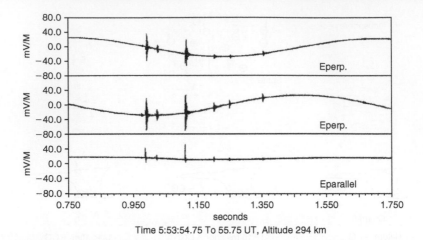

Figure 7.21   Vector electric fields over a thunderstorm. [After Kelley et al. (1990). Reproduced with permission of the American Geophysical Union.]

mode wave. Four rockets were launched by Cornell University over thunderstorms from Wallops Island (Kelley et al., 1990; Siefring and Kelley, 1991; Baker et al., 1996; Kelley et al., 1997). One of the most intriguing, and as yet not understood, features observed in these launches is a parallel electric field accompanying each strike. An example of electric fields observed in the ionosphere at 294 km altitude is shown in Fig. 7.21. The field strength is as high as 50 mV/m, but it is a far cry from the 5 V/m seen the same distance away along the ground in the earth-ionosphere waveguide (Kelley and Barnum, 2009; Vlasov and Kelley, 2009). The dissipated energy heats the mesospheric electrons. On one occasion, keV electrons were detected over a thunderstorm-rich hurricane with an associated electric field burst (Burke et al., 1992). The parallel electric field component observed by the rockets rides on the leading edge of the wave front, which travels at nearly the speed of light. These parallel fields may be associated with an ac conductivity, $\sigma_{ac} = j\omega\varepsilon$, associated with the leading edge. Large-amplitude waves launched from the ground have produced particle acceleration and heating at midlatitudes at HF frequencies (Djuth et al., 1999; Kagan et al., 2000; Vlasov et al., 2005), which may be a similar process.

7.9   Nonlinear Mesospheric Waves

7.9.1   Observations

Another new and interesting mesospheric phenomenon was first observed over Haleakala, Hawaii, on October 10, 1993 (Taylor et al., 1995). A summary of the event is shown in Fig. 7.22. The feature was first detected coming in from

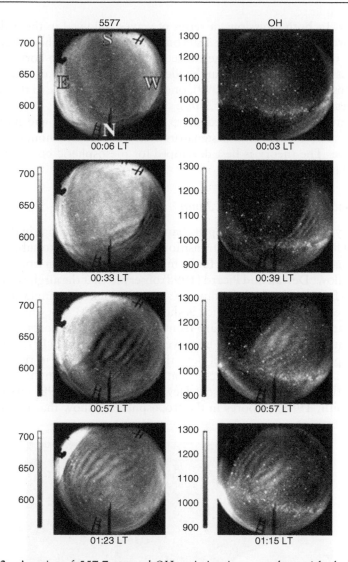

Figure 7.22 A series of 557.7 nm and OH emission images taken with the Utah State University Imager on October 10, 1993, in Hawaii (Taylor et al., 1995). The images show a front approach from the northwest. Note the inverse relationship between the two emissions. [After Taylor et al. (1995). Reproduced with permission of the American Geophysical Union.]

the northwest at 23:30 LT. By 0:30 LT the principal features of this event were apparent. There was a sharp front approaching from the northwest which, in the 5577 emission (at 96 km), brought with it a sharp decline in the airglow intensity, whereas in the OH emission (at 87 km), it brought a sharp increase in the intensity. In both emissions, a series of waves followed in the wake of the front. The waves themselves were 180° out of phase between the two emissions. The feature eventually covered the entire sky, from horizon to horizon in every direction. As many as 10 crests can be identified. Very similar events have been reported subsequently (Garcia, 1999), one of which was visible to the naked eye in the OH emission (Smith et al., 2003).

These features have been called sharp fronts, wall waves, and mesospheric bore waves. The former two designators refer to a guided nonlinear step or solitary wave. The theory is best developed for mesospheric bores and will be emphasized here.

7.9.2   Analogy to a Hydraulic Jump

A bore, or hydraulic jump, in an open channel is an abrupt change in the depth of the channel fluid in response to a transition from supercritical to subcritical flow and is analogous to the shock that forms in a gas in the transition from supersonic to subsonic flow. Dewan and Picard (1998) observed that the airglow image data of the sharp front in the Aloha campaign showed remarkable similarities to river channel bores, an example of which is shown in Fig. 7.23. They have advanced a theory to try to attribute the sharp fronts to "mesospheric bores."

Bores in an open channel have been well studied. Lighthill (1979) provides a good description of open channel bore theory. The problem in relating atmospheric bores to open channel bores is that there is no channel in the atmosphere, so what is the guiding mechanism? Dewan and Picard (1998) cite examples (e.g., see Smith, 1988) of bores in the troposphere to show that internal bores

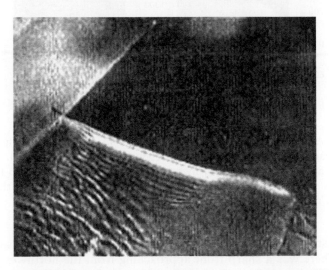

Figure 7.23   A tidal bore on the River Mersey from the air (Dewan and Picard, 1998) from (Tricker, 1965). [After Dewan and Pickard (1998). Reproduced with permission of the American Geophysical Union.]

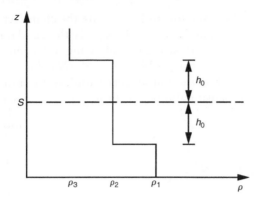

Figure 7.24   The highly idealized density profile used by Dewan and Picard (1998). [After Dewan and Picard (1998). Reproduced with permission of the American Geophysical Union.]

can and do occur. The channels in these cases are hypothesized to be layers of stable air surrounded immediately above and below by less stable layers. Layers of temperature inversion could provide for such a situation. In explaining the Aloha front, they believe there must have existed a horizontal waveguiding channel between the two layers (say, around 90–94 km) in which a bore propagated. As the bore theoretically passed overhead, it produced symmetrical oscillations about the center of the channel. The upper layer (5577) was pushed higher, making it cooler, less dense, and presumably less bright. The lower layer (OH) was pushed symmetrically lower, making it warmer, more dense, and brighter.

Dewan and Picard use the very simplified density profile depicted in Fig. 7.24, which they claim is the simplest guiding structure that can support an internal bore. The ρ_1 and ρ_3 layers are assumed to be semi-infinite in extent. The ρ_3 layer is of finite thickness $2h_0$. They argue that any oscillation on the $\rho_3 - \rho_2$ surface will be mirrored in the $\rho_3 - \rho_1$ surface about the plane of symmetry S, where there is no vertical motion. Thus, S acts as if it were the rigid bottom of an open channel bore. Dewan and Picard state that the equations governing open channel bores will hold for this simple model if the acceleration due to gravity, g, is replaced by the buoyant acceleration g' defined as

$$g' = \frac{\rho_2 - \rho_3}{\rho_2} g$$

Figure 7.24 is equivalent to a Brunt-Väisälä profile with a step.

7.9.3   Nonlinear Simulation of Mesospheric Bores

Seyler's (2005) two-dimensional nonlinear model for the mesosphere depends on the following parameters: N_0, N_1, h, L_x, L_z, and A. $N_1 < N_0$ are Brunt-Väisälä frequencies associated with inversion layers guiding the resulting waves; h is the characteristic scale of the inversion and L_x, L_z are the characteristics of a ducted gravity wave with initial amplitude A. If N_0^{-1} is the fundamental time scale and h is the fundamental length scale, the following independent dimensionless parameters then characterize the problem: $\delta = N_1/N_0$, $\tilde{L}_x = L_x/h$, $\tilde{L}_z = L_z/h$, and, $\tilde{A} = A/N_0 h^2$, which is the initial dimensionless amplitude of the stream function, φ. The nondimensional model is then

$$\frac{\partial}{\partial t} \nabla_\perp^2 \varphi + \left[\varphi, \nabla_\perp^2 \varphi\right] = -\frac{\partial \theta}{\partial x} + \mu \nabla_\perp^4 \varphi$$

$$\frac{\partial}{\partial t} \theta + [\varphi, \theta] = N^2(z) \frac{\partial \varphi}{\partial x}$$

where θ is the potential temperature,

$$[A, B] = \frac{\partial A}{\partial x} \frac{\partial B}{\partial z} - \frac{\partial B}{\partial x} \frac{\partial A}{\partial z}$$

and μ is a dimensionless kinematic viscosity.

For the numerical solutions, Seyler used a smooth approximation to a sharp profile, $N^2(z) = \delta^2 + \exp(-z^4)$, where $\delta = N_1/N_0$. The initial conditions on φ and θ were chosen to be of the form

$$\psi(x, z, 0) = A \sin\left(\frac{2\pi x}{L_x}\right) \sin\left(\frac{n\pi z}{L_z}\right), \quad n = 1, 2, \Lambda$$

and,

$$\theta(x, z, 0) = -\frac{2\pi A N_1}{L_z} \sin\left(\frac{2\pi x}{L_x}\right) \sin\left(\frac{n\pi z}{L_z}\right), \quad n = 1, 2, \Lambda$$

This initial condition is related to a ducted wave since it is a superposition of oblique right-upward and right-downward propagating gravity waves with a frequency below the background cutoff frequency N_1.

Table 7.1 provides a summary for 8 of the numerical simulations investigated. All of the runs listed in the table form bores, although in run 8, a single peak was formed and, strictly speaking, this should be considered a solitary wave. The wavelength λ_x was determined approximately ¾ of the way through the runs and the phase speed was averaged over the last half of the runs. The initial wavelength at the time of peak formation is considerably smaller.

Figure 7.25a is a set of contour plots of the stream function given at four uniformly spaced times for run 1. In the upper left panel at time 200, the distortion of the initial "ducted" gravity wave is evident. At time 400, shown in the upper right panel, three peaks have formed. Peak formation continues to occur in the bottom left and right panels at times 600 and 800, respectively. A maximum of 8 peaks was generated for this case. The propagation is from left to right and

Table 7.1 Parameters Used in the Seyler (2005) Numerical Bore Simulator

Param.	Run 1	Run 2	Run 3	Run 4	Run 5	Run 6	Run 7	Run 8
δ	0.2	0.1	0.25	0.15	0.2	0.25	0.1	0.25
L_x	30	50	30	30	30	40	60	60
L_z	1.5	2.0	1.0	2.0	3.0	3.0	5	5
A	0.14	0.25	0.1	0.2	0.28	0.28	0.3	0.6
Bore speed	0.54	0.55	0.60	0.58	0.7	0.80	0.57	1.1
λ_x	18	16	12	25	25	35	25	—
# of peaks	8	13	9	11	6	2	9	1

$$\psi(x, z)$$

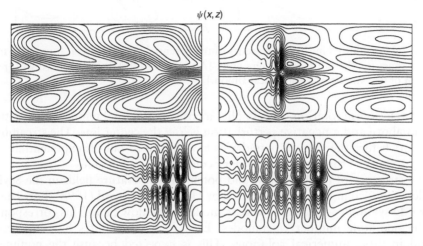

Figure 7.25a Solution of the two-dimensional equations for the parameters in Table 7.1 corresponding to run 1. Plotted in (a) are contours of the stream function at four uniformly spaced times in the sequence: upper left, upper right, lower left, and lower right. The horizontal axis is the x direction and the vertical axis is the z direction. The parameters for the run are $L_x = 30$, $d = 1.5\pi$, $\mu = 0.001$, $N_0 = 1$, $h = 1$, $N_1 = 0.2$, and $A = 0.14$. The closed contours just above the center line correspond to a flow rotation in the clockwise direction and the contours just below the center line correspond to counterclockwise rotation. [After Seyler (2005). Reproduced with permission of the American Geophysical Union.]

one should note that because of the periodic boundary conditions the waves that go out the right boundary emerge from the left boundary.

Figure 7.25b is the potential temperature perturbation for the same times shown in Fig. 7.25a. Note that almost all of the potential temperature perturbation occurs in the layer while the flow can extend well beyond the layer. The closed contours just above the center line correspond to a flow rotation in the clockwise direction and the contours just below the center line correspond to

$$\theta(x, z)$$

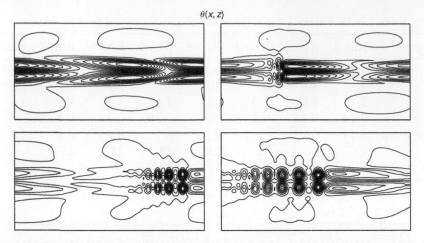

Figure 7.25b Plotted in (b) are contours of the potential temperature at four uniformly spaced times for the same parameters used. [After Seyler (2005). Reproduced with permission of the American Geophysical Union.]

counterclockwise rotation. This implies a positive potential temperature perturbation above and a negative perturbation below the center line. The streamline contours before the vertical boundaries have begun to fall off exponentially so that the effect of the periodic boundary conditions is negligible.

The structure of the bore flow and potential temperature fields in the Seyler model is consistent with that proposed by Dewan and Picard (1998). Bores have only odd symmetry (i.e., area varicose mode) in the flow field. This corresponds to the odd-node number eigenmodes. Even eigenmode bores have not been found in these numerical solutions. This is expected because the nonlinear steepening process resulting from flow advection of the potential temperature is not significant for even parity solutions. A property found in the simulations that differs from that assumed by Dewan and Picard (1998) is the constancy of the wavelength, or, perhaps more properly, the separation between crests. Seyler finds that the wavelength decreases from the front to the back of the bore. The variation in peak separation can be large, as in runs 5 and 7 or considerably smaller, as in runs 1, 2, 3, and 4, and depends significantly on the time at which the separation is measured.

The linear bore phase speed is close to that predicted by linear theory—namely,

$$c = -\frac{2hN_0}{\pi}\left(1 - \frac{4}{\pi^2}|k|h\right)$$

For example, using $k = 2\pi/18 \approx 0.35$ in run 1, the linear model phase speed is $c = 0.55$. This is very close to the simulation value for the bore speed, which is 0.54.

In order to compare these results with observations, we need to revert to physical space and time units need to be introduced. To set the characteristic spatial and time scales, let the layer thickness be 1.5 km and the peak Brunt period, in inverse radians, is 25 s using data from Garcia (1999). From Table 7.1, corresponding to run 1, the wavelength of the longest undulation is 1.5 km $\times$ 18 $= 27$ km. The bore formation time is 25 s $\times$ 300 $\approx 2h$. The phase speed is $c = 0.54 \times N_0 h \approx 32$ m/s. These values are typical of what is observed for mesospheric bores.

It is important to note that the bore formation process considered by Seyler (2005) differs from that in Dewan and Picard (2001) who consider the bore formation results from steepening of a "hydraulic jump" created by an accelerating piston, whereas Seyler only considered ducted gravity wave initial conditions. Nonetheless, the analysis and numerical solutions presented by Seyler show that mesospheric bores are consistent with nonlinear internal gravity waves trapped within a thermal inversion layer, as proposed by Dewan and Picard (1998). The nonlinear solutions are consistent with observations of mesospheric bores with respect to propagation speed, wavelength, number of peaks, and formation time. The nonlinear model predicts additional properties of mesospheric bores that are potentially testable but have yet to be observed, such as the vertical structure of the flow and potential temperature field.

References

Alcala, C. M., and Kelley, M. C. (2001). Nonturbulent layers in polar summer mesosphere: 2. Application of wavelet analysis to VHF scattering. *Radio Sci.* **36**(5), 891–903.

Alcala, C. M., Kelley, M. C., and Ulwick, J. C. (2001). Nonturbulent layers in polar summer mesosphere: 1. Detection and characterization of sharp gradients using wavelet analysis. *Radio Sci.* **36**(5), 875–890.

Baker, S. D., Kelley, M. C., and Wheeler, T. F. (1996). Thunderstorm III 4-Megahertz Burst-Mode Data Acquisition System. *J. Spacecraft Rockets* **33**(1), 92–95.

Barrington-Leigh, C. P., Inan, U. S., and Stanley, M. (2001). Identification of sprites and elves with intensified video and broadband array photometry. *J. Geophys. Res.* **106**(A2), doi: 10.1029/2000JA000073.

Batchelor, G. K. (1959). Small-scale variation of convected quantities like temperature in turbulent fluid, Part 1. General discussion and the case for small conductivity. *J. Fluid Mech.* **5**, 113.

Blix, T. A. (1988). In situ studies of turbulence in the middle atmosphere by means of electrostatic ion probes. *REP. NDRE/PUBL-88/1002*, Norw. Def. Res. Estab., Kjeller.

Bolgiano, R. (1968). The general theory of turbulence: Turbulence in the atmosphere. In *Wind and Turbulence in the Stratosphere, Mesosphere, and Ionosphere*, K. Rawer (ed.). North-Holland, Amsterdam, 371–400.

Burke, W. J., Aggson, T. A., Maynard, N. C., Hogey, W. R., Hoffman, R. A., Holzworth, R. H., Liebrecht, C., and Rodgers, E. (1992). Effects of a lightning discharge observed by the DE-2 satellite over Hurricane Debbie. *J. Geophys. Res.* 97, 6359.

Cho, J. Y. N., and Kelley, M. C. (1993). Polar mesosphere summer radar echoes: Observations and current theories. *Rev. Geophys.* 31, 243–265.

Cho, J. Y. N., Hall, T., and Kelley, M. (1992a). On the role of charged aerosols in polar mesosphere summer echoes. *J. Geophys. Res.* 97, 875–886.

Cho, J. Y. N., Kelley, M. C., and Heinselman, C. J. (1992b). Enhancement of Thomson scatter by charged aerosols in the polar mesosphere: Measurements with a 1.29-GHz radar. *Geophys. Res. Lett.* 19, 1097–1100.

Cho, J. Y., Sulzer, M. P., and Kelley, M. C. (1998). Meteoric dust effects on D-region incoherent scatter radar spectra. *J. Atmos. Solar-Terr. Phys.* 60, 349–357.

Dewan, E. M., and Picard, R. H. (1998). Mesospheric bores. *J. Geophys. Res.* 103(D6), 6295–6305.

Dewan, E. M., and Picard, R. H. (2001). On the origin of the mesospheric bores. *J. Geophys. Res.* 106(D3), 2921–2927.

Djuth, F. T., Bernhardt, P. A., Tepley, C. A., Gardner, J. A., Kelley, M. C., Broadfoot, A. L., Kagan, L. M., Sulzer, M. P., Elder, J. H., Selcher, C., Isham, B., Brown, C., and Carlson, H. C. (1999). Large airglow enhancements produced via wave-plasma interactions in sporadic E. *Geophys. Res. Lett.* 26(11), 1557–1560, doi:10.1029/1999GL900296.

Ecklund, W. L., and Balsley, B. B. (1981). Long-term observations of the arctic mesosphere with the MST radar at Poker Flat, Alaska. *J. Geophys. Res.* 86, 7775–7780.

Fukunishi, H., Takahashi, Y., Kubota, M., Sakanoi, K., Inan, U. S., and Lyons, W. A. (1996). Elves: Lightning-induced transient luminous events in the lower ionosphere. *Geophys. Res. Lett.* 23, 2157.

Gadsden, M. (1990). A secular change in noctilucent cloud occurrences. *J. Atmos. Terr. Phys.* 52, 247–251.

Gadsden, M., and Schröder, W. (1989). *Noctilucent Clouds*. New York: SpringerVerlag.

Garcia, F. J. (1999). *Atmospheric Studies Using All-Sky Image of Airglow Layers*. Ph. D. Thesis, Cornell University, Ithaca, NY.

Gelinas, L. J., Lynch, K. A., Kelley, M. C., Collins, S., Baker, S., Zhou, Q., and Friedman, J. (1998). First observation of meteoritic charged dust in the tropical mesosphere. *Geophys. Res. Lett.* 25(21), 4047–4050.

Gerken, E. A., Inan, U. S., and Barrington-Leigh, C. P. (2000). Telescopic imaging of sprites. *Geophys. Res. Lett.* 27, 2637.

Hagfors, T. (1992). Note on the scattering of electromagnetic waves from charged dust particles in a plasma. *J. Atmos. Terr. Phys.* 54, 333–338.

Havnes, O., de Angelis, U., Bingham, R., Goertz, C. K., Morfil, G. E., and Tsytovich, V. (1990). On the role of dust in the summer mesopause. *J. Atmos. Terr. Phys.* 52, 637–643.

Havnes, O., Trøim, J., Blix, T., Mortensen, W., Naesheim, L. I., Thrane, E., and Tønnesen, T. (1996). First detection of charged dust particles in the earth's atmosphere. *J. Geophys. Res.* 66, 1985.

Hill, R., GibsonWilde, D., Werne, J., and Fritts, D. (1999). Turbulence-induced fluctuations in ionization and application to PMSE. *Earth Planets Space* 51, 499–513.

Huaman, M. M., and Kelley, M. C. (2002). "Mesosphere: Polar Summer Mesopause," in *Encyclopedia of Atmospheric Science.* James R. Holton, John A. Pyle, and Judith A. Curry, eds. Academic Press: Elsevier Science, London, 2171–1279.

Hunten, D. M., Turco, R. P., and Toon, O. B. (1980). Smoke and dust particles of meteoric origin in the mesosphere and stratosphere. *J. Atmos. Sci.* 37, 1342.

Inan, U. S., Bell, T. F., and Rodriguez, J. V. (1991). Heating and ionization of the lower ionosphere by lightning. *Geophys. Res. Lett.* 18, 705.

Kagan, L. M., Kelley, M. C., Garcia, F., Bernhardt, P.A., Djuth, F. T., Sulzer, M. P., and Tepley, C. A. (2000). The structure of electromagnetic wave-induced 557.7 nm emission associated with a sporadic E event over Arecibo. *Phys. Rev. Lett.* 85(1), 218–221.

Kelley, M. C., and Barnum, B. (2009). An explanation for parallel field pulses observed over thunderstorms. *J. Geophys. Res.*, submitted.

Kelley, M. C., and Ulwick, J. (1988). Large and smallscale organization of electrons in the highlatitude mesosphere: Implications of the STATE data. *J. Geophys. Res.* 93, 7001.

Kelley, M. C., Ding, J. G., and Holzworth, R. (1990). Intense ionospheric electric and magnetic field pulses generated by lightning. *Geophys. Res. Lett.* 17(12), 2221–2224.

Kelley, M. C., Farley, D. T., and Röttger, J. (1987). The effect of cluster ions on anomalous VHF backscatter from the summer polar mesosphere. *Geophys. Res. Lett.* 14, 1031–1034.

Kelley, M. C., Seyler, C. E., and Larson, M. (2009). Evidence from observations of Space Shuttle plume transport and two-dimensional turbulence theory that the Great Siberian Impact was due to a coment. *Geophys. Res. Lett.*, submitted.

Kelley, M. C., Baker, S. D., Holzworth, R. H., Argo, P., and Cummer, S. A. (1997). LF and MF observations of the lightning electromagnetic pulse at ionospheric altitudes. *Geophys. Res. Lett.* 24(9), 1111–1114.

La Hoz, C. (1992). Radar scattering from dusty plasmas. *Phys. Scr.*, 45, 529–534.

La Hoz, C., Havnes, O., Naesheim, L. I., and Hysell, D. L. (2006). Observations and theories of Polar Mesospheric Summer Echoes at a Bragg wavelength of 16 cm. *J. Geophys. Res.* 111, D04203, doi:10.1029/2005JD006044.

Larsen, M. F. (2002). Winds and shears in the mesosphere and lower thermosphere: Results from four decades of chemical release wind measurements. *J. Geophys. Res.* 107, 1215, doi:10.1029/2001JA000218.

Lighthill, J. (1979). *Waves in Fluids.* Cambridge University Press, New York.

Lübken, F. J., and von Zahn, U. (1991). Thermal structure of the mesopause region at polar latitudes. *J. Geophys. Res.* 96(D11), 20,841–20,857.

Lübken, F. J., Fricke, K. H., and Langer, M. (1996). Noctilucent clouds and the thermal structure near the arctic mesopause in summer. *J. Geophys. Res.* 101, 9489.

Lübken, F. J., Lehmacher, G., Blix, T., Hoppe, U. P., Thrane, E., Cho, J., and Swartz, W. (1993). First in situ observations of neutral and plasma density fluctuations within a PMSE layer. *Geophys. Res. Lett.* 20(20), 2311–2314.

Muraoka, Y., Sugiyama, T., Kawahira, K., Sato, T., Tsuda, T., Fukao, S., and Susumu, K. (1988). Cause of a monochromatic inertia gravity wave breaking observed by the MU radar. *Geophys. Res. Lett.* 15(12), 1349–1352.

Nicolls, M. J., Kelley, M. C., Varney, R. H., and Heinselman, C. J. (2008). Spectral observations of polar mesospheric summer echoes at 33 cm (450 MHz) with PFISR. *J. Atmos. Solar-Terr. Phys.*, in press.

Nordberg, W., Katchen, L., Theon, J., Smith, W. S. (1965). Rocket observations of the structure of the mesosphere. *J. Atmos. Sci.* 22, 611–622.

Orlanski, I., and Bryan, K. (1969). Formation of the thermocline step structure by large-amplitude internal gravity waves. *J. Geophys. Res.* 74, 6975–6983.

Pasko, V. P., Inan, U. S., and Bell, T. F. (1997). Sprites produced by quasielectrostatic heating and ionization in the lower ionosphere. *J. Geophys. Res.* 102, 4529.

Pasko, V. P., Stanley, M. A., Mathews, J. D., Inan, U. S., and Wood, T. G. (2002). Electrical discharge from a thundercloud top to the lower ionosphere. *Nature* 416, 152–154.

Pedersen, A., Trøim, J. and Kane, J.A. (1970). Rocket measurements showing removal of electrons above the mesopause in summer at high latitude. *Planet. Space Sci.* 18, 945.

Pfaff, R., Holzworth, R., Goldberg, R., Freudenreich, H., Voss, H., Croskey, C., Mitchell, J., Gumbel, J., Bounds, S., Singer, W., and Latteck, R. (2001). Rocket probe observations of electric field irregularities in the polar summer mesosphere. *Geophys. Res. Lett.* 28(8), 1431–1434.

Rapp, M., and Lübken, F. J. (2004a). On the nature of PMSE: Electron diffusion in the vicinity of charged particles revisited. *J. Geophys. Res.* 108(D8), 8437, doi:10.29/2002JD002857.

———. (2004b). Polar mesosphere summer echoes (PMSE): Review of observations and current understanding. *Atmos. Chem. Phys.* 4, 2601–2633.

Reid, G. C. (1990). Ice particles and electron "bite-outs" at the summer polar mesopause. *J. Geophys. Res.* 95, 13,891–13,896.

———. (1997). The nucleation and growth of ice particles in the upper mesosphere. *Adv. Space Res.* 20(6), 1285–1291.

Roble, R. G., and Dickinson, R. E. (1989). How will changes in carbon dioxide and methane modify the mean structure of the mesosphere and thermosphere? *Geophys. Res. Lett.* 16, 1441.

Roble, R. G., and Ridley, E. C. (1994). A thermosphere-ionosphere-mesosphere-electrodynamics general circulation model (TIMEGCM): Equinox solar cycle minimum simulations (30–500 km). *Geophys. Res. Lett.* 21, 417–420.

Rosinski, J., and Snow, R. H. (1961). Secondary particulate matter from meteor vapors. *J. Meteorol.* 18, 736–745.

Röttger, J., Rietveld, M. T., La Hoz, C., Hall, T., Kelley, M. C., and Swartz, W. (1990). Polar mesosphere summer echoes observed with the EISCAT 933 MHz radar and the CUPRI 46.9 MHz radar, their similarity to 224 MHz radar echoes and their relation to turbulence and electron density profiles. *Radio Sci.* 25, 671–687.

Røyrvik, O., and Smith, L. (1984). Comparison of mesospheric VHF radar echoes and rocket probe electron concentration measurements. *J. Geophys. Res.* 89, 9014.

Schröder, W. (2001). Otto Jesse and the investigation of noctilucent clouds 115 years ago. *Bulletin of the American Meteorological Society* 82(11), 2457–2468.

Seyler, C. E. (2005). Internal waves and undular bores in mesospheric inversion layers. *J. Geophys. Res.* 110, D09S05, doi:10.1029/2004JD004685.

Siefring, C. L., and Kelley, M. C. (1991). Analysis of standing wave patterns in VLF transmitter signals: Effects of sporadic E-field layers and in situ measurements of low electron densities. *J. Geophys. Res.* **96**, 17,813–17,826.

Smith, R. K. (1988). Travelling waves and bores in the lower atmosphere. The "Morning Glory" and related phenomena. *Earth Sci. Rev.* **25**, 267–290.

Smith, S. M., Taylor, M. J., Swenson, G. R., She, C. Y., Hocking, W., Baumgardner, J., and Mendillo, M. (2003). A multi-diagnostic investigation of the mesospheric bore phenomenon. *J. Geophys. Res.* **108**, 13–30.

Stevens, M. H., Gumbel, J., Englert, C. R., Grossman, K. U., Rapp, M., and Hartogh, P. (2003). Polar mesospheric clouds formed from space shuttle exhaust. *Geophys. Res. Lett.* **3**, 10, 1546, doi:10.1029/2003GL017249.

Stevens, M. H., Meier, R. R., Chu, X., DeLand, M. T., and Plane, J. M. C. (2005). Antarctic mesospheric clouds formed from space shuttle exhaust. *Geophys. Res. Lett.* **32**, L13810, doi:10.1029/2005GL023054.

Strelnikova, I., Rapp, M., Raizada, S., and Sulzer, M. (2007). Meteor smoke particle properties derived from Arecibo incoherent scatter radar observations. *Geophys. Res. Lett.* **34**, L15815, doi:10.1029/2007GL030635.

Su, H. T., Hsu, R. R., Chen, A. B., Wang, Y. C., Hsiao, W. S., Lai, W. C., Lee, L. C., Sato, M., Fukunishi, H. (2003). Gigantic jets between a thundercloud and the ionosphere. *Nature* **423**, 974.

Taylor, M. J., Turnbull, D. N., and Lowe, R. P. (1995). Spectrometric and imaging measurements of a spectacular gravity wave event observed during the ALOHA93 campaign. *Geophys. Res. Lett.* **22**, 2848–2852.

Thomas, G. E. (1991). Mesospheric clouds and the physics of the mesopause region. *Rev. Geophys.* **29**, 553–575.

———. (1996a). Global change in the mesospherelower thermosphere region: Has it already arrived? *J. Atmos. Terr. Phys.* **58**, 1629.

———. (1996b). Is the polar mesosphere the miner's canary of global change? *Adv. Space Res.* **18**, 149.

Tricker, R. A. R. (1965). *Bores, Breakers, Waves and Wakes.* Elsevier Science, New York.

Ulwick, J. C., Baker, K. D., Kelley, M. C., Balsley, B. B., and Ecklund, W. L. (1988). Comparison of simultaneous MST radar and electron density probe measurements during STATE. *J. Geophys. Res.* **93**(D6), 6989–7000.

Vlasov, M. N., and Kelley, M. C. (2009). Electron heating and airglow emission due to lightning effects on the ionosphere. *J. Geophys. Res.*, submitted.

Vlasov, M. N., Kelley, M. C., and Gerken, E. A. (2005). On the energy distribution of suprathermal electrons produced by HF heating in the F_2 region. *J. Atmos. Solar-Terr. Phys.* **67**(4), 405–412.

von Zahn, U., and Bremer, J. (1999). Simultaneous and commonvolume observations of noctilucent clouds and polar mesosphere summer echoes. *Geophys. Res. Lett.* **26**(11), 1521–1524.

Whipple, F. J. W. (1930). The great Siberian meteor and the waves, seismic and aerial, which it produced. *Q. J. R. Meteorol. Soc.* **56**(236), 287–304.

Yokoyama, T., Yamamato, M., Fukao, S., Takahashi, T., and Tanaka, M. (2005). Numerical simulation of midlatitude ionospheric Eregion based on SEEK and SEEK2 observations. *Ann. Geophys.* **23**, 2377–2384.

Yu, J. R., and She, C. Y. (1995). Climatology of a midlatitude mesopause region observed by a lidar at Fort Collins, Colorado (40.6° N, 105° W). *J. Geophys. Res.* **100**(D4), 7441–7452.

Zinn, J., and Drummond, J. (2005). Observations of persistent Leonid meteor trails: 4. Buoyant rise/vortex formation as mechanism for creation of parallel meteor train pairs. *J. Geophys. Res.* **110**, A04306, doi:10.1029/2004JA010575.

8 High-Latitude Electrodynamics

In this chapter we study the macroscopic motion of the high-latitude ionospheric plasma in the plane perpendicular to the magnetic field lines. Since the magnetic field is nearly vertical, this corresponds to the horizontal motion of plasma. At the large scales (>100 km) considered here, the electric force in (2.36b) dominates the pressure gradient and gravitational forces so only an imposed electric field and a neutral wind need be considered in the perpendicular plasma motion. In order to understand the characteristics and the sources of the imposed electric field, we first deal briefly with the relationships between electric fields and currents that exist in the ionosphere, outer magnetosphere, and solar wind. These fields and currents are coupled along the earth's magnetic field. Following this, we discuss the observed characteristics of electric fields and currents in the ionosphere and their relationships to the magnetic field topology throughout the ionosphere, the magnetosphere, and the solar wind system.

8.1 Electrical Coupling Between the Ionosphere, Magnetosphere, and Solar Wind

8.1.1 General Relationships

We begin by considering two regions: one in the ionosphere and one in the magnetosphere. Below about 200 km the ionosphere is a resistive medium, and in Chapter 2 we showed that the electric field $\mathbf{E}$ and electric current $\mathbf{J}$ are related by the equation

$$\mathbf{J} = \sigma \cdot (\mathbf{E} + \mathbf{U} \times \mathbf{B}) \tag{8.1}$$

where σ is the ionospheric conductivity tensor and $\mathbf{U}$ and $\mathbf{B}$ are the neutral wind velocity and magnetic field, respectively. In that chapter we also pointed out that above about 2000 km, the magnetospheric plasma is essentially collisionless and the equations governing $\mathbf{E}$, $\mathbf{J}$, and the plasma velocity $\mathbf{V}$ are

$$\mathbf{E} + \mathbf{V} \times \mathbf{B} = 0 \tag{8.2}$$

The Earth's Ionosphere: Plasma Physics and Electrodynamics

and

$$\rho d\mathbf{V}/dt = -\nabla p + \rho \mathbf{g} + \mathbf{J} \times \mathbf{B} \tag{8.3}$$

Taking the cross product of (8.3) with $\mathbf{B}$, the current density perpendicular to $\mathbf{B}$ is given by

$$\mathbf{J}_\perp = (1/B^2)(\rho \mathbf{B} \times d\mathbf{V}/dt + \rho \mathbf{g} \times \mathbf{B} + \mathbf{B} \times \nabla p) \tag{8.4}$$

The gravitational term in (8.4) can usually be neglected, so we note that in the high-altitude magnetosphere the cross-field current is controlled by pressure gradients and space or time-dependent flow and not directly by the electric field. Electrical coupling between the regions is described by the equation for current continuity

$$\nabla \cdot \mathbf{J} = 0 \tag{8.5}$$

and Faraday's law

$$\nabla \times \mathbf{E} + \partial \mathbf{B}/\partial t = 0 \tag{8.6}$$

which must apply throughout the system. The boundaries of these regions can be defined rather loosely for our purposes, since only a qualitative treatment of magnetosphere-ionosphere coupling is being undertaken. The region between 200 and 2000 km is a transition zone, but for most purposes the plasma motion there is also given by (8.2). Throughout this chapter we assume that the magnetic field lines connecting these regions are electric equipotentials. This assumption breaks down drastically in the lower magnetosphere on auroral zone field lines where electrons and ions are accelerated by parallel electric fields. This source region for auroral arcs is discussed in Chapter 9.

The space above 2000 km can be divided into two topologically different regions by the magnetic field geometry described in Chapter 1 (see Fig. 1.13). At the highest latitudes, the magnetic field lines extend either to the magnetopause and subsequently into the magnetosheath and solar wind, or far down the magnetotail into the boundary layer that lies just inside the magnetopause. Field lines extending to the magnetosheath are "open"—that is, they have one foot on the earth and the other connected to the interplanetary magnetic field (IMF). Field lines extending to the boundary layer and inner magnetosphere are "closed"—that is, they have both feet on the earth even though the field line may be extremely long. In the magnetosheath and in the boundary layer, the plasma is flowing rapidly antisunward and is driven by the expanding solar atmosphere. In the inner magnetosphere the plasma flow is dependent on the plasma pressure and upon both internally and externally applied electric fields.

8.1.2   A Qualitative Description of Convection for Southward IMF

Some fundamental properties of the coupling of electric fields, currents, and energy in the ionosphere, magnetosphere, and solar wind system can be understood as follows. Consider the case where magnetic field lines in the ionosphere connect to a southward IMF. This defines an area at high latitudes in the ionosphere called the polar cap. The geometry is shown schematically in Fig. 8.1a. This interaction is described qualitatively in Chapter 1, which the reader is encouraged to revisit at this time. Since the solar wind plasma is collisionless and expands radially outward from the sun, the electric field in the solar wind vanishes, while in the earth's frame it is given by $\mathbf{E}_{sw} = -\mathbf{V}_{sw} \times \mathbf{B}_{sw}$. For a southward IMF the field will have a component pointing from dawn to dusk. The electric potential across the connected field lines will be applied across the magnetosphere and will map down to the polar cap ionosphere, where a dawn-to-dusk directed ionospheric electric field, $\mathbf{E}_I$, will also result. This electric field drives the ionospheric F-region plasma in the antisunward direction at a speed $\mathbf{V}_I = \mathbf{E}_I \times \mathbf{B}_I/B_I^2$. The magnetic flux density is higher in the ionosphere than in the solar wind and, since the equipotential surfaces converge, the electric field in the ionosphere will be larger than in the solar wind. Typical numbers are

$$B_I/B_{sw} = 50{,}000\,\text{nT}/5\,\text{nT} = 10^4$$

$$E_I/E_{sw} = 50\,\text{mV}\,\text{m}^{-1}/2\,\text{mV}\,\text{m}^{-1} = 125$$

$$V_I/V_{sw} = 1\,\text{km}\,\text{s}^{-1}/400\,\text{km}\,\text{s}^{-1} = 2.5 \times 10^{-3}$$

Returning to (8.1) and ignoring the neutral wind in the ionosphere for the present, the electric field will drive a current at ionospheric heights given by

$$\mathbf{J} = \sigma \cdot \mathbf{E}_I \tag{8.7}$$

The Pedersen component of this ionospheric current is parallel to $\mathbf{E}_I$ and so is such that $\mathbf{J} \cdot \mathbf{E} > 0$. In this interaction, then, the ionosphere is a load and we must ascertain where the electrical energy originates. Referring to Fig. 8.1a, suppose that the solar wind slows down slightly when it is in contact with the region shown and that we can ignore the gravity and pressure gradients. Then (8.4) shows that $\mathbf{J}_{\perp} = (\rho/B^2)\mathbf{B} \times d\mathbf{V}/dt$ is antiparallel to $\mathbf{E}_{sw}$—that is, $\mathbf{J}_{sw} \cdot \mathbf{E}_{sw} < 0$. The solar wind acts as an MHD generator, feeding energy to the ionosphere as a load. Note that in this region the $\mathbf{J} \times \mathbf{B}$ force is directed back toward the sun, which shows consistency with the concept that the solar wind is slowing down due to the interaction. These same principles hold in any electromotive generator in which kinetic energy is converted to electrical energy. Finally, since the interaction region where the solar wind slows down is bounded, $\mathbf{J}_{sw}$ has a divergence which provides the field-aligned currents feeding the load.

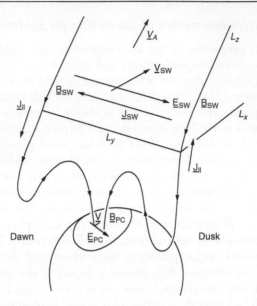

Figure 8.1a   Schematic representation of the magnetic connection between the solar wind dynamo and the ionospheric load. Note that $\mathbf{J} \times \mathbf{B}$ in the solar wind is toward the sun.

Our goal here is to make a simplified but interesting estimate of the percentage change of the solar wind velocity as it streams by the earth. We argue that what controls this slowdown in part is the conductivity of the polar cap ionospheres. This is characterized by the magnetic field line-integrated Pedersen conductivity, Σ_P, which is the electrical load on the MHD generator.

The total current enclosed in the polar cap is $I_{pc} = \Sigma_P L_{pc} E_{pc}$, where L_{pc} is the size of the polar cap and E_{pc} is the dawn-to-dusk component of the polar cap electric field. Since Σ_P can be very different in the two hemispheres at solstice, we treat each polar cap separately. I_{pc} must be closed in the solar wind. Its value is given by the expression $I_{sw} = J_{sw} L_T L_z$, where L_T is the distance the field line moves before it reconnects in the magnetotail and L_z is the distance along the IMF in which an Alfvén wave travels in this time. For a "square" polar cap we have $L_z = V_A \delta t$, where $\delta t = (L_{pc} B_{pc})/E_{pc}$, since the polar cap velocity is E_{pc}/B_{pc}. Likewise, $L_T = V_{sw} \delta t$.

By Kirchoff's current law, $I_{pc} = I_{sw}$ and thus,

$$J_{sw} = \frac{\Sigma_P L_{pc} E_{pc}}{L_T L_z}. \tag{8.8}$$

The current density can also be expressed as

$$J_{sw} = (\rho/B_{sw}) \frac{dV_{sw}}{dt}. \tag{8.9}$$

Equating Eqs. (8.8) and (8.9) yields

$$\Sigma_P L_{pc} E_{pc} = \frac{\rho}{B_{sw}} \frac{\delta V_{sw}}{\delta t} (V_A \delta t)(V_{sw} \delta t).$$

We should note that by letting dt in the time derivative equal δt, we are assuming the slowdown occurs at a large spatial scale. Now, using $V_A = \sqrt{B_{sw}^2/\mu_0 \rho}$ and letting $\eta_A = \mu_0 V_A$ be the intrinsic impedance of a magnetized plasma at low frequencies, we can write

$$\frac{\delta V_{sw}}{V_{sw}} = \Sigma_P \eta_A \left(\frac{E_{pc}^2}{B_{pc}} \frac{B_{sw}}{E_{sw}^2} \right).$$

However, if magnetic field lines are equipotentials, then E^2/B is a constant anywhere along the field line, since $E \propto \Phi/\delta l$ and $B \propto 1/\delta l^2$, where Φ is the constant potential and δl is the distance between two field lines. Thus, the term in brackets goes to 1 and we get

$$\frac{\delta V_{sw}}{V_{sw}} = \Sigma_P \eta_A. \tag{8.10a}$$

This simple yet elegant result demonstrates how a planetary magnetospheric/ionospheric system interacts with a stellar wind and the controlling influence of the polar cap conductivity. For the summer polar cap, $\Sigma_P = 5\Omega^{-1}$ and, taking $\eta_A = 0.1\Omega$, we find a 50% slowdown. This equation implies that the winter polar cap electric field should be larger than the summer field. This implication is suggested by data such as that in Fig. 8.1b in which the polar cap in winter exceeds the summer value in agreement with Eq. (8.10a).

More sophisticated studies by Siscoe et al. (2002) and Hill (1984) find the following relationship between the polar cap potential and interplanetary parameters:

$$\Phi_{pc} = \frac{57.6 E_{sw} P_{sw}^{1/3} D^{4/3} F(\theta)}{P_{sw}^{1/2} D + 0.0125 \xi \Sigma_P E_{sw} F(\theta)} \tag{8.10b}$$

where Φ_{pc} is the polar cap potential drop, E_{sw} is the electric field in the upstream solar wind ($E_{sw} = |\mathbf{V}_{sw} \times \mathbf{B}_{sw}|$), P_{sw} is the ram pressure exerted by the solar wind ($P_{sw} = \rho_{sw} V_{sw}^2$), D is the earth's dipole field normalized to 1 for the present-day value, $F(\theta)$ is a function of the clock angle of the IMF to account for the geometry of reconnection (here $F(\theta) = \sin^n(\theta/2)$, so $F(\theta) = 0/1$ for IMF pure northward or southward), ξ is a dimensionless coefficient between 3 and 4 that depends on the geometry of currents in the ionosphere (Crooker and Siscoe, 1981), and Σ_P is the height-integrated Pedersen ionospheric conductivity (assumed to be uniform

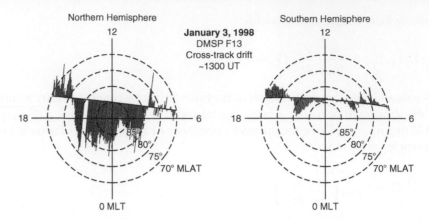

Figure 8.1b An example of the cross-track plasma drift velocity measured by DMSP satellites on consecutive polar passes. The left panel shows the drift in the northern hemisphere and the right panel shows the drift in the southern hemisphere. Ten degrees of latitude is about 1 km/s on this scale. The winter drift shows significantly more structure than the summer drift and is at least 3–4 times the average summer drift. B_z as measured by the ACE satellite was southward during this time, with a value of about −0.8 nT during the northern pass and about −2.2 nT during the southern pass. (Figure courtesy of M. J. Nicolls.)

for simplicity's sake). In the preceding equation, E_{sw} is expressed in mV/m, P_{sw} in nPa, Σ_P in S, and Φ_{pc} in kV. As in the simple conceptual approach above, the polar cap ionospheric electric field (proportional to Φ_{pc}) can be inversely proportional to Σ_P. This holds true whenever E_{sw} or Σ_P is large enough to cause a saturation of the polar cap potential, limiting it to values of the order of 250 kV (Hairston et al., 2003).

Weimer (see Appendix B) has developed a climatological model of the cross polar cap potential based on DMSP data. Inputs involve solar wind parameters. In Fig. 8.1c, the Hill-Siscoe equation (8.10b) is compared to the Weimer model using TIMEGCM to determine Σ_p (Weimer, 2001). The agreement is remarkable except when B_z is almost exactly northward. For the high solar wind velocity and pressure in this event during B_z north, a PCP of almost 50 keV was observed.

From the viewpoint of ionospheric physics, it is important to understand how the currents from the generator link up to the load currents. Dividing **J** into components perpendicular ($\perp$) and parallel ($\parallel$) to the magnetic field and setting the divergence of **J** to zero yields

$$\nabla \cdot \mathbf{J} = \nabla_\perp \cdot \mathbf{J}_\perp + \partial J_\parallel \partial s = 0 \tag{8.11}$$

where s is distance along the magnetic field line. Repeating and slightly extending the discussion in Section 2.4, we may integrate this equation over the field line

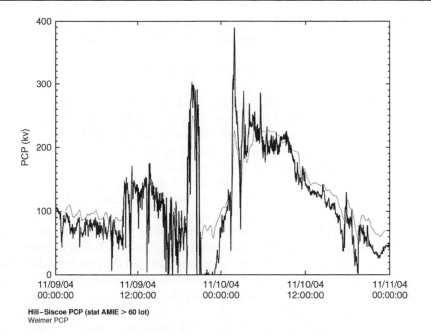

Figure 8.1c Weimer Convection Model PCP (thin line), along with the Hill-Siscoe polar cap potential prediction (thick curve). [After Kelley et al. (2009). Reproduced with permission of Elsevier Science Ltd.]

distance Δs where the ionospheric currents flow (200 to 80 km). Assuming that no current flows out of the bottom of this region, we obtain an expression for the field-aligned current (which is approximately vertical in the polar cap) at the top of the region:

$$J_{||} = \int_{\Delta s} (\nabla_\perp \cdot \mathbf{J}_\perp) ds \tag{8.12}$$

Since the electric field is independent of the variable s, we can substitute (8.7) in (8.12) and remove the electric field and divergence operator from the integral to yield the expression

$$J_{||} = \nabla_\perp \cdot (\mathbf{\Sigma}_\perp \cdot \mathbf{E}_I) \tag{8.13}$$

where $\mathbf{\Sigma}_\perp$ is the height-integrated horizontal conductivity tensor. Breaking this expression into Hall and Pedersen components, we finally have

$$\begin{aligned} J_{||} = \Sigma_p(\nabla_\perp \cdot \mathbf{E}_I) + \mathbf{E}_I \cdot \nabla_\perp \Sigma_p + \Sigma_H [\nabla_\perp \cdot (\mathbf{E}_I \times \mathbf{s})] \\ + (\mathbf{E}_I \times \mathbf{s}) \cdot \nabla_\perp \Sigma_H \end{aligned} \tag{8.14}$$

We showed in Chapter 3 that, for an electrostatic field, $\nabla_\perp \cdot (\mathbf{E}_1 \times \mathbf{s})$ is very small in the F region and above (incompressible flow), and thus

$$J_\parallel = \Sigma_P(\nabla_\perp \cdot \mathbf{E}_I) + \mathbf{E}_I \cdot \nabla_\perp \Sigma_P + \left(\mathbf{E}_I \times \hat{s}\right) \cdot \nabla_\perp \Sigma_H \tag{8.15}$$

This expression shows explicitly that field-aligned currents are intimately related to spatial variations in the ionospheric electric field and conductivity. It is important to realize, however, that the field-aligned current in (8.15) is driven by some divergence of current in the generator, but modified by conductivity gradients and polarization field in the ionosphere.

Next, consider the electric field in the closed field line region of the magnetosphere. (We are still considering the southward IMF case.) In Chapter 1 we saw that the magnetic field lines in the boundary layer and the plasma sheet are distorted from a dipole shape to produce a magnetic tail extending away from the sun. This magnetic geometry has a tension (see Chapter 2) that exerts a force on the plasma. Together with the pressure gradient and the potential difference applied across the magnetosphere by the flowing solar wind, these forces produce motion of the magnetospheric plasma on closed field lines and an associated dawn-to-dusk magnetospheric electric field in the tail. Further, since the gradient and tension forces are equivalent to the $\mathbf{J} \times \mathbf{B}$ force, the geometry requires the existence of electric currents and vice versa. Figure 8.2a shows the configuration of electric fields and currents in the ionosphere and magnetosphere. The tail (or neutral sheet) current, $\mathbf{J_T}$, flows across the tail near the magnetic equatorial plane to support the curl implied by the stretched magnetic field geometry. The tail current is closed primarily in the magnetosheath by currents flowing on the magnetopause (which are not shown in the figure). The ring current, $\mathbf{J_R}$, to first order flows in closed loops around the earth at distances between 2 and 10 earth radii (R_e) and is driven primarily by pressure gradient forces. Any divergences in these currents must be closed by field-aligned currents that enter the ionosphere. The portion of the tail and ring current that is closed through the ionosphere is called the partial ring current, J_{PR}. These currents, the so-called Region 2 currents, are labeled R_2 in the figure and link the inner magnetosphere with the auroral oval near its equatorward edge. Region 1 currents, labeled R_1, link the poleward portion of the auroral oval and the polar cap to the magnetosheath, solar wind, or the boundary layer plasma near the magnetopause. Figure 8.2b shows how the magnetospheric electric field, $\mathbf{E}_m$, maps to the ionosphere on closed field lines.

The ionosphere is not a passive element in this circuit because some of the hot plasma in the magnetosphere can move along the magnetic field and strike the atmosphere, producing significant ionization. This particle precipitation produces the discrete and diffuse auroral airglow and can play a dominant role in determining the ionospheric conductivity in the eclipsed ionosphere. Notice that the electric field mapping causes the sign of the electric field to be reversed (e.g., Fig. 8.2b) so that in the ionospheric auroral zone the electric field $\mathbf{E}_a$ is directed from dusk to dawn, which is opposite from the polar cap electric field. The resulting ionospheric plasma flow and electric field patterns are shown in Fig. 8.3a. The flow is made up of antisunward flow at the highest latitudes resulting from the connection of the open magnetic field lines to the solar wind electric field discussed earlier. The return sunward flow in the auroral zones results from the electric field E_a, which in turn is determined by the potential difference across the closed field line portion of the magnetosphere.

Figure 8.3b shows the electric and magnetic fields measured in a polar pass of the S3-2 satellite field. The electric field detector had a higher spatial resolution but otherwise the measured electric field looks remarkably similar to the magnetic field. It is easy to show that the Poynting flux is almost always downward and consistent with Σ_P values for the polar cap and auroral oval (see the next section).

The electric field reversal at the polar cap boundary is associated with the so-called Region 1 field-aligned or Birkeland currents (R_1 in Fig. 8.2a) after the scientist who first postulated their existence. They are closed at one end by currents in the magnetosheath. In the summer polar cap, a large fraction of the Region 1 currents close across the conducting polar cap. Note from examination of Fig. 8.2 that the ionospheric and internal magnetospheric currents and associated electric fields all have the same direction at any one place, designating a load, whereas in the magnetosheath the electric field and current have opposite signs, as required for a generator. In fact, the entire system is analogous to a magnetohydrodynamic (MHD) generator, where the solar wind in the magnetosheath is the flowing conductor connected by the Region 1 Birkeland currents to the ionosphere-magnetosphere system, which is the load. Region 2 currents are connected to Region 1 currents by another $(\rho/B^2)\mathbf{B} \times d\mathbf{V}/dt$ current, since the magnetic field increases as the plasma from the tail nears the earth and E/B decreases. Here $\mathbf{J} \times \mathbf{B}$ is antisunward (observe that $J_\perp$ flows counter to $E_\perp$ in the equatorial plan of Fig. 8.2b).

8.1.3 Energy Transfer

Further insight into the methods by which energy is transmitted and converted in this system can be obtained from the idealized model of Region 1 currents and the closure current across the summer polar cap, shown in Fig. 8.4. Here, two parallel current sheets of thickness dx are oriented in the y-z plane and carry equal and opposite current densities in the z direction, which is along the magnetic field. These current sheets represent the Region 1 currents connecting the MHD generator in the solar wind or magnetosheath plasma to the ionosphere on the dawn and dusk sides of the polar cap, respectively. The current sheets are closed in the polar cap ionosphere, represented here as a resistive medium of vertical extent h having a uniform conductivity, σ_p, perpendicular to the current sheets. Associated with the current sheets are a magnetic field (δB) and an electric field (δE). Now consider a surface S_1 that is bounded by a rectangular loop 1,

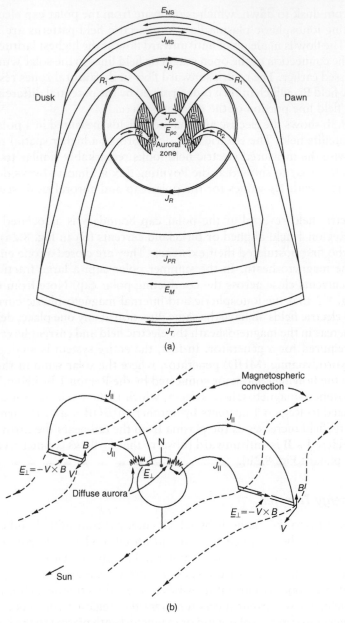

(a)

(b)

Figure 8.2 (a) Schematic diagram of the currents and electric fields that exist as a result of the extended magnetic field in the tail of the magnetosphere and the interaction between the solar wind and the earth's magnetic field. (b) Three-dimensional view of the electric and magnetic field geometry on auroral zone flux tubes.

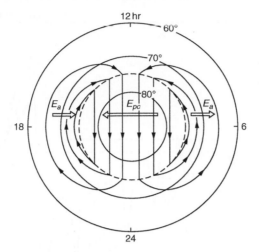

Figure 8.3a Representation of ionospheric electric fields in the northern hemisphere polar cap and auroral zone, as well as the plasma flow due to those fields.

encompassing a width w in the current sheet and extending a distance $d/2$ in each direction perpendicular to the current sheet. We further take $d \gg dx$. For such a loop, the steady-state integral form of Ampère's law can be written

$$\oint_1 \delta\mathbf{B} \cdot d\mathbf{l} = \mu_0 \iint_{S_1} \mathbf{J} \cdot d\mathbf{a}$$

When the surface S_1 is far from the magnetosheath and ionospheric ends of the current sheet, we may assume that the magnetic perturbation, δB_x, is zero at the surface edges. We may also assume from symmetry that δB_y along part (a) of the loop is equal and opposite to δB_y along part (h) of the loop. Thus, evaluating both sides of the previous equation gives

$$2\delta B_y w = \mu_0 J_z w dx$$

Hence,

$$\delta B_y = \mu_0 J_z dx / 2 \tag{8.16}$$

Note that the magnetic perturbation amplitude is independent of the distance from an infinite current sheet. Thus the magnetic perturbations from the two equal and opposite current sheets shown in Fig. 8.4 will add together in the region between the two sheets and exactly cancel each other in the regions outside. The result will be a magnetic perturbation, $\delta B_y = \mu_0 J_z \, dx$, confined completely to the region between the current sheets.

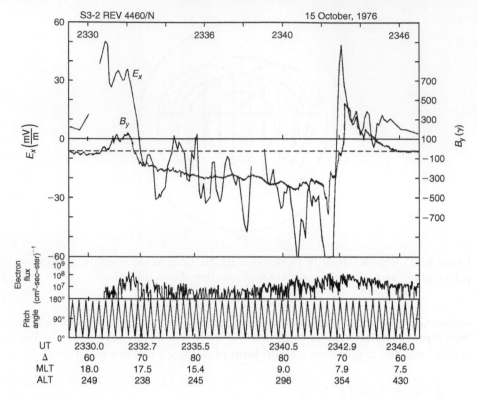

Figure 8.3b   S3-2 observations from a north polar pass on October 15, 1976. The dashed line is the baseline for B_y. [After Smiddy et al. (1980). Reproduced with permission of the American Geophysical Union.]

Now consider the ionospheric load. In this resistive medium the horizontal current from dawn to dusk is $J = \sigma_p \delta E_x h$. In a steady state the current entering the ionosphere in current sheet 1, $J_z\,dx$, must be equal to the total horizontal current in the vertical extent h of the ionosphere. Thus, making use of $\delta B_y = \mu_0 J_z dx$ and the fact that $\Sigma_p = \sigma_p h$, we have

$$\delta B_y = \mu_0 \Sigma_p \delta E_x \tag{8.17a}$$

This equation may be rewritten in the equivalent forms

$$\delta E_x = \delta B_y / \mu_0 \Sigma_p \tag{8.17b}$$

$$\delta B_y / \delta E_x = \mu_0 \Sigma_p \tag{8.17c}$$

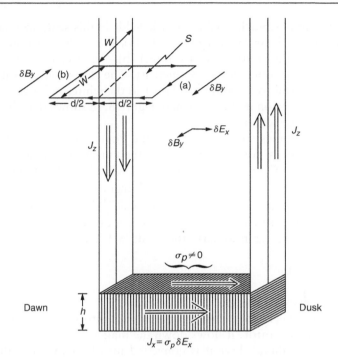

Figure 8.4 Schematic representation of field-aligned currents and closure currents in the polar cap ionosphere and the electric and magnetic fields associated with them.

or

$$\Sigma_p = \delta B_y / \mu_0 \delta E_x \qquad (8.17d)$$

These expressions show that for a uniform Σ_p the electric and magnetic fields at altitudes away from the ionospheric load should be highly correlated. DE-2 satellite observations have shown some events with just such correlations at a 99% confidence level (Sugiura et al., 1982), although Smiddy et al. (1980) have also shown examples where the correlation breaks down. The latter authors have interpreted their results in terms of the $E_I \cdot \nabla_\perp \Sigma_p$, term in (8.14) and (8.15), which must be ignored to derive the relationship above.

One may consider the solar wind MHD generator as a voltage source characterized by the electric field δE, which is attached to a resistive load characterized by Σ_p. Then (8.17a) yields the perturbation magnetic field as a function of the load conductivity. This viewpoint may be applicable for the large-scale magnetosphere-ionosphere interaction (Lysak, 1985). Alternatively, the generator may be viewed as a current source. There is some evidence that the latter viewpoint is more accurate at smaller scales (≤ 100 km), since the summer polar cap has much smaller electric fields in this scale size regime than does the winter polar cap (Vickrey et al., 1985). For such a current source, $\delta \mathbf{B}$ would be fixed by the source and, from (8.17b), the electric field would then be determined by the ionospheric conductivity. The electric field would then be inversely proportional to Σ_p and thus, larger in the winter polar cap, where the lack of solar illumination makes Σ_p small. Figure 8.1b shows that this idea may be true, even at large scales.

Further insight comes from considering the Poynting flux. In the geometry of Fig. 8.4, $\delta \mathbf{E} \times \delta \mathbf{H}$ is downward between the current sheets and the energy input is $(\delta E_x \delta B_y / \mu_0)$ W/m^2. This energy must be dissipated as Joule heat in the ionosphere at the rate of $W = \mathbf{J} \cdot \mathbf{E} = \sigma_p \delta E_x^2$. Integrating over the vertical extent of the ionosphere yields a dissipation rate of $(\Sigma_p \delta E_x^2)$ W/m^2. Since the Poynting flux yields the power flow into the region per unit area, we may equate the two expressions, and once again we have the result (8.17c):

$$\delta B_y / \delta E_x = \mu_0 \Sigma_P$$

The two approaches are therefore self-consistent.

To summarize, mechanical energy is converted into electromagnetic energy in the solar wind generator. It flows down the magnetic field lines to the ionosphere as Poynting flux, where it is converted into heat by Joule dissipation. For the typical parameters of $\delta E_x = 50$ mV/m and $\delta B_y = 500$ nT, we can estimate the Poynting flux to be $\delta E_x \delta B_y / \mu_0 = 0.02$ W/m$^2 = 20$ ergs/cm$^2 \cdot$ s. This is a substantial amount of energy, roughly 10^{11} W over the whole region. It is important to notice also that an energy flux of 20 ergs/cm$^2 \cdot$ s is very large compared to typically observed energy fluxes in auroral particle precipitation except for extremely intense localized auroral arcs. In fact, the Joule heat input is the primary reason that the thermosphere has a local temperature maximum in the high-latitude region, which competes with the solar photon-driven temperature maximum that occurs near the subsolar point.

8.1.4 Additional Complexities

Before leaving this introductory discussion, we emphasize several important considerations to keep in mind when examining the observations of high-latitude ionospheric plasma motion in the next section. First, the qualitative discussion presented previously is centered around a direct connection between the earth's magnetic field and the IMF. The subsequent communication of the interplanetary electric field to the ionosphere and the magnetosphere gives rise to a two-cell convection pattern as shown in Fig. 8.3a. This process, which produces antisunward flow on open field lines if the IMF has a component in the southward direction, was first described by Dungey (1961). Axford and Hines (1961) showed that a similar motion of the plasma at high latitudes would result if solar wind momentum was transferred across the magnetopause without any direct connection between the ionosphere and the solar wind magnetic fields. This process, called "viscous interaction," produces a relatively narrow region of antisunward-flowing plasma just inside the magnetopause. This region is now called the equatorial magnetospheric boundary layer, and in this region the anti-sunward plasma flow occurs on closed field lines. The ionospheric flow pattern we described above for the electrical connection model (often called reconnection) does not differ dramatically for a theory in which viscous interaction produces the plasma flow in the ionosphere and thus the electric field pattern. The major difference is that the antisunward convecting plasma is on closed instead of open field lines. Second, when the interplanetary magnetic field is northward, the viscous interaction idea does not change very much but connection to the interplanetary magnetic field and the solar wind potential is drastically different. Since B_z is northward roughly one-half of the time, it is not surprising that some of the flow patterns shown below are quite different from the idealized two-cell pattern discussed thus far.

We also need to discuss the physics at the inner boundary of the region dominated by magnetospheric processes. We know from Chapters 3 and 5 that the electric field pattern on field lines within 3 or 4 earth radii is controlled primarily by the atmosphere. How does this transition take place? As the plasma in the magnetosphere flows toward the earth, it encounters an increasing magnetic field strength. Since the first adiabatic invariant is conserved (see Section 2.5.2), the perpendicular energy of the plasma increases. Since the gradient and curvature drifts of these particles depend on both their energy and charge, a zonal charge separation occurs with positive charges at dusk and negative charges at dawn. This creates an electric field pointed from dusk toward dawn in the inner magnetosphere, which tends to cancel out the applied dawn-dusk electric field. The inner magnetosphere is therefore shielded from the magnetospheric electric field and the plasma flows around this region. This shielding only operates on long time scales, however, and fluctuations of the external field with periods shorter than eight hours or so can penetrate (see Chapter 3).

Since the magnetospheric electric field is reduced to almost zero earthward of the ring current, the electric field due to the rotation of the earth becomes comparable to the magnetospheric field near this boundary. When this source of the plasma motion is included, the ionospheric flow paths acquire a corotating component and look like those shown in Fig. 8.5. At latitudes below about 50° (not shown on this figure), the plasma undergoes circular convection paths around the earth at the corotation speed. On the dusk side near 60° latitude, the corotation and the magnetospheric electric fields oppose each other, leading to complex flow trajectories that involve a stagnation point, marked S. At still higher latitudes, the two-cell convection pattern is preserved. The corresponding plasma flow paths in the equatorial plane of the inner magnetosphere might look like those shown in Fig. 8.6. Near the earth in the shaded region, plasma flows in concentric circles. This region is called the plasmasphere because it contains a cool, dense plasma. This region has a high plasma content since, once per

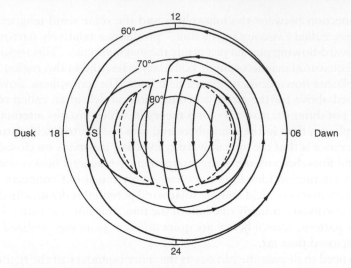

Figure 8.5   Ionospheric flow paths at high latitudes including the effect of corotation of the plasma with the earth. The diagram is fixed with respect to the sun.

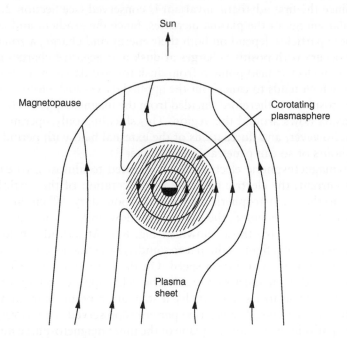

Figure 8.6   Contours of plasma flow velocity and equipotentials in the magnetic equatorial plane. The concentric circles indicate corotating plasma deep in the magnetosphere. The diagram is fixed with respect to the sun.

day, the associated flux tubes fill from below with ionospheric plasma produced on the dayside of the earth (see Chapter 5). To first order, the plasmasphere lies just inside the ring current and the plasma sheet field lines. Outside of this region, the flow is more or less toward the sun in the equatorial plane. Where a flow line meets the magnetopause, it is assumed either to make contact with the interplanetary field and flow back over the top of the magnetopause or to flow back down the flanks of the tail in the boundary layer. Figures 8.5 and 8.6 are in the sun-fixed frame.

On the magnetic field lines in contact with the corotating plasmasphere, the ionosphere to first order corotates with the earth just as the atmosphere does. In Chapters 3 and 5 we worked in this earth-fixed frame where the corotation field vanishes and discussed the ionospheric "weather." This involved electric fields and consequent plasma motions in the rotating frame, which had typical magnitudes of 1–10 mV/m and velocities of 20–200 m/s. For reference, in the equatorial plane the earth's rotation speed at the surface is 434 m/s, while the corotation speed at $L = 4$ in the equatorial plane is near 2 km/s.

8.2 Observations of Ionospheric Convection

Our knowledge of the large-scale motion of the high-latitude F-region plasma has come from satellite, rocket, balloon, and ground-based radar and magnetic field measurements. These and other measurement techniques have also been used extensively to study smaller-scale features of the plasma density and plasma motion associated with the aurora. This topic is dealt with in Chapter 10. No single measurement can provide a complete description of the large-scale motion of F-region plasma. Over a 24-hour period, for example, a satellite measurement can be repeated 10 to 15 times over the entire high-latitude range in the iono-sphere but only in a very limited local time region. Over the same time period a ground-based radar measurement can be made over the entire local time region but in a limited latitude range. A description of the high-latitude plasma motion is therefore made up from a synthesis of these complementary measurements, taken over a period of many years.

For a time-independent system, $\nabla \times \mathbf{E} = -\partial \mathbf{B}/\partial t = 0$ and the electric field may be described by an electrostatic potential ϕ such that $\mathbf{E} = -\nabla \phi$. The electric field is perpendicular to a line of constant potential and is also perpendicular to the $\mathbf{E} \times \mathbf{B}$ motion of the plasma. Thus, the low temperature plasma flows along lines of constant potential and a pattern of equipotentials also represents the plasma flow pattern. Each measurement of electric field or plasma velocity therefore provides a signature of some portion of the convection pattern. These measure-ments show that the convection pattern is highly dependent on the orientation and the magnitude of the IMF. Recall from Chapter 1 that when discussing the IMF, its direction can be described in a number of ways. The z component can be negative or positive and the IMF is often referred to as southward or northward, respectively. Similarly, the y component can be directed opposite to the earth's revolution about the sun or parallel to it, which is referred to as positive or negative, respectively. Finally, it is worth noting that the tendency of the IMF to assume a garden hose–like spiral in the solar wind means that, when the IMF is directed away from the sun, the y component is positive and when the field line is directed toward the sun, B_y is negative. Thus, the terms "toward" and "away" are sometimes used in the literature to describe the sign of the IMF y component. One final note concerns the presentation of observational data at high latitudes. Since the earth's magnetic field lines can be highly stretched on the nightside and slightly compressed on the dayside, a dipole representation is not very informative. Thus, it is frequently the practice to describe a point in the high-latitude region with two parameters. One is the L value, defined as the equatorial crossing point (in units of R_E) of the magnetic field line passing through the point. The second is the magnetic local time (MLT), defined using the angle between the earth-sun line and the plane containing the magnetic axis of the earth and a line from the center of the earth to the point. Frequently, the L value is expressed in terms of an invariant latitude Λ where

$$\Lambda = \cos^{-1}\left(1/\sqrt{L}\right)$$

An L value of 4, then, corresponds to $\Lambda = 60°$. In a centered dipole field, the latitude at which a magnetic field line passed through the surface of the earth would be equal to the invariant latitude. The MLT is almost always expressed in hours. It is also quite common to express the date as a five-digit number, called the Julian day. In this format the first two digits of the number denote the year while the last three digits denote which day of the year (assume that January 1 is day 1).

The most radical differences in the convection pattern can be seen by comparing signatures when the IMF is southward with signatures when the IMF is northward. The discussion following is divided into these two categories. However, it should be emphasized that substantial variability exists in the convection pattern, and our understanding of the nature of the controlling factors is still evolving.

8.2.1 Observations During Southward IMF

Figure 8.7a shows vectors representing the plasma velocity perpendicular to the magnetic field at points along a satellite track as the spacecraft passes through the high-latitude convection pattern during a period of southward IMF. The regions of sunward and antisunward convection are separated by well-defined reversals near dawn and dusk and the signature of the two-cell convection pattern is indicated by the dashed lines. The auroral zone is coincident with the sunward flow regions as described above. If we arbitrarily assign a zero value for the electrostatic potential at low latitudes, data of this nature can be integrated along the satellite track to produce a representative electrostatic potential distribution,

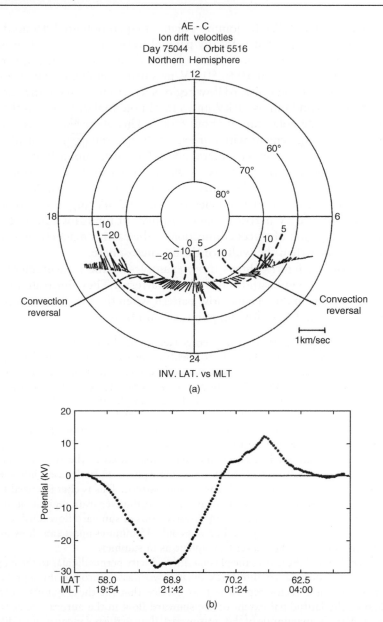

Figure 8.7 (a) A satellite flight across the high-latitude convection pattern provides drift velocity profiles, which are shown along with the inferred convection pattern. (b) The potential distribution resulting from this convective flow pattern shows maxima and minima at the polar cap boundaries and a total potential difference of about 60 kV across the polar cap. [After Heelis and Hanson (1980). Reproduced with permission of the American Geophysical Union.]

which is shown in Fig. 8.7b. By joining points of equal potential deduced from the lower plot with a line that is locally parallel to the observed flow, it is possible to deduce the flow pattern, as shown by the dashed lines in the upper plot. Typical electric field values of 10 to 50 mV/m (corresponding to flow velocities of 200 to 1000 m/s) in the sunward flow regions lead to maxima in the potential of the order of plus and minus 30 kV and a total potential drop across the open field line region of the polar cap of about 60 kV. This observed convection pattern is in qualitative agreement with our expectations from earlier consideration of the solar wind-magnetosphere interaction. However, only a fraction of the magnetospheric potential drop across a distance equal to the dimension of the magnetosphere appears across the ionospheric polar cap. For example, using our previous estimate for the solar wind electric field of 2 mV/m, we obtain a total potential of about 500 kV across a magnetosphere of typical width $40R_e$. This indicates that the region of direct connection of the earth's magnetic field to the IMF is much smaller than the width of the magnetosphere.

The next series of figures shows more examples of the measurements used in the derivation of high-latitude convection patterns. They illustrate the most important variations in the plasma flow pattern when the IMF has a southward component ($B_z < 0$). The following points should be noted.

1. *Average auroral zone pattern.* The average electric field pattern in the auroral oval determined from 500 h of balloon electric field measurements is presented in Fig. 8.8a. The dominant variation is diurnal with the meridional component considerably larger than the zonal component. When collated with respect to high and low K_p values, the pattern does not change much in shape but the amplitude of the dominant feature decreases by about a factor of 2 as K_p varies from 6 to 0.

2. *Seasonal dependence.* In Fig. 8.8b the plasma motion in a direction almost parallel to the earth-sun line is shown for two dawn-dusk satellite passes. The figure illustrates that the plasma velocity is considerably more structured in the winter (Southern) hemisphere (top panel) than in the summer hemisphere. In both cases, however, the data are consistent with a two-cell convection pattern. This is characterized by two large-scale reversals (shown by the heavy arrows) that separate regions of antisunward flow (dawn-to-dusk electric field) from regions of sunward flow (dusk-to-dawn electric field). As shown in Fig. 8.8b, passes only 45 minutes apart can show average fields twice as large in the winter hemisphere as in summer.

3. *Variability.* Figure 8.8c shows the plasma drift velocity perpendicular to the magnetic field measured along a satellite track during two passes through the high-latitude ionosphere using a vector representation. Notice that for these quite similar IMF conditions, the latitudinal extent of the sunward flow in the auroral zones is quite different. Also the magnitude of the antisunward flow in the polar cap is quite different in these two cases.

4. *Response to B_y.* Figure 8.8d again shows the ion drift velocity vector perpendicular to the magnetic field for two passes through the high-latitude ionosphere. In the first pass (top panel) B_y is negative, and in the bottom panel B_y is positive. Notice that when B_y is positive, the sunward flow from the dusk side passes across local noon in the auroral zone before flowing antisunward. When B_y is negative, the sunward flow from the dawn side passes across local noon before flowing antisunward.

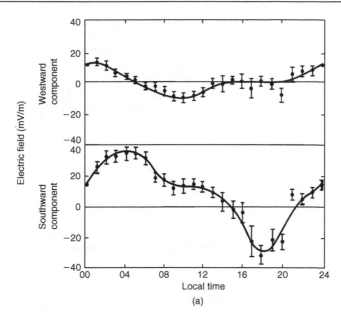

Figure 8.8a Hourly averages of auroral zone electric field data from balloon flights. The error bars are standard deviations of the means, and the solid curves are empirical fits to the data, satisfying the constraint imposed by $\nabla \times \mathbf{E} = 0$ that the 24-hour average westward field be zero. [After Mozer and Lucht (1974). Reproduced with permission of the American Geophysical Union.]

These observations show that the ionospheric conductivity, the internal state of the magnetosphere, and the external state of the interplanetary environment can all affect the high-latitude ionospheric convection pattern. However, when the IMF has a southward component, a two-cell convection pattern made up of predominantly antisunward convection at the highest latitudes with sunward convection occurring at lower latitudes is almost always observed.

In principle, a number of different radar stations and satellites operating at the same time can provide a simultaneous signature of the convection pattern at different local times. Alternatively, if some degree of stability in the convection pattern is assumed, a single radar site can sample all local times over a substantial latitudinal width of the pattern during a day. These techniques and the previously mentioned statistical syntheses of data bases are currently being used to elucidate details of the seasonal dependence in the convection pattern and the factors determining the convection speed. A climatological model by Weimer (1995, 2001), driven by solar wind parameters, provides a synthesis of the existing data. The dependence on B_y is well documented for that portion of the convection pattern on the sunward side of the dawn-dusk meridian and is shown schematically in Fig. 8.9. The convection pattern is most easily characterized by one small cell and one large cell. The large cell has an almost circular perimeter and the small

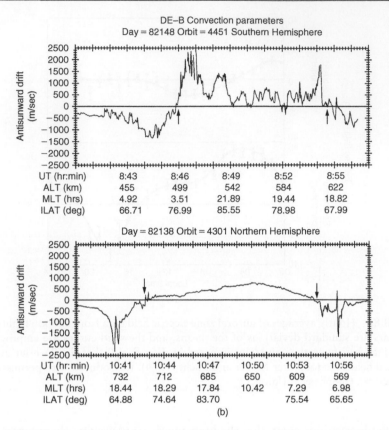

Figure 8.8b   The antisunward component of the plasma velocity, measured in both the Northern and Southern Hemispheres, is plotted as a satellite moves across the auroral zone and polar cap.

cell is crescent-shaped. The latter surrounds a portion of the circular cell. The placement of the small and large cells depends to first order on the sign of B_y. When the IMF has a southward component, the signature of a two-cell convection pattern is seen well beyond the dawn-dusk meridian into the nightside. The nightside convection pattern, however, does have some distinctive geometric characteristics that are related to the field-aligned and horizontal currents, which are described later. Figure 8.10 illustrates the variety of convection geometries that are seen.

8.2.2 Observations During Northward IMF

A significant departure from the previously described convection patterns is seen when the IMF has a northward component. In this case the convection velocity is usually much more structured and of smaller magnitude. When the identification

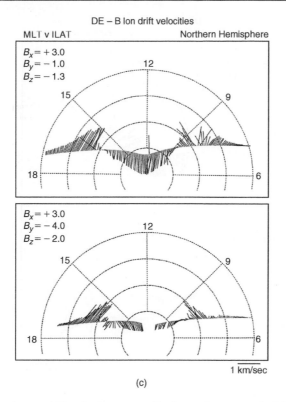

DE – B Ion drift velocities

MLT v ILAT   Northern Hemisphere

B_x = + 3.0
B_y = − 1.0
B_z = − 1.3

B_x = + 3.0
B_y = − 4.0
B_z = − 2.0

1 km/sec

(c)

Figure 8.8c   The plasma drift velocity perpendicular to the magnetic field measured during two satellite passes through the high-latitude ionosphere is represented here by drift velocity vectors.

of an organized convection pattern is possible, it is characterized surprisingly by a sunward flow component (a dusk-to-dawn electric field) in the central polar cap. In addition, the entire region of significant plasma motion is confined to much higher latitudes than in the southward IMF case. Figure 8.11 shows several examples of the dawn-to-dusk electric field measured at high latitudes when the IMF had a northward component. The shaded regions show the existence of a dusk-to-dawn field and thus sunward flow deep in the polar cap. Quite frequently, the flow is extremely structured at small scales, and observations can look more typically like those shown in Fig. 8.12a. Despite the difficulties associated with the recognition of convection features during northward IMF, a qualitative description is possible, as summarized in Fig. 8.12b. When the IMF has a northward component, four convection cells can be identified in the dayside hemisphere. Two high-latitude cells circulate in a manner that produces sunward convection at the highest latitudes. The size and orientation of these cells depend to first order on the magnitude and sign of the y component of the IMF. When B_y is very small, these two cells are located approximately on either

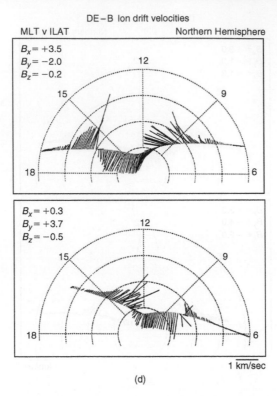

Figure 8.8d Measurements of the vector plasma drift velocity in the dayside high-latitude ionosphere show that the antisunward flow can be preferentially biased toward the dawn or dusk directions depending on the sign of B_y. (Parts (b)–(d) courtesy of R. A. Heelis and W. B. Hanson.)

side of the noon-midnight meridian and are of equal size. When B_y is positive, the clockwise circulation in the high-latitude dawn cell tends to expand into the dusk side, and the anticlockwise cell virtually disappears. Expansion of the dawn cell is apparent as B_y becomes positive. The lower-latitude convection cells circulate in the manner expected during a southward IMF. Their geometry and position show no strong dependence on the IMF y component and the total potential drop across these cells rarely exceeds $10 \, \text{kV}$. It may well be that this part of the convection pattern is driven by the viscous interaction described earlier.

These patterns correspond to a steady state, which is probably never attained. One experiment has revealed how the polar cap dynamically changes from B_z south to B_z north. Three rockets were flown during just such a transition. It is well known that the polar cap shrinks when B_z turns north. These data suggest that this occurs when an auroral arc surges sunward, bringing sunward convection behind it. Figure 8.13 shows the convection pattern along a flight trajectory of one rocket directed toward the southeast. As the arcs surged sunward at

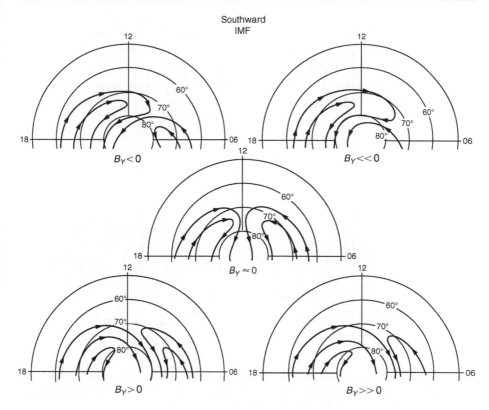

Figure 8.9 Schematic representation of the day-side high-latitude convection pattern showing its dependence on the y component of the IMF when B_z is south. [After Heelis (1984). Reproduced with permission of the American Geophysical Union.]

high velocity, the convection changed from antisunward to sunward (Berg et al., 1994). Thus, the polar cap/auroral oval boundary seems to shrink in discrete steps as arcs propagate across the polar cap.

8.3 Simple Models of Convection in the Magnetosphere

In our simple considerations of magnetosphere-solar wind coupling, the ionospheric plasma motion reflects the motion of the magnetospheric plasma to which it is magnetically connected. At the highest latitudes, antisunward convection is associated with similar flow in the magnetosheath, while sunward convection at lower latitudes is associated with flow in the plasma sheet. This relationship between plasma flows in different regions makes the concept of moving magnetic field lines or frozen-in flux extremely useful. However, if, as we suggested, the

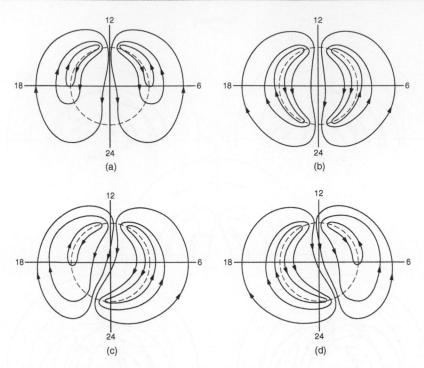

Figure 8.10 The nightside convection pattern can have a variety of geometries. Patterns a and d are most frequently observed. [After Heelis and Hanson (1980). Reproduced with permission of the American Geophysical Union.]

magnetic field lines associated with plasma flow in the polar cap are open and those associated with plasma flow in the auroral zones are closed, then this concept must break down wherever the plasma flows across the boundary between these two regions. When this breakdown occurs, the term "merging" or "reconnection" is frequently used to describe the phenomenon. A description of the plasma processes involved in a merging or reconnection region is beyond the scope of this book. It is convenient, however, to use these terms when describing moving magnetic field lines that change their identity either from an open to a closed topology or from one open topology to another.

8.3.1 Models for Southward IMF

We have seen that many ionospheric observations indicate that the IMF exercises significant control over the plasma convection pattern. From examining the energetic electron precipitation patterns at high latitudes and the direct detection of solar cosmic rays in the polar ionosphere, it is generally believed that the anti-sunward flow in the polar cap during a southward IMF exists on open magnetic field lines. The sunward flow in the auroral zone that completes the two-cell

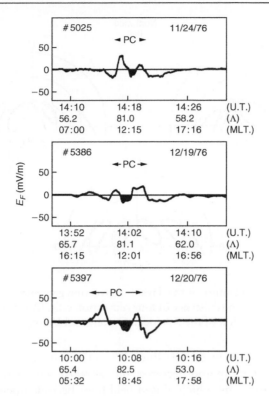

Figure 8.11 Three examples of the high-latitude, dawn-to-dusk component of the electric field in the ionosphere for a northward IMF. The shaded regions indicate sunward plasma flow. [After Burke et al. (1979). Reproduced with permission of the American Geophysical Union.]

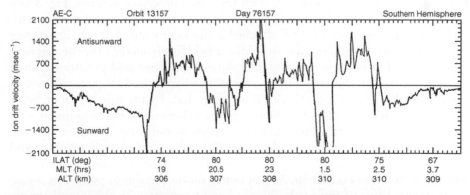

Figure 8.12a High-latitude plasma drift velocity in the ionosphere shown for a northward IMF. The flow is extremely structured and does not indicate a simple two-cell convection pattern, as is often the case during southward IMF. [After Heelis and Hanson (1980). Reproduced with permission of the American Geophysical Union.]

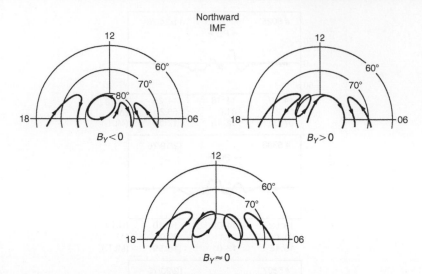

Figure 8.12b   The main feature of the dayside convection geometry when the IMF has a northward component is the existence of four convection cells. [After Heelis et al. (1986). Reproduced with permission of the American Geophysical Union.]

convection pattern is on closed field lines. When the sunward flow of the auroral zone crosses the polar cap and becomes antisunward, we use the term *merging* to describe the process by which closed field lines become open and connected to the interplanetary magnetic field (Dungey, 1961). The term "reconnection" describes the process by which open magnetic field lines are joined in the magnetotail to produce a closed field line. The reconnection region must correspond in the ionosphere to the region where antisunward flow crosses the polar cap boundary and becomes sunward. With these two processes in mind, Fig. 8.14a shows that a southward-pointing IMF field line (A-B) breaks and merges with the earth's field at point N1. Still attached to the solar wind, that field line moves across the polar cap and eventually rejoins a field line from the other hemisphere at point N2. During this time the interplanetary electric field penetrates down into the ionosphere, where it drives an antisunward flow. This picture is not to scale since the solar wind velocity is so large that, by the time the foot point in the ionosphere traverses the polar cap, the other end of the field line is very far away from the earth. Hence, a long magnetic tail forms on the nightside. The time history of convecting plasma or flux tubes in the magnetosphere and ionosphere during the circulation of plasma in one cell of the two-cell convection pattern is shown in Fig. 8.14b. As just noted, on the dayside at point 1, a closed magnetic field line of the earth "breaks" and connects with the interplanetary magnetic field to produce an open field line. Not shown is the magnetic field line from the south polar region that connects to the "other half" of the interplanetary field line. This point may alternatively be viewed as the location at

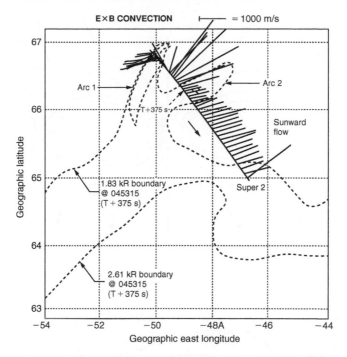

Figure 8.13 E × B drift velocities along the footprint of Super 2 deduced from the double-probe electric field data. The heavy and dashed lines denote the boundaries of 2.61 and 1.83 kR luminosity at 5577, respectively, at 0453:15 UT ($T + 375$ s). An arrow marks the location of the Super 2 footprint at that time. The rocket flew toward the southeast as illustrated by the arrow parallel to the trajectory. [After Berg et al. (1994). Reproduced with permission of the American Geophysical Union.]

which magnetospheric plasma moves from a region located on closed magnetic field lines to a region on open field lines. After merging takes place, the magnetospheric plasma moves antisunward along the path labeled 1, 2, and 3. Small arrows show the corresponding convective path of the plasma in the ionosphere as it moves from 1' to 2' to 3'. At point 3, open field lines from the Northern and Southern Hemispheres reconnect in the equatorial plane and subsequently convect sunward in the equatorial plasma sheet as closed field lines. Again, one may alternatively view point 3 as a place where magnetospheric plasma makes a transition from being an open field line to being a closed field line. From either viewpoint, the subsequent convective paths of the plasma in the ionosphere and magnetosphere, respectively, are 3', 4', 5', 6', 1' and 3, 4, 5, 6, 1. Note in this figure that the ionospheric flow is shown on a relatively enlarged sphere so that its features can be seen.

These basic elements of a two-cell convection pattern can also be attributed to the viscous interaction process mentioned earlier (Axford and Hines, 1961). In this process the momentum of the magnetosheath plasma is transmitted across

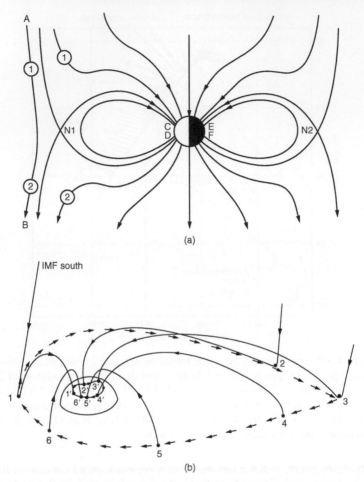

Figure 8.14 (a) Schematic diagram showing dayside merging (NI) and nightside reconnection (N2) in the noon-midnight meridian. (Courtesy of D. P. Stern.) (b) Time history of convecting flux tubes that result from connection of a southward IMF to the earth's magnetic field. The convection along path 1, 2, 3 is on open field lines, while the path from 3, 4, 5, 6 to 1 is on closed field lines that connect to the southern hemisphere.

the magnetopause by waves and diffusion, creating an effective viscosity. Plasma near the equatorial plane, which is just inside the magnetopause, the so-called equatorial boundary layer, is set in motion antisunward. A return flow toward the sun occurs on lower latitude field lines and is driven by back pressure in the magnetosphere near midnight. The ionospheric convection signature of this process is qualitatively identical to that created by direct connection with the solar wind magnetic field or merging, except that even the antisunward convective paths in the magnetosphere are on closed field lines. Figure 8.15a shows

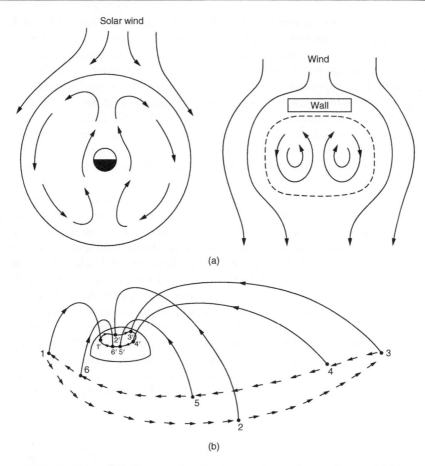

(a)

(b)

Figure 8.15 (a) Schematic diagram showing how a viscous interaction could drive magnetospheric convection, along with a fluid analogy. (Courtesy of D. P. Stern and S. Lantz.) (b) Time history of convecting flux tubes resulting from a viscous interaction. Here all the flux tubes are closed and also connect into the southern hemisphere, which is not shown.

the convection cycle in the magnetosphere for the viscous interaction mechanism along with an analogous process in fluid dynamics. Convection paths in the magnetosphere and ionosphere are shown in Fig. 8.15b for the viscous interaction driver. In this case there are no merging and reconnection regions but simply locations at which convection on closed field lines changes direction from sunward to antisunward at point 1 and from antisunward to sunward at point 3. The sunward convection from 4 to 5 to 6 is identical to that shown in Fig. 8.14, but the antisunward flow in the magnetosphere now exists on closed field lines in the equatorial plane, as shown by the arrows connecting 1 to 2 to 3.

It can be seen from a comparison of the primed paths in Figs. 8.14 and 8.15 that observations of just the ionospheric convection signature cannot distinguish between these two convection mechanisms. However, simultaneous observations of the convection signature and the energetic particle environment indicate that the magnetic field topology is often open. Likewise, theoretical considerations of the physics involved in a viscous interaction indicate that the potential associated with this mechanism is usually less than 10 kV. Observed cross-polar cap potentials in excess of 60 kV and the dependence of the convection geometry on the y component of the solar wind magnetic field strongly suggest that the merging process dominates over viscous interaction when the IMF has a southward component.

Although a rigorous discussion of the way in which the solar wind and earth's magnetic fields interconnect is beyond the scope or requirements of this book, it is possible to understand the basic properties of the convection geometry from rather simple considerations. We start by assuming that field lines interconnect when they have antiparallel components in a plane perpendicular to the earth-sun line (the y-z plane). Figure 8.16 shows views from above the north pole of one of the earth's magnetic field lines that is connected to the IMF when the IMF y component is nonzero. In the situation for negative B_y, once merging has taken place the newly open field lines have extremely large curvature near the merging point. The resulting tension in the Northern Hemisphere open field lines will produce a duskward component in the plasma flow in the dayside magnetosphere. A corresponding duskward plasma flow will exist in the dayside polar cap of the northern hemispheric ionosphere. In the Southern Hemisphere

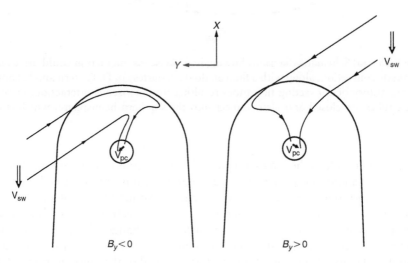

Figure 8.16 Schematic illustration of the magnetic field line orientation producing different flow directions in the polar cap when B_y is positive versus the situation where B_y is negative.

the plasma flow at the magnetopause and in the ionosphere will be directed toward dawn when B_y is negative. Similar arguments can be applied to the situation when B_y is positive. In this case a dawnward flow component will be seen in the northern ionospheric polar cap flow. This rather simple idea can account for the B_y dependence of the ionospheric flow direction seen near local noon in Fig. 8.9 and for the dependence of the flow speed at the dawn and dusk edges of the polar cap on B_y (Heppner, 1972). This dependence is such that in the Northern Hemisphere a larger flow speed is seen near the dawn side of the polar cap when B_y is positive and a larger flow speed is seen at the dusk side of the polar cap when B_y is negative. Also note that this simple model predicts that the B_y dependences seen in the Northern Hemisphere appear in the opposite sense for the Southern Hemispheres, a fact borne out by observation.

8.3.2 Models for Northward IMF

The pattern of the high-latitude circulation cells during northward IMF can again be understood by the rather simple concepts of merging and frozen-in flux used in the case of southward IMF. As we shall see, a number of different magnetic field topologies might exist during times of northward IMF. Experimental and theoretical investigations in this area are still being undertaken and we will describe here only some of the basic concepts involved.

One way in which the earth's magnetospheric field may connect to a northward IMF is shown in Fig. 8.17. Here magnetic field lines that extend into the tail of the magnetosphere have the right orientation for merging to occur with a northward IMF at position 1. Subsequent antisunward motion of open field lines to points 2 and 3 produces antisunward motion in the ionosphere from 1' to 3'. At this

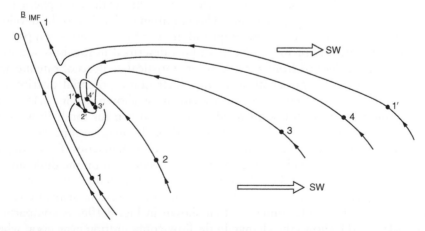

Figure 8.17 Time history of convecting flux tubes (connecting to a northward IMF) that produce a dominant high-latitude convection cell. This must be combined with the pattern for viscous interaction shown in Fig. 8.15 to produce the observed four-cell pattern.

point the open or closed state of the flux tube becomes an important issue. In Fig. 8.17 the field line remains open and convects down the tail and toward the earth-sun line until it assumes the position of the original field tube (1'-1). The associated convection is labeled 1, 2, 3, 4, 1 in which the antisunward flow toward the earth-sun line at the magnetopause corresponds to sunward flow 3', 4', 1' in the ionosphere. Alternative or additional convection paths in which field tubes reconnect at point 3 and subsequently flow sunward in a plasma sheet that thickens along the earth-sun line have been proposed to explain the possible existence of closed magnetic field lines at the highest latitudes. It should be emphasized that the precise location of the transition from open to closed field lines in the ionosphere can be quite variable within the picture we have drawn.

The convective motion described above exists in addition to the viscous interaction mechanism described earlier, and we must invoke both of them to explain the four-cell convection pattern that is most frequently observed. We note that the tendency of the flow of the inner two cells to be preferentially directed toward dawn or dusk, depending on whether B_y is positive or negative, can again be explained by the effective "tension" in connection to the IMF (see Fig. 8.16).

8.4 Empirical and Analytical Representations of High-Latitude Convection

As we shall see in Chapter 9, knowledge of the high-latitude ionospheric convection pattern is required to effectively calculate the distribution and composition of plasma in the F region and to determine the F-region neutral wind field and Joule heating rate. For such studies it is necessary to know the ionospheric flow velocity at all points at any given time. This requirement has led to the development of several empirical and semi-empirical models for the convection pattern designed to mimic one or more of the features described in previous sections. These models are derived from a synthesis or statistical analysis of satellite and radar data. A global representation of the electrostatic potential distribution derived from satellite electric field measurements is shown in Fig. 8.18. This empirical model is valid for southward B_z and is shown for the IMF with B_y negative (a) and positive (b). It depicts several of the major properties of the convective flow that we have previously noted in individual cases. For example, the ion flow or electric field is larger on the dawn side than on the dusk side in the Northern Hemisphere when B_y is positive (i.e., the potential lines are closer together). Figure 8.18b also shows the existence of a crescent-shaped convection cell on the dusk side similar to that shown in Fig. 8.10d. A comparison of Fig. 8.18a and b shows the change in the flow configuration near noon when B_y changes sign, as depicted in Fig. 8.8d. Data such as this can be useful in

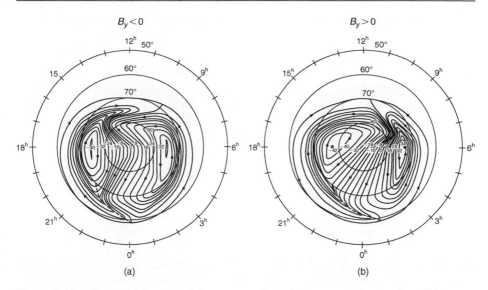

Figure 8.18 Electrostatic potential contours derived from a synthesis of satellite measurements of the ionospheric electrostatic potential when B_y is negative (a) and when it is positive (b). [After Heppner and Maynard (1987). Reproduced with permission of the American Geophysical Union.]

computer modeling studies if numerical values of the potential are obtained over a latitude-local time grid.

More detailed distributions of the electrostatic potential in the sunward flow regions have been obtained from ground-based radars that probe this region of the ionosphere. Many years of data from these radars can be sorted by latitude and local time as well as by magnetic activity and solar wind conditions. Tabulated values of the coefficients of functional fits to these data are then available. Figure 8.19 shows examples of these average convection patterns derived from radars at Chatanika, Alaska, and Millstone Hill. Although data such as these do not cover the entire high-latitude region, they do provide a means of assessing the effects of magnetic activity on the auroral zone flow, a task that is not easily accomplished using satellite data.

Analytical models use a collection of mathematical expressions to specify the electrostatic potential as a function of position in the high-latitude F region. The comparative ease with which they can be incorporated into computer code makes them the most often used in F-region modeling studies, and here we provide illustrative examples. With this technique the polar cap is usually designated as a circle and outside this circle the potential θ decays via an inverse power law sine wave. That is,

$$\phi = \phi_0 \left(\sin \theta / \sin \theta_0\right)^n \tag{8.18a}$$

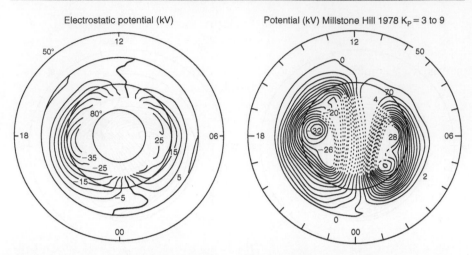

Figure 8.19   Radar data from Chatanika and Millstone Hill used to derive the convection pattern in the auroral zone (sunward-flow regions). [After Foster (1983) and Oliver et al. (1983). Reproduced with permission of the American Geophysical Union.]

Here θ is magnetic co-latitude, ϕ_0 is the potential at the polar cap boundary where $\theta = \theta_0$, and the index n typically takes values between -2 and -4. Inside the polar cap boundary, Volland (1975) showed that if

$$\phi = \phi_0 \, (\sin \theta / \sin \theta_0) \sin(H) \tag{8.18b}$$

where H is the local hour angle (MLT), then the lines of electrostatic equipotential depict convective flow that is antisunward inside the polar cap. This model can be modified to allow ϕ_0 to take different values on the dawn- and dusk-side polar cap boundaries and to allow the origin of the potential coordinate system to be displaced from the magnetic pole. It has been used extensively in F-region modeling studies where the offset between the geomagnetic and geographic poles has also been taken into account.

To reproduce such observations as the crescent-shaped cell in the dusk sector and the predominance of eastward or westward flow at local noon, some modifications to the mathematical model can be made. These include the removal of any local time dependence of the potential in certain local time sectors and the removal of the latitudinal discontinuity in the potential gradient at the polar cap boundary. A variety of convection geometries are then available with realistic field-aligned current densities similar to those described in Section 8.5. The challenge of producing mathematical models that adequately describe the effects of changing B_y when the IMF is southward and the four-cell convection pattern when the IMF is northward still remains.

In our description in previous sections we have shown how the IMF exercises a large degree of control on the geometry or shape of the convection cells. This geometry is related to the distribution of electrostatic potential around the polar cap boundary and within the polar cap itself. These are the two fundamental properties addressed by empirical models. To date, our description has been only qualitative or empirical. There is, however, convincing evidence that merging or direct connection with the solar wind dynamo via the IMF is an important contributor to the flow in the polar cap. In this case we may reasonably ask what effect interplanetary conditions have on the magnitude of the drift velocities or, alternatively, on the size of the polar cap and the potential drop across it. In this area an analytical model can be extremely useful.

A model developed by Siscoe (1982) describes the Region 1 and Region 2 field-aligned currents introduced in Section 8.1 as two concentric rings. A potential ϕ is assumed to be distributed sinusoidally in local time around the region 1 ring. Then consideration of the conservation of magnetic flux in the ionosphere and the energy dissipated in the Region 2 current loop can be used to show that, in an equilibrium situation, the radius of the Region 1 circle—that is, the polar cap radius r—is related to the potential by the expression

$$r = r_0 \phi^{3/16} \tag{8.19}$$

Here r_0 depends on the ionospheric conductivity and the width of the auroral zone, but these quantities may be considered as constants to first order. The limited data that exist do not disagree with this relationship, but the exponent on the potential is extremely small and thus, for potential differences exceeding about 20 kV, the polar cap radius is almost independent of the magnitude of ϕ. Magnetic storms and substorms (discussed briefly in the next section) are known to cause expansions of the polar cap and the entire convection pattern, but even a qualitative description of this behavior has not yet been accomplished.

The polar cap potential drop is, however, found to be a strong function of the interplanetary magnetic field magnitude and orientation. From rather simple considerations, one might expect that the potential difference depends on the area over which the IMF and geomagnetic field interconnect and the efficiency with which the electric field in the interplanetary medium is transferred across that area. Hill (1975) approached this problem by determining the dissipative component of the solar wind electric field at the magnetopause (i.e., the component parallel to the magnetopause current) that separates an internal earth's magnetic field, B_2, from an external field, B_1 (the IMF). He called this field the merging region field and denoted it by E_j. He then showed that this field and the relative orientation of the magnetic fields on either side of the magnetopause would determine the effective potential transmitted across the boundary to the ionosphere.

The electric field in the merging region can be expressed as

$$E_j = E_0 \frac{\alpha^2 (\alpha - \cos \theta)^2}{1 + \alpha^2 - 2\alpha \cos \theta} \tag{8.20}$$

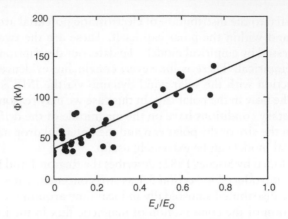

Figure 8.20 Measured polar cap potential difference, ϕ, plotted as a function of the quantity E_j/E_0 defined in (8.20). The potential increases as the IMF becomes more southward but a potential still exists when the solar wind electric field parallel to the magnetopause current is zero. [After Reiff et al. (1981). Reproduced with permission of the American Geophysical Union.]

where $\alpha = B_1/B_2$, θ is the angle between $\mathbf{B}_1$ and $\mathbf{B}_2$, and E_0 depends on the solar wind density and the magnetic field magnitudes. The total potential drop Φ across the polar cap due to the merging field is simply $\Phi = E_j d$, where d represents the length of the magnetopause merging region. One would therefore expect that, when the IMF has no component perpendicular to the sheet current at the magnetopause (and thus $E_j = 0$), Φ should also be zero. Observations of the polar cap potential difference are plotted in Fig. 8.20 against values of E_j calculated for the measured solar wind and the earth's magnetic fields at different times. The E_j is normalized to E_0, which depends on the plasma density in the solar wind and the earth's magnetic field. As the solar wind magnetic field changes, the observed potential does not go to zero when $E_j = 0$. This additional potential difference may be related to electric fields communicated by processes other than merging or to a time response of the magnetosphere-ionosphere system that does not allow it to change instantaneously when the IMF changes.

Figure 8.21 shows the results of studying the time response of the ionosphere-magnetosphere system to a change in the north-south component of the IMF. It shows the polar cap potential difference measured as a function of time after the IMF turns northward. The data suggest that relatively large potentials, associated with a two-cell convection pattern, tend to persist for 1 or 2 hours following a transition from southward to northward IMF. After this time quite small polar cap potential differences are observed. Bearing in mind the uncertainties in defining the polar cap, these small values are consistent with a multicell convection pattern for a northward IMF. It is interesting to note that this time delay is consistent with the time required to reverse a two-cell convection pattern in the neutral gas that is established during a southward IMF (see Chapter 9). Thus, a dynamo

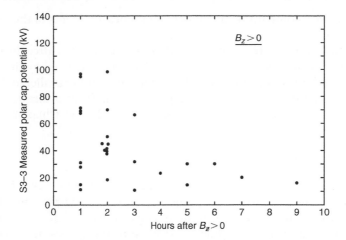

Figure 8.21 The polar cap potential decreases slowly after the IMF turns northward. It finally reaches a value between 20 and 15 kV that is perhaps attributable to the viscous interaction process. [After Wygant et al. (1983). Reproduced with permission of the American Geophysical Union.]

electric field might be produced in the ionosphere at high latitudes by a flywheel effect during transitions from previously stable IMF orientations. In such a case, $\mathbf{J} \cdot \mathbf{E} < 0$ in the ionosphere and the Poynting flux is upward. Rocket data indicate that shrinkage of the polar cap for B_z north occurs in discrete steps accompanied by new auroral arc generation poleward of the instantaneous auroral oval (Berg et al., 1994). Alternatively, this decay may be the time constant required to dissipate stored magnetic energy by Joule heating at ionospheric altitudes (Kelley et al., 2009).

The observations of the flow geometry when the IMF has a northward component are not sufficiently advanced to make quantitative analysis possible. As mentioned earlier, it is likely that processes other than merging (e.g., viscous interaction) may be important in producing the ionospheric motion in this case, and advances in this area are occurring at the time of this writing. An excellent climatological model for magnetospheric convection has been developed by Weimer (1995, 2001).

8.5 Observations of Field-Aligned Currents

Field-aligned currents are essential to the linkage between the solar wind-magnetosphere system and the ionosphere. The ionosphere is not a passive element in the electric field mapping process between these regions and, in fact, the electric field can be modified at the driver and throughout the system by the ionosphere. This is particularly true in the auroral zone, where the ionosphere and magnetosphere are linked and the hot plasma sheet particles can determine the ionospheric conductivity through particle precipitation. Study of these currents yields information comparable in importance to measurements of the convective motions.

Field-aligned currents, frequently called Birkeland currents after the man who first postulated their existence (see Section 8.1), can be detected by the magnetic perturbation they produce. A current flowing in a direction parallel to the earth's magnetic field will produce a magnetic perturbation δB in a direction perpendicular to the field. From the diagnostic equation (2.23c) in Chapter 2, the two quantities are related by

$$\mu_0 J_{||} = (\nabla \times \delta B)_{||} \tag{8.21}$$

The magnetic disturbances can be observed by satellite and rocket-borne magnetometers that measure three mutually perpendicular components of the disturbance vector. A measurement along a single trajectory does not allow the determination of a curl, so in practice some approximations are necessary to derive $J_{||}$.

Field-aligned currents frequently, but not always, occur as sheets whose dimensions in one horizontal direction greatly exceed those in the other (see Fig. 8.4). The magnetic disturbance is then predominantly parallel to the sheet, and Fig. 8.22 shows schematically the disturbance that would be measured on passing perpendicularly through the sheet. The current density is given by (8.21), which simplifies to $\mu_0 J_{||} = (\partial/\partial x)(\delta B_y)$, and for the dimensions shown in Fig. 8.22 the disturbance would correspond to an outward current sheet of density $0.8 \ \mu A/m^2$. Most observations are not, of course, made with the convenience of such perpendicular crossings of the current sheet, so care must be taken in the arbitrary application of (8.21). In general, the full perturbation vector is measured, so in many cases only a rotation of a coordinate system is required to determine $J_{||}$. A statistical analysis of satellite magnetometer data of this kind has enabled a systematic picture of field-aligned current directions, magnitudes, and locations to be derived. As in the case of electric field or plasma flow observations, there exist some relatively stable characteristics in the patterns of field-aligned currents that

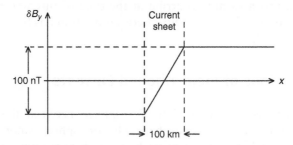

Figure 8.22   A uniform $0.8 \ \mu A/m^2$ current sheet in a plane perpendicular to the paper and occupying the 100-km extent shown here produces a perturbation change of $100 \ nT$ in the magnetic field.

persist at all times. In addition, the interplanetary magnetic field magnitude and orientation introduce changes in the location and configuration of the current systems. This should not be surprising, considering the intimate relationships between currents and electric fields that were derived in Section 8.1. Since we have established the basic magnetic field topologies associated with the electric field configuration and given a qualitative description of the drivers that produce a dependence on the IMF orientation, we will confine ourselves here to a description of the field-aligned current morphology. In the final section we will describe the behavior of the horizontal currents.

8.5.1   Current Patterns for a Southward IMF

When the IMF has a southward component, Fig. 8.23 shows the most stable features of the field-aligned current system seen during times (a) when the internal state of the magnetosphere is quiet and (b) when it is more disturbed. The following features should be noted:

1. There exist essentially two concentric field-aligned current rings approximately the same as those predicted from the simple considerations of Section 8.1.
2. The inner ring, which may be considered to contain the driving currents for ionospheric convection, has current directed into the ionosphere in the morning hemisphere and away from the ionosphere in the evening hemisphere. These are the Region 1 currents, identified in Section 8.1.

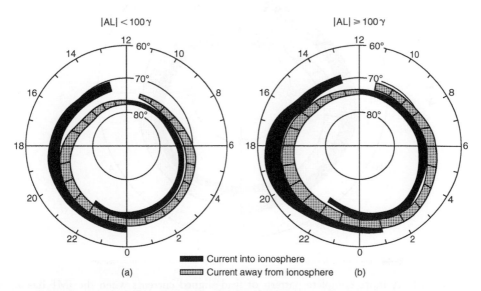

Figure 8.23   Two overlapping rings of field-aligned currents always exist at high latitudes. They occur at higher latitudes and have smaller latitudinal extent during quiet times. The inner ring is termed Region 1 current and the outer ring Region 2. [After Iijima and Potemra (1978). Reproduced with permission of the American Geophysical Union.]

3. The outer ring, which results from the feedback between the ionosphere and the magnetospheric plasma sheet, has the opposite current direction in the morning and evening hemispheres from that of Region 1. These are the Region 2 currents identified in Section 8.1.

4. In the magnetic local time sector from 2200 to 2400 hours, three regions of field-aligned current are formed by the overlap of the two rings seen at other local times.

5. During magnetically active periods there is a slight expansion of both current rings to lower latitudes. The region of overlap then increases to occupy the MLT region from 2200 to 0100 hours. The main characteristics of the current system remain unchanged.

These properties of the field-aligned current system are common to both hemispheres. However, there is an additional current system that shows a strong dependence on the IMF and has a dominant polarity that for strong B_y is opposite in the two hemispheres. This additional current system is called the cusp current and, as shown in Fig. 8.24, it lies just poleward of the Region 1 currents near local noon. When the IMF y component is small, the cusp current system consists

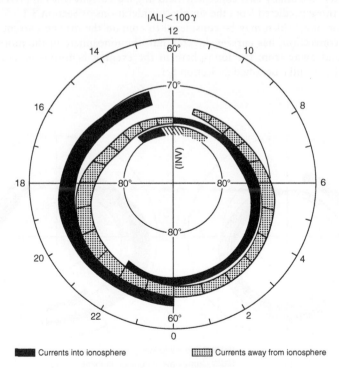

Figure 8.24 A more complete pattern of field-aligned currents when the IMF has a southward component and $B_y = 0.1$. An additional field-aligned current system resides near local noon. These "cusp" currents have a dominant polarity determined by the y component of the IMF. [After Iijima and Potemra (1976). Reproduced with permission of the American Geophysical Union.]

of two localized field-aligned current regions symmetrically located about local noon in both hemispheres. The current directions are opposite to the adjacent Region 1 currents, being downward into the ionosphere in the afternoon and upward away from the ionosphere in the morning. Change in the sign of B_y significantly affects this distribution. Observations made in both the Northern and Southern Hemispheres show that when $B_y > 0$, the cusp region current is extremely asymmetric and is predominantly upward in the Northern Hemisphere and downward in the Southern Hemisphere. When $B_y < 0$, the current is predominantly downward in the north and upward in the south.

8.5.2 Current Patterns for a Northward IMF

When the IMF has a northward component, the distribution of field-aligned currents changes quite dramatically (as does the electric field distribution). The two-ring pattern of field-aligned currents is retained, but they are seen to move to higher latitudes and are significantly reduced in intensity. Perhaps more significant is a reorganization of the cusp currents in such a way that they expand over most of the high-latitude area enclosed by the Region 1 currents and become as intense as the Region 1 and Region 2 currents themselves. This new cusp current system has been called the "NBZ" current system to denote the northward-directed B_z component that exists during its occurrence. The NBZ current system again shows a strong dependence on the sign of B_y, and Fig. 8.25 shows schematically its location and distribution in the Southern Hemisphere. Except for their dramatic redistribution in area, the NBZ currents show essentially the same behavior as the cusp currents, being predominantly downward in the Southern Hemisphere when B_y is positive and upward when B_y is negative. Conversely, we expect the NBZ current to be upward in the Northern Hemisphere when B_y is positive and downward when B_y is negative. Because the currents are out of

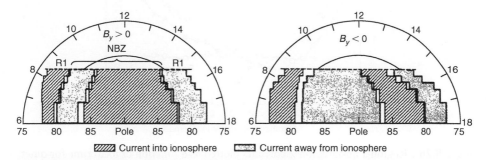

Figure 8.25 Cusp currents in the southern hemisphere, shown for the case when the IMF has a northward component. The cusp currents expand to fill the polar cusp and have been called NBZ currents. Their dominant polarity is determined by the y component of the IMF. [After Iijima et al. (1984). Reproduced with permission of the American Geophysical Union.]

balance a net current flows into one pole and out the other. This current must close in the solar wind in the direction perpendicular to the ecliptic (Kelley and Makela, 2002).

8.5.3 Dependence on Magnetic Activity, IMF, and Season

When the IMF has a southward component, some qualitative descriptions of the ionospheric field-aligned currents are available in terms of their dependence on the degree of magnetic activity, on the sign of B_y, and on season. In all of these descriptions, the field-aligned current density is regarded as being uniform in each of the Region 1 and Region 2 rings, so a magnitude and a local time distribution can be ascribed to them. A comparison of the upper and lower panels of Fig. 8.26 shows that the local time distribution is only slightly affected

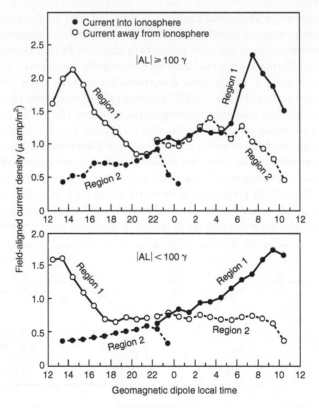

Figure 8.26   Region 1 and Region 2 currents plotted as a function of local time for quiet and disturbed geomagnetic conditions. The Region 1 currents have maxima near 14.00 and 8.00 h where they exceed the Region 2 currents. Elsewhere the Region 1 and Region 2 currents are about the same magnitude and show little variation with local time. [After Iijima and Potemra (1978). Reproduced with permission of the American Geophysical Union.]

by magnetic activity. The maximum Region 1 current intensities occur between about 0800 and 1000 h on the morning side and between about 1400 and 1600 h on the afternoon side. At these times the Region 1 currents are about a factor of 2 to 3 greater than the corresponding Region 2 currents. During equinox and winter, the Region 1 and Region 2 currents show relatively little dependence on local time except for the two Region 1 current maxima. In the summer, however, it is found that the Region 1 currents can exceed the Region 2 currents at all local times, although the region of maximum difference remains on the dayside. The Region 2 currents show a much higher degree of variability with magnetic activity than do the Region 1 currents. This can lead to situations on the nightside in which the Region 2 currents exceed the Region 1 currents during disturbed periods.

The magnitude of the Region 1 currents shows a strong dependence on the interplanetary magnetic field, as might be expected from arguments similar to those applied for the polar cap potential difference. Figure 8.27 shows the dependence of the morningside and afternoonside Region 1 currents on the magnitude of the solar wind electric field. These data show that the field-aligned currents increase as the electric field associated with the IMF increases from zero. Similar conclusions concerning the polar cap potential difference were drawn from Fig. 8.20, so it is not surprising that the driving currents should show the same behavior. As in the case of the polar cap potential difference, the Region 1 current density does not reduce to zero when the solar wind electric field reduces to zero. This observation supports the existence of a mechanism in addition to merging (perhaps viscous interaction) that also provides driving currents for ionospheric convection.

8.6 Horizontal Currents at High Latitudes

Equations (8.7) and (8.12) represent the fundamental relationships between horizontal and field-aligned currents and the electric field and conductivity in the ionosphere. The discussion of the dependence of the ionospheric conductivity on altitude given in Chapter 2 shows that the region over which substantial currents flow perpendicular to the magnetic field lines is restricted to the range from about 90 to about 130 km in the sunlit ionosphere and may extend up to 300 km when no appreciable local ionization source is present. This current flow is almost horizontal at high latitudes and produces a magnetic signature that can be observed on the ground. It has been shown, however, that ground magnetometers respond primarily to Hall currents and thus cannot be directly converted to electric fields (Piddington, 1962).

Ground-based magnetometers function somewhat differently from those on satellites, but their output, three mutually perpendicular components of the magnetic perturbation from a normal steady baseline, is the same. These magnetic field perturbations are usually resolved along the geographic north, east, and

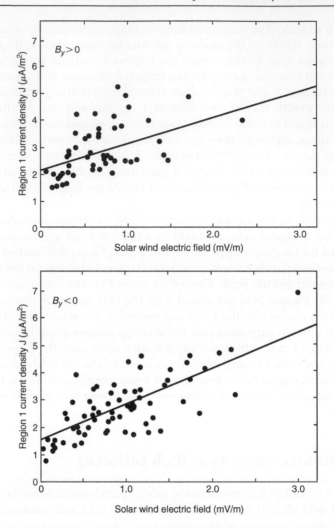

Figure 8.27   Region 1 currents plotted against the solar wind electric field [$V_{sw}(B_y^2 + B_z^2)^{1/2} \sin(\theta/2)$] for $B_y > 0$ and $B_y < 0$. The Region 1 currents show a dependence on the solar wind electric field similar to that of the polar cap potential drop. Notice that the Region 1 currents do not reduce to zero when the solar wind electric field does, possibly indicating a viscous interaction. [After Iijima and Potemra (1982). Reproduced with permission of the American Geophysical Union.]

vertically down directions and are denoted by H, D, and Z components, respectively. Sometimes a geomagnetic coordinate system is used, in which case the symbols X, Y, and Z denote the magnetic perturbations. It is impossible to derive the true horizontal ionospheric current distribution uniquely from ground magnetic perturbations, since they are a superposition of contributions from the horizontal ionospheric currents, field-aligned currents, distant currents in the magnetosphere, and currents induced in the earth's surface. For these reasons the ground magnetic perturbations are usually expressed in terms of "equivalent" ionospheric currents. The study of magnetic perturbations and their interpretations as current systems in the earth and in space is extremely complex and we will not discuss this topic in detail. However, magnetic perturbations are used to describe phenomena such as magnetic storms and substorms and to derive indices such as DST, K_p, and AE that describe the magnetic activity in the earth's environment. It is therefore necessary to discuss the meaning of these indices and the nature of the magnetic measurements. This discussion is located in Appendix B.

Essentially two techniques are utilized to derive the equivalent overhead horizontal current flowing in a thin shell near 100 km altitude. One method calculates the magnetic signature on the ground from a current flowing east-west in a small horizontal cell which in turn flows into field-aligned currents at the edges of the cell (Kisabeth and Rostoker, 1971). The field-aligned currents are assumed to flow along dipole magnetic field lines and subsequently close in the equatorial plane. By a "best-fit" process, this technique yields the total three-dimensional current system (horizontal and field-aligned) that produces the measured ground magnetic perturbations.

The other, more widely used method expresses the overhead current $\mathbf{J}$ in terms of a divergence-free component sometimes called the equivalent current $\mathbf{J}_e$ and a curl-free component sometimes called the potential current $\mathbf{J}_p$. The potential current can be viewed as the closing current for the field-aligned currents. For vertical magnetic field lines it can be shown that the field-aligned and potential current circuit produces no magnetic perturbation at the ground. Further, if the ionospheric conductivity is uniform it can be shown that the potential current is the Pedersen current. In this case, since other magnetic effects are small, the equivalent current will largely represent the horizontal ionospheric Hall current. Since it is divergence-free, the equivalent current can be expressed in terms of a current function Γ such that

$$\mathbf{J}_e = \mathbf{r} \times \nabla \Gamma \tag{8.22}$$

where $\mathbf{r}$ is the unit radial vector in a coordinate system with the origin at the center of the earth.

No current flows near the ground, so the magnetic perturbation there, $\delta \mathbf{B}$, can be expressed in terms of a magnetic potential φ such that

$$\delta \mathbf{B} = -\nabla \varphi \tag{8.23}$$

There are straightforward mathematical relationships that uniquely relate Γ and φ so that the horizontal equivalent current can be derived from the potential function φ. The function φ is determined by taking the divergence of (8.23) and solving the two-dimensional Poisson equation

$$\nabla^2 \varphi = -\nabla \cdot \delta \mathbf{B} \tag{8.24}$$

The right-hand side of this equation is given by a global distribution of ground observations with suitable interpolation at each grid point.

These two quite different techniques yield very similar results for the horizontal equivalent ionospheric current. In principle, only the equivalent current can be derived from ground magnetic perturbations, and some model of ionospheric conductivity is required to proceed further. However, many features of the equivalent current system are useful in establishing self-consistency among the electrodynamic parameters and we will now describe these characteristics.

Our purpose here is to describe the dependence of the equivalent current distribution on magnetic activity and the interplanetary magnetic field orientation. In this way a self-consistent picture of the electrodynamic parameters $\mathbf{J}_\perp, \mathbf{J}_{||}$, and $\mathbf{E}_\perp$ can be established. The effects of magnetic activity on the horizontal equivalent current manifest themselves principally in small-scale features that we will discuss later. The large-scale features of the horizontal current and its dependence on the IMF are shown in Fig. 8.28. A two-cell pattern of current flow exists for all orientations of the IMF. In each of the pictures, the current flows in closed loops around a focus labeled "$+$," located on the nightside near midnight, and around a focus labeled "$-$," located on the dayside near noon. The sense of circulation of the current is clockwise around the $+$ focus and anticlockwise around the $-$ focus. When the IMF has a southward component ($B_z < 0$) the current is much larger and the two-cell pattern is more well defined than when the IMF is northward. Examining the southward IMF case in more detail, we note that at the low-latitude extremes of the current system there exist bands of current flowing, respectively, eastward on the dusk side and westward on the dawn side. These current bands are collocated with the diffuse auroral zones and are called the eastward and westward electrojets. Note that the direction of the electrojet currents is opposite to that of the F-region plasma flow.

The fact that a two-cell current pattern is retained for all orientations of the IMF indicates that some element of the current pattern may be independent of the IMF. Statistical analysis of ground-based magnetograms shows this to be the case, and this portion of the current system is called the S_q^p ("solar quiet polar") current system. Detailed analysis shows that there may also exist semidiurnal contributions to this current that may be attributable to lunar variations, as is the case in the equatorial region.

For a southward IMF during times of relatively low magnetic activity (i.e., $K_p < 3$), the total current system is made up of the S_q^p system plus another system called the DP2 system; the DP2 current system is made up of the DPY and DPZ

Equivalent current system

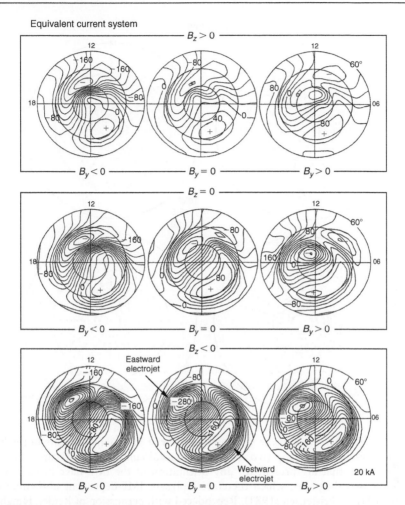

Figure 8.28 Contours showing the horizontal equivalent currents for different orientations of B_z and B_y. The horizontal equivalent current shows many of the features of the high-latitude convection pattern. In fact, the features differ only where substantial field-aligned currents and conductivity gradients exist. [After Friis-Christensen et al. (1985). Reproduced with permission of the American Geophysical Union.]

systems, which are strongly dependent on the y and z components of the IMF, respectively. The features of the IMF-dependent DPY current system dominate the horizontal current distribution and can be clearly seen if we subtract the IMF-independent contribution from equivalent current systems measured when B_y has a substantial positive component and when it has a negative component. Figure 8.29 shows the result of such an exercise, in which it can be seen that the IMF-dependent system is approximately zonal and flows across the local noon

B_z South

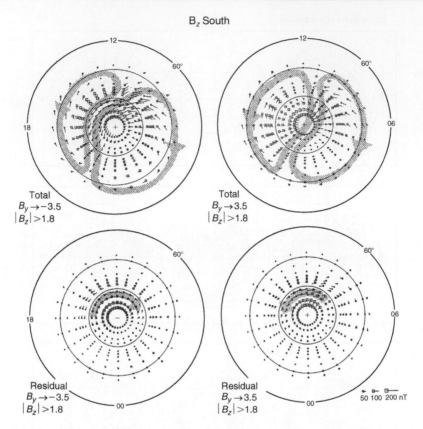

Figure 8.29 The arrows indicate the flow direction of horizontal currents in the high-latitude ionosphere. In the top two diagrams the total currents are indicated. Features of these currents that are present for any orientation of the IMF have been removed in the lower two diagrams, which therefore represent IMF-dependent horizontal currents. [After Friis-Christensen (1981). Reproduced with permission of Reidel, Hingham, Massachusetts.]

meridian. This "DPY current" flows from dawn to dusk when B_y is positive and from dusk to dawn when B_y is negative. The DPY current produces a rotation and an asymmetry in the current vortices so that the high-latitude sunward-flowing current is directed toward the prenoon sector for both polarities of the IMF. When the z component of the IMF is very small the two-cell horizontal current system is preserved, but the intensity of the current is much less than that existing when the IMF is strongly southward. As B_y changes from negative to positive, the line of symmetry for the current vortices rotates clockwise as it does for a southward IMF. However, when B_z is very small the degree of rotation is much larger, leading to currents that flow almost parallel to the dawn-dusk meridian when B_y is positive.

When the IMF has a strong northward component, the horizontal currents associated with closure of the field-aligned NBZ current system can be observed from satellites but, from the ground, it is only observable in the summer polar cap. What is detectable on the ground is a low-intensity version of the two-cell current system that exists when B_z is negative or very small. In addition, there is evidence for an antisunward current in the center of the polar cap.

References

Axford, W. I., and Hines, C. O. (1961). A unifying theory of high-latitude geophysical phenomena and geomagnetic storms. *Can. J. Phys.* **39**(10), 1433.

Berg, G. A., Kelley, M. C., Mendillo, M., Doe, R., Vickrey, J., Kletzing, C., Primdahl, F., and Baker, K. D. (1994). Formation and eruption of sun-aligned arcs at the polar cap-auroral oval boundary. *J. Geophys. Res.* **99**(A9), 17,577.

Burke, W. J., Kelley, M. C., Sagalyn, R. C., Smiddy, M., and Lai, S. T. (1979). Polar cap electric field structures with a northward interplanetary magnetic field. *Geophys. Res. Lett.* **6**, 21.

Crooker, N. U., and Siscoe, G. L. (1981). Birkeland currents as the cause of the low-latitude asymmetric disturbance field. *J. Geophys. Res.* **86**, 11,201.

Dungey, J. W. (1961). Interplanetary magnetic field and the auroral zones. *Phys. Rev. Lett.* **6**, 47.

Foster, J. C. (1983). An empirical electric field model derived from Chatanika radar data. *J. Geophys. Res.* **88**, 981.

Friis-Christensen, E. A. (1981). High latitude ionospheric currents. In *Exploration of the Polar Upper Atmosphere* (C. S. Deehr and J. A. Holtet, eds.). Reidel, Hingham, Massachusetts.

Friis-Christensen, E. A., Kamide, Y., Richmond, A. D., and Matsushita, S. (1985). Interplanetary magnetic field control of high latitude electric fields and currents determined from Greenland magnetometer data. *J. Geophys. Res.* **90**, 1325.

Hairston, M. R., Hill, T. W., and Heelis, R. A. (2003). Observed saturation of the ionospheric polar cap potential during the 31 March 2001 storm. *Geophys. Res. Lett.* **30**(6), 1325, doi:10.1029/2002GL015894.

Heelis, R. A. (1984). The effects of interplanetary magnetic field orientation on dayside high latitude convection. *J. Geophys. Res.* **89**, 2873.

Heelis, R. A., and Hanson, W. B. (1980). High latitude ion convection in the nighttime F-region. *J. Geophys. Res.* **85**, 1995.

Heelis, R. A., Reiff, P. H., Winningham, J. D., and Hanson, W. B. (1986). Ionospheric convection signatures observed by DE-2 during northward interplanetary magnetic field. *J. Geophys. Res.* **91**, 5817.

Heppner, J. P. (1972). Polar cap electric field distributions related to the interplanetary magnetic field direction. *J. Geophys. Res.* **77**, 4877.

Heppner, J. P., and Maynard, N. C. (1987). Empirical high latitude electric field models. *J. Geophys. Res.* **92**, 4467.

Hill, T. W. (1975). Magnetic merging in a collisionless plasma. *J. Geophys. Res.* **80**, 4689.

———. (1984). Magnetic coupling between solar wind and magnetosphere: Regulated by ionospheric conductance? (abstract). *Eos Trans. AGU*, **65**(45), 1047.

Iijima, T., and Potemra, T. A. (1976). Field-aligned currents in the dayside cusp observed by Triad. *J. Geophys. Res.* **81**, 5971.

———. (1978). Large-scale characteristics of field-aligned currents associated with substorms. *J. Geophys. Res.* **83**, 599.

———. (1982). The relationship between interplanetary quantities and Birke-land current densities. *Geophys. Res. Lett.* **9**, 442.

Iijima, T., Potemra, T. A., Zanetti, L. J., and Bythrow, P. F. (1984). Large-scale Birkeland currents in the dayside polar region during strongly northward IMF: A new Birkeland current system. *J. Geophys. Res.* **89**, 7441.

Kelley, M. C., and Makela, J. J. (2002). B_y-dependent prompt penetrating electric fields at the magnetic equator. *Geophys. Res. Lett.* **29**(7), doi: 10.1029/2001GL014468.

Kelley, M. C., Crowley, G., and Weimer, D. R. (2009). Comparison of the Hill-Siscoe polar cap potential theory with the Weimer and AMIE models. *J. Atmos. Solar-Terr. Phys.*, in press.

Kisabeth, J. L., and Rostoker, G. (1971). Development of the polar electrojet during polar magnetic substorms. *J. Geophys. Res.* **76**, 6815.

Lysak, R. L. (1985). Auroral electrodynamics with current and voltage generators. *J. Geophys. Res.* **90**, 4178.

Mozer, F. S., and Lucht, P. (1974). The average auroral zone electric field. *J. Geophys. Res.* **79**, 1001.

Oliver, W. L., Holt, J. M., Wand, R. H., and Evans, J. V. (1983). Millstone Hill incoherent scatter observations of auroral convection over $60° < \Lambda < 75°$. 3. Average patterns versus K_p. *J. Geophys. Res.* **88**, 5505.

Piddington, J. H. (1962). A hydromagnetic theory of geomagnetic storms. *Geophys. J. Roy. Astro. Soc.* 7, 183–193.

Reiff, P. H., Spiro, R. W., and Hill, T. W. (1981). Dependence of polar cap potential drop on interplanetary parameters. *J. Geophys. Res.* **86**, 7639.

Siscoe, G. L. (1982). Polar cap size and potential: A predicted relationship. *Geophys. Res. Lett.* **9**, 672.

Siscoe, G. L., Erickson, G. M., Sonnerup, B. U. Ö., Maynard, N. C., Schoendorf, J. A., Siebert, K. D., Weimer, D. R., White, W. W., and Wilson, G. R. (2002). Hill model of transpolar potential saturation: Comparisons with MHD simulations. *J. Geophys. Res.* **107**(A6), 1075, doi:10.1029/2001JA000109.

Smiddy, M., Burke, W. J., Kelley, M. C., Saflekos, N. A., Gussenhoven, M. S., Hardy, D. A., and F. J. (1980). Effects of high-latitude conductivity on observed convection electric fields and Birkeland currents. *J. Geophys. Res.* **85**, 6811.

Sugiura, M., Maynard, N. C., Farthing, W. H., Heppner, J. P., Ledley, B. G., and Cahill, L. J. (1982). Initial results on the correlation between the electron and magnetic fields observed from the DE 2 satellite in the field-aligned current regions. *Geophys. Res. Lett.* **9**, 985.

Vickrey, J. F., Livingston, R. C., Walker, N. B., Potemra, T. A., Heelis, R. A., and Rich, F. J. (1985). On the current-voltage relationship of the magnetospheric generator at small spatial scales. *Geophys. Res. Lett.* **13**, 495.

Volland, H. (1975). Models of the global electric fields within the magnetosphere. *Ann. Geophys.* **31**, 159.

Weimer, D. R. (1995). Models of high-latitude electric potentials derived with a least error fit of spherical harmonic coefficients. *J. Geophys. Res.* **100**, 19,595.

————. (2001). An improved model of ionospheric electric potentials including substorm perturbations and application to the Geospace Environment Modeling November 24, 1996, event. *J. Geophys. Res.* **106**(A1), 407–416.

Wygant, J. R., Torbert, R. B., and Mozer, F. S. (1983). Comparison of S3-2 polar cap potential with the interplanetary magnetic field and models of magnetopause reconnection. *J. Geophys. Res.* **88**, 5727.

9 Ionospheric Response to Electric Fields

In this chapter we describe some additional features of the high-latitude ionosphere that distinguish it from the lower-latitude regions. We have defined the high-latitude ionosphere as that region of latitudes in which (at least some of the time) the plasma flows in magnetic flux tubes that either have only one foot on the ground or are so extended that the effect of connecting to the conjugate hemisphere is negligible on the local plasma. From our knowledge of the high-latitude convection pattern, we know that such a definition can include the ionosphere at invariant latitudes above about 50°. In this region the ionospheric plasma may be subject to "boundary conditions" in the outer magnetosphere that allow rapid expansion along the magnetic field lines. We also know that within this region the plasma velocities perpendicular to the magnetic field can be of the order of 1 km/s. This gives the bulk thermal plasma some very different properties from those found at lower altitudes, and it is necessary to consider the details of ionospheric plasma motion, both parallel and perpendicular to the magnetic field, in order to understand them. The high plasma velocities that occur in this region also greatly affect neutral atmospheric dynamics.

9.1 Ionospheric Effects of Parallel Plasma Dynamics

9.1.1 Ionospheric Composition at High Latitudes

The effects of plasma motion along the magnetic field can be seen most dramatically in the ionospheric composition near an altitude of 1000 km as a function of latitude. At this altitude the ionosphere has essentially three ion species: O^+, H^+, and He^+. Figure 9.1 shows the quiet-time average distribution of these ion species for satellite passes across the polar region from the dayside ($\sim$1635 magnetic local time) to the nightside ($\sim$0446 magnetic local time). The field-aligned flow velocities and number flux of the light ions are also given. We obtained the data near equinox with the ISIS II satellite, which was located at 1400 km altitude. These data show quite abrupt decreases in the H^+ and He^+

The Earth's Ionosphere: Plasma Physics and Electrodynamics

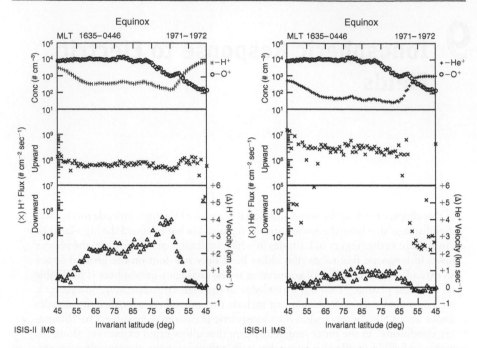

Figure 9.1 The average equinox distribution of O^+, H^+, and He^+ at 1400 km plotted in the upper panels shows that the reduced concentration of light ions at high latitudes is accompanied by significant outward flow velocities. Quiet magnetic conditions pertained in the data sets used to construct these curves. [After Hoffman and Dodson (1980). Reproduced with permission of the American Geophysical Union.]

concentrations between 55° and 65° invariant latitudes to very low, almost constant concentrations near $100\,\mathrm{cm}^{-3}$ at higher latitudes. In the same latitude sector the O^+ concentration is nearly constant on the dayside and shows only a gradual change at night. This signature in the light ion species may be seen at all altitudes where the species H^+ or He^+ is detectable and is a persistent feature at all local times. It is called the light ion trough. The latitude gradient in the H^+ concentration defines the equatorward edge of the light ion trough and the magnitude of the gradient may change significantly as a function of local time and season. Its position is also strongly dependent on magnetic activity. As we shall see in Section 9.2, the magnetic dependence is a straightforward consequence of the variability of convection paths of the plasma that, in fact, determine where the trough will be formed. This large area of depressed light ion concentrations relative to the O^+ concentration can be produced only by a field-aligned motion of the light ions relative to O^+. Satellite-borne ion mass spectrometers have observed this relative motion at altitudes below 2000 km as well as at much higher altitudes. The light ion velocities are described as the "polar wind." The velocity of the light ions can, in fact, be supersonic at times. The existence of the polar wind and the depressed light ion concentrations at high latitudes indicates that in this region the light ion population of the ionosphere is usually not in diffusive equilibrium along the magnetic field lines.

9.1.2 Hydrodynamic Theory of the Polar Wind

In Chapters 2 and 5 we discussed the forces that contribute to motion of the ionospheric plasma along the magnetic field. In these discussions we included the generation of an electric field with a component along the magnetic field lines that was created internally by the plasma. This "ambipolar" field does not drive a current parallel to the magnetic field but merely serves to equalize the forces on the ions and electrons so they move together. In the inner magnetosphere the field lines are closed and relatively short in length. This allows the flux tubes to fill to a condition of equilibrium on a time scale of a day or so. Under normal conditions, then, the plasma is produced during the daytime and tends to flow slowly up into the inner magnetosphere. At night the flow is reversed and plasma tends to flow back into the ionosphere from above, replacing the ionospheric plasma against recombination losses. Charge exchange converts O^+ to H^+ and vice versa at the top of the ionosphere. The crucial difference between the high- and low-latitude cases is that, in the former, the flux tubes are either open or so long that no semblance of diffusive equilibrium exists. In effect, whatever plasma is produced by sunlight or particle influx merely expands into a near vacuum.

Below about 2000 km, we learned that collision frequencies and gyrofrequencies are sufficiently high that, for most problems, the acceleration term may be neglected in the equations of motion. In this region the plasma is said to be collision dominated, meaning that the ions undergo several collisions while moving through one scale height. A plasma is collision dominated if

$$(V_i/H_i) \ll \nu_i$$

where V_i is the field-aligned ion velocity and H_i and ν_i are the ion scale height and appropriate collision frequency, respectively. However, since the upper boundary condition is crucial at high latitudes, we cannot make this approximation or restrict ourselves to altitudes below 2000 km. Note that it is not our intent here to provide all the mathematical details involved in the theoretical formation of parallel ion flow in such a situation; we only outline the initial steps and give some physical insight into the phenomenon based on experiments and simple model calculations.

In a collision-dominated plasma with multiple ion species, where one ion may move relative to another at a high speed that varies with altitude, the advective derivative in (2.5) must be retained. The velocity of the jth ion species then can be found from the hydrodynamic equation (2.22b):

$$\rho_j(\mathbf{V}_j \cdot \nabla)\mathbf{V}_j = -\nabla p_j + \rho_j \mathbf{g} + n_j q_j(\mathbf{E} + \mathbf{V}_j \times \mathbf{B}) - \sum_k \rho_j \nu_{jk}(\mathbf{V}_j - \mathbf{V}_k) \quad (9.1)$$

We drop the $\partial/\partial t$ term under an assumption of steady-state flow conditions in time. In the equation of motion for the electrons, the terms containing the electron mass can be neglected, as can the acceleration and advective derivative terms. Then, as shown in Chapter 5, the ambipolar electric field component parallel to $\mathbf{B}$ is given by

$$E_{||} = (1/n_e q_e)\nabla_{||}p_e \quad (9.2)$$

Remember that since q_e is a negative number, $E_{||}$ is opposite in direction to $\nabla_{||}p_e$; that is, $E_{||}$ points upward in the topside ionosphere. If we now let s be measured opposite to the magnetic field, the component of the equation of motion in the s direction for the jth ion species in the Northern Hemisphere is

$$V_{js}\left(\frac{\partial V_{js}}{\partial s}\right) + \frac{1}{n_e m_j}\frac{\partial p_e}{\partial s} + \frac{1}{n_j m_j}\frac{\partial p_j}{\partial s} - (\mathbf{g} \cdot \hat{B}) + \sum_{j \neq k} \nu_{jk}(V_{js} - V_{ks}) = 0 \quad (9.3)$$

Inspection of (9.3) reveals the forces acting on a minor ion. From right to left they arise from friction brought about by collisions with the other ion species, the component of gravity along the magnetic field, the partial pressure gradient in the minor ion species itself, and the ambipolar electric field generated between the electrons and the major ions. The final term on the left is the advective derivative. The upward electric force must ultimately become small as the light ions become the dominant species. Then on closed magnetic field lines the light ion plasma can approach or attain a state of quasi-diffusive and hydrodynamic equilibrium. While this situation can occur along the relatively short, closed magnetic field lines that exist in the midlatitude ionosphere, it may not be achievable along the open or highly distended field lines in the high-latitude ionosphere. A continuous outward flow of the minor light ions, H^+ and He^+, can therefore occur at either subsonic or supersonic velocities in a manner consistent with the forces.

A description of this ion motion can be obtained by considering the equations of continuity, motion, and energy for each species. These equations make up a closed system that is generally solved numerically (see Schunk, 1977, and Schunk and Nagy, 2000, for a full description of these equations), given a boundary condition at the top of the ionosphere near 3000 km. If we apply some simplifying assumptions, however, a description of the fundamental physics that is operating can be obtained. Consider the motion of a minor ion species, j, embedded in a major ion species, i, where the ith species is in diffusive equilibrium in an isothermal topside ionosphere. In such a case the electron and major ion gas is distributed along the magnetic field tube according to the expression

$$n_e = n_{e0} \exp[-(s - s_0)/H_p],$$ (9.4)

where n_{e0} is the electron density at some reference altitude s_0, H_p is the plasma scale height, and the variation of gravity with altitude has been neglected (Rishbeth and Garriott, 1969). The pressure for each species may be written as $p_j = n_j k_B T_j$. If we also assume that the altitude is sufficiently high for neutral collisions to be neglected, the minor ions collide only with the major ions and (9.3) can be rewritten as

$$\nu_{ji} V_{js} + V_{js}\left(\frac{\partial V_{js}}{\partial s}\right) = (\mathbf{g} \cdot \hat{B}) + \left(\frac{k_B T_e}{m_j H_p}\right) - \left(\frac{k_B T_j}{m_j}\right)\left(\frac{1}{n_j}\right)\left(\frac{\partial n_j}{\partial s}\right)$$ (9.5)

where we allow $T_e \neq T_i$ or T_j but do not allow variation with altitude. Neglecting the perpendicular motion of the plasma, the steady-state continuity equation for species j can be written from (2.22a) in the form

$$A(P_j - L_j) = (\partial/\partial s)(An_j V_{js})$$ (9.6)

where A is the cross-sectional area of a magnetic flux tube, P_j is the production term, and L_j is the loss term. Integrating this equation along the magnetic flux tube from the reference altitude to any point s, we find

$$(1/Q_j)(\partial Q_j/\partial s) = (1/n_j)(\partial n_j/\partial s) + (1/V_{js})(\partial V_{js}/\partial s) + (1/A)(\partial A/\partial s)$$ (9.7)

where

$$Q_j = \int_{s_0}^{s} A(P_j - L_j)ds$$ (9.8)

Substituting for $(1/n_j)(\partial n_j/\partial s)$ in (9.5), we arrive finally at the equation

$$\left(V_{js}^2 - V_t^2\right)(1/V_{js})(\partial V_{js}/\partial s) = -\nu_{ji} V_{js} + \mathbf{g} \cdot \hat{B} - (k_B T_e/m_j H_p)$$
$$- V_t^2[(1/Q_j)(\partial Q_j/\partial s) - (1/A)(\partial A/\partial s)],$$ (9.9)

where the thermal speed of the jth species is given by

$$V_t = (k_B T_j/m_j)^{1/2}.$$ (9.10)

This first-order differential equation for V_{js} has a singularity where $V_{js} = V_t$. This point is commonly called the supersonic transition in fluid theory. In a multiple ion species plasma the plasma sound speed is not given by V_{js} but for convenience here, we to refer to $V_{js} > V_t$ as representing the supersonic flow of the jth species. Figure 9.2 shows the family of solutions for (9.9), using the hydrogen Mach number $M_j = V_{js}/V_t$ plotted as a function of altitude. The physically meaningful solutions are those with $M_j > 1$ at all altitudes and one solution (marked as curve A) passing through the transitional point at $M_j = 1$ for which the flow is subsonic at low altitudes and supersonic at higher altitudes. A similar analysis has been done for the solar wind in the pioneering work by Parker (1958).

The solutions of these hydrodynamic equations are now sophisticated enough that the details of heat transfer and energy balance may be considered in addition to the momentum and density distribution of various ion species. In such models the acceleration of minor ion species with mass less than the mean ion mass can occur in which ion-ion collisions are sufficient to produce frictional heating and anisotropic ion temperatures. In the high-latitude ionosphere, both H^+ and He^+ can be accelerated through the O^+ gas and be subjected to this heating process. The existence of different temperatures and temperature distributions in the ion species can induce motion in these species that (depending on the ion mass) either resists or assists their outward flow.

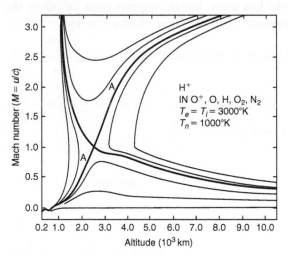

Figure 9.2 Solutions of the hydrodynamic equations for the hydrogen ion Mach number as a function of altitude in a gas containing O^+ ions as well as the neutral species O, H, O_2, and N_2. Curve A provides a transition from subsonic to supersonic flow. In this figure, u is the hydrogen ion bulk velocity and c is the hydrogen velocity given by (9.10). [After Banks and Holzer (1969). Reproduced with permission of the American Geophysical Union.]

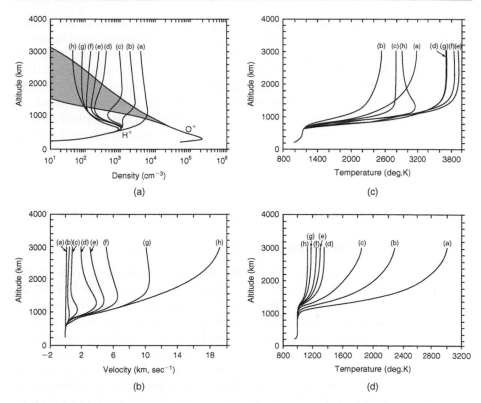

Figure 9.3 Attitude distribution of (a) O^+ and H^+ densities, (b) H^+ velocity, (c) H^+ temperature, and (d) O^+ temperature from models of the polar wind. The curves labeled (a)–(h) correspond to assumed H^+ outflow velocities at 3000 km ranging from 0 to about 20 km/s. [After Raitt et al. (1975). Reproduced with permission of Pergamon Press.]

Figure 9.3 shows some typical results of calculations performed assuming different outflow velocities for H^+ at 3000 km. Notice in Fig. 9.3b that the H^+ velocity increases rapidly in the region between 600 and 1400 km where O^+ is the dominant ion (see Fig. 9.3a). This flow is induced by the O^+-electron polarization field. The flow of H^+ through O^+ produces frictional heating of the H^+ gas to temperatures above that of the O^+ ions in this region. At higher altitudes the distribution of the species is controlled by the specified H^+ outflow velocity, that is, by the boundary condition placed on the solution. It shows that, at 3000 km, the predicted H^+ polar wind can be supersonic since velocities in excess of 10 km/s are associated with temperatures less than 4000 K.

Two important considerations should be kept in mind when considering the usefulness of this hydrodynamic description of the polar wind. First, the polar wind does not simply result from an open field-line geometry of field tubes with very large volumes. The minor ions H^+ and He^+, for example, receive very little acceleration if the electron population is very cold, as it might be in the winter polar region. Second, the theoretical descriptions of the polar wind are dependent on some boundary condition that is usually applied at the top of the modeled region near 3000 km, where the ionosphere becomes essentially collisionless. The effects of the acceleration processes and the distribution of the ion species below this altitude are dependent on their boundary conditions. Ultimately, boundary conditions must be consistent with conditions in the outer magnetosphere, where the field tubes are in contact with a much lower-density plasma having a variety of possible energy distributions that differ significantly from those existing at lower altitudes.

9.2   Ionospheric Effects of Perpendicular Plasma Dynamics

9.2.1   The Role of Horizontal Transport

In Chapter 8 we described in some detail the convective motion of the ionospheric plasma that results from an electric field of solar wind origin. The large-scale convection pattern at high latitudes is fixed with respect to the earth-sun line and the magnetic dipole axis. In this coordinate system the convection pattern is dependent on latitude and longitude (also referred to as magnetic local time). The plasma moves both toward and away from the sun and into and out of the auroral zone. Since the solar radiation and auroral particle precipitation are the most important ionization sources, the plasma convection velocity is thus one of the most important factors affecting the plasma distribution and temperature at all altitudes.

In addition to the ionospheric motion imposed from the magnetosphere, the earth's atmosphere and portions of its plasma environment corotate about the geographic axis. We thus have a rather complicated situation in which one source of motion and ionization is most easily specified in a geographic reference frame and another is most easily specified in a geomagnetic reference frame. During the course of a day the geomagnetic pole rotates around the geographic pole and, in a coordinate system fixed with respect to the sun, there is therefore a universal time dependence (in addition to latitude and local time dependence) in both the magnetospheric convection velocity and the auroral zone ionization rate. At high latitudes the relatively high ion velocities produce frictional heating due to collisions between the ions and the neutral gas. These collisions also affect the rate at which ions recombine with the neutral gas. We will deal with these collisional effects later. We will not deal in detail with the production and loss processes for plasma in the ionosphere but we will describe the convection effects in sufficient detail that their effects on the plasma composition, distribution, and temperature will be recognized.

To understand the net plasma distribution at high latitudes, we will consider a frame of reference in which the earth-sun line and the geographic pole are fixed. This frame is essentially fixed in inertial space since its motion, consisting of rotation around the sun once per year, is negligible compared to the convective motion of the plasma during a one-day cycle. In this reference frame the plasma has a corotation component that is independent of local time and a magnetospheric component that is dependent on local time and universal time as the magnetic pole rotates around the geographic pole. Figure 9.4 shows several different plasma convection paths resulting from this complex addition of corotation and magnetospheric convection. Each path is traced for a 24-hour period to illustrate that the paths do not necessarily close and are certainly quite different from the convection paths illustrated in Chapter 8, in which the motion of the coordinate system was removed. The starting points for these three trajectories may be roughly summarized as follows: (a) dusk convection cell, (b) polar cap, and (c) dawn convection cell. Also shown in each portion of the figure is the position of the solar terminator in the Northern Hemisphere during winter, summer, and equinox. Notice in this inertial frame that the plasma can move in and out of sunlight as well as in and out of the auroral zone. A complex distribution of plasma production, transport, and loss, having local time and universal time dependences, results from this motion.

With the goal of understanding some of the consequences of this plasma motion, we will consider some large-scale features observed in the high-latitude F region. Figure 9.5 shows the total plasma concentration and the perpendicular components of the plasma drift velocity observed by a satellite traversing the high-latitude ionosphere. The plot is made in a rotating frame such that a corotating plasma would register zero velocity. In the time interval labeled B, the zonal flow velocity indicated by the upper curve is weak and to the west. At the point where the curve crosses the dashed line, the plasma is flowing toward the west with the same velocity at which the earth is rotating toward the east at that latitude. This means that the dusk terminator is moving westward at the same velocity as the plasma, and thus, the plasma sees solar conditions that are independent of time.

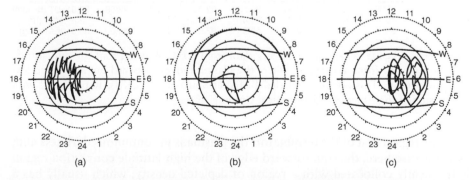

Figure 9.4 Several complex trajectories that plasma may undergo in a 24-hour period due to displacement of the geomagnetic and geographic poles. [After Sojka et al. (1979). Reproduced with permission of the American Geophysical Union.]

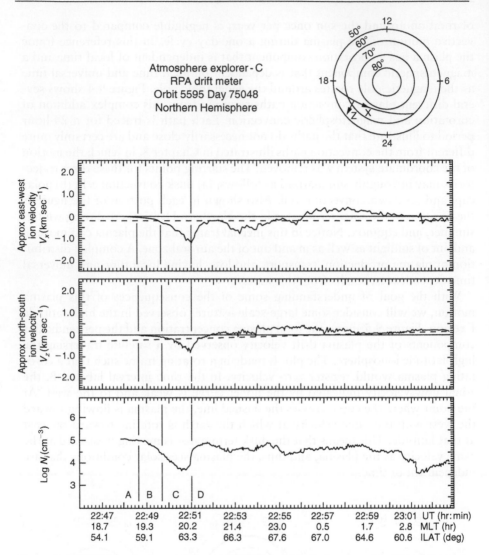

Figure 9.5 Plasma velocity and density data from AE-C showing that the midlatitude trough begins to form in a region where the plasma moves eastward but at a speed less than corotation. This region is labeled B. [After Spiro et al. (1978). Reproduced with permission of the American Geophysical Union.]

In the region east of the terminator, the plasma is in continuous darkness and, as is the case here, the equatorward edge of the high-latitude convection region is frequently collocated with a region of depleted density, which usually has a sharp poleward edge and a more gradual decline at its equatorward side. The total plasma concentration depletion can exceed an order of magnitude in the

F region and is called the midlatitude trough. Examination of the plasma drift characteristics within the premidnight trough in Fig. 9.5 shows that the plasma depletion lies within a region in which there is a transition from either eastward or corotating plasma drift equatorward of the auroral zone to westward drift in the auroral zone. At such a transition zone the plasma flows very slowly, and chemical recombination will have more time to reduce the ion concentration than in regions where the plasma moves more quickly or where auroral zone ionization exists.

Figure 9.6 shows a schematic diagram of a modeled set of convection paths in the evening trough region that are similar to the data shown in Fig. 9.7. On the right-hand side the results of a simple chemical decay model are given for the plasma content, taking into account its motion along these paths. The results confirm that plasma stagnation in the nighttime region just equatorward of the auroral zone is a likely candidate for the quiet-time trough formation. We discuss the storm-time trough in Chapter 10.

The convection paths giving rise to the trough in the premidnight sector result quite naturally from adding the essentially eastward corotation, dominating the flow characteristics at low latitudes, and the essentially westward flow from the magnetospheric field, dominating the flow characteristics at high latitudes. Figure 9.7 shows the results of adding these two convection sources in a magnetic coordinate frame. In this figure the premidnight trough results from the very slow flow in the nighttime region of the trajectory labeled II. The illustrated flow pattern, of course, has considerable variations because of the universal time effects mentioned earlier and the temporal changes in the magnetospheric electric field. These variations can explain the variety of observed latitudinal profiles of

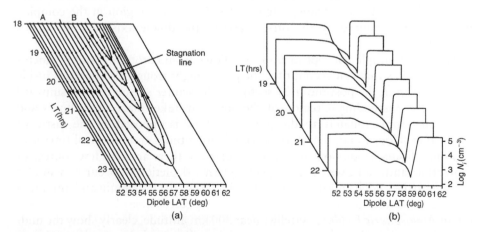

Figure 9.6 Model convection paths having the same properties as those observed in Fig. 9.5 are shown in (a). The resulting number density profile due to the almost stagnant flow is shown in (b). [After Spiro et al. (1978). Reproduced with permission of the American Geophysical Union.]

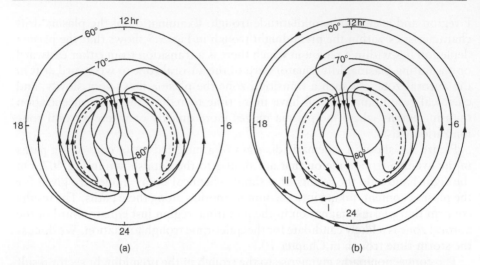

Figure 9.7 The model convection shown in Fig. 9.5 can be created by the addition of a simple two-cell convection pattern shown in (a) with a corotation velocity. The result shown in (b) produces typical trough flow characteristics along the contour labeled II. [After Spiro et al. (1978). Reproduced with permission of the American Geophysical Union.]

the midlatitude trough. Note however that a midlatitude trough is observed in regions other than the premidnight sector, so it is unlikely that the slow flow convection trajectories described here are the only source of trough formation.

At other local times the midlatitude trough may be due to low plasma densities produced elsewhere and transported to the observation region in the complex flow pattern produced by the offset between the dipole axis and the earth's rotation axis.

The rather simple concept of opposing flows from corotation and magnetospheric sources that can account for the premidnight trough also pertains inside the polar cap in the postmidnight sector. In the winter season the antisunward flow in the postmidnight region of the polar cap takes place in darkness and is directed approximately to the west. This flow is opposed by the eastward corotation velocity and, therefore, at certain longitudes and universal times can produce plasma stagnation. In the dark polar cap there are very few sources of ionization, and, thus, we might expect a plasma depletion to occur in a similar manner to the midlatitude trough. Figure 9.8 shows the maximum and minimum O^+ concentrations in the southern winter hemisphere. These observations, made on an *Atmospheric Explorer* satellite near 300 km altitude, clearly show the midlatitude trough in the O^+ concentration and, as discussed above, a large plasma depletion at latitudes above the auroral zone in the postmidnight polar cap. This polar cap depletion is called the high-latitude or polar hole and is the site of the lowest plasma concentration in the winter high-latitude ionosphere. It is a

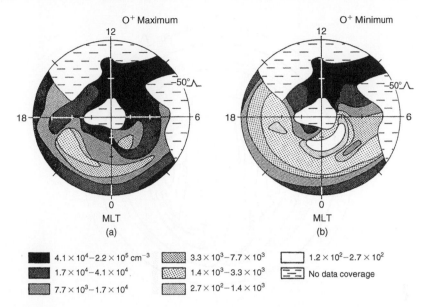

O⁺ Maximum

O⁺ Minimum

████ $4.1 \times 10^4 - 2.2 \times 10^5 \, cm^{-3}$	▓▓▓ $3.3 \times 10^3 - 7.7 \times 10^3$	☐ $1.2 \times 10^2 - 2.7 \times 10^2$
████ $1.7 \times 10^4 - 4.1 \times 10^4$	▒▒▒ $1.4 \times 10^3 - 3.3 \times 10^3$	⊟ No data coverage
▓▓▓ $7.7 \times 10^3 - 1.7 \times 10^4$	░░░ $2.7 \times 10^2 - 1.4 \times 10^3$	

Figure 9.8 Examination of the O⁺ concentration of the high-latitude winter ionosphere near 300 km shows two regions of plasma depletions that may result from plasma convection: the midlatitude trough seen near 60° invariant latitude and between 1800 and 2400 h, and the high-latitude hole seen near 80° invariant latitude between 2300 and 0530 MLT hours. [From Brinton et al. (1978). Reproduced with permission of the American Geophysical Union.]

persistent feature in the winter, appears sporadically at equinox, and almost never appears in the summer. This occurrence pattern can easily be attributed to the movement of the solar terminator, which places this location always in darkness in the winter, partially in darkness at equinox, and completely in sunlight during the summer. Figure 9.9 summarizes some characteristics of the total ion concentration and its relationships to a typical high-latitude convection pattern.

9.2.2 Ion Heating Due to Collisions

When ionic and neutral particles collide, the amount of energy exchanged depends on the relative velocity between the two particles. At high latitudes this relative velocity can become quite high due to the magnetospheric convection electric field. Then this energy exchange can significantly affect the ion temperature, the ion composition, and even the neutral wind.

The ion temperature at high latitudes is determined principally by frictional heating, which occurs whenever a relative velocity exists between the ion and neutral gases, and by heat exchange with the neutral gas and electron gas. The former can be described equally well as Joule heating due to the Pedersen current

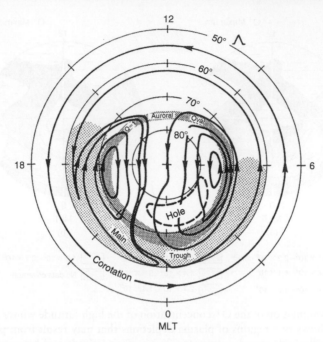

Figure 9.9   The plasma stagnation that may produce the depletions seen in Fig. 9.8 occur when the magnetospheric convection pattern and the corotation velocity are combined, as shown here. [After Brinton et al. (1978). Reproduced with permission of the American Geophysical Union.]

$J_P = \sigma_P E$. When all of these effects are considered, the ion temperature can be expressed as (St.-Maurice and Hanson, 1982):

$$T_i = T_{eq} + (m_n \phi_{in}/3k_B \varphi_{in})|V_i - V_n|^2 \tag{9.11}$$

where

$$T_{eq} = T_n + [(m_i + m_n)\nu_{ie}/m_i \nu_{in}\varphi_{in}](T_e - T_i). \tag{9.12}$$

Here the dimensionless parameters ϕ_{in} and φ_{in} depend on the nature of the collisional interactions between the ions and the neutral gas, but above about 200 km, where O^+ collides principally with atomic oxygen, both of these parameters are approximately unity.

These expressions show that the ion temperature will increase from its equilibrium value whenever a relative velocity exists between the ion and neutral gases. This relationship between the ion drift velocity and the ion temperature can easily be seen at high latitudes, where the magnetospheric electric field can rapidly produce ion velocities that greatly exceed the neutral gas velocity. One such example is shown in Fig. 9.10, where an extremely good correlation between

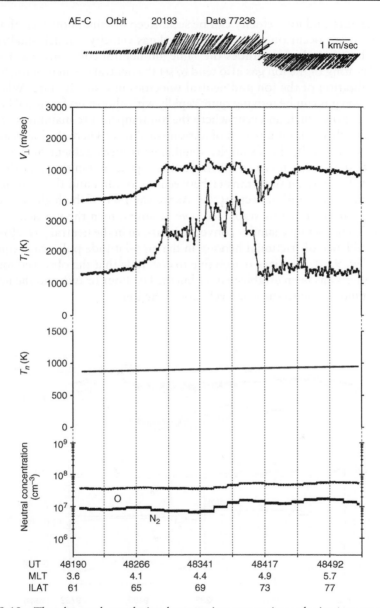

Figure 9.10   The observed correlation between ion convection velocity (top two panels) and the ion temperature is extremely good evidence for Joule heating in the region of sunward flow. [After St.-Maurice and Hanson (1982). Reproduced with permission of the American Geophysical Union.]

ion temperature and ion velocity can be seen. Keep in mind, however, that such a correlation can occur only when the neutral gas velocity is much smaller than the ion velocity. At high latitudes the same ion-neutral collisions that produce frictional heating of the ion gas also tend to set the neutral gas in motion, leading to an equalization of the ion and neutral velocities in a steady state. When this occurs, the ion gas can be moving quite rapidly with almost no frictional heating. Figure 9.11 shows such an event where the ion temperature maintains its equilibrium value throughout a region of antisunward flow where the ion velocity is in excess of 1 km/s. In this example, simultaneous measurement of the neutral gas velocity shows that it is also moving at high speed and in the same direction as the ions. The momentum transfer between the plasma and the neutral gas is described in more detail in Section 9.3. As we shall see, if the high-latitude ion convection pattern remains stable for several hours, then the ion-neutral collisions can, in fact, set up a similar convection pattern in the neutral gas. However, if we see evidence of frictional heating in the high-latitude polar cap region, as shown in Fig. 9.10, then it is reasonable to conclude that the observed ion convection pattern has been established within the last hour or two and the neutrals are not yet moving with the same velocity as the ions.

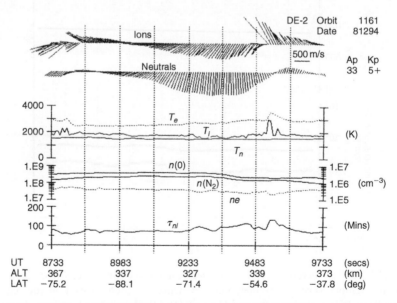

Figure 9.11 Under stable conditions the ion and neutral velocities can become comparable, as observed in the polar cap here by DE-2. Then, little Joule heating occurs, even when the ion velocity is large. [After Killeen et al. (1984). Reproduced with permission of the American Geophysical Union.]

9.2.3 Velocity-Dependent Recombination

In addition to frictional heating, the relative energy with which the ion and neutral particles collide can affect the chemical reaction rates of the ions. This is particularly true in the ionospheric F region, where the rates for the charge exchange reactions are extremely sensitive to the relative energy of the reacting particles. If these rates are enhanced, then the plasma density will decrease since NO^+ and O_2^+ recombine quite rapidly because of dissociative recombination. The plasma ion composition will also change if O^+ is converted to molecular species via the reactions

$$O^+ + N_2 \rightarrow NO^+ + N \quad \text{(with rate } k_1) \tag{9.13a}$$

$$O^+ + O_2 \rightarrow O_2^+ + O \quad \text{(with rate } k_2) \tag{9.13b}$$

Laboratory studies of these reactions express their rates in terms of an effective temperature. The experiments are performed by passing one gas through the other in a drift tube in a manner quite similar to the drift of ions through the neutrals in the ionosphere. The most important reaction rate in the F region is the ion-atom exchange between O^+ and N_2. The rate coefficient of this reaction can be written in the form

$$k_1 = \Big[1.53 \times 10^{-12} - 5.92 \times 10^{-13}(T_{eff}/300)$$

$$+ 8.6 \times 10^{-14}(T_{eff}/300)^2\Big]\text{cm}^3/\text{s}, \quad T_{eff} = 0.636T_i + 0.364T_n.$$

To determine the effective temperature for the ionosphere, we must take into account the relative velocity of the neutral gas with respect to the ions. If we then express the relative ion-neutral velocity in terms of the electric field in the neutral frame such that

$$|\mathbf{V}_{i\perp} - \mathbf{U}_\perp| = E'_\perp/B,$$

then the effective temperature is given by

$$T_{eff} = T_i + 0.33E'^2_\perp \tag{9.14}$$

where $E'_\perp$ is expressed in millivolts per meter (Schunk et al., 1975). We see, therefore, that the rate coefficient, k_1, increases strongly with the ion-neutral velocity difference. This highly sensitive variation in the charge exchange rate can be reflected quite dramatically, both in the absolute ion concentration and in the relative ion composition at high latitudes. Figure 9.12 shows an example of this effect in the high-latitude F region. Near 250 km, O^+ is the dominant ion by

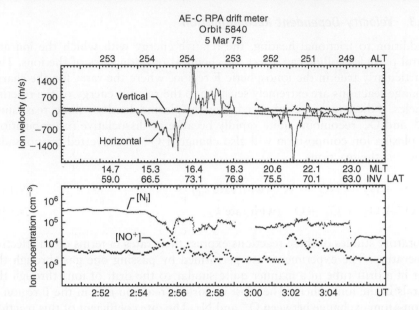

Figure 9.12 The upper panel shows the vertical and horizontal convection velocities measured by the AE-C satellite. The lower panel shows the total ion concentration comprising O^+ and the molecular ions, as well as the NO^+ concentration itself. Notice that when the horizontal ion drift is large, the relative abundance of NO^+ increases due to the enhancement of the rate of reaction (9.13a). (Figure courtesy of R. Heelis and W. Hanson.)

at least an order of magnitude over NO^+ for almost the entire pass. In regions of large drift, however, we observe an increase in the relative concentration of NO^+ and a decrease in the total ion concentration. The decrease in total ion concentration is due to the very rapid recombination of NO^+.

9.2.4 Positive and Negative Ionospheric Storms

Energy, momentum, and Joule heating in the high latitude ionosphere all contribute to what are known as ionospheric storms. As the names indicate, ionospheric plasma densities at midlatitudes may be higher or lower than the norm in various phases of the magnetic storms, which are the primary context in which ionospheric storms are described. The term ionospheric storm refers to global ionospheric effects as opposed to Convective Equatorial Ionospheric Storms (CEIS), described in Chapter 4, and that are confined to the low latitude sector.

Positive storms can be due to a variety of mechanisms (Fuller-Rowell and Rees, 1981; Huang et al., 2005). Strong neutral winds blowing equatorward out of the high latitude zone will tend to push plasma away from the earth and its neutral atmosphere along the magnetic field lines. This in turn reduces recombination of O^+, which requires interaction with a neutral molecule for efficient recombination, thereby increasing the plasma density. Similarly, penetrating zonally eastward electric fields associated with an increase in the dawn-to-dusk interplanetary electric field in the earth's reference frame (IMF B_z south conditions) causes an uplift of the daytime equatorial ionosphere, removing ions from the recombination zone and allowing for further production of plasma in sunlight. This uplift eventually leads to an enhanced fountain effect and an expansion of the equatorial anomaly (equatorial arcs) to higher latitudes. If, in turn, this plasma is ripped away from the midlatitude zone by meridional penetrating fields, a SED (Storm Enhanced Density) event may occur, leading to channels of dense plasma stretching even over the polar zone (Foster and Burke, 2002; Vlasov et al., 2003; Foster et al., 2005).

Negative storms are thought to be due to changes in the composition of the neutral atmosphere at thermosphere/ionosphere heights. This in turn is due to heating and upwelling of the lower atmosphere with its molecular-rich composition. In turn, this decreases the $[O]/[N_2]$ ratio and enhances the recombination of the O^+ plasma, resulting in a negative ionospheric storm. The molecular-rich atmosphere can be transported for vast distances by disturbance winds (Fuller-Rowell and Rees, 1981). Large scale TIDs can also affect the density as the associated oscillating winds drive the ionosphere into the neutral atmosphere in one phase, decreasing the local density, while in the equatorward wind phase, to first order, the plasma merely rises without changing its content, something of a half-wave rectifier effect.

9.3 Electrodynamic Forcing of the Neutral Atmosphere

9.3.1 $J \times B$ Forcing

The fact that the plasma is often in motion with high velocities in the polar region has important consequences for thermospheric dynamics. We cannot hope to treat this fascinating topic in the detail it deserves but will only outline some of its most important aspects.

In Chapters 3 and 5 we pointed out that the plasma acts as a drag on the thermosphere at low and midlatitudes. We showed that this drag force was equivalent to a $J \times B$ force on the neutral gas where

$$J = \sigma \cdot (U \times B).$$

If this current is not divergence free, we found that electric fields build up and usually reduce the total current with the result that the $J \times B$ force decreases. Neutral winds in such a case therefore create electric fields via a *dynamo* process.

At high latitudes, the electric fields are impressed from outside the ionosphere/thermosphere system and the possibility for a *motor* arises. In fact, the origin of the force on the neutrals is again the $J \times B$ force, only in the high-latitude case the expression

$$J = \sigma \cdot (E + U \times B)$$

is usually dominated by the impressed electric field, E. In the F region, where the conductivity tensor σ is diagonal, the force becomes

$$F = \sigma_p E' \times B$$

where E' is the electric field in the neutral frame of reference. A data set taken in the evening auroral oval, which shows the importance of electrodynamic forcing, is presented in Fig. 9.13a and b. Figure 9.13a is an all-sky camera picture showing the location of several chemical releases made from the same rocket. The rocket was launched from Poker Flat, Alaska, at 1810 local time on February 28, 1978. Figure 9.13b is a schematic diagram that describes the various features in the photograph. The two visible barium (Ba) ion clouds have been driven by the $E \times B$ drift at high velocity toward the west, since E was the usual northward electric field at that time period. The two circular strontium (Sr) neutral clouds have hardly moved since their release (which was at the same time and altitude (≥ 200 km) as the ion clouds). However, the trimethyl aluminum (TMA) trail shows a high-velocity region streaming in the same direction as the ion cloud at about one-third the plasma cloud velocity. More quantitative data from the same experiment are given in Fig. 9.14. The three solid tracks in the figure show sequential locations of the barium ion clouds and are labeled I205, I268, and I214, corresponding to the three ion (I) cloud altitudes in kilometers. Minute markers show the temporal progression of the three clouds, which moved parallel to the auroral oval at about 1 km/s. The two dashed lines marked N131 and N140 show the neutral (N) trail locations at the peak in the two neutral velocity profiles, which occurred at 131 and 140 km, respectively. Notice that the southern trail moved very nearly parallel to the $E \times B$ direction with a velocity equal to about 27% of the velocity of the ion cloud nearest in horizontal location. The northern trail had a speed equal to 44% of the nearest ion cloud velocity and a direction rotated slightly poleward of the $E \times B$ velocity. The high-altitude neutral clouds (N210, N278, and N209) moved much more slowly and had a considerably different direction of motion from the $E \times B$ direction.

Some of these results may be explained in a straightforward manner. First, note that since the E field maps uniformly with altitude, the main altitude dependence in F is in the σ_P term. The simultaneous Chatanika radar data showed that this term (primarily due to σ_P) peaked at 140 km altitude, in excellent agreement with the observed peaks in the neutral wind profiles. A crude estimate of the time

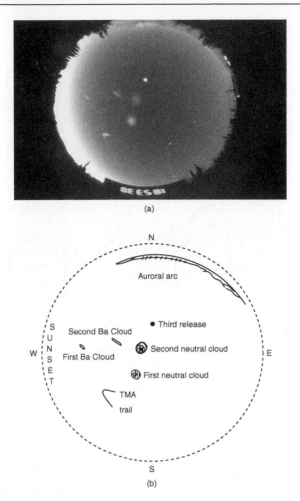

Figure 9.13 (a) An all-sky camera photo showing the position of ion and neutral clouds in the BaTMAn experiment several minutes after the second barium release. The bright dot [see (b)] is the third barium release at the detonation time. (b) Key for identifying features in part (a).

required to accelerate the neutrals may be derived as follows. The acceleration of the neutral gas by the electromagnetic force is given by an expression of the form

$$\rho(\partial u/\partial t) \cong |\mathbf{J} \times \mathbf{B}| = \sigma_P E B$$

where u is the zonal velocity and E is the meridional electric-field component. Since from (2.40b) the F-region Pedersen conductivity is given as approximately

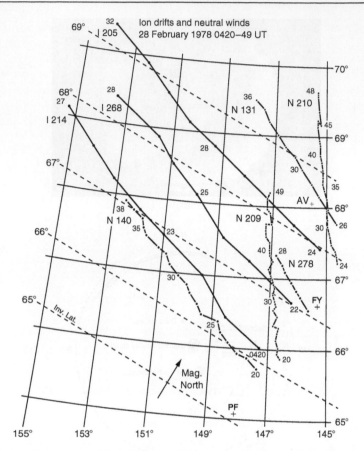

Figure 9.14 Ground tracks and altitudes of neutral (N) and ion clouds (I) of the various releases during the BaTMAn experiment shown in Fig. 9.13. [After Mikkelsen et al. (1981a). Reproduced with permission of the American Geophysical Union.]

$\sigma_P = nm_i v_{in}/B^2$, this relationship may be written as

$$\delta u/\delta t = (nv_{in}/n_n)(E/B)$$

where we have assumed that the ion mass and the neutral mass are the same. Now, since v_{in} is proportional to the neutral density n_n, the latter cancels out, and we have the interesting result that the neutral acceleration depends only on the plasma density n. Using $v_{in} \cong 5 \times 10^{-10} n_n$ (see Appendix B), we can estimate the time for acceleration to a velocity, $\delta u = (1/3)(E/B)$, to be $\delta t_{1/3} \cong 10^{10}/15n$. For a peak density of $n = 3 \times 10^5$ cm^{-3}, as was the case in the Alaskan experiment, $\delta t_{1/3} \cong 2000$ s, which is quite short.

This estimate ignores many other terms in the equation of motion, but it does show that $J \times B$ forcing is significant. Viscosity is particularly important in the dynamics since it spreads the strong height-dependent $J \times B$ forcing to other altitudes. Mikkelsen et al. (1981a, b) have studied this event with a two-dimensional numerical model using the local forcing deduced from the radar and rocket data. They found good agreement with the calculated and observed zonal wind (see Fig. 9.15). Their model included the pressure gradient, a curvature effect due to the shape of the auroral oval and advection of momentum in the vertical and meridional directions. The agreement was not quite as good between the observations and the model wind in the meridional component. Furthermore, when applied to a second data set, obtained three days later on March 2, 1978, the agreement was poor in both components. The primary difference in the two events was the location of the forcing in longitude. In the February 28 data, shown in Figs. 9.13–9.15, a substorm occurred near College, Alaska, and local forcing was an adequate approximation. On March 2, however, the substorm was far to the east of College. Since the observed winds were very similar, Mikkelsen et al. concluded that advection of momentum in the zonal direction was crucial and that a three-dimensional model was essential for progress in understanding electrodynamic forcing of the upper atmosphere.

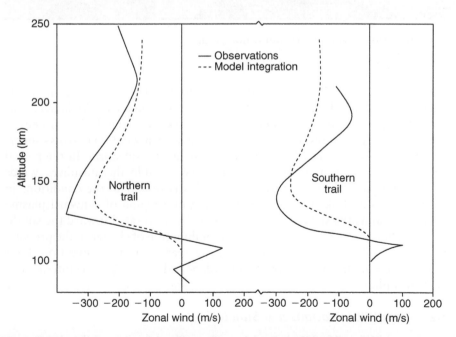

Figure 9.15 Comparison of the observed zonal wind profiles at the northern and southern positions on February 28 with the results of the modeling. [After Mikkelsen et al. (1981b). Reproduced with permission of the American Geophysical Union.]

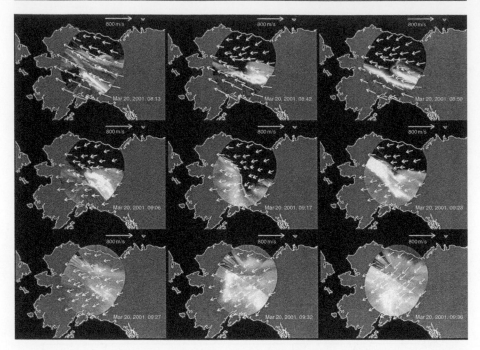

Figure 9.16   All-sky wind observations, along with images of the aurora. (Figure courtesy of M. Conde.)

Developments in optical technology now permit Doppler shift measurements of thermospheric winds at up to 48 different locations in an all-sky image. This "all-sky Fabry-Perot" instrument has shown remarkable variability in the neutral wind near auroral features. Figure 9.16 shows just such winds via arrows superposed on an all-sky image obtained during a magnetic substorm. In the period of 0813–0906 nT (evening over Alaska), very sizeable thermospherc winds were blowing parallel to the auroral oval. This is similar to the conditions in Figs. 9.13–9.15 and likely represents ion drag in a region of sunward plasma convection. But as the substorm developed, the wind rotated toward the southwest. This change most likely was due to a combination of change in the pressure gradients due to Joule heating and ion drag due to antisunward convection closer to magnetic midnight. We simply do not understand neutral/plasma interactions on our own planet.

9.3.2 Global Observations and Simulations

Heppner and Miller (1982) collected a number of barium neutral cloud measurements in the thermosphere and compared them to several convection models. One such comparison is shown in Fig. 9.17. The wind vectors are superimposed on a representative convection pattern in Fig. 9.17a at the observed location in

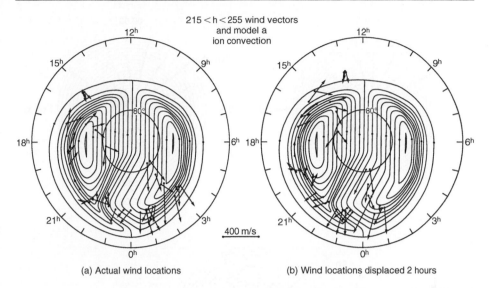

Figure 9.17 (a) Compilation of barium neutral cloud velocities for a number of local times superimposed on a plasma convection model. In (b) the same data are plotted 2 hours earlier in local time. [After Heppner and Miller (1982). Reproduced with permission of the American Geophysical Union.]

Fig. 9.17b. The wind locations were arbitrarily shifted by two hours to take into account the sluggish response of the neutrals (e.g., see the preceding calculation of δt). The vectors show a much clearer relationship to the ion flow in Fig. 9.17b. One obvious difference between the plasma flow and the measured winds, even in Fig. 9.17b, is that the strong neutral flow seems to continue at latitudes well below the auroral oval. This occurs because there is no comparable force to "turn" the neutral wind when the flow leaves the sunward convection part of the two-celled plasma flow pattern. Notice that something of a flywheel effect will occur in this case since the wind in the subauroral region can create electric fields via the F region dynamo process discussed in Chapters 3 and 5. This effect has been suggested by Gonzales et al. (1978) to explain some of the anomalous midlatitude electric field observations in the evening sector (see Chapter 5).

Although the primary electrodynamic effect seems to be momentum transfer via the $\mathbf{J} \times \mathbf{B}$ force, changes in pressure due to Joule heating of the neutral gas discussed in Section 9.2 also affect the neutral atmospheric dynamics. This effect may be more important in producing vertical upwelling and neutral composition changes than in generating strong neutral winds.

Three-dimensional models of neutral wind forcing due to realistic electric field patterns are now available that include all of the important physical processes. An example of such a calculation is given in the series of plots in Fig. 9.18, which are polar diagrams of the wind velocity at 240-km altitude for various

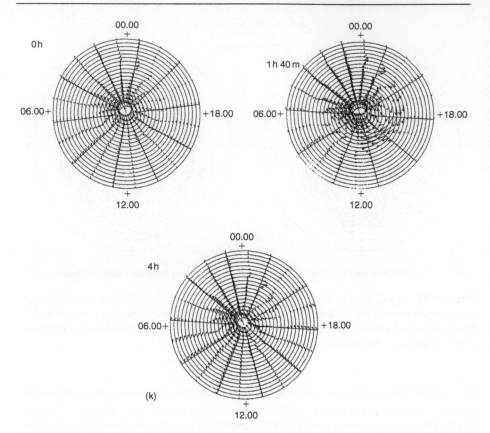

Figure 9.18 Plot of wind vectors from 50° latitude to the pole at 240-km altitude during a simulated substorm. Plots are shown at intervals of 0, 1 h 40 min, and 4 h. Scale: an arrow the same length as 2° in latitude corresponds to 160 m/s. [After Fuller-Rowell and Rees (1981). Reproduced with permission of Pergamon Press.]

time intervals after a model magnetic substorm that starts at $t = 0$ and lasts for 2 h. The initial plot at $t + 20$ min is representative of the solar forced winds. By $t + 100$ min a clear vortex has formed in the dusk sector where wind speeds exceed 300 m/s. Very low velocities are found in the morning hours where the substorm effects are counteracting the solar forcing. Two hours after the storm ended, the pattern returned virtually to the initial state.

Winds in the E region for the same simulation are shown in Fig. 9.19 at $t + 100$ min. Note that the scale differs by a factor of four relative to the F region calculation. The wind vectors show a clear double-cell pattern with a line of symmetry running from 0300 to 1500. The symmetry of the two vortices is due to the lower value of pure solar forcing at E-region heights. The rotation of the pattern away from a noon-midnight symmetry is due to the importance of the Hall conductivity in the $\mathbf{J} \times \mathbf{B}$ force.

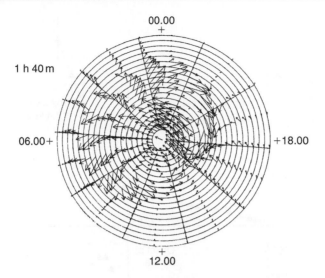

Figure 9.19 Plot of wind vectors from 50° latitude to the pole at 120-km altitude during a simulated substorm. The plot shown is at a time of 1 h 40 min after the substorm onset. Scale: an arrow the same length as 2° in latitude corresponds to 40 m/s. [After Fuller-Rowell and Rees (1981). Reproduced with permission of Pergamon Press.]

With computing power increasing quickly and becoming more readily available, a number of the three-dimensional models now exist. Models such as those discussed here will prove quite valuable in developing further insights into the electrodynamic forcing of the neutral atmosphere.

9.4 Particle Acceleration in the Topside Ionosphere

Remarkably, the earth's topside ionosphere is a powerful particle accelerator. This is manifested most obviously by one of nature's most spectacular phenomena: the aurora. We can only hope to scratch the surface of ion and electron acceleration phenomena in this text but at least some introduction is possible.

9.4.1 Parallel Electric Fields in the Upper Ionosphere

Low-altitude rocket experiments gave the first evidence for the existence of large parallel electric fields via observations of anticorrelated electron and ion fluxes (Mozer and Bruston, 1966) and mono-energetic electron spectra peaked at 6 keV (Evans, 1968). The first definitive detection came from observation of upward barium ion acceleration using shaped, charged injections (Haerendel et al., 1976) and from in situ parallel electric field observations from a satellite (Mozer et al., 1977). Figure 9.20 shows the altitude of the tip of a barium-shaped charge

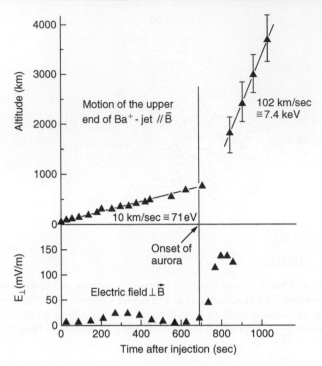

Figure 9.20 Upward velocity of the upper end of the barium ion jet (January 11, 1975) (upper panel) and transverse electric field projected to the 100 km level as derived from the low altitude portion of the ion jet (lower panel).

injection. The velocity change corresponds to passage through an acceleration zone where **E** points upwards antiparallel to **B** in the Northern Hemisphere.

Such fields are unexpected due to the high conductivity parallel to the magnetic field. Early theories appealed to anomalous resistivity supported by ion acoustic or ion cyclotron waves. The idea was that the electrons would collide with waves rather than other particles, creating an effective collision frequency, ν^*. Then $\sigma^* = ne^2/m\nu^*$ would decrease, allowing $E_{||} = J_{||}/\sigma_0^*$ to become large. This is an attractive idea since both ion cyclotron and ion acoustic waves are unstable to large $J_{||}$ (Kindel and Kennel, 1971). Another attraction was that the plasma density decreases faster than does the area of the magnetic flux tube. Thus, for a given $J_{||} = neV_d$, where V_d is the parallel differential drift of ions and electrons, V_d can become quite large a few thousand kilometers above the auroral ionosphere. When V_d exceeds the electron thermal speed, the plasma is unstable; that is, for

$$V_d = \frac{J_{||}}{ne} > V_e^{th} \qquad (9.15)$$

where V_e^{th} is the electron thermal velocity, the plasma is unstable to generation of Langmuir waves. This is the instability threshold for $T_e = T_i$, but for $T_e \gg T_i$, the plasma is unstable for

$$V_d^2 > C_s^2 = \frac{k(T_e + T_i)}{M} \tag{9.16}$$

where C_s is the ion acoustic speed. This threshold is much lower than (9.15). In this case, sound waves are generated and could lead to an anomalous resistivity.

However, it turns out that many other exotic plasma processes occur around and even below this threshold and it seems very likely that other nonlinear phenomena dominate the physics before anomalous resistivity comes into play. We cannot hope to do this nonlinear physics justice but end this section with data from the FAST satellite, illustrating the complexity of the parallel electric field acceleration zone (Frey et al., 2001). Figure 9.21 shows typical examples of electron acceleration by both upward (left-hand panel) and downward parallel

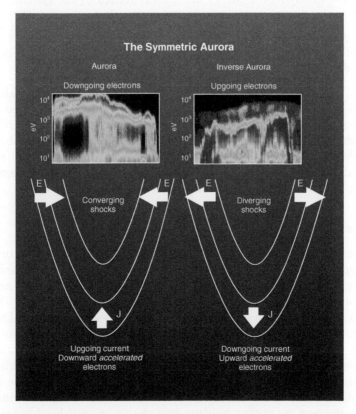

Figure 9.21 Energetic electrons and ions for opposite types of accelerating structures that have scales of about 3 km. (Figure courtesy of C. Carlson.)

electric fields. These parallel electric field components are accompanied by the large (nearly 1 V/m) perpendicular electric fields illustrated in the figure. As indicated by the closed potential contours, these large perpendicular fields do not map to the ionosphere below. Ions are accelerated in opposite directions by the parallel electric field.

It is important to note in these figures that $J_{||}$ and $E_{||}$ are such that $\mathbf{J}_{||} \cdot E_{||} > 0$. Applying to Poynting's theorem, consider a volume capped above the acceleration zone and below the E region and with sides parallel to **B**. Since the current closes in the E region, no Poynting flux escapes and the Poynting flux into the top must be equal to the Joule dissipation somewhere inside the volume. An interesting way to think about the auroral acceleration zone, then, is as follows. For some modest level of Poynting flux at the top, the ionospheric dissipation $\mathbf{J} \cdot \mathbf{E}' = \sigma_P |\mathbf{E}'|^2$ can cope with the input. In a sunlit ionosphere this is possible, since Σ_p is many mhos. At night some drizzle precipitation can also keep Σ_p elevated. But as the magnetosphere increases its energy input, something must give. Generation of a finite parallel electric field at high altitude has two great advantages for satisfying the energy dissipation needed by the system:

1. if $\mathbf{J} \cdot \mathbf{E} > 0$ exists in regions other than the E region and electrical energy is directly converted into mechanical energy in another volume; and
2. the resulting downward accelerated electrons increase Σ_p in the E region where Joule dissipation can increase.

As shown by both the barium ion injection experiments mentioned above and the satellite particle detectors (Shelley et al., 1976), ions are accelerated upward into the magnetosphere by the same $E_{||}$ accelerating electrons downward. This is an important ionospheric source of O^+ ions for magnetospheric plasma, which was thought to be entirely made up of H^+ ions for many years. In the next section we describe other sources of ionospheric particle injection into the magnetosphere, beginning at lower altitudes where O^+ is more prevalent.

9.4.2   Ion Outflows and Perpendicular Ion Acceleration

An important source of mass and energetic ions for the magnetosphere originates in the auroral F-region ionosphere. There appears to be three mechanisms for transporting energetic ions upward. Type 1 upflows are friction driven and operate in the presence of strong convection and ion-neutral collisions (St.-Maurice and Schunk, 1979). Evidence for this kind of upflow exists from satellite observations (St.-Maurice et al., 1976) and from ISR EISCAT observations (Lockwood et al., 1987). Type 2 ion upflows are always found in association with soft electron precipitation. In this case the soft electron precipitation heats the F region electron gas, which expands upward, carrying the ions along through an ambipolar electric field (Wahlund and Opgenoorth, 1989; Wahlund et al., 1992). In both cases the upflow is driven by a change in scale height of either the ion or electron gas and the observed upflow is at the transition between regions with two different scale heights. Since the gravitational binding energy of O^+ is about 9 eV, heating either the electron or ion gas to even a few eV will not eject O^+ into the magnetosphere. In fact, downflow due to a transition to a smaller ionospheric scale height is also observed.

Surveys of ion outflow below 1000 km have been performed using DE-2 (Loranc et al., 1991) and Hi-Lat (Tsunoda et al., 1989). The Loranc et al. study focused on the altitude range of 200–800 km altitude. They confirmed the prenoon cleft/cusp as the largest source of ion fluxes and found upward fluxes capable of supplying the cleft ion fountain. They also found upward drift velocities exceeding 250 m/s at altitudes above 600 km and smaller upward drift velocities at lower altitudes. The Tsunoda et al. study examined 76 orbits of Hi-Lat observations at 800 km altitude and found the mean ion outflow velocity to be 700 m/s. Furthermore, their events were characterized by upward field-aligned currents; intense, soft electron precipitation; and convection velocity shears. The DE-2 data were re-examined by Seo et al. (1997) in the altitude range of 850–950 km to investigate the details of ionospheric plasma parameters and soft electron precipitation. They found that the upward ion velocity correlated best with electron temperature ($r = 0.97$) and less well with ion temperature ($r = 0.91$), where r is the correlation coefficients. They concluded that soft electron precipitation and the associated electron heating, expansion, and ambipolar electric field were the most common sources for ion outflow, but that frictional heating contributes as well.

The third source of hot O^+ ions is wave-induced ion acceleration associated with field-aligned currents. This mechanism is easily identified by the existence of transversely accelerated ions, sometimes referred to as ion conics. The full analysis involves plasma physics beyond the scope of this text but a few comments are in order. Figure 9.22 shows rocket data from a flight over the dayside auroral zone (P. M. Kintner, personal communication, 2002). The rocket was launched from Norway, overflew the island of Spitzbergen, and landed in the Arctic Ocean. Initially the rocket was located on field lines that threaded the plasma sheet and which are most likely closed deep in the tail of the magnetosphere. This region is characterized by the 1–5 keV electron fluxes in the upper panel. Abruptly the payload enters the region of highly structured, intense 0.2–1 keV electrons and energetic ions that originated in the magnetosheath, the shocked dayside solar wind that connects directly to the ionosphere in the cusp. In the expanded plots below, we see intense O^+, H^+, and low frequency plasma waves in this region. Even the O^+ ions in this region are energetic enough to escape earth's gravitational pull. Furthermore, these field lines are swept over the polar cap and the oxygen ions certainly become part of the magnetosphere. This region is very likely a significant source of magnetospheric ions, many of which are oxygen.

The clear correlation of energized ions and low frequency waves shown in the lower panel suggests ion energization by waves. Current thinking is that a combination of ion acoustic, lower hybrid, and ion cyclotron waves are responsible for this energization. The latter two waves, in particular, preferentially accelerate

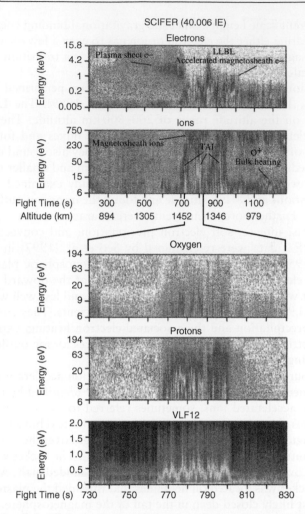

Figure 9.22 Panel 1: Flight overview of electrons from 0.005 to 15.8 keV. Note the boundary between closed and "open" field lines at about 710 s. Panel 2: Flight overview of ions from 6 to 750 eV. Panel 3: O^+ ions from 6 to 194 eV showing the major transversely accelerated ion (TAI) event at 765–800 s. Panel 4: H^+ ions from 6 to 194 eV showing the major TAI event at 765–800 s. Panel 5: Electric field (VLF 12) from 20 Hz to 2 kHz showing the major TAI event at 765–800 s. (Figure courtesy of P. M. Kintner.)

ions perpendicular to the magnetic field. Subsequent work suggests a connection with small-scale turbulence resulting from the decay of inertial Alfvén waves. The decay yields wave structures with scale lengths the order of an ion gyroradius, which can rapidly accelerate ions transversely (Klatt et al., 2005; Chaston et al., 2006).

9.5 Summary

In this chapter we have discussed a variety of phenomena observed in the high-latitude ionosphere. Knowledge of the current and prior convective history of the plasma and the magnetic field topology encountered by that plasma is essential to understanding or explaining many phenomena.

When considering plasma near the F peak, the local solar zenith angle, the time at which the flux tube encountered the auroral ionization source, the spatial variation of the drift velocity along the convection path, and the length of time the drift allows the plasma to reside in a production region must all be taken into account. The magnetic field topology is a minor consideration. The midlatitude trough, for example, can be understood in this manner. Above the F peak, the magnetic field topology is an important consideration, as is the plasma temperature. They will both affect the field-aligned motion of the plasma and its topside distribution. The light ion trough can be understood with these considerations. Progress toward a more complete understanding of the high-latitude ionospheric plasma will undoubtedly require multipoint measurements that provide some information about the time history of the plasma motion and the ionization sources it encounters. Some data sets of this nature are available from simultaneous measurements made by rockets, satellites, and ground-based instrumentation, but they must be integrated into a study that includes a versatile computer model in order to sort out the most important processes.

At the top of the ionosphere, dramatic particle acceleration zones exist. At 500-1000 km, ions are transversely accelerated far beyond escape speed, supplying the magnetosphere with hot oxygen ions. Above this height, near 5000 km, both ions and electrons are accelerated by parallel electric fields, forming the classic electron-induced aurora and precipitating oxygen ions and protons. The upward parallel fields also expel hot ions into the magnetosphere and create "black" aurora.

The effect of plasma dynamics on the neutral atmosphere can also be quite dramatic. Numerous measurements have shown extremely strong winds in the high-latitude thermosphere. Simultaneous plasma velocity measurements show clearly that these winds are driven by electric fields that are impressed from above through interactions between the earth's magnetosphere and the solar wind. This control of the earth's upper atmosphere by interplanetary processes is a fascinating example of the interaction between a flowing plasma and a planetary atmosphere.

References

Banks, P. M., and Holzer, T. E. (1969). Features of plasma transport in the upper atmosphere. *J. Geophys. Res.* 74, 6304.

Brinton, H. C., Grebowsky, J. M., and Brace, L. H. (1978). The high-latitude winter F region at 300 km: Thermal plasma observations from AE-C. *J. Geophys. Res.* 83, 4769.

Chaston, C. C., Genot, V., Bonnell, J. W., Carlson, C. W., McFadden, J. P., Ergun, R. E., Strangeway, R. J., Lund, E. J., and Hwang, K. J. (2006). Ionospheric erosion by Alfvén waves. *J. Geophys. Res.* **111**, A03206, doi:10.1029/2005JA011367.

Evans, D. S. (1968). The observations of a near mono-energetic flux of auroral electrons. *J. Geophys. Res.* **73**, 2315.

Foster, J. C., and Burke, W. J. (2002). SAPS: A new characterization for subauroral electric fields. *Eos Trans. AGU* **83**, 393.

Foster, J. C., Coster, A. J., Erickson, P. J., Rideout, W., Rich, F. J., Immel, T. J., and Sandel, B. R. (2005). Redistribution of the stormtime ionosphere and the formation of a plasmaspheric bulge. *Inner Magnetosphere Interactions: New Perspectives from Imaging*, Geophys. Monogram Series, American Geophysical Union Press, Washington, D.C., 277.

Frey, H. U., Mende, S. B., Carlson, C. W., Gérard, J.-C., Hubert, B., Spann, J., Gladstone, R., Immel, T. J. (2001). The electron and proton aurora as seen by IMAGE-FUV and FAST. *Geophys. Res. Lett.* **28**(6), 1135.

Fuller-Rowell, T. J., and Rees, D. (1981). A three-dimensional simulation of the global dynamical response of the thermosphere to a geomagnetic substorm. *J. Atmos. Terr. Phys.* **43**, 701.

Gonzales, C. A., Kelley, M. C., Carpenter, L. A., and Holzworth, R. H. (1978). Evidence for 3 magnetospheric effect on mid-latitude electric fields. *J. Geophys. Res.* **83**, 4399.

Haerendel, G., Rigger, E., Valenzuela, A., Föppl, H., Stenback-Nielsen, H. C., and Wescott, E. M. (1976). First observation of electrostatic acceleration of barium ions into the magnetosphere. *European Programmes on Sounding-Rocket and Balloon Research in the Auroral Zone*, Rep. ESA-SP115, European Space Agency, Neuilly, France.

Heppner, J. P., and Miller, M. L. (1982). Thermospheric winds at high latitudes from chemical release observations. *J. Geophys. Res.* **87**, 1633.

Hoffman, J. H., and Dodson, W. H. (1980). Light ion concentrations and fluxes in the polar regions during magnetically quiet times. *J. Geophys. Res.* **85**, 626.

Huang, C.-S., Foster, J. C., Goncharenko, L. P., Erickson, P. J., Rideout, W., and Coster, A. J. (2005). A strong positive phase of ionospheric storms observed by the Millstone Hill incoherent scatter radar and global GPS network. *J. Geophys. Res.* **110**, A06303, doi:10.1029/2004JA010865.

Killeen, T. L., Hays, P. B., Carignan, G. R., Heelis, R. A., Hanson, W. B., Spencer, N. W., and Brace, L. H. (1984). Ion-neutral coupling in the high latitude F-region: Evaluation of ion heating terms from Dynamics Explorer-2. *J. Geophys. Res.* **89**, 7495.

Kindel, J. M., and Kennel, C. F. (1971). Topside current instabilities. *J. Geophys. Res.* **76**, 3055.

Klatt, E. M., Kintner, P. M., Seyler, C. E., Liu, K., MacDonald, E. A., and Lynch, K. A. (2005). SIERRA observations of Alfvénic processes in the topside auroral ionosphere. *J. Geophys. Res.* **110**, A10S12, doi:10.1029/2004JA010883.

Lockwood, M., Bromage, B. J. I., Horne, R. B., St.-Maurice, J.-P., Willis, D. M., and Cowley, S. W. H. (1987). Non-maxwellian ion velocity distributions observed using EISCAT. *Geophys. Res. Lett.* **14**, 111.

Loranc, M., Hanson, W. B., Heelis, R. A., and St.-Maurice, J. P. (1991). A morphological study of vertical ionospheric flows in the high-latitude F-region. *J. Geophys. Res.* **96**, 3627.

Mikkelsen, J. S., Jorgensen, T. S., Kelley, M. C., Larsen, M. F., Pereira, E., and Vickrey, J. F. (1981a). Neutral winds and electric fields in the dusk auroral oval. 1. Measurements. *J. Geophys. Res.* **86**, 1513.

Mikkelsen, J. S., Jorgensen, T. S., Kelley, M. C., Larsen, M. F., and Pereira, E. (1981b). Neutral winds and electric fields in the dusk auroral oval. 2. Theory and model. *J. Geophys. Res.* **86**, 1525.

Mozer, F. S., and Bruston, P. (1966). Observation of the low latitude acceleration of auroral protons. *J. Geophys. Res.* **71**, 2201.

Mozer, F. S., Carlson, C. W., Hudson, M. K., Torbert, R. B., Parady, B., Yateau, J., and Kelley, M. C. (1977). Observations of paired electrostatic shocks in the polar magnetosphere. *Phys. Rev. Lett.* **38**, 292.

Parker, R. N. (1958). Dynamics of the interplanetary gas and magnetic fields. *Astrophys. J.* **128**, 664.

Raitt, W. J., Schunk, R. W., and Banks, P. M. (1975). A comparison of the temperature and density structure in high and low speed thermal proton flows. *Planet. Space Sci.* **23**, 1103.

Rishbeth, H., and Garriott, O. K. (1969), *Introduction to Ionospheric Physics*. Int. Geophys. Ser. **14**. Academic Press, New York.

St.-Maurice, J.-P., and Hanson, W. B. (1982). Ion frictional heating at high latitudes and its possible use for an in situ determination of neutral thermospheric winds and temperatures. *J. Geophys. Res.* **87**, 7580.

St.-Maurice, J.-P., and Schunk, R. W. (1979). Ion velocity distributions in the high-latitude ionosphere. *Rev. Geophys. Space Phys.* **17**, 243.

St.-Maurice, J.-P., Hanson, W. B., and Walker, J. C. G. (1976). Retarding potential analyzer measurement of the effect of ion-neutral collisions on the velocity distribution in the auroral ionosphere. *J. Geophys. Res.* **81**, 5438.

Schunk, R. W. (1977). Mathematical structure of transport equations for multispecies flows. *Rev. Geophys. Space Phys.* **15**, 429.

Schunk, R. W., and Nagy, A. (2000). *Ionospheres, Physics, Plasma Physics and Chemistry*. Cambridge University Press, Cambridge, UK.

Schunk, R. W., Raitt, W. J., and Banks, P. M. (1975). Effect of electric fields on the daytime high-latitude E and F regions. *J. Geophys. Res.* **80**, 3121.

Seo, Y., Horwitz, J. L., and Caton, R. (1997). Statistical relationships between high-latitude ionosphere F region/topside upflows and their drivers: DE 2 observations. *J. Geophys. Res.* **102**(A4), 7493–7500.

Shelley, E. G., Sharp, R. D., and Johnson, R. G. (1976). Satellite observations of an ionospheric acceleration mechanism. *Geophys. Res. Lett.* **3**, 654.

Sojka, J. J., Raitt, W. J., and Schunk, R. W. (1979). Effect of displaced geomagnetic and geographic poles on high-latitude plasma convection and ionospheric depletions. *J. Geophys. Res.* **84**, 5943.

Spiro, R. W., Heelis, R. A., and Hanson, W. B. (1978). Ion convection and the formation of the P90, mid-latitude F-region ionization trough. *J. Geophys. Res.* **83**, 4255.

Tsunoda, R. T., Livingston, R. C., Vickrey, J. F., Heelis, R. A., Hanson, W. B., Rich, F. J., and Bythrow, P. F. (1989). Dayside observations of thermal ion upwellings at 800 km altitude: An ionospheric signature of the cleft ion fountain. *J. Geophys. Res.* **94**, 15,277.

Vlasov, M., Kelley, M. C., and Kil, H. (2003). Analysis of ground-based and satellite observations of F-region behavior during the great magnetic storm of July 15, 2000. *J. Atmos. Solar-Terr. Phys.*, 1223–1234.

Wahlund, J.-E., and Opgenoorth, H. J. (1989). EISCAT observations of strong ion outflows from the F-region ionosphere during auroral activity: Preliminary results. *Geophys. Res. Lett.* **16**, 727.

Wahlund, J.-E., Opgenoorth, H. J., Häggström, I., Winser, K. J., and Jones, G. O. L. (1992). EISCAT observations of topside ionospheric ion outflows during auroral activity: Revisited. *J. Geophys. Res.* **97**, 3019.

10 Instabilities and Structure in the High-Latitude Ionosphere

The high-latitude sector is extremely rich in plasma instabilities and other processes that act to create structure in the ionosphere. Our primary interest lies in the horizontal variation of plasma density and electric fields. Since the magnetic field is nearly vertical, the horizontal structure is equivalent to variations perpendicular to the magnetic field. We use the generic term *structure*, since terms such as *waves* or plasma density *irregularities* conjure up specific sources. In fact, a very long list of processes contributes to the generation of horizontal structures in the plasma density and velocity field of the high-latitude ionosphere. It is essentially impossible to treat the topic in its entirety, since our understanding is developing very quickly. Our hope instead is to give a reasonable hint at the breadth of the phenomena involved and, within the various subtopics, to treat a few important processes in some detail. As was the case in earlier chapters, we start with the F region and follow with the E region in the second portion of the chapter.

10.1 Planetary and Large-Scale Structures in the High-Latitude F Region

In this section we review the physical processes that create horizontal variations in the high-latitude plasma density at the largest scales. We somewhat arbitrarily define the planetary scale to be larger than 500 km and large-scale processes to have perpendicular wavelengths in the range $30\,\text{km} \leq \lambda \leq 500\,\text{km}$. In these regimes production, loss, and transport dominate the ionospheric processes.

Ionospheric plasma instabilities are not very important in this scale size regime (although some local ionospheric processes do contribute at the lower end). Of course, this is not to say that *magnetospheric* plasma instabilities are unimportant, since they control some of the precipitation and turbulence in the flow patterns which are impressed on the ionosphere. Detailed study of the associated magnetospheric physics is beyond the scope of the present text, although we do refer in passing to some of the most important processes that occur.

The Earth's Ionosphere: Plasma Physics and Electrodynamics

10.1.1   Convection and Production as Sources of Planetary Scale Structure in the High-Latitude Ionosphere

We have already discussed variations of plasma flow and plasma density variations at planetary scales in Chapters 8 and 9. We briefly summarize the material here for completeness. The velocity field applied to the ionosphere from the solar wind-magnetosphere interaction has a number of discernible patterns at planetary scales. However, which of these patterns applies at a given time is highly dependent on conditions in the interplanetary medium. The most crucial parameter is the sign of the north-south component (B_z) of the interplanetary magnetic field (IMF). When the IMF is southward for any extended time (e.g., tens of minutes) the classic two-celled convection pattern is imposed on the ionosphere. The magnetic field lines that thread the high-latitude ionosphere spread out enormously as they recede from the earth. This means that the pattern in the ionosphere represents a focused version of the electrodynamic processes that create the electric field pattern. Unfortunately, the topological mapping that occurs is quite complicated, since some field lines are open and some are closed.

For steady B_z south, the two-cell ionospheric flow pattern is more or less fixed with respect to the sun-earth line. The earth and its neutral atmosphere rotate under this plasma flow pattern once a day (ignoring tidal forcing and acceleration of the neutrals by the plasma). This creates a planetary scale, diurnally varying plasma flow field in the ionosphere. Now, even with B_z held south, the classic symmetric two-celled flow shifts with respect to the sun-earth line as the other components of the IMF vary (particularly the component parallel to the earth's orbit, B_y). This shifting of the flow field can occur within one or two Alfvén travel times from ionospheric altitudes to the generator. Likewise as B_z and the velocity of the solar wind change, the rate of energy transfer to the magnetosphere varies from minute to minute and a flow field which is highly variable in space and time results.

When B_z changes sign, the major source of energy transfer ceases but other effects take over. A viscous interaction seems to create a small two-cell pattern and the connection of field lines to the IMF in regions of the magnetosphere far from the ecliptic plane also creates multiple cells with planetary scales (Burke et al., 1979).

To gauge the effect of these complex flow fields on plasma content we first need to discuss the processes which create and destroy plasma on planetary scales. The most important source is photoionization by sunlight. In a nonrotating frame this region is bounded by the terminator, which moves across the polar region on a seasonal basis. On the dark side of this line, recombination rapidly destroys plasma below 200 km but only very slowly erodes the F-peak region. The time constant is roughly 1 hour at 300 km.

Now when the planetary scale flow is imposed on this source and loss pattern, it is clear that solar plasma can be transported for vast distances into and clear through regions of total darkness before recombination can play much of a role.

In this way, planetary scale structuring of the plasma occurs which is much more complex and interesting than the simple terminator effect would be.

A more subtle process can create planetary scale depletions of plasma as well. Since the dipole magnetic axis of the earth is offset by $11°$ from the rotation axis and the plasma flow is organized by the magnetic geometry, in the winter time some convection patterns have flux tubes that are never illuminated by sunlight. Then, very deep plasma depletions can occur due to recombination, yielding peak plasma densities as low as 10^3 cm^{-3} with He$^+$ the dominant ion (see Fig. 9.1).

The two-cell convection pattern is associated with another planetary scale plasma source: impact ionization by particle precipitation in the auroral oval. From a visual perspective, this band of light around the polar region expands and thickens with increasing B_z southward and shrinks when B_z is northward. Much of the plasma in this oval is created so low in the atmosphere (≤ 200 km) that it is short-lived. Nonetheless, the lowest energy precipitating particles produce plasma high enough in altitude to create an important F-layer plasma source, particularly in winter.

10.1.2 Some Effects of Plasma Transport and Loss on the Large-Scale Horizontal Structure of the Ionosphere

One of the most fascinating aspects of the high-latitude ionosphere is its interaction with the various magnetospheric regions to which it is connected by magnetic field lines. In portions of the ionosphere which are not sunlit, the influx of precipitating particles is one of the dominant sources of the ionospheric plasma; the other major source is transport from either a sunlit region or a region where particles are precipitating. One might at first suspect that solar-produced plasma would display very little structure in the F region, since the ionization is long lived and the source is smoothly varying. However, since the flow field varies drastically in time and space, even solar-produced plasma may become horizontally structured. It is, in fact, difficult to separate particle precipitation zones from electric field patterns, since they are intimately related. We concentrate on the latter in this section and then follow with some comment on precipitation in the next section. As discussed in Chapters 8 and 9, perpendicular electric fields occur throughout the auroral zone and polar cap. Their existence results in transport of F region plasma from production zones to areas where one might not expect to find much plasma at all. A good example is found in the winter polar cap. F region plasma is not produced by sunlight at all in this region and there is only a very weak particle input, called "polar rain," when B_z is northward. However, observations show considerable structured plasma in the polar cap. The red light emission of atomic oxygen at 630 nm due to recombination is sufficiently large to make this plasma visible to image-intensified camera systems. In Chapter 4 we showed that localized low-density regions could be observed as an absence of these emissions and pointed out that these airglow depletions were due to equatorial spread F wedges or bubbles. At high latitudes it is the enhancements of plasma that are detected and which can be tracked by optical techniques. The all-sky camera photographs in Fig. 10.1 show just such a plasma patch marching across the field of view over Thule, Greenland. These structures are of the order of 1000 km across.

In Section 10.2 we study the plasma instabilities that may develop on the edges of these regions of enhanced plasma density, but here we are interested in their origin and lifetime. This is determined by both recombination and diffusion. As noted by Rishbeth and Garriott (1969) and discussed in Chapter 5, a Chapman-like F layer subject to diffusion, gravity, and loss will preserve its shape and decay at a rate given by $e^{-\beta(z_0)t}$, where $\beta(z_0)$ is the recombination rate at the altitude z_0 of the peak in the plasma density. The decay time constant is about 1 hour. Numerical calculations verify this behavior (Schunk et al., 1976; Schunk and Sojka, 1987). The typical dawn-dusk electric field across the polar cap with B_z south is 25 mV/m, which corresponds to 500 m/s velocity. In 1 h then a convecting patch can move 1800 km, just over 16° of latitude. Clearly, in such a rapid flow pattern both the dayside auroral oval and the sunlit dayside of the high-latitude zone can supply plasma to the dark polar cap regions with only modest loss due to recombination, gravity, and parallel diffusion.

Diffusion of plasma perpendicular to **B** is much more complex and interesting than parallel diffusion, and we discuss this in Section 10.3.2. A quick estimate, however, shows that large structures will easily survive transport over long distances without perpendicular diffusive loss. The time scale for perpendicular diffusion, τ_D, is given by

$$\tau_D = (k^2 D_\perp)^{-1}$$

For the F region the classical diffusion coefficient has an upper limit given by the ion diffusion coefficient $D_{i\perp}$, which has typical values in the range 10–50 m^2/s. Even for $D_{i\perp} = 50$ m^2/s, a 10 km wavelength structure has a time constant of 14 h. The evolution of large-scale features is therefore controlled

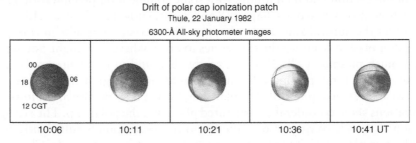

Drift of polar cap ionization patch
Thule, 22 January 1982

6300-Å All-sky photometer images

| 10:06 | 10:11 | 10:21 | 10:36 | 10:41 UT |

Figure 10.1 All-sky (155° field of view) 6300-Å images at 5-minute intervals illustrating large-scale patch structure and drift in the polar cap. The dawn-dusk and noon-midnight meridians are projected into the images at a height of 250 km. [After Weber et al. (1986). Reproduced with permission of the American Geophysical Union.]

by gravity, production, loss, and transport rather than classical perpendicular diffusion.

We still need to discuss how the patches shown in Fig. 10.1 are formed. We postpone for the moment structured sources of ionization to consider only electric field effects. For example, large-scale ($k \sim 1000$ km) organization of the F-region plasma can occur when the flow field has spatial variations in that same scale. Such plasma structure can even evolve out of a uniform sunlit region if different portions of the plasma move into dark areas at different speeds. As shown in Fig. 2.2, this advection effect can cause the local plasma density to vary if

$$\partial n/\partial t = -(\mathbf{V} \cdot \mathbf{\nabla}_{\perp} n) \neq 0 \tag{10.1}$$

even when the production and loss terms vanish. Equation (10.1) holds for an incompressible fluid, a valid approximation for the F region. For sunlit conditions $\mathbf{\nabla}_{\perp} n$ is determined by the solar depression angle, and the typical perpendicular gradient scale length $L_{\perp} = [(1/n)(dn/dx)]^{-1}$ is several hundred kilometers. In (10.1) $\mathbf{V} = \mathbf{E} \times \mathbf{B}/B^2$ where $\mathbf{E}$ is the applied electric field. A graphic illustration of plasma structuring by a flow field is shown in Fig. 10.2. Here an initially cylindrical plasma blob is placed in a spatially varying flow characteristic of the duskside auroral oval. With time, the structure becomes very elongated in the magnetic east-west direction. Since the flow is incompressible, the elongation results in a very short scale size in the meridional direction. If the flow field is turbulent, as it usually is, patches may be formed from eddies in the flow and be carried off by the mean flow into the polar cap.

Plasma loss through recombination can also create structure when coupled to a spatially varying flow field. This process was discussed in detail in Chapter 9, where the midlatitude plasma trough and the polar hole phenomena were shown to occur when the plasma remains for a long time on flux tubes with no sunlight and no particle precipitation. This can occur on flux tubes that flow west at the same speed as the earth rotates (the trough) or that circulate around a vortex that is entirely in darkness (the polar hole). Then even a slow decay rate is sufficient to deplete the plasma density.

The midlatitude nightside trough can become particularly deep during magnetic storms, when very large electric fields have been observed to build up at the equatorward edge of the plasma sheet-ring current system in the magnetosphere (Smiddy et al., 1977). This electric field points radially outward from the earth in the equatorial plane. Mapped to the ionosphere, this localized electric field is poleward and causes intense ionospheric flow toward the dusk terminator. In fact, the largest ionospheric electric field ever reported—350 mV/m—occurred in just such an event (Rich et al., 1980). Satellite data taken during such an event are reproduced in Fig. 10.3a. The top panels show the intense localized electric field (250 mV/m) near $L = 4$ (60° invariant latitude) with magnetic field, density, and energetic particle influx data plotted following. The large pulse corresponds to a potential drop of about 25 keV across the boundary. Notice that

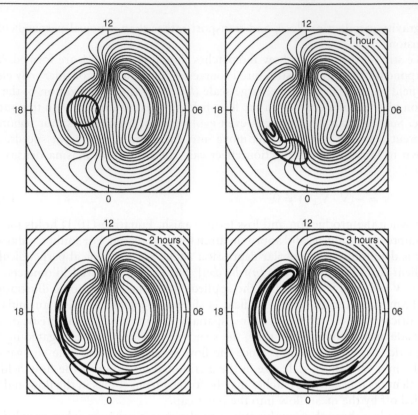

Figure 10.2 Distortion of a circular blob of ionization as it convects from the polar cap through the auroral zone. The first panel shows the initial conditions and the assumed convection model. [After Robinson et al. (1985). Reproduced with permission of the American Geophysical Union.]

the ionospheric plasma composition changes from H^+ at lower latitudes to O^+ at higher latitudes. At ionospheric altitudes the width of this high-flow region is between 150 and 200 km. These meridional scales are in the intermediate range but, through shear instabilities, can create larger scale longitudinal structures, as discussed in the next section. Modest small field-aligned currents were detected during this pass (Providakes et al., 1989).

These large poleward electric field events have come to be called subauroral ion drifts (SAID) due to their signature in drift meter instruments (Spiro et al., 1979). There are two ways to think about their origin. The combination of $E \times B$, gradient, and curvature drifts yield eastward zonal pressure gradients deep in the magnetosphere and thus radially inward currents. These currents close through poleward electric fields in the low conductivity nighttime subauroral ionosphere, creating large poleward electric fields. The second viewpoint is that these same factors cause ions to precipitate equatorward of electrons

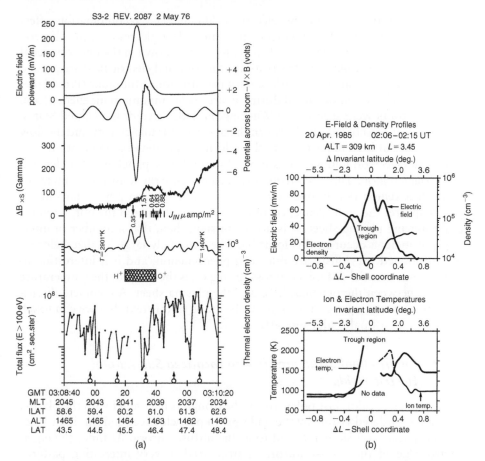

Figure 10.3 (a) Electric field, magnetic field, thermal electron, and energetic electron data obtained on the S3-2 satellite near the intense poleward electric field for revolution 2087 on May 2, 1976. The transition from H⁺ to O⁺ (shaded region) is coincident with the strong electric field. The calculated field-aligned current (J_{IN}) is displayed under the magnetometer trace with arrows indicating current into (downward arrow) or out of (upward arrow) the ionosphere. [After Rich et al. (1980). Reproduced with permission of the American Geophysical Union.] (b) Millstone Hill radar-measured electric field and electron density profiles at an altitude of 309 km (top) and corresponding electron and ion temperature profiles (bottom). Both plots use a coordinate system centered at the maximum in the electric field. [After Providakes et al. (1989). Reproduced with permission of the American Geophysical Union.]

(typically found in low-altitude satellite observations, e.g., Gussenhoven et al., 1983), yielding a poleward electric field. Although these viewpoints must in some sense be equivalent, most if not all theoretical models use the pressure gradient approach.

In the storm-time trough the electric field-enhanced recombination effect discussed in Chapter 9 then acts to help create a deep decrease in the electron density in this sector. An example of such an event is shown in the upper panel of Fig. 10.3b, where the plasma density and the meridional electric field are plotted from data taken during an azimuth scan of the Millstone Hill radar (Providakes et al., 1989). This event occurred during a large solar minimum magnetic storm in which auroras were seen in the Washington, D.C. area of the United States. There is a clear anticorrelation between the observed plasma density and the flow velocity that is at least in part due to the velocity-dependent recombination rate. The ion and electron temperatures plotted in the lower panel are also elevated in the trough region. This deep trough is in some sense the ionospheric image of the interface region that bounds two quite different plasma populations in the magnetosphere: the cool, dense, corotating plasmasphere and the hot, tenuous, rapidly moving plasma sheet. It is not surprising, then, that a number of interesting phenomena occur at this location. The extremely large localized electric fields mentioned previously are one such phenomenon, and it is interesting to note that the potential difference across the region is the order of the temperature difference between the plasmasphere and the plasma sheet. A complete discussion of this interesting region is beyond the scope of this text, but we can touch on a few topics of particular interest.

10.1.3 Longitudinal Structures Due to Localized Sub-Auroral Electric Fields

The ionospheric region just poleward of the trough is in contact with the plasma sheet and is rendered visible by the widespread particle precipitation that causes the atmospheric emissions referred to as the diffuse aurora. At times the equatorward edge of the diffuse aurora is structured in very interesting patterns (Lui et al., 1982; Kelley, 1986). Two photographs taken in visible light from the DMSP satellite (800 km altitude) that illustrate this effect are shown in Fig. 10.4a. Note that the equatorial edge of the diffuse aurora is scalloped in a highly nonlinear fashion. Typical observed wavelengths of this feature range from 200 to 800 km. This seems to be a clear example of ionospheric F-region structuring due to a magnetospheric process. The argument is as follows. Since the plasma sheet is the source of diffuse auroral precipitation, if the inner edge of the plasma sheet is distorted the light emissions and the production of plasma (see Section 10.1.3) in the ionosphere will mirror that distortion. Consequently, the F-region plasma will take on a horizontal structure with the same scale size. These relationships have been verified by simultaneous observations of the undulations by radar and optical means (Providakes et al., 1989). This longitudinal structure will be superimposed on the latitudinal structure of the trough itself. There is evidence that the undulations on the equatorward edge of the auroral oval seen in high-altitude photographs (e.g., Fig. 10.4) occur at the time when the large localized electric fields exist and that the latter drive

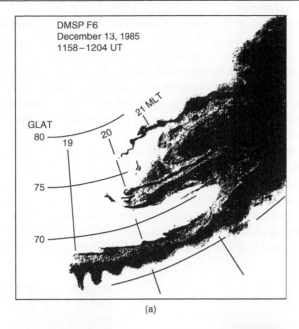

(a)

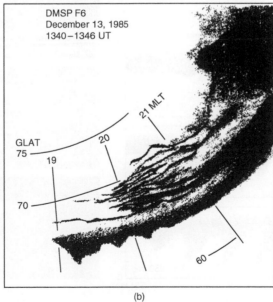

(b)

Figure 10.4a Outlines of auroral images photographed from the DMSP F6 satellite in the Northern Hemisphere at around (a) 1202 UT and (b) 1345 UT on December 13, 1985. In (a), giant (type 2) undulations are seen on the equatorward boundary of the diffuse aurora near 19 MLT. The undulations in (b) are thought to be type 2 undulations in the early stages of development. [After Yamamoto et al. (1994). Reproduced with permission of the American Geophysical Union.]

some magnetospheric instability (e.g., Kelvin-Helmholtz) that causes undulations (Kelley, 1986). Yamamoto et al. (1994) have simulated this process with the results shown in Fig. 10.4b. The similarity with the data is striking. Another suggestion is that kinetic drift waves driven by the ion temperature gradient in the magnetosphere are the source of these features (Lewis et al., 2005). The latitudinally elongated smaller-scale structures of sub-arcs in the bottom panel of Fig. 10.4a suggest a set of thin field-aligned current sheets coming in and going out of the ionosphere.

Radar observations of plasma at the poleward edge of the ionospheric trough also show that it is often structured. One example of the latitudinal structure of this region is given in Fig. 10.5, which shows the development and meridional motion of enhancements in the F-region plasma density. Initially there is a dense E-region layer beneath the F-region structure, indicating that it is collocated with precipitating electrons. As the plasma blob drifted equatorward, the plasma density became quite low in the E region below it. A plasma trough separates

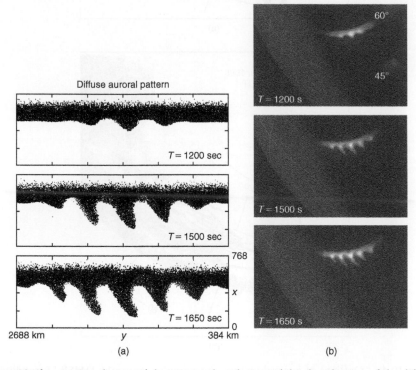

(a)   (b)

Figure 10.4b   (a) Simulation of the temporal evolution of the distribution of the diffuse aurora on the poleward edge of the trough. (b) A three-dimensional picture of giant undulations with imitative color, obtained by projecting the computer-produced auroral pattern (a) onto the globe surface. [After Yamamoto et al. (1994). Reproduced with permission of the American Geophysical Union.] See Color Plate 27.

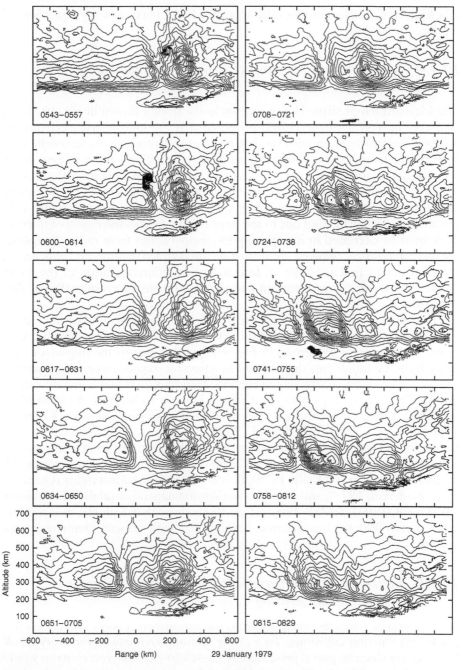

Figure 10.5 Sequence of radar electron density contour maps from 0543 to 0829 UT. January 29, 1979. [After Weber et al. (1985). Reproduced with permission of the American Geophysical Union.]

this high-latitude plasma from the solar-produced plasma in the equatorward region.

10.1.4 Temperature Enhancements in the Trough and Stable Auroral Red Arcs

The localized temperature enhancements in the trough shown in Fig. 10.3b are also very interesting. Cole (1965) proposed that the ionospheric electrons are in good thermal contact with the hot magnetosphere along the magnetic field lines. In the trough, where the electron density is low, this input of heat raises the electron temperature more than, say, in the region poleward of the trough, where the heat input is also large but the electron density is high. The enhanced ion temperatures may be explained by a combination of thermal exchange with electrons and Joule heating due to the large electric field in the region of interest. Care must be exercised in interpreting ion temperatures from radar data, since the ion composition changes, shear flow on the sampled volume, and anisotropic ion distribution functions can all lead to incorrect interpretation of the width of the incoherent scatter spectrum and the T_i estimate that results (Providakes et al., 1989).

The high electron temperature in the trough leads to the phenomenon of stable auroral red (SAR) arcs. The emission is $O(^1D)$, the same as that which occurs in recombination as illustrated in Fig. 10.1, and is due to the impact of electrons from the tail of this distribution function on oxygen atoms. Very intense events can be visible to the naked eye (M. Mendillo, personal communication, 2005). These emissions usually reach from horizon to horizon. An example is presented in Fig. 10.6a, using a camera located in western New York State at approximately $L = 3.5$ (Nicolls et al., 2005). The red arc stretches from horizon to horizon but is not as unstructured as previous publications seem to imply. Possible explanations for the undulations on the poleward edge include the $\nabla n \times \nabla T$ instability (Hudson and Kelly, 1978), the current convective instability (Ossakow and Chaturvedi, 1979), and possibly the thermomagnetic instability (Erukhimov and Kagan, 1994). DMSP data in Fig. 10.6b show that high electron temperature (over 6000 K), low electron density, and high velocity characterize the red arc.

10.1.5 Horizontal Plasma Variations Due to Localized Plasma Production and Heating

When energetic particles precipitate into the atmosphere, energy loss occurs primarily via ionizing collisions. To a crude first approximation, for every 35 eV lost, one ion-electron pair is produced. A single kilo-electron-volt electron would thus yield about 30 pairs. The type of ion produced depends on the altitude where the collision occurs, since the atmospheric constituents vary with height. The initial particle energy determines how deeply it can penetrate into the atmosphere.

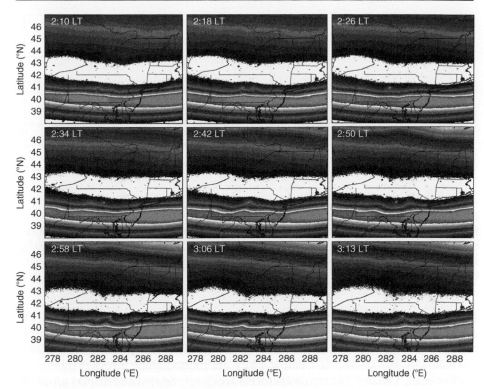

Figure 10.6a   Series of SAR arc images observed from Ithaca, NY, on October 28–29, 2000. The arc is centered at 40° geographical latitude, while the edge of the diffuse aurora is near 43°N. [After Nicolls et al. (2005). Reprinted with permission of the Institute of Electrical and Electronics Engineers, Inc. (© 2005, IEEE).] See Color Plate 28.

For F-region physics we are most concerned with "soft" particles—that is, electrons with energy less than about 500 eV—since they deposit their energy at the highest altitudes. Vertical transport by diffusion is crucial, however, since even soft electrons produce plasma at relatively low altitudes. Electron heat conduction is also very important because secondary electrons are much more energetic than the ionospheric electrons and may heat the latter quickly, depending on the ratio of two electron populations. This can create a redistribution of plasma at high altitudes which mirrors the horizontal precipitation pattern.

As discussed briefly in Chapter 1, soft electrons most copiously precipitate in the dayside cusp region, where magnetosheath plasma is in direct contact with the ionosphere. This occurs in the dayside auroral oval at invariant latitudes in the range 70–74°. Our interest here is not so much in the ionization process, which is more the topic of classical aeronomy (see, for example, Banks and Kockarts, 1973), as in the resulting horizontal variation in plasma content and structure. Strong evidence that the equatorward boundary of structured plasma in the

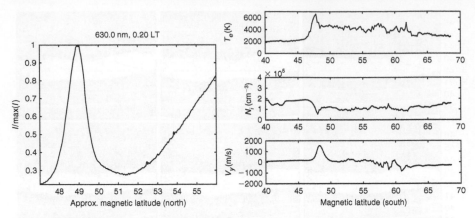

Figure 10.6b Cut across one of the images (left) along with DMSP-measured electron temperature, ion density, and cross-track drift (right). [After Nicolls et al. (2005). Reproduced with permission of the Institute of Electrical and Electronics Engineers, Inc. (© 2005, IEEE).]

dayside cusp region is collocated with the equatorward boundary of 300 eV electron flux was given by Dyson and Winningham (1974). Rocket data from the cusp region are even more striking. In Fig. 10.7 the plasma density and the total electron energy flux below 1 keV are plotted along a rocket trajectory (Kelley et al., 1982b). The two waveforms are clearly related and display an outer scale in the range $\lambda_\perp = 75\text{--}100$ km. The plasma spectrum of the rocket data in Fig. 10.8 exhibits an outer scale of 75 km and a power law for smaller scales. For $k_0 \leq (2\pi/30)$ rad/km, the power spectrum displays a power law dependence of the form, $k^{-1.9}$.

This good correlation between plasma density and electron flux is at first glance somewhat surprising, since the altitude of the observation is about a hundred kilometers higher than the altitude where most of the ionization occurs. This correlation, however, has been confirmed in another dayside auroral oval rocket experiment using the results of several independent plasma density instruments, so there is no question of contamination in the probe response. The particle flux and plasma density data from this latter flight are shown in Fig. 10.9 and clearly show a collocation of structured F-region plasma and particle precipitation zones. Roble and Rees (1977) have made a time-dependent model for the soft electron flux case in which the typical energy is the order of 100 eV. We may use this calculation to investigate the time scale required to build up a density increase of 1.5×10^5 cm^{-3} at about 375 km as required by the data in Fig. 10.7. Roble and Rees found a local production rate of 300 cm^{-3} s^{-1} for a total energy input of 1 erg/cm$^2 \cdot$ s. For 5 ergs, then, as indicated by the particle detector data, the rate will be about 1.5×10^3 cm^{-3} s^{-1}. Roughly 100 s is therefore required. The enhancements in the data of Fig. 10.9 have been quantitatively modeled

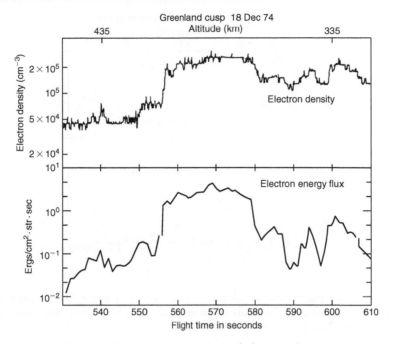

Figure 10.7 Simultaneous in situ measurements of electron density and precipitating auroral energy flux in the polar cusp. [After Kelley et al. (1982a). Reproduced with permission of the American Geophysical Union.]

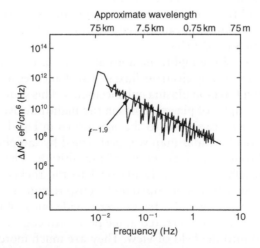

Figure 10.8 Spectrum of irregularities obtained from analysis of the electron density measurements shown in Fig. 10.7. [After Kelley et al. (1982a). Reproduced with permission of the American Geophysical Union.]

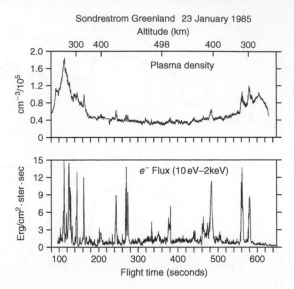

Figure 10.9   Local plasma density along a rocket trajectory in the dayside auroral oval along with precipitating electron fluxes. [After LaBelle et al. (1989). Reproduced with permission of the American Geophysical Union.]

by LaBelle et al. (1989) assuming a steady electron flux for about 2 minutes. Auroral arcs are very dynamic but this time scale is not at all unreasonable in the cusp, and we conclude that structured soft electron input can create a very irregular F-layer plasma in the dayside auroral oval. Conversely, the hard spectrum studied by Roble and Rees (1977), which is typical in the heart of the nighttime auroral oval, had a production rate of only $100 \, cm^{-3} \, s^{-1}$ at 400 km even for a total energy input of 8 ergs/$cm^2 \cdot$ s.

The poleward edge of the nighttime auroral oval, on the other hand, is also characterized by intense soft electron fluxes (Tanskanen et al., 1981) that are capable of producing F-region plasma enhancements. This source may be related to the numerous examples of plasma blobs in the nighttime oval detected by the Chatanika radar which are similar to the one shown in Fig. 10.10a. A spectral analysis of this particular radar map was performed by sampling the data with a horizontal cut through the data set at 350 km altitude. The power spectrum is given in Fig. 10.10b and is nearly identical to the rocket power spectrum shown in Fig. 10.8, which was obtained in the cusp region. These plasma blobs convect into and through the radar beam over a wide range of local times. Such enhancements could have been created at the poleward edge of the auroral oval and then convected into the field of view. They are much more common in the Alaskan sector at solar maximum, which may be due to the larger scale height of the neutral atmosphere and the corresponding increase in F-region plasma production when soft particles precipitate (J. Vickrey, personal communication, 1982).

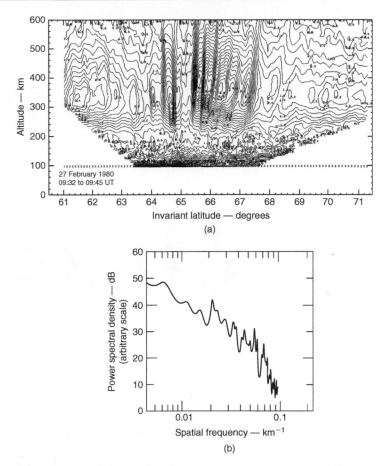

Figure 10.10 (a) Altitude/latitude variation of electron density in the midnight sector auroral zone measured by the Chatanika radar. (b) Spectrum of electron density irregularities obtained by analyzing the radar measurements of the latitudinal variations of electron density at 350 km altitude shown in (a). In this plot the x-axis corresponds to $1/\lambda$, not $2\pi/\lambda$, as is usual in this text. [After Kelley et al. (1982a). Reproduced with permission of the American Geophysical Union.]

We turn now to a discussion of polar cap aurora. When B_z is northward, the auroral oval contracts and particle precipitation are much reduced in the auroral zone. However, auroral activity actually increases in the polar cap region during B_z north conditions as illustrated in Fig. 10.11a. There seem to be two basic states of the polar cap when B_z is north, an ordered multicell flow pattern and a chaotic flow (see Chapter 8 for details). In the ordered case, a very long sun-aligned auroral arc can form in the polar cap. An example is shown in Fig. 10.11b, which is a photograph of the polar region taken on the *Dynamics Explorer 1* satellite. The circular auroral oval is linked from day to night by a sun-aligned

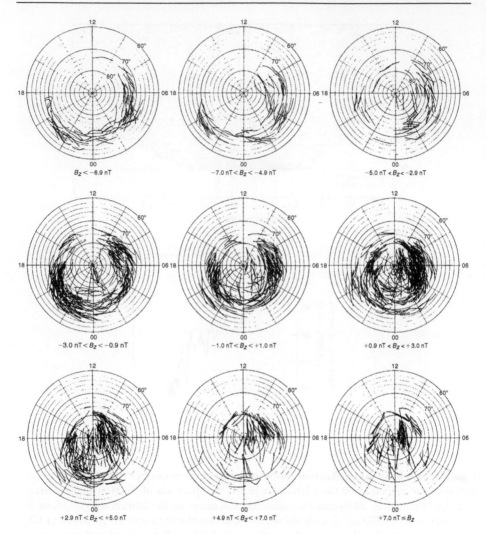

Figure 10.11a   Mass plots of arcs in corrected geomagnetic coordinates for different values of B_z. [After Lassen and Danielsen (1978). Reproduced with permission of the American Geophysical Union.]

auroral emission. The entire feature is referred to as a theta aurora. The particle energies associated with these auroras are in the "soft" category (<100 eV), and both the plasma production and auroral airglow emissions therefore occur at lower F-region or upper E-region altitudes.

Other experiments show that sun-aligned arcs seem to occur in a region of shear in the plasma flow. In fact, the data suggest that electrons precipitate in a region where the flow vorticity has a positive sign (Burke et al., 1982). This

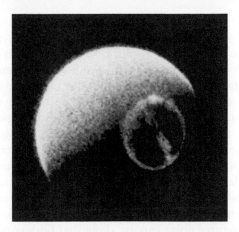

Figure 10.11b Image of the auroral oval and a transpolar arc in the southern hemisphere at 0022 UT on May 11, 1983. The combination of continuous luminosities around the auroral oval and a transpolar arc can produce a pattern of luminosities which, when viewed from polar latitudes, is reminiscent of the Greek letter θ. Such auroral distributions are generally referred to as "theta" auroras. The dayglow and auroral emissions observed in this image are predominantly form neutral atomic oxygen at 130.4 and 135.6 nm. (Figure courtesy of L. A. Frank, J. D. Craven, and R. L. Rairden, University of Iowa.)

may be understood as follows. The flow vorticity, a measure of velocity shear, is defined as the curl of the velocity field:

$$\mathbf{W} = \nabla \times \mathbf{V} \tag{10.2}$$

As usual, let the z direction be downward parallel to the magnetic field and the $\hat{a}_x$ direction along the arc. For $E_z = 0$, the plasma flow velocity $\mathbf{E} \times \mathbf{B}/B^2$ is independent of z. If we consider a flow field parallel everywhere to the arc and in the direction $\hat{a}_x$, then the vorticity has only a z component,

$$\mathbf{W} = -(\partial V_x/\partial y)\hat{a}_z = W_z\hat{a}_z \tag{10.3a}$$

Suppose for simplicity that we take the height-integrated Pedersen conductivity to be uniform in the horizontal plane. This would occur for a weak arc or for an arc in the sunlit hemisphere. Then, because $\mathbf{V} = \mathbf{E} \times \mathbf{B}/B^2$, we have

$$J_z = \Sigma_P(\partial E_y/\partial y) = \Sigma_P B(\partial V_x/\partial y)$$

and thus

$$J_z = -\Sigma_P B W_z \tag{10.3b}$$

This equation shows that J_z is out of the northern hemisphere ionosphere when W_z is positive and into the ionosphere when W_z is negative. Since precipitating electrons carry current away from the ionosphere, this relationship suggests that an arc should occur when the vorticity is positive. This dependence has indeed been reported for polar cap arcs by Burke et al. (1982). It is of interest also to note that, from Poisson's equation, this result is equivalent to the statement that electron precipitation occurs into regions of net negative charge density (see Section 2.4), which seems counterintuitive but is correct for the usual case that the ionosphere acts as an electrical load. The multitude of polar arcs shown in Fig. 10.11a suggests a turbulent flow with many sign changes of the vorticity. This vorticity in the magnetospheric flow or solar wind creates charge separation in a low impedance source, which leads to field-aligned currents. These currents then become unstable at some altitude, which results in particle acceleration via parallel electric fields. The experimental situation is complicated by the conductivity gradients which are created by the particle precipitation. In fact, detailed observations near a winter polar cap arc show that both conductivity gradients and structured electric fields are important in the horizontal current divergence (Weber et al., 1989).

So far we have discussed the production of F-region plasma in the cusp (dayside oval), in the polar cap, and at the edge of the nightside auroral oval. In the heart of the nightside oval the situation is very chaotic, due in part to the role of substorm activity in the midnight sector. The series of *Dynamics Explorer 1* images of the auroral oval in Fig. 10.12 shows some of the dynamical features in a typical substorm. As a crude first approximation there are three classes of oval precipitation, which may be described by their optical and plasma signatures. The diffuse aurora is characterized by a widespread, nearly uniform particle influx from the plasma sheet. Since the diffuse aurora mirrors the hot plasma sheet, it is also characterized by the same flow pattern, which is roughly zonally westward before midnight and zonally eastward after midnight. Imbedded in this plasma are regions of discrete auroral arcs that are usually aligned east-west and are often associated with potential drops along the magnetic field which accelerate electrons into the atmosphere. This acceleration zone has been located by satellite and rocket techniques at altitudes ranging from 2000 to 8000 km. This process occurs throughout the auroral oval. The auroral arcs in the dayside oval are associated with acceleration zones having several hundred to a thousand volts of potential drop. In the midnight sector the potential is higher and most of the plasma in this nighttime sector is produced by accelerated electrons striking neutrals in the E region. The intense E-region plasma density increase shown in Fig. 1.5 is of this type. Quiet auroral arcs are usually east-west aligned, that is, much smaller in latitudinal extent than in longitude. The active auroral forms associated with substorms comprise a third rough class of auroras.

Other than the obvious comment that more plasma is produced in a bright aurora, active auroral forms are often extremely contorted. The examples in Fig. 1.4 and Fig. 10.12 show this very clearly. The plasma density produced by particle precipitation in the E and F regions will then also be quite structured. So many factors contribute to the horizontal structuring in the nightside oval

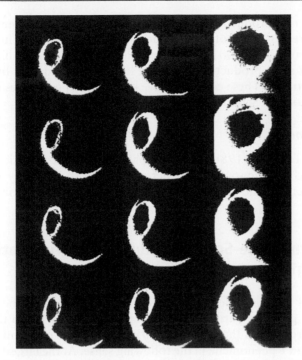

Figure 10.12 Twelve consecutive images at ultraviolet wavelengths 123–160 nm record the development of an auroral substorm. The sequence begins at 0529 UT on April 2, 1982 (bottom left image), as the NASA/GSFC spacecraft *Dynamics Explorer 1* first views the auroral oval from the late evening side of the dark hemisphere at low northern latitudes near apogee (3.65 earth radii altitude) and then from progressively greater latitudes as the spacecraft proceeds inbound over the auroral oval toward perigee. In each panel, time increases upward and from left to right. The poleward bulge at onset of the auroral substorm is observed beginning at 0605 UT (fourth frame at upper left). In successive 12-minute images the substorm is observed to expand rapidly in latitude and longitude. (Figure courtesy of L. A. Frank, J. D. Craven, and R. L. Rairden, University of Iowa.)

that it is often difficult to separate out cause and effect in that sector. In some cases the character of the density profile itself gives a clue to the origin or at least the time history of the plasma. For example, if the E-region density is high, production must be occurring simultaneously, since recombination is very fast at low altitudes. Likewise, if electron temperature data are available (for example, from an incoherent scatter radar), they may be used to determine the history of F-region plasma. If precipitation is occurring, T_e will be well elevated over the ion and neutral temperatures. If the long-lived F-region plasma has been produced elsewhere and has been advected into the field of view, however, the electron temperature will have had time to equilibrate with the other gases and a low T_e is expected.

We conclude that production and heating of plasma by long-lived soft particle precipitation creates much of the F-region horizontal structure in the auroral oval at scales from 30 to 100 km. Larger-scale features are more likely due to a combination of solar production, chemical losses, and transport by large-scale electric field patterns. These concepts have been verified in a series of computer model calculations by Schunk and Sojka (1987).

10.1.6   Summary

We agree with Tsunoda (1988) that ionospheric plasma instabilities per se have very little to do with ionospheric structure at large to planetary scales. Convection, production, and loss are sufficient to create a marvelously complex ionosphere.

10.2   Intermediate-Scale Structure in the High-Latitude F Region

The ordering of high-latitude plasma at scales above 30 km has been described in detail in the previous part of this chapter. Here we look into the processes which create structure in the range 0.1–30 km. There is a close analogy here to equatorial spread F phenomena in the following sense. At the largest scales in the equatorial case aeronomical processes such as production, recombination, and electrodynamic transport act in concert with neutral wind phenomena such as gravity waves to create the large-scale patterns. The intermediate-scale structuring then proceeds primarily via the generalized Rayleigh-Taylor instability. At high latitudes, we have seen that aeronomical processes are further supplemented by structured particle precipitation and structured large-scale flow patterns. It should not be surprising then that intermediate-scale structuring in the auroral zone and polar cap is also more complicated than at the magnetic equator. In a turbulent neutral fluid, viscosity eventually limits the scale sizes where structure occurs to values larger than some wavelength corresponding to the so-called viscous cutoff. In a plasma there are many more degrees of freedom in the system and new sources of structure may arise at various wavelength scales in the medium. In Sections 10.2.1 and 10.2.2 we concentrate on mechanisms that generate structure at intermediate scales, scales that lie between the size of the auroral oval and the size of auroral arcs. Since the linear theory for plasma instabilities in this range appeals to diffusive damping as a limiting mechanism, we spend some time discussing cross-field plasma diffusion and images in Section 10.2.3.

10.2.1   The Generalized $E \times B$ Instability at High Latitudes

Before delving into the differences, some similarities to equatorial spread F are to be noted. Figure 10.13 shows three sets of plasma density patterns detected along

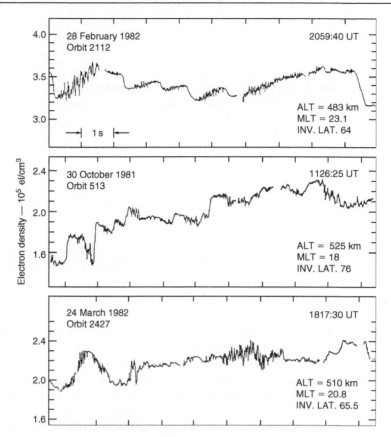

Figure 10.13   Examples of the preferential structuring of high-latitude plasma density enhancements on gradients of a particular sign. [Adapted from Cerisier et al. (1985). Reproduced with permission of the American Geophysical Union.]

the flight path of the *AUREOL-3* satellite (Cerisier et al., 1985). Notice that the structure seems to occur preferentially on only one sense of the gradients. In Fig. 10.14 we have turned one of the vertical equatorial spread F-rocket profiles on its side for comparison with the horizontal satellite data in Fig. 10.13. The similarity is striking in that both profiles show a strong tendency for irregularity development on a preferred direction of the density gradient. The equatorial case is more violent with many irregularities pushing through to the stable side of the gradient.

Gravity plays very little role in high-latitude phenomena, since **g** is essentially parallel to **B**. We therefore concentrate our analysis on the electric field-driven aspects of the instability in this chapter. We refer to the process as a generalized **E** × **B** instability and include neutral winds, electric fields, and field-aligned

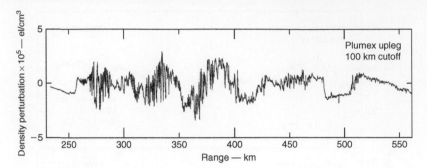

Figure 10.14   A vertical plasma density profile during equatorial spread F "tipped on its side" to compare with Fig. 10.13. [After Kelley et al. (1982a). Reproduced with permission of the American Geophysical Union.]

currents. As noted earlier, when $(\mathbf{E}' \times \mathbf{B})$ has a component parallel to ∇n—that is, when

$$(\mathbf{E}' \times \mathbf{B}) \cdot \nabla n > 0 \tag{10.4}$$

the system is unstable. Referring back to Fig. 4.9a, we see that in this case perturbation electric fields develop when a Pedersen current flows perpendicular to the zero-order density gradient in the presence of a small disturbance. In effect, a high-density region polarizes in such a way that it has a slower drift velocity than the background plasma. A high-density region therefore lags behind and seems to grow with respect to the background density as it drifts "down" the gradient. Low-density regions move in the opposite direction and seem to grow with respect to the background. Indeed, based on the *AUREOL-3* satellite measurements of the electric field, the event in the upper panel of Fig. 10.13 was such that the instability condition in (10.4) was satisfied, since a 12 mV/m eastward zonal electric field component was measured and the density gradient was almost certainly poleward. We say "almost certainly" because only one component of ∇n is measurable from a polar-orbiting satellite. Cerisier et al. (1985) have studied six events of this type and found four of them to have (10.4), to the satisfied, to the best of their knowledge, given the ambiguity in ∇n. It seems clear that the $\mathbf{E} \times \mathbf{B}$ process can and does occur in the high-latitude sector, and most of the discussion of the generalized Rayleigh-Taylor instability in Chapter 4 is directly applicable at high latitudes. However, there is one important caveat to raise. The satellite measures only $\mathbf{E}$, not $\mathbf{E}' = \mathbf{E} + \mathbf{U} \times \mathbf{B}$, and the role of the neutral wind is just not known in most experiments.

Usually one argues that because $|\mathbf{E} \times \mathbf{B}/B^2| > |\mathbf{U}|$ at high latitudes, the neutral wind plays only a minor role. However, at the boundaries of the auroral oval, where the plasma flow changes direction, electrodynamically driven neutral winds will not change direction as quickly as the plasma flow and could act as a source for instability. In the midnight sector, for example, we expect a southward neutral wind driven by ion drag in the polar cap. At the convection boundary E is in the meridian plane and thus plays no role in the stability of the poleward plasma density gradient which often exists at that boundary. However, if, as argued previously, U is in the southward direction, $\mathbf{E}' \times \mathbf{B}$ is poleward, and thus the poleward wall of the trough could be unstable to the generalized $\mathbf{E} \times \mathbf{B}$ process.

Concerning details of the $\mathbf{E} \times \mathbf{B}$ instability, there are some aspects of this process that differ from its development at equatorial and middle latitudes. Tsunoda (1988) has written an excellent review of these phenomena and we only hit the high points here. One feature involves the relatively high E-region Pedersen conductivity that exists at high latitudes. This affects both the growth rate and the loss rate of plasma structure. For example, a simple expression (Vickrey and Kelley, 1982) for the linear growth rate of the one-dimensional $\mathbf{E} \times \mathbf{B}$ instability for waves perpendicular to the gradient is of the form

$$\gamma = \frac{E_0'}{BL} \left(\frac{M-1}{M} \right) - k^2 D_\perp \qquad (10.5)$$

where E_0'/B is the component of $\mathbf{E}' \times \mathbf{B}/B^2$ parallel to ∇n, L the inverse gradient scale length, $D_\perp$ the height-averaged perpendicular diffusion coefficient, and $M = \left(\Sigma_P^E + \Sigma_P^F \right)/\Sigma_P^E$. In the definition of M, Σ_P^F is the field-line-integrated Pedersen conductivity in the F region and Σ_P^E is the field-line-integrated Pedersen conductivity in the E region. If $\Sigma_P^E > \Sigma_P^F$, which is almost always the case in the auroral oval due to particle precipitation, the E region tends to short out the perturbation electric field, δE, produced in an $\mathbf{E} \times \mathbf{B}$ instability and the growth rate becomes small. This is easy to understand because "growth" in the case of the $\mathbf{E} \times \mathbf{B}$ instability is just due to advection of high-density plasma down a gradient and low-density plasma up a gradient. The rate of change of the density is given by

$$\partial n/\partial t = -\delta \mathbf{V} \cdot \nabla n \cong \gamma \delta n \qquad (10.6)$$

where $\delta \mathbf{V} = \delta \mathbf{E} \times \mathbf{B}/B^2$. If the perturbation charges that set up δE are shorted out by current flow to the E region, the growth rate becomes small. The diffusive damping term also depends on the E region but in a subtle fashion that we postpone until Section 10.2.3. Tsunoda (1988) has compared M values for the auroral case with those for equatorial spread F and for large midlatitude barium releases. For equatorial spread F, he quotes values in the range 10 to 10^4, while for low-altitude barium releases M ranges from 60 for a 1 kg release to 470 for a 48 kg release. Auroral F-region plasma enhancements, on the other hand, have M values less than 1.2 for reasonable E-region densities $\left(\Sigma_P^E = 5 \text{ mho} \right)$. Even for nighttime conditions with no E layer $M \leq 10$ for F-region densities as high as 10^6 cm^{-3}. On this basis alone it is clear that local plasma instabilities may be less dominant in the auroral zone than at other latitudes.

The shorting process discussed previously is accomplished by field-aligned currents due to the E region, which shorts out both the driving fields, δE, and the ambipolar electric fields (see Section 10.2.3) that limit diffusion across **B**. In addition to these structure-related currents, there are also large-scale field-aligned currents that link the magnetospheric and solar wind generators to the ionospheric load and that were discussed in some detail in Chapter 8. Here we investigate the role that such applied currents play with regard to the $E \times B$ instability. When field-aligned currents are included, the process is referred to in the literature as the current convective instability (Ossakow and Chaturvedi, 1979).

In brief, when the ion and electron species have different drift velocities along the magnetic field line (and thus a net parallel current is carried by the thermal plasma), the $E \times B$ instability is modified in such a way that even plasma gradients stable to the $E \times B$ instability may in principle be unstable to the current convective process. In the case of the pure $E \times B$ process the most unstable wave has $k_{\parallel} = 0$. For the current convective instability to operate, a finite $k_{\parallel}$ is required.

To understand why this instability occurs we isolate the effect of an upward (northern hemisphere) field-aligned current acting alone as shown in Fig. 10.15. The finite $k_{\parallel}$ is exaggerated to make the effect easier to illustrate. We assume a density perturbation of the form

$$\tilde{n}(s, t) = \delta n' e^{i(ks - \omega t)}$$

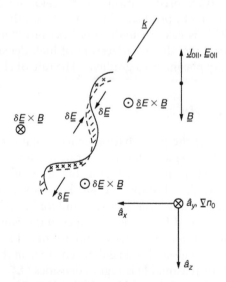

Figure 10.15 Electrostatic wave with finite (exaggerated) $k_{\parallel}$ propagating in the presence of a field-aligned current. If a zero-order density gradient exists pointed into the page the wave is unstable.

where s is a distance measured in the direction parallel to $\mathbf{k}$. The perturbation vector $\mathbf{k}$ is confined to the x-z plane, while ∇n is in the y direction. Because both elements σ_0 and σ_P of the tensor conductivity are proportional to the density, it follows that they will depend on s in a similar manner,

$$\sigma_0(s, t) = \sigma_0 + \delta\sigma_0 e^{i(ks-\omega t)}$$

$$\sigma_P(s, t) = \sigma_P + \delta\sigma_P e^{i(ks-\omega t)}$$

Our goal is to determine the perturbation electric field δE_k that is parallel to $\mathbf{k}$, since we are dealing with an electrostatic wave,

$$E_k(s, t) = \delta E_k e^{i(ks-\omega t)}$$

We now solve for δE_k in terms of the initial perturbation in δn by setting $\nabla \cdot \mathbf{J} = 0$. In the x-z plane,

$$\mathbf{J} = \sigma_0 E_z \hat{a}_z + \sigma_P E_x \hat{a}_x$$

For now, we take the zero-order perpendicular electric field to be zero. Since the zero-order parallel electric field is upward in the case illustrated by Fig. 10.15, we have

$$\mathbf{J} = \left(\sigma_0 + \delta\sigma_0 e^{i(k_x x + k_z z - \omega t)}\right)\left(-E_{\|} + \delta E_k \sin\theta e^{i(k_x x + k_z z - \omega t)}\right)\hat{a}_z$$

$$+ \left(\sigma_P + \delta\sigma_P e^{i(k_x x + k_z z - \omega t)}\right)\left(\delta E_k \cos\theta e^{i(k_x x + k_z z - \omega t)}\right)\hat{a}_x$$

Evaluating $\nabla \cdot \mathbf{J}$, dropping second-order terms, and setting the result equal to zero yields

$$\left(-ik_z \delta\sigma_0 E_{\|} + ik_z \sigma_0 \delta E_k \sin\theta + ik_x \delta\sigma_P \delta E_k \cos\theta\right)e^{i(k_x x + k_z z - \omega t)} = 0$$

Substituting $k_z = k \sin\theta$ and $k_x = k \cos\theta$ and solving for δE_k,

$$\delta E_k = E_{\|}\delta\sigma_0 \sin\theta \Big/ \left(\sigma_0 \sin^2\theta + \sigma_P \cos^2\theta\right)$$

which may be written

$$\delta E_k + E_{\|}\left(\frac{\delta\sigma_0}{\sigma_0}\right)\frac{\sin\theta}{\sin^2\theta + \left(\sigma_P/\sigma_0\right)\cos^2\theta} \tag{10.7}$$

Our goal is not to carry out a full algebraic analysis of the process, which is quite messy, but rather to gain some physical insight. To proceed, however, we need one result of the detailed linear analysis made by Ossakow and Chaturvedi (1979). They linearized the continuity equation, the momentum equation, and $\nabla \cdot \mathbf{J} = 0$ and showed that, for maximum growth,

$$\theta = \frac{k_{||}}{k_{\perp}} \left[\frac{\nu_i}{\Omega_i} \left(\frac{\Omega_i}{\nu_i} + \frac{\Omega_e}{\nu_e} \right)^{-1} \right]^{1/2}$$

Substituting $\sigma_P = (ne^2 \nu_i / M\Omega_i^2)$ and $\sigma_0 = (ne^2 / m\nu_e)$ for the zero-order conductivities in this expression yields

$$\theta = \left[\frac{\sigma_P}{\sigma_0} \left(1 + \frac{\Omega_i \nu_e}{\Omega_e \nu_i} \right) \right]^{-1/2} \cong \left(\frac{\sigma_P}{2\sigma_0} \right)^{1/2}$$

where in the last step we use the fact that for typical F-region conditions $\Omega_i \nu_e / \Omega_e \nu_i \approx 1$. Substituting this result into (10.7) and using the fact that $\theta \cong \sin\theta$ and $\cos\theta \cong 1$,

$$\delta E_k = \delta\sigma_0 / \sigma_0 \left[\sqrt{2} E_{||} / 3 \left(\sigma_P / \sigma_0 \right)^{1/2} \right]$$

and, finally, because $\sigma_0 \propto n$, we have for maximum growth

$$\delta E_k = (0.47) (\delta n / n) E_{||} (\sigma_P / \sigma_0)^{1/2} \tag{10.8}$$

This electric field is very nearly perpendicular to $\mathbf{B}$, since $k_{\perp} \gg k_{||}$ and, referring to Fig. 10.15, we see that it is such that δE_k is in the $\hat{a}_x$ direction when $\delta n / n > 0$. The perturbation $\mathbf{E} \times \mathbf{B}$ drift in this phase of the wave is thus equal to $+(\delta E_k / B)\hat{a}_y$, which means that a high-density region moves down the gradient to lower-density regions and therefore grows in relative amplitude. Therefore, the plasma is unstable.

In this same configuration, suppose a background perpendicular zero-order electric field $\mathbf{E}_{0\perp} = E_{0\perp}\hat{a}_x$ existed in addition to the zero-order parallel electric field $E_{0||}$. A review of the $\mathbf{E} \times \mathbf{B}$ instability theory presented in Chapters 4 and 5 shows that the growth rate is given by

$$\gamma_{EB} = -E_{0\perp} / BL = +V_{0y} / L$$

That is, if the perpendicular electric field is in the $+\hat{a}_x$ direction, the system is stable, while if it is in the $-\hat{a}_x$ direction, instability occurs.

We now may compare the growth rate of the pure current convective instability to the classical $\mathbf{E} \times \mathbf{B}$ process. The latter we write in the form

$$\gamma_{EB} = E_{0\perp} / BL = V_{0\perp} / L$$

where $V_{0\perp}$ is the magnitude of the zero-order drift parallel to ∇n. Using (10.5), which can be written $\gamma = (\delta E_k / BL)(\delta n / n)$ and substituting δE_k from (10.8) yields

$$\gamma_{cc} = \frac{\sqrt{2}}{3} \frac{E_{||}}{BL(\sigma_P/\sigma_0)^{1/2}}$$

The usual practice is to compare these two growth rates in terms of $V_{0\perp}$ and the zero-order parallel drift velocity difference of the ions and electrons, which may be written

$$V_{||} = \sigma_0 E_{||}/ne$$

Then

$$\frac{|\gamma_{EB}|}{|\gamma_{cc}|} = \frac{3}{\sqrt{2}} \frac{V_{0\perp} B(\sigma_P \sigma_0)^{1/2}}{V_{||} ne}$$

This result makes it easy to compare the two effects. First, we note that if both processes are unstable, (10.9) will show which one is more important. Second, if $E_{0\perp}$ is in the stable configuration while $J_{||}$ is destabilizing, instability is still predicted if (10.9) is less than one. To estimate the magnitude of $J_{||}$ required to overcome, say, a stabilizing 10 mV/m perpendicular electric field ($V_{0\perp} = 400$ m/s), we use $V_{||} = j_{||}/ne$ and set $n = 5 \times 10^4$ cm^{-3} to find $J_{||} \geq 7 \, \mu A/m^2$. This is a sizable current but is not out of the question for the auroral zone.

Using the same expressions for σ_P and σ_0 as used previously, $(\sigma_P \sigma_0)^{1/2} = (\Omega_e \nu_i / \Omega_i \nu_e)^{1/2}$ and Eq. (10.10) becomes,

$$\frac{|\gamma_{EB}|}{|\gamma_{cc}|} = \frac{3}{\sqrt{2}} \frac{V_{0\perp}}{V_{||}} \left(\frac{\Omega_e \nu_i}{\Omega_i \nu_e} \right)^{1/2} \tag{10.9}$$

Finally, once again, we note that the quantity in parentheses is approximately equal to unity in the F region, and we have

$$\gamma_{EB}/\gamma_{cc} \cong 2V_{0\perp}/V_{||}$$

which is identical to the result usually quoted. A complete analysis (Ossakow and Chaturvedi, 1979; Vickrey et al., 1980) yields the following expression for the local growth rate of the current convective instability including the possible existence of $E_{0\perp}$:

$$\gamma_{cc} = \frac{(-1/L)\left[(-E_{0\perp}/B)(\nu_{in}/\Omega_i) + V_{||}\theta_{max}\right]^{1/2}}{(\Omega_i/\nu_{in} + \Omega_e/\nu_{ei})\theta_{max}^2 + \nu_{in}\Omega_i}$$

$$- \frac{(k_\perp^2 \nu_{ei})}{\Omega_e \Omega_i} C_s^2 - \frac{k_{||}^2 C_s^2}{\nu_{in}} \left\{ 1 + \left(\nu_{in}^2/\Omega_i^2 \right) \Big/ \left[(\nu_{ei}\nu_{in}/\Omega_e\Omega_i) + \theta_{max}^2 \right] \right\} \tag{10.10}$$

where $L^{-1} = 1/n(dn/dy)$, C_s is the ion acoustic speed, ν_{in} and ν_{ei} are the ion-neutral and electron-ion collision frequencies, Ω_e and Ω_i are the electron and ion gyrofrequencies, $E_{0\perp}$ is the component of the perpendicular electric field in the $\nabla n \times B$ direction, and the ratio of parallel to perpendicular wave numbers for maximum growth, θ_{max}, is given by

$$\theta_{max} = -\left(\frac{E_{0\perp}}{BV_{||}}\right)\left(\frac{\nu_{in}}{\Omega_i}\right) \pm \left[\left(\frac{E_{0\perp}}{BV_{||}}\right)^2\left(\frac{\nu_{in}}{\Omega_i}\right)^2 + \frac{\nu_{in}/\Omega_i}{\Omega_i/\nu_{in} + \Omega_i/\nu_{en}}\right]^{1/2}$$

$$(10.11)$$

If we let $V_{||}$ go to zero in (10.11) and choose the negative sign, then $\theta_{max} = 0$. This result yields the flute mode ($k_{||} = 0$) $E \times B$ instability. Indeed, if we set $\theta_{max} = 0$ in (10.10), it reduces to the growth rate of the gradient drift instability. Thus, as already noted, a field-aligned current can serve to destabilize a plasma configuration that is otherwise stable to the $E \times B$ instability, or to enhance the growth rate of an already unstable situation. Choice of the negative sign in (10.11) corresponds to a damped acoustic mode.

The linear local theory described here has a preferential tendency to create unstable conditions at low plasma densities, since for a fixed $J_{||}$, $V_{||}$ is inversely proportional to n. The region above the F peak is therefore a region of high growth rate. However, the generalized process could very easily be *stable* on the same magnetic field line at or below the F peak. It is very clear then that a local theory is not very suitable. Such nonlocal effects in general reduce the growth rate. Note also that E-region shorting is every bit as important to the current convective instability as it is to the $E \times B$ instability and is not included in (10.10). Also, the growth rate given in (10.10) uses a perpendicular diffusion coefficient corresponding to a nonconducting E region, which may also underestimate the damping. Taken together, these negative aspects have led to the conclusion that although it is an interesting physical process, the current convective instability can only rarely overcome the stabilizing effects of an unfavorable $E \times B$ geometry. Finally, it should be noted that the instability only occurs when the current is carried by thermal plasma and does not apply when energetic electrons carry the bulk of an upward current.

Erukhimov and Kagan (1994) proposed a thermal instability mechanism (which they named thermomagnetic) in which a strong field-aligned electric field is set up by conversion of the field-perpendicular electric field by a geomagnetic field-perpendicular magnetic perturbation. The induced field-aligned electric field is prevented from shorting out by electron inertia and is sustained by the plasma drift. Such type of electric field conversion takes place in stationary Alfvén wave (Knudsen, 1996). The induced field-aligned electric field is much stronger than usual and leads to the thermal instability generation discussed in (Erukhimov et al., 1982). The E-region analogue of the latter, based on ion heating, is developed by Kagan and Kelley (2000). The main idea is that temperature fluctuations provide a feedback that enhances the electric field in plasma depletions and therefore lead to further plasma forcing out from the plasma-depleted regions. Similar to the current-convective processes, a finite field-aligned wave vector is essential. The instability criterion and the growth rate are

$$V_d V_{\parallel} \left(1 + V_d \frac{d(\ln V_d)/dy}{\delta_e \nu_e}\right) > \frac{k_B(T_e + T_i)}{m_e} \lambda_p^2 k_{\parallel} k,$$

$$\gamma = \left(\delta_{eff} \nu_e + V_d \frac{d(\ln V_d)}{dy}\right) \frac{\nu_e}{\nu_{en}},$$

where

$$\lambda_p = c\sqrt{\frac{m_e}{4\pi e^2 n}}$$

is the wavelength of the plasma wave, $\delta_{eff} \nu_e$ involves the heating loss caused by collisions and the field-aligned electron thermal conductivity, V_d is the plasma drift velocity in $\mathbf{E} \times \mathbf{B}_0$ fields, and $V_{\parallel}$ is the field-aligned electron velocity.

Analyses of coordinated scintillation observations using orbital satellite beacons and the EISCAT radar by Kagan et al. (1995) showed that the thermomagnetic instability is the most probable candidate for F-region 1 km-scale irregularities inside the trough. It is interesting that the same type of field-aligned electric field associated with stationary Alfvén waves seems to be responsible for ion outflows from auroral ionosphere that are observed above stable auroral arcs and are accompanied by strong electron temperature enhancements (Kagan and St.-Maurice, 2005).

10.2.2 Turbulent Mixing as an Alternative to Plasma Instabilities

It is well known that the magnetospheric flow pattern is seldom laminar but rather is usually somewhere between turbulent and extremely turbulent (Crowley and Hackert, 2001; Earle and Kelley, 1993). The effect of this turbulence is to mix any existing density gradient due to solar production or particle impact ionization regardless of the sign of the gradient. We can estimate the turbulence level necessary to accomplish a higher level of density structure than that provided by the $\mathbf{E} \times \mathbf{B}$ process as follows. We assume that the layer has the appropriate sign for instability, but, of course, this will happen only 50% of the time. The simple form of the linear growth rate for the $\mathbf{E} \times \mathbf{B}$ instability is γ_{EB}, given previously. If that same gradient is structured by a spatial electric field pattern characterized by some wave number spectrum $E(k)$, then we may use the continuity equation

$$\partial n/\partial t = -\mathbf{V} \cdot \nabla n$$

to define a mixing growth rate $\gamma_m(k)$, which is a function of k:

$$\gamma_m(k) = E(k)/BL$$

To compare the mixing growth rate to the linear $\mathbf{E} \times \mathbf{B}$ growth rate, which is independent of k, we must integrate $\gamma_m(k)$ over some portion of the turbulent spectrum. Then

$$\Gamma_m = \left[\int_{\Delta k} E^2(k) \, dk \right]^{1/2} \Big/ BL \qquad (10.12)$$

is the appropriate growth rate provided Δk is chosen properly. Since the important wavelength range is for $k > k_L = 2\pi/L$, we integrate from k_L to ∞. The ratio of turbulent mixing to the plasma instability source is then

$$R = \Gamma_m/\gamma_{EB} = (1/E_{0\perp}) \left(\int_{k_L}^{\infty} E^2(k) dk \right)^{1/2} = \frac{E_{rms}}{E_{0\perp}} \qquad (10.13)$$

The numerator is the rms electric field between k_L and $\infty(E_{rms})$. Turbulent mixing is important when $R \geq 1$—that is, when the mixing growth rate is comparable to the linear growth rate of the plasma instability. The mixing process will even dominate the physics on an unstable gradient if the rms electric field in the range Δk exceeds $E_{0\perp}$, the quasi-dc component perpendicular to ∇n. For a stable gradient the mixing effect will, of course, be the dominant source and R need not be greater than 1. Furthermore, since turbulent fields are supplied by a low impedance magnetosphere generator, they will not be shorted out by the E region.

Surprisingly, very few electric field spectra exist in the literature. One example from the dawn auroral oval is presented in Fig. 10.16. The upper plot comes from periods of the flight near aurorae arcs (note that the scale is broken). The spectra show evidence for two turbulent processes, one operating at long wavelengths yielding a $k^{-5/3}$ power law and one which injects turbulence in the ionosphere at the scale of individual auroral arcs. The latter process seems to result in a k^{-3} power law form for large $k > k_b \approx 20\,\text{km}^{-1}$. These two power laws are in agreement with predictions of two-dimensional turbulence theory (Kintner and Seyler, 1985). The k-space integrals corresponding to these two regimes yield rms electric fields of 10 and 9 mV/m, respectively. Such values for E_{rms} indicate a significant role for turbulent mixing, since $E_{rms}/E_{0\perp} \approx 0.5$ for this event.

The fact of the matter is that auroral arcs are inextricably intertwined with structured electric fields. At the altitude of the auroral acceleration zone ($\approx$ 3000 − 5000 km), in fact, the electric field is turbulent in a much larger region than that of the electron beams themselves. The acceleration zone is imbedded in this turbulent plasma which partially extends to ionospheric heights. It seems

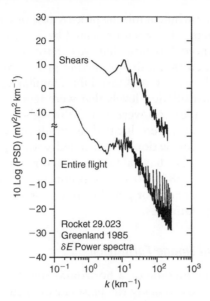

Figure 10.16 Wave number spectra of the electric field from a rocket flight in the dayside auroral oval. The lower plot represents a large spatial region, while the upper plot shows smaller-scale spectra near auroral arcs. The latter are imbedded in the region sampled to produce the lower plot. Note that the scales are shifted. [After Earle and Kelley (1993). Reproduced with permission of the American Geophysical Union.]

clear that any ionospheric plasma gradient *must* be mixed by those applied turbulent electric fields.

The ionosphere is not passive in this context, however. Detailed study has shown that the magnetosphere behaves as a current generator at these scales (Vickrey et al., 1986). This in turn implies that the turbulent electric field and flow velocity are inversely proportional to the ionospheric conductivity. This is evident when one compares winter and summer hemispheres (e.g., see Fig. 8.8b). The winter ionosphere has a much more structured velocity field than does the summer ionosphere. One hypothesis is that the solar wind-driven flow in the magnetosphere has its own turbulent characteristics that are moderated by the gross E-region conductivity due to solar lighting conditions, diffuse auroral precipitation, and polar rain. Superposed on this are localized regions of turbulence associated with auroral arcs that create plasma and strongly mix it at the same time. At this scale the E region itself becomes dominated by the precipitation and the conductivity, and the flow field and plasma content are intertwined in a fundamental way. A review (Kintner and Seyler, 1985) discussing the interrelationship between plasma fluid turbulence and plasma density variations in the ionosphere has been published. Although no direct conclusions were reached, a framework was provided for this study and progress is still needed in this area.

Further evidence for a turbulent mixing process at scales on the border between "large" and "intermediate" has been presented by Tsunoda et al. (1985). They studied a highly structured F region that was characterized by a stable $\mathbf{E} \times \mathbf{B}$ situation. They concluded that a driven mixing process was occurring but one that was *not* explained by the $\mathbf{E} \times \mathbf{B}$ instability. Similarly, referring to the data in Fig. 10.1, the high scintillation levels that were reported in association with plasma patches in the polar cap were found throughout the patches. That is, kilometer-scale irregularities were located on gradients with either sign, even though the mean patch convection is associated with a particular $\mathbf{E} \times \mathbf{B}$ flow direction. Scintillation studies have also shown that regions of plasma shear are very irregular (Basu et al., 1986). We therefore conclude that impressed turbulent electric fields of magnetospheric origin play an important role in creating ionospheric structure at high latitudes.

10.2.3   Diffusion and Image Formation

As noted previously, the lifetime of a horizontal structure in the high-latitude ionosphere depends on how quickly the plasma can diffuse across the magnetic field. Also, whether a particular plasma instability is stable or unstable at a given k value depends on whether some positive contribution to the growth rate exceeds diffusive damping due to terms of the form $-k^2 D_\perp$. In Chapter 5 we pointed out that parallel diffusion which operates with a time scale $(k^2 D_{||})^{-1}$ is determined by the ion diffusion coefficient and that

$$D_{||} = 2D_i$$

The factor of 2 comes from the fact that an ambipolar electric field builds up when electrons attempt to diffuse away quickly parallel to $\mathbf{B}$ due to their small mass. In effect, each electron has to drag a heavy ion with it along $\mathbf{B}$, and the *plasma* diffusion coefficient is then determined by the ions.

Perpendicular to $\mathbf{B}$, an analogous process occurs. The equations of conservation of momentum (2.22b) for each species in the moving reference frame of the neutral wind are

$$0 = -k_B T_j \nabla n + n M_j \mathbf{g} + n q \mathbf{E}' + n q_j (\mathbf{V}'_j \times \mathbf{B}) - n M_j v_{jn} \mathbf{V}'_j$$

where we have taken $d\mathbf{V}'_j / dt = 0$. Let $\mathbf{B}$ be in the z direction and neglect the effect of gravity and neutral winds. Then in the directions perpendicular to $\mathbf{B}$,

$$0 = -k_B T_j (\partial n / \partial x) + n q_j E_x + n q_j V_{jy} B - n M_j v_{jn} V_{jx}$$

$$0 = -k_B T_j (\partial n / \partial y) + n q_j E_y - n q_j V_{jx} B - n M_j v_{jn} V_{jy}$$

Substituting for V_{jy} in the second equation and substituting into the first yields an expression involving only V_{jx}:

$$V_{jx}\left(1 + \frac{q_j^2 B^2}{M_j^2 v_{jn}^2}\right) = \frac{-k_B T_j}{M_j v_{jn}}\frac{1}{n}\frac{\partial n}{\partial x} + \frac{q_j}{M_j v_{jn}}E_x$$

$$- \frac{q_j B k_B T_j}{M_j^2 v_{jn}^2}\frac{1}{n}\frac{\partial n}{\partial y} + \frac{q_j^2 B^2}{M_j^2 v_{jn}^2}E_y$$

Similarly, we can derive

$$V_{jy}\left(1 + \frac{q_j^2 B^2}{M_j^2 v_{jn}^2}\right) = \frac{-k_B T_j}{M_j v_{jn}}\frac{1}{n}\frac{\partial n}{\partial y} + \frac{q_j}{M_j v_{jn}}E_y$$

$$- \frac{q_j B k_B T_j}{M_j^2 v_{jn}^2}\frac{1}{n}\frac{\partial n}{\partial x} + \frac{q_j^2 B^2}{M_j^2 v_{jn}^2}E_x$$

Remembering that the $\mathbf{E} \times \mathbf{B}$ and the diamagnetic drift velocities can be written

$$\mathbf{V}_{Ej} = \mathbf{E} \times \mathbf{B}/B^2, \quad \mathbf{V}_{Dj} = -\frac{k_B T_j}{q_j B^2 n}\nabla n \times \mathbf{B}$$

the equations for the perpendicular velocity can be recombined, giving

$$\mathbf{V}_{j\perp}\left(1 + \frac{q_j^2 B^2}{M_j^2 v_{jn}^2}\right) = -\frac{k_B T_j}{M_j v_{jn} n}\nabla_\perp n + \frac{q_j}{M_j v_{jn}}\mathbf{E}_\perp$$

$$+ \frac{q_j^2 B^2}{M_j^2 v_{jn}^2}\mathbf{V}_D + \frac{q_j B^2}{M_j^2 v_{jn}^2}\mathbf{V}_E$$

Defining the perpendicular diffusion coefficient for motion antiparallel to a density gradient by an expression of the form $\mathbf{V} = -D(\nabla n/n)$ yields,

$$D_{j\perp} = k_B T_j/M_j v_{jn}\left(1 + \Omega_j^2/v_{jn}^2\right) \tag{10.14a}$$

Now for F-region altitudes $\Omega_j \gg v_{jn}$, and this expression becomes

$$D_{j\perp} = (k_B T_j/M_j)\left(v_{jn}/\Omega_j^2\right)$$

Finally, since the species gyroradius is given by the expression

$$r_{gj} = \left(k_B T_j / M_j\right)^{1/2} \Omega_j^{-1} \tag{10.14b}$$

the perpendicular diffusion coefficient of some species j may be written

$$D_{j\perp} = r_{gj}^2 \nu_j \tag{10.14c}$$

where r_{gj} is the gyroradius and ν_j is the collision frequency. Intuitively this makes sense, since a particle will randomly walk one gyroradius between each collision. Since r_{gj} is much larger for ions than for electrons, the ions tend to diffuse more rapidly down a density gradient which exists perpendicular to the magnetic field than do the electrons. However, when this occurs an electric field builds up parallel to ∇n, so the ion motion across the field lines is retarded and the electron motion is enhanced. The result (derived following) is that an isolated plasma structure diffuses across a magnetic field with a diffusion coefficient equal to the low electron diffusion coefficient.

The preceding discussion assumes that the F-region plasma is the only plasma in the system. However, if an E region exists due to production of plasma by sunlight or production by auroral particle precipitation, the ambipolar electric field discussed previously will be shorted out and the ions may then diffuse rapidly across the magnetic field. The electrons still cannot move across $\mathbf{B}$, but they *can* move along the magnetic field to complete the circuit through the E region and therefore neutralize the ambipolar electric field. The entire process is illustrated schematically in Fig. 10.17a. The converging arrows in the E region represent the ambipolar electric field, which maps down to the E region due to the high conductivity parallel to $\mathbf{B}$. Those arrows also show the direction of the E region ion current that completes the circuit. The inner and outer sets of arrows show the electron flow direction. The line through the F-region structure represents the vertical magnetic field line.

A quantitative description of this diffusion process must allow for altitude variations in all the relevant quantities. Vickrey and Kelley (1982) have performed such calculations for a model ionosphere and computed the typical effective diffusion coefficient plotted in Fig. 10.17b as a function of the ratio of E region to F-region conductivities. The effective (field-line-integrated) diffusion coefficient increases from roughly 1 m²/s to more than 10 m²/s as the E-region conductivity increases. Also plotted is the e-folding time, $\tau_\perp = (k_\perp^2 D_\perp)^{-1}$, for a horizontal structure with a characteristic perpendicular wavelength of 1 km. Even for high E-region conductivity the diffusion time scale is several hours for such a structure, and it could survive considerable transport around the high-latitude zone. In this particular calculation, the F layer had a Chapman distribution characterized by the parameters listed in the plot. Kelley et al. (1982b) used this semiquantitative approach to calculate the relative amplitude of kilometer-scale structures produced at various places in the auroral oval which were then allowed to

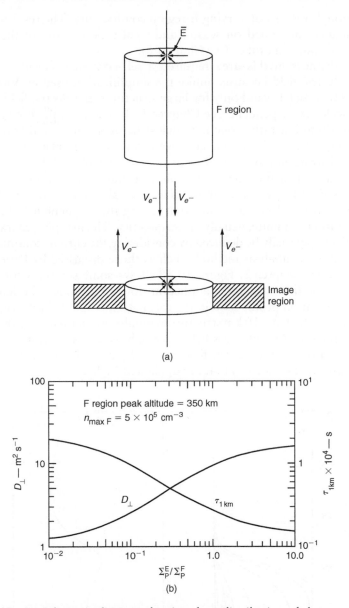

(a)

(b)

Figure 10.17 (a) Schematic diagram showing the redistribution of electrons along the magnetic field during the image formation process as well as the ambipolar electric field. The arrows in the image region also show the direction of the ion Pedersen velocity. (b) Dependence of cross-field diffusion rate $D_\perp$ and lifetime of 1 km scale irregularities $\tau_{1\,km}$ on the ratio of E to F region Pedersen conductivities. The F-layer altitude and peak density are fixed. [After Vickrey and Kelley (1982). Reproduced with permission of the American Geophysical Union.]

convect through regions of varying E-region conductivity. The resulting amplitude distribution depended on season and local time because of the peculiar trajectories of some flux tubes (see Chapter 8).

Further advances in this area required a self-consistent calculation of the ambipolar electric field because, unlike the simplification used by Vickrey and Kelley (1982), which is valid only for large structures, the electric field mapping process is scale size dependent (see Chapter 2). In addition, the E region is not a passive medium but rather one in which structures can form that mirror the F region irregularities. These "images" were first pointed out in the context of barium cloud striation physics by Goldman et al. (1976).

In the remainder of this section we follow a unified description of diffusion that includes F and E-region coupling (Heelis et al., 1985). First, it should be emphasized that the electrical conductivity along the magnetic field lines is not infinite but rather is a finite, usually large, quantity. The mapping characteristics of electric fields can easily be deduced by considering the current continuity equation and employing analysis methods such as those discussed by Farley (1959) and presented in Chapter 2. Figure 10.18 shows some solutions for different scale size electric field structures applied at 500 km altitude. The ionospheric plasma concentration profile used in the calculation is shown in the right panel. As expected, we see that a 10 km structure maps almost unattenuated throughout the E and F regions. At scale sizes below 5 km, however, significant attenuation occurs and can render the electric field essentially zero below 200 km for fields with scale sizes of 1 km or less that are applied at high altitudes.

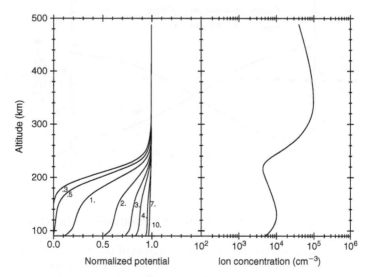

Figure 10.18 The variation in electrostatic potential from different scale size electric fields is shown in the left panel. The right panel shows the ionospheric ion concentration profile. The potentials were applied at 500 km altitude. (Figure courtesy of R. Heelis.)

The importance of electric field mapping in the ionosphere lies in the fact that electric fields created locally in a region of poor horizontal conductivity can map to an altitude where the horizontal conductivity is much larger. The presence of an electric field in a conducting region will drive a horizontal current that will tend to short the applied electric field. The electric field can then be maintained at the generator only if it can supply the current required in the conducting region. Let us consider the forces acting on a plasma concentration structure in an ionosphere with a vertical magnetic field. First, a pressure gradient force will exist on both the ions and the electron gases. But, as shown previously, owing to the different masses of these particles and their correspondingly different collision frequencies and gyrofrequencies, these two species will move at different rates. Any tendency for separation of the ions and electrons in this manner will produce an electric force.

The horizontal motion of the electrons and ions subject to these forces can be determined by manipulation of the steady-state equations of motion for each species. Rather complex expressions result if all the terms in these equations are retained. This is the usual practice in computer models, but in order to illustrate the dominant physics we consider only the horizontal ion motion and assume that electron-ion collisions have a negligible effect. At high latitudes we may assume $g \times B = 0$, in which case the horizontal motions are not affected by gravity. Then from (2.36b)

$$V_{ix} = (\sigma_{Pi}/en)\, E - (D_{i\perp}/n)\, \nabla_x n \tag{10.15}$$

Here x again denotes the direction of the gradient, and we have set $n_i = n_e = n$ due to the quasi-neutrality condition. The ion Pedersen conductivity, or σ_{Pi}, and the ion perpendicular diffusion coefficient, $D_{i\perp}$, are given in general form by

$$\sigma_{Pi} = \left(e^2 n v_{in}\right) \Big/ M_i \left(\Omega_i^2 + v_{in}^2\right)$$

$$D_{i\perp} = \left(k_B T_i v_{in}\right) \Big/ M_i \left(\Omega_i^2 + v_{in}^2\right)$$

Similarly, the horizontal electron velocity can be expressed as

$$V_{ex} = -(\sigma_{ei}/en)E - (D_{e\perp}/n)\nabla_x n + R V_{ix} \tag{10.16}$$

where the factor R is given by

$$R = \frac{\Omega_e^2 v_{ie}/v_{in} + v_{ei}/v_e}{\Omega_e^2 + v_e^2}$$

and arises because ion-electron collisions are not always negligible for the electrons. Finally, Eqs. (10.15) and (10.16) can be combined to yield an expression for the local horizontal Pedersen current,

$$J_P = [(1 - R)\sigma_{Pi} + \sigma_{Pe}]E - e[(1 - R)D_{i\perp} - D_{e\perp}]\nabla_x n \tag{10.17}$$

The magnitude of ionospheric structure and its time evolution are related by the equations of continuity and momentum, both of which contain the plasma velocity. If we consider an ionospheric structure not subject to any production, then the continuity equation for the ions may be written as

$$\partial n/\partial t = (-L) - \nabla \cdot (nV_{i\perp}) - \nabla \cdot (nV_{i\parallel})$$

Here the ion velocity has been expressed in terms of its components perpendicular ($\perp$) and parallel ($\parallel$) to the magnetic field. If we further simplify the problem by considering electric fields produced by the structure itself and neglect any parallel ion motion produced by the structure, then the ion velocity is given by (8.15) and the continuity equation can be rewritten in the form

$$\partial n/\partial t = -L - \nabla \cdot (\sigma_{Pi}/e)E + \nabla \cdot (D_{i\perp}\nabla n)$$

At this point it is a common practice to "linearize" the equations by expressing each of the plasma properties as the sum of a background value and a small perturbation value due to the existence of the structure. Then all terms containing products of perturbation values are ignored, since they are much smaller than the other terms. If we denote all background values with the superscript "o," assume a horizontally stratified background ionosphere and ignore losses, then the continuity equation may be written finally as

$$\partial n\partial t = - \left(\sigma_{Pi}^{o}/e\right) \nabla_x E + D_{i\perp}^{o} \nabla_x^2 n \tag{10.18}$$

Now consider two extreme situations. First, suppose the local electric field associated with the plasma structure in the F region is completely shorted ($E = 0$) by mapping to a highly conducting E region. Investigation of (10.18) shows that for $E = 0$ the classical diffusion equation results with $D_\perp = D_{i\perp}$, the ion diffusion coefficient. This is the maximum possible value of $D_\perp$ under the circumstances considered. In the other limit, suppose that the structure cannot drive a Pedersen current because it maps to an E region that is an insulator. Then we may solve for the electric field obtained by setting the current in (10.17) to zero and substitute the result in (10.18). In the F region, where $R \ll 1$, the electric field from (10.17) becomes

$$E_x = \frac{e \left(D_{i\perp}^{o} - D_{e\perp}^{o}\right)}{\sigma_{iP}^{o} + \sigma_{eP}^{o}} \nabla_x n$$

Substituting this result into (10.18) yields

$$\frac{\partial n}{\partial t} = \left[D_{i\perp}^{o} - \sigma_{iP}^{o} \left(\frac{D_{i\perp}^{o} - D_{e\perp}^{o}}{\sigma_{iP}^{o} + \sigma_{eP}^{o}}\right)\right] \nabla_x^2 n$$

Since in the F region $\sigma_{iP}^o >> \sigma_{eP}^o$, this reduces to

$$\partial n/\partial t = D_{e\perp}(\nabla_x^2 n)$$

In this case therefore the velocity and associated decay rate of the structure will be the minimum possible under the circumstances considered and will be characterized by the electron diffusion coefficient $D_{e\perp}$.

From the previous discussion it should be clear that the decay rate of an F-region irregularity depends not only on its scale size and the local ionospheric conditions but also on the mapping properties of the electric field and on the conductivity of the regions throughout which the field maps. In our description we considered two extreme examples where the electric field was easily determined. In practice, however, the electric field will be determined by applying the current continuity condition on an intermediate state. This yields slightly more complicated equations but with essentially the same characteristics.

One further point to consider is the effect that the current produced by a structure in the F region might have in the E region. This can be an important consideration because the relatively high collision frequencies in the E region mean that the plasma is compressible. If this compressional force can overcome the horizontal diffusion forces that oppose it in the E region, then a structure can form there that mimics the one in the F region. In order to be significant, this structure formation process in the E region must also overcome the high rate of dissociative recombination that exists at such altitudes. We now investigate the conditions under which such an "image" structure can be formed.

The image formation process can be understood from Fig. 10.17a for a pure decaying cylindrical irregularity in which the only electric field is due to the ambipolar effect. For a large enough structure, the E field maps as shown and E-region ions are gathered together in the center region because of their Pedersen drifts. The requirement of charge neutrality is met by electrons flowing down **B** to the E region. The net result is that the E-region plasma density (ions and electrons) increases at the center. That is, an *image* high-density region forms in the center and two low-density regions form outside. The image amplitude will eventually be limited by recombination in the E region, which is proportional to n^2. Notice that the plasma-gathering process can be discussed in terms of compressibility of plasma in the E region in response to the applied electric field. The F region plasma, on the other hand, is virtually incompressible perpendicular to **B**.

Although interesting in their own right, we have not explained the role images play in F-region diffusion. The E-region plasma density gradient due to an image drives an ion current opposite in direction to the mapped electric field. The net field-aligned current from the F region is therefore decreased, and the F-region diffusion coefficient is lowered, tending more toward the low value of $D_\perp$ which pertains for an insulating E region. Now, since the mapping process is itself scale size dependent, we conclude that the effective diffusion coefficient of high-latitude plasma irregularities may also be scale size dependent if the E-region density is neither too high nor too low. Vickrey et al. (1984) provided evidence that images can be formed on off-equatorial field lines in contact with a structural F region during a convective ionospheric storm.

10.3   Small-Scale Waves in the High-Latitude F Region

For this purpose we define small-scale wavelengths to be less than 100 m. Research in this area has not progressed as rapidly as it has in the equatorial zone. One reason is that the observational data base using radars is not very large. In the equatorial case the Jicamarca and Altair radars can easily scatter from irregularities with wave numbers perpendicular to **B**, since the geometry is favorable. Since these waves have much larger growth rates than waves with finite $k_{\parallel}$, they dominate the spectrum. At the high-latitude radar sites, it is geometrically impossible to obtain the appropriate backscatter angle in the F region using VHF and UHF systems.

F-region radar observations are possible at HF frequencies, since ionospheric refraction can be used to bend the radar signals and attain a scattering geometry nearly perpendicular to **B**. Successful measurements of this type were first made systematically in Scandinavia (Villain et al., 1985) and from a transmitting station in Goose Bay, Labrador, Canada (Greenwald et al., 1983). The observation volume for the latter site is located over the Sondre Stromfjord, Greenland, incoherent scatter radar station. The results show that echoes are indeed received in the tens of meters wavelength range but a detailed explanation does not yet exist for the scattering structures. The Super Darn HF radar system uses these structures and it is very surprising that we do not understand them better.

Clues concerning the origin of these waves come from the Doppler shift of the returned HF signal. During one event simultaneous measurements were made with the incoherent scatter radar at Sondre Stromfjord. The latter can be used independently to determine the plasma drift velocity projected along the HF radar line of sight. The two data sets are superimposed in Fig. 10.19 and show quite good agreement. The implication is that the irregularities are frozen into the plasma flow—that is, that the phase velocity of the structures is small in the plasma reference frame.

The long-wavelength instability processes discussed in Section 10.2 do have this low-phase velocity property. However, it is easy to show that at a wavelength of 10 m, the **E** × **B** instability is stable. For example, we may let $E_{0\perp} = 20$ mV/m, $L = 10$ km, and $D_{\perp} = 1$ m^2/s, which is on the small side, and solve the equation

$$\gamma = mE_{0\perp}/LB - k_m^2 D_{\perp} = 0$$

for the marginally stable wave number k_m. This yields a wavelength $\lambda_m = 30$ m. A larger diffusion coefficient will only make the value of λ_m larger.

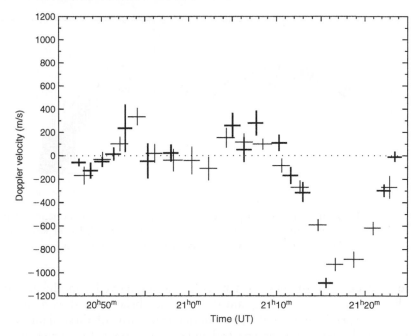

Figure 10.19 Comparison of time series data for HF (heavy crosses) and incoherent scatter (light crosses) velocity estimates calculated for the returns received from the nearby volumes. [After Ruohoniemi et al. (1987). Reproduced with permission of the American Geophysical Union.]

Unless some additional instability arises at small scales, a possibility we address soon, the only way waves with $\lambda \le 10$ m can arise is through a cascade of structure from long to short λ. There is a curious difference between high- and low-latitude irregularities in this regard. As discussed in Chapter 4, convective ionospheric storms evolve in such a way that a very steep spectrum (k^{-5}) evolves for $\delta n^2(k)/n^2$ when $\lambda \le 100$ m. This spectral range looks very much like a viscous subrange in neutral turbulence and has been termed the diffusive subrange. No analogy to the viscous or diffusive subrange has been reported at high latitudes.

Measurements on satellites and rockets do show that the spectrum of plasma density fluctuations varies as k^{-n} in the wavelength regime from tens of meters to tens of kilometers (Dyson et al., 1974; Sagalyn et al., 1974; Cerisier et al., 1985), where typically $1.5 \le n \le 2.5$. As discussed in Chapter 4, this spectral form has also been reported for equatorial measurements and barium cloud striations. In both bottomside spread F and low-altitude barium striations such spectra were shown to be due to wave steepening and not to turbulence (Costa and Kelley, 1978; Kelley et al., 1979). There has been no evidence to date that steepening occurs at high latitudes; rather, turbulent processes appear to dominate. An example from a rocket flight is reproduced in Fig. 10.20. The

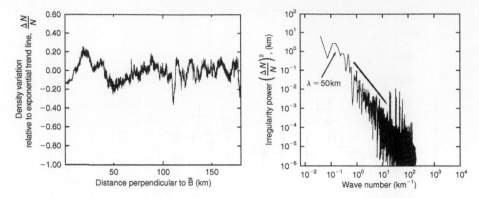

Figure 10.20 Horizontal variations in electron density relative to an exponential dependence with altitude (top) and power spectrum of the data (bottom). [After Kelley et al. (1980). Reproduced with permission of the American Geophysical Union.]

altitude dependence of the plasma density has been removed from the data by a detrending technique, and the quantity $\nabla n/n$ has been plotted as a function of distance perpendicular to **B**. The Fourier transform of these data is also shown and indicates a power spectral index n of about 1.6, which is consistent with a Kolmogorov ($n = 5/3$) spectrum (the straight line has a 5/3 slope). These auroral rocket data were obtained within minutes of the poleward surge of a magnetospheric substorm. The signal level was above the noise for $\lambda \geq 60$ m.

This unexplained difference between the "q-machine" plasma at low latitudes and the violent high-latitude ionosphere is perplexing. Our instincts suggest that field-aligned currents matter at high latitudes and break the transition to damped diffusion waves. Perhaps the current convective instability dominates the cascade process. Bohm diffusion always seems to occur in the laboratory and perhaps this is the answer here as well. The latter is thought to be related to field-aligned currents and ion cyclotron waves. Some evidence for the latter exists in very high $J_{\parallel}$ cases.

Information on even shorter-wavelength waves (≤ 10 m) comes almost entirely from rocket measurements, due to the high data rates required. Three distinct types of irregularities were observed during a rocket flight into the recovery phase of a magnetospheric substorm on April 2, 1970 (Bering et al., 1975; Kelley et al., 1975; Kelley and Carlson, 1977). At the edge of an auroral arc, as detected by onboard particle detectors, intense (5 mV/m) oxygen electrostatic ion cyclotron waves were observed in data from the electric field wave receiver. A spectrum from the rocket results is presented in Fig. 10.21. Existence of such waves is consistent with auroral backscatter measurements of cyclotron waves in the ionosphere. Both data sets indicate that the waves occur in a very localized area. The waves detected on the rocket were observed in conjunction with

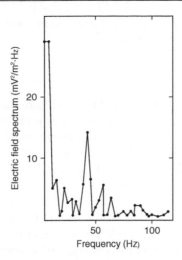

Figure 10.21 Detection of a pure oxygen cyclotron wave near the edge of an auroral arc. [After Kelley et al. (1975). Reproduced with permission of the American Institute of Physics.]

an intense upward field-aligned current. Hydrogen ion cyclotron waves are commonly observed above 5000 km (Kintner et al., 1978) but observations of oxygen cyclotron waves such as these are more rare.

Kelley and Carlson (1977) also showed evidence for less intense and more broadband electrostatic waves in regions of intense velocity shear and field-aligned currents at the edge of the same auroral arc. The velocity shear measured by the dc electric field detector in the same region was $20\,s^{-1}$, which is very high and implies a change in the plasma flow velocity comparable to the sound speed in one ion gyroradius. Less intense irregularities in the same wavelength regime were observed equatorward of the arc but not above the arc itself or poleward of it. The dc electric field was such that the ionospheric plasma convection had an equatorward component in the whole region probed, which suggests that the structures may have been formed at the arc boundary and were transported equatorward. Similar spectra have been reported in the dayside auroral oval by Earle (1988) and Earle and Kelley (1993).

The possible origins of ionospheric structure with scales in the range of a few meters ($\approx r_{gi}$) to a few tens of meters are quite numerous and no unifying mechanism is likely to explain all of the observations. It is very likely that some energy cascades into this wavelength range from larger scales in a manner similar to neutral turbulence. In a plasma the situation is made much more complicated by the possible generation of wave modes which compete with or modify the cascade concept.

Of the several possible free energy sources available, most theoretical effort has gone into calculations of the generation of electrostatic ion cyclotron waves by field-aligned currents and, as noted previously, evidence does exist for their generation. This wave mode arises when

$$V_i^{th} \leq J_{||}/ne \leq V_e^{th}$$

which corresponds to parallel current due to a differential drift between ions and electrons which falls between the ion and electron thermal speeds. In the lower ionosphere collisions must be taken into account, while higher up collisionless theory is adequate. The threshold parallel current density for O^+ and NO^+ cyclotron waves is plotted in Fig. 10.22 for a reference ionospheric profile and for a wavelength corresponding to $(kr_{gi}) = \sqrt{2}$. The calculation includes collisions. The required current densities are higher than the average observed field-aligned currents but not higher than some of the largest reported examples of $J_{||}$ (e.g., Burke et al., 1980). Thus, oxygen cyclotron waves should occur in the ionosphere but are rare and may be restricted to the edges of auroral arcs and/or regions of strong velocity turbulence.

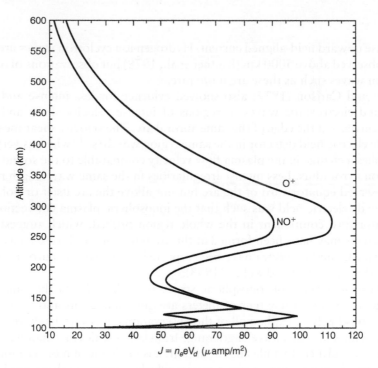

Figure 10.22 Threshold currents required to excite O^+ and NO^+ EIC waves in O^+ and NO^+ plasmas, respectively. The parameters used are $k\rho_i = (2)^{1/2}$ and $k_{||}/k_\perp = 0.06$. [After Satyanarayana and Chaturvedi (1985). Reproduced with permission of the American Geophysical Union.]

Although most theoretical effort has involved current-driven modes, Ganguli et al. (1985) have pointed out that intense shears can also generate ion cyclotron waves. Since shears and intense field-aligned currents are often collocated [see Eq. (10.3)], it is not yet clear which free energy source is most important. Mixing and turbulent cascade will occur as well. Basu et al. (1986) have also shown a tendency for intense irregularities at large scales to arise in regions of velocity shear. An ionospheric Kelvin-Helmholtz instability may occur due to these shears (Keskinen et al., 1988). Finally, it seems very likely that turbulence in the auroral acceleration zone will propagate via an Alfvén wave mode to ionospheric altitudes and create considerable structure in the plasma flow velocity and the plasma density (Knudsen et al., 1992).

10.4 E-Region Layering at High Latitudes

Plasma layers are also common at high latitudes and are reasonably common in the central polar cap. The upper panel of Fig. 10.23 shows a very nice example detected with the Sondre Stromfjord radar in Greenland. The dark altitude-extended regions are due to auroral particle impact. The layer stands out very

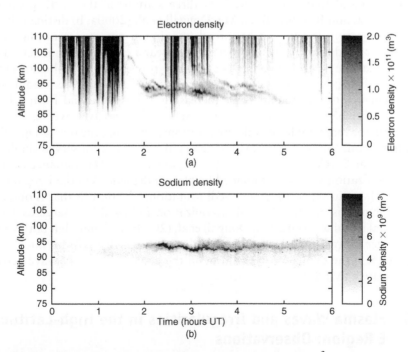

Figure 10.23 Electron and neutral sodium number densities (m^{-3}) as functions of altitude and time on December 11, 1997. [After Heinselman et al. (1998). Reproduced with permission of the American Geophysical Union.]

well between 90 and 95 km. The lower panel is the sodium atom density measured using sodium lidar and is clearly related to the plasma layer. The most likely explanation for such effects is that ionized sodium recombines to yield the atom layer (Cox and Plane, 1998). There seems to be little evidence that auroral particles release sodium from a dust layer (Heinselman et al., 1998), as has been suggested (von Zahn et al., 1987). For example, no precipitation occurred but a strong Na layer developed. The magnetic field is almost vertical at high latitudes, so the midlatitude wind-shear mechanism is not very effective in gathering the plasma. The small off-vertical angle, however, is very important because of large-magnitude perpendicular electric fields. For example, a southward electric field will cause a small Pedersen drift downward, which is a strong function of altitude. As the particles drift down perpendicular to B, their velocity decreases due to the increasing collision frequency. As they slow down, a layer develops with time. The abrupt onset in Fig. 10.23 could be due to advection of the plasma cloud into the view of the radar.

In the polar cap neither of the mechanisms involving horizontal electric fields and/or neutral winds will work because the vertical velocity is proportional to cosine of a dip angle, which becomes 0.034 for the dip of 88° compared to 0.17 for the dip angle of 80°. Based on observations made with the Canadian Advanced Digital Ionosonde (CADI) at three stations in the north polar cap (Alert, Eureka, and Resolute Bay), MacDougall et al. (2000a, b) distinguish two types of ionization layers: height-spread (maximum occurrence during 20:00 LT in winter) and thin (maximum occurrence near 12:00 LT in summer). They proposed that the height-spread Es is caused by transport of metallic ions from the dayside E region to the central polar cap lower F region. The metallic ions then sediment into an exponentially increasing atmosphere and form a layer. They may also be converged by gravity waves. The gravity wave signature is clearly seen for many high-latitude observations, including the one in Fig. 10.23 for Sondrestrom (dip is 80°). MacDougall et al. (2000a, b) observed similar Es undulations at Eureka and Resolute Bay on January 1, 1995. Enhancements in E-region ionization were accompanied by plasma depletions in the F region (see their Fig. 1), again suggesting some sedimentation. The thin Es that is observed only in summer is produced by photoionization followed by charge exchange with neutral metallic atoms. MacDougall et al. (2000a, b) found that both winter and summer Es types were associated with positive IMF By. Why this is so is yet not clear. One explanation could be in the enhanced gravity wave activity during positive IMF By.

10.5 Plasma Waves and Irregularities in the High-Latitude E Region: Observations

Due to the high neutral density at E-region heights, plasma production, diffusion, and recombination all proceed very rapidly. The large-scale horizontal organization of E-region plasma is therefore dominated by the spatial character of solar and particle precipitation sources. In the vertical direction, the energy distributions of the photon or particle impact ionization sources are important. Since we are more interested in dynamical processes in this text, we do not treat production and loss processes in any detail. The reader is referred to texts by Rishbeth and Garriott (1969) and Banks and Kockarts (1973). Gravity waves are somewhat less important as a source of vertical structuring at high latitudes due to the large dip angle of the magnetic field.

Just as is the case at the magnetic equator, the two primary plasma instabilities sources of structure in the auroral E region are the two-stream and gradient drift instabilities. The radar echoes are referred to as type 1 and type 2, based on the spectral width and mean Doppler shift. Although it is often thought that these two designation types correspond to each other, it seems this is not the case. These instabilities have been discussed in great detail in Chapter 4. There are some major differences in the character of the ionosphere at high latitudes, however, that affect the relative importance of these two processes. In addition, there is evidence for a third radar mode (type 3) and for a special type of type 1 associated with anomalous electron heating by the electrojet waves (type 4). The electric field in the polar cap and auroral zone is applied from "above" and is appreciably larger and more widespread than in the equatorial region. The average auroral electric field, for example, has a strong diurnal component with an amplitude of 30 mV/m and maxima near 0500 and 1800 LT (Mozer and Lucht, 1974). Fields as high as 50 mV/m are common in both the oval and polar cap, and numerous measurements exceeding 100 mV/m have been reported. The latter corresponds to a drift velocity of 2000 m/s, which is a value five times higher than the threshold of the two-stream instability.

The plasma density structure is much more variable in the auroral zone than in the equatorial region. However, in the auroral region the magnetic field is inclined at about 10° to the vertical, and therefore a vertical electron density gradient has a density scale length perpendicular to the magnetic field that is appreciably larger than the same vertical gradient would have at the magnetic equator. This in turn means that the gradient drift instability tends to have a lower growth rate at high latitudes for the same size vertical gradients even though the plasma drifts are larger. In the auroral region there are also important horizontal density gradients due to spatial variations in the particle precipitation that must be considered but that are difficult to measure.

The dynamic nature of both electric fields and density gradients makes analysis of auroral E-region instabilities even more challenging than their equatorial counterpart. On the other hand, a wealth of radar data does exist at E-region heights. These data have been supplemented by results from a number of rocket flights and a fairly complete experimental picture has emerged. In this chapter we concentrate on the plasma instabilities which operate in the auroral E region. In the next few sections some of the salient experimental data are presented. This is followed by a theoretical discussion in which additional data are introduced where appropriate.

Since auroral currents can be quite large, the term *auroral electrojet* is often used. The "jet" in this case refers to the relatively localized height range of the current rather than a narrow latitudinal range which characterizes the equatorial jet as well.

10.5.1 Radar Observations

In the equatorial electrojet most radar observations resolve into two classes: type 1 irregularities, which display a narrow Doppler spike that is offset from zero by the acoustic frequency $\omega_A = kC_s$; and type 2 irregularities, which display a broad Doppler spread centered on the frequency corresponding to the line-of-sight electron drift velocity. Both of these echo types have their counterparts in the auroral case. However, as summarized in a review by Fejer and Kelley (1980), there are various Doppler signatures associated with the "radar aurora." Since that review was written, considerable progress has been made in theoretical and experimental studies aimed at sorting out the origins of the various spectral signatures.

The four Doppler spectral types are shown in Fig. 10.24. The spectrum in the upper panel is very much like a narrow type 1 echo at the magnetic equator that locks onto the acoustic frequency. The second spectrum is very broad with a Doppler shift of less than C_s and is not unlike the equatorial type 2 case.

In the third panel a very narrow spectrum is shown that has a Doppler velocity that is much less than the acoustic speed and which corresponds to a Doppler shift of about 70 Hz. The signals in this mode are very strong and thus when the spectra are normalized the narrow spike becomes a dominant feature. These waves were first reported by Fejer et al. (1984b) and have inspired considerable theoretical interest. The implication of the narrow spike is that a very coherent wave is present. The 70 Hz Doppler shift is about 40% above the gyrofrequency of O^+, which has led to a number of studies dealing with the excitation of oxygen cyclotron waves in the upper E region or lower F region. We return to this topic in the discussion section following.

We turn now to the fourth type of Doppler signature. High-latitude Doppler spectra often display such large "type 1" Doppler shifts that it was originally thought that in the high-latitude case, 3 m waves did not reach a limiting phase velocity at C_s. The Doppler drift was therefore assumed to be equal to the line-of-sight electron drift speed, a result in agreement with the linear theory of two-stream irregularities. This interpretation is very much in doubt now, and a considerable literature exists on anomalous electron heating due to electrojet turbulence. This topic is taken up separately in some detail in Section 10.5.2. As we shall see, the result is that the type 4 waves might still saturate at a phase velocity equal to C_s but that the sound velocity is much higher than expected.

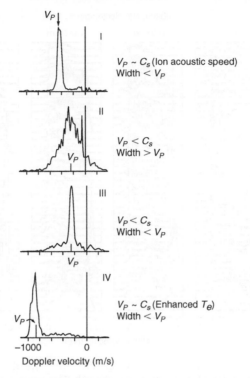

Figure 10.24   The four types of radar spectra that are observed in the auroral electrojet. (Figure courtesy of J. Providakes.)

An example showing the complexity of the radar aurora is given in Fig. 10.25. Consecutive spectra separated by about 6 s are shown from four different ranges. Many examples of broad type 2 Doppler shifts were obtained during this time period, but in addition very strong sporadic echoes were seen which have very large Doppler shifts. The spectra look like type 1 events but have shifts much larger than expected for model values of the temperature and thus the acoustic speed at E region altitudes. These are type 4 echoes.

Since the coherent scatter radar data are complicated to interpret, it is important to determine the zero-order plasma properties and the k spectrum of the waves independently in order to distinguish and understand the various possible plasma instability mechanisms. As in the equatorial case, rocket measurements have been used to add to our knowledge of auroral electrojet instabilities. These data are discussed next.

10.5.2   Rocket Observations of Auroral Electrojet Instabilities

The plot in Fig. 10.26 is a composite from six different rocket wave electric field measurements in very different background dc electric field conditions

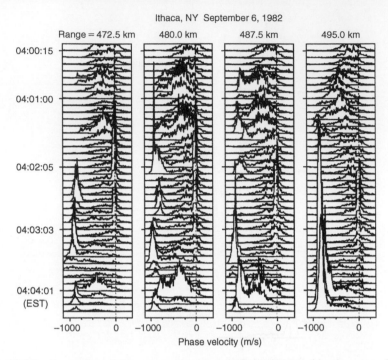

Figure 10.25 Examples of the temporal and spatial variations of backscattered power and Doppler velocity at 50 MHz in the auroral oval. The spectra were normalized to the same value and the integration time was about 6 s. [After Providakes et al. (1985). Reproduced with permission of the American Geophysical Union.]

(Pfaff et al., 1984; Pfaff, 1985). The frequency-altitude-intensity format used here shows where the waves occur and their relative intensity. In the bottom right-hand panel ($E_{dc} = 10$ mV/m) there is no detectable signal. As the dc electric field strength increases the layer thickness and the intensity both increase dramatically. Near the top of the layer a discrete narrow band emission is observed in some cases (e.g., panels a through d). This feature is also apparent in the sequence of discrete spectra in Fig. 10.27, which correspond to the experiment in panel b of Fig. 10.26. At a height of 120 km a very narrow band signal was detected which, except for the exact value of the center frequency, is nearly indistinguishable from the topside equatorial electrojet spectrum plotted in Fig. 4.30. The value of the center frequency in this coherent feature can vary considerably from flight to flight. This has been explained through the dependence of the Doppler shift of the signal on the rocket velocity vector, the plasma velocity vector, and the acoustic speed in a given flight (Pfaff, 1985). The data seem to indicate a very turbulent "heart" to the electrojet with coherent structures in the upper regions. This description is very similar to that of the daytime equatorial electrojet rocket data discussed in Chapter 4.

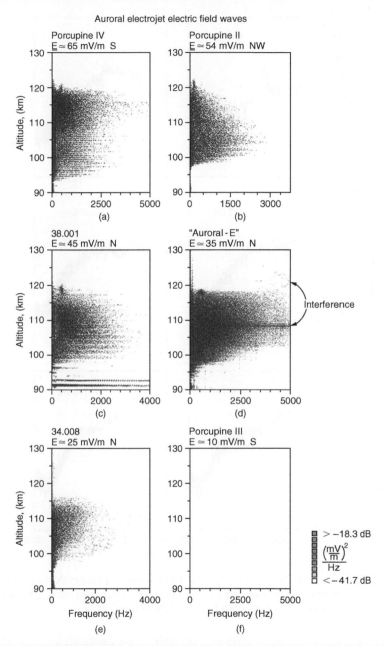

Figure 10.26 Frequency-height sonograms of electric field wave measurements made during six different auroral zone experiments. The gray levels are identical for all six sonograms and the dc electric field is noted for each panel. [After Pfaff et al. (1984). Reproduced with permission of the American Geophysical Union.]

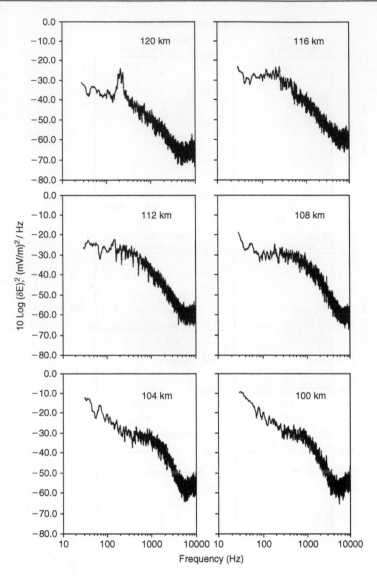

Figure 10.27   Six electric field spectra spaced 4 km apart as a rocket descended through the electrojet. The spectra each represent 1.092 s of data. [After Pfaff et al. (1984). Reproduced with permission of the American Geophysical Union.]

Most rocket measurements, including those shown here, emphasize waves with $\lambda \leq$ tens of meters. This is also true in the equatorial electrojet. This limitation is due to the high velocity at which most rockets penetrate the E region. Like the Condor experiments (see Chapter 4), the *Auroral-E* rocket experiment was dedicated to low-altitude studies and thus the rocket spent more time in the electrojet (Pfaff, 1985). This allows longer-wavelength waves to be studied. The electric field data from the latter experiment are shown in Fig. 10.28e. The electric field signal is modulated by the rocket spin. The envelope of this modulation is smoothly varying above about 102 km. Below this height the spin signal is modulated by the geophysical phenomenon of interest here. Analysis indicates that these modulations correspond to roughly a 300 m horizontal wavelength for the electric field fluctuation. These waves occurred in a region of vertical plasma density gradient (see Fig. 10.28a) and of northward dc electric field. This, of course, is the proper orientation for instability of the gradient drift wave. Notice that the waves are localized in altitude even though the near-vertical magnetic field allows very efficient coupling along **B**. We return to a discussion of this problem and these same data following.

10.5.3 Simultaneous Data Sets

As in the equatorial case, significant advances are possible when more than one measurement technique is applied in a given experiment. The data in Fig. 10.28 are of this type, since the incoherent scatter radar at Chatanika was used to provide the electron density profile shown in panel a as well as the ambient electric field. Armed with these data, the vertical profile of $V_e - V_i$ was calculated (panel b) and the linear theory (see Section 10.6) was used to calculate the growth rate as a function of wave number in panel c. When compared to the electric field wave data in panels d and c, it is clear at once that high-frequency (high k value) waves are observed in the upper electrojet, where the two-stream term dominates. However, the long-wavelength waves are present in the bottomside gradient region, where the quantity $V_d/(l + \psi)$ is below the threshold value

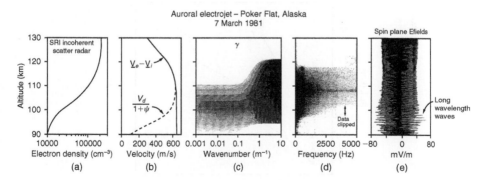

Figure 10.28 Growth rate modeling for the combined two-stream and gradient drift instability and comparison to the *Auroral-E* data. The model assumes a stable, vertical electron density gradient as measured by the SRI incoherent scatter radar and a 35 mV/m ambient electric field. The electron density is assumed to be uniform in horizontal extent. The in situ measured electric field data are shown in the two panels to the far right. [After Pfaff et al. (1984). Reproduced with permission of the American Geophysical Union.]

(C_s) required for two-stream instability. It is clear from this study that both the gradient drift and the two-stream condition play a role in the auroral electrojet.

A controversy over interpretation of the phase velocity of meter-wavelength waves at high latitudes has been resolved by simultaneous coherent scatter radar and incoherent scatter radar observations. The STARE coherent scatter Doppler data at first were interpreted as representing the projection of $(\mathbf{E} \times \mathbf{B}/B^2)$ on the radar line of sight. This interpretation agrees with linear theory, since (see Chapter 4) that theory predicts

$$\omega \cong \mathbf{k} \cdot \mathbf{V}_D/(1 + \Psi) \tag{10.19}$$

but disagrees with the equatorial electrojet observations, which show that $\omega/k = C_s$ at any angle to the current. In view of recent equatorial data showing that Doppler shift velocities match the nonisothermal ionacoustic speed rather than the isothermal (Chapter 4), an application of these equatorial results to high latitudes is now in progress with added consideration for elevated electron and ion temperatures and electron heat flows (L. M. Kagan, personal communication). On the other hand, $\mathbf{E} \times \mathbf{B}/B^2$ drifts as large as 2000 m/s are not uncommon at high latitudes, so it was not unreasonable to assume that the coherent scatter Doppler shift yields the projection of the $\mathbf{E} \times \mathbf{B}$ velocity.

Some of the first evidence that E-region temperatures might be very high came from incoherent scatter observations in Alaska (Schlegel and St.-Maurice, 1981). An example is shown in Fig. 10.29, where T_e near 110 km is about three times the neutral gas temperature. The dc electric field in this event was measured by the same radar to be 85 mV/m, which clearly must have been driving very intense

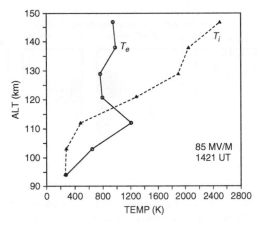

Figure 10.29 Typical electron (solid line) and ion (dashed line) temperature profiles in the polar E region in the presence of a high electric field (southward, 85 mV/m in this example) measured at 1421 UT on November 13, 1979. [After Schlegel and St.-Maurice (1981). Reproduced with permission of the American Geophysical Union.]

waves in the electrojet (e.g., see Fig. 10.26). For reference, the sound speed for a mean ion mass of 31 amu, $T_e = 1200$ K, and $T_i = 500$ K is 900 m/s.

Simultaneous observations of the plasma drift velocity and the wave phase velocity were made at EISCAT in conjunction with the STARE coherent scatter radar (Nielsen and Schlegel, 1985). The data are compared in Fig. 10.30 for (a) the linear "fluid" model and (b) the "acoustic" model. The acoustic model is clearly in better agreement with the data. The implication is that the waves travel at the speed of $C_s \cos\theta$, which contradicted the conventional wisdom for equatorial observations that the waves travel at the speed of C_s independent of θ. But as discussed in Chapter 4, there is a way to reconcile the results at both latitudes. Nielsen and Schlegel (1985) also found that the acoustic speed is an increasing function of plasma drift and thus the applied electric field. This is also strong evidence for heating.

The $C_s \cos\theta$ condition was further supported by simultaneous rocket electric field and 30 MHz radar observatories in Alaska (Bahcivan et al., 2005). The radar used a sophisticated interferometer that allowed cross field of view as well as radial velocities to be determined. They used an empirical relationship between T_e and E developed by Nielsen and Schlegel (1985) to determine C_s. This work also synthesized our understanding of the relationship between type 1, 2, and 3 waves. They found that at high latitudes the narrow type 1 echoes are only seen when the radar is looking nearly parallel or antiparallel to the electron drift velocity. The Doppler shift is then C_s. When there is a reasonable angle to the drift velocity, the spectrum widens, but the mean Doppler shift traces $C_s \cos\theta$. It seems that as far as strong radar echoes are concerned, at high latitudes type 1 echoes are driven by the two-stream instability. Type 4 echoes are simply due to nonisothermal ion-acoustic waves at elevated electron temperatures.

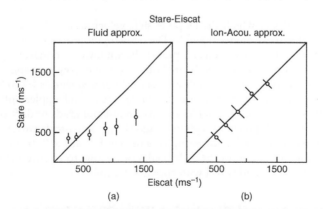

Figure 10.30 Magnitude and direction of the electron drift velocity estimated from the STARE Doppler velocity measurements, applying (a) the cosine relationship and (b) the ion-acoustic approach, as a function of the electron drift velocity determined by the incoherent scatter facility EISCAT. [After Nielsen and Schlegel (1985). Reproduced with permission of the American Geophysical Union.]

From these results it seems clear that the type 4 echo occurs when strong electron heating occurs in the electrojet due to the waves themselves. The anomalous heating mechanism is not entirely understood at present but there seems little doubt that it is occurring (Providakes et al., 1988). This interesting nonlinear plasma problem is treated further following.

10.5.4   Summary

About 40 years after the first detection and theoretical work on the Farley-Buneman instability, and with combinations of incoherent and coherent scatter radars as well as simultaneous rocket electric field and coherent scatter radar, it seems the experimental description is finally self-consistent at equatorial and high latitudes. At high latitudes we have the following summary:

1. The two-stream instability dominates as a radar echo source.
2. The spectra are narrow when the line of sight is near the electron drift velocity and shifted by C_s.
3. At large angles, the main Doppler is $C_s \cos \theta$ and the spectra are broad.
4. For large electric fields, the electron temperature is enhanced and C_s rises.
5. The gradient drift instability occurs, as indicated by rocket experiments, but is less important.

At equatorial latitudes we have the following summary:

1. Rocket data indicate strong, horizontally propagating, narrow band, two-stream waves at the top of the electrojet where the gradient vanishes. These waves cannot be detected directly by ground-based radars.
2. Vertical echoes from this pure two-stream region (i.e., perpendicular to the electron drift) have broad spectrum with an average Doppler shift of zero which equals $C_s \cos \theta$.
3. There are altitudes at which the Doppler shift velocities match nonisothermal ion-acoustic speed and are higher than C_s. The two-stream processes change from being super-adiabatic at lower electrojet altitudes to becoming isothermal at higher altitudes via a transitional process defined by inelastic electron energy exchange. The latter is important at low frequencies and disappears at high frequencies $\geq 150\,\mathrm{MHz}$.
4. The ubiquitous observation of narrow (type 1) spectra with Doppler shifts equal to C_s is due to the simultaneous instability of two-stream and gradient drift processes and the subsequent development of high-amplitude large (kms)-scale waves. The result is that somewhere in the radar scattering volume the $(\mathbf{E} \times \delta \mathbf{E}) \times \mathbf{B}$ drift exceeds C_s.
5. Below this region, pure gradient drift modes occur due to electric fields and winds having type 2 spectra and travel with speeds $\leq V_d \cos \theta$.

10.6   Linear Auroral Electrojet Wave Theories

The local linear theory for the gradient drift and two-stream instabilities in the auroral electrojet does not differ very much from its counterpart in the equatorial case (see Chapter 4). In this section we delineate some of the additional factors that arise at high latitudes. By "local" in this context we mean that we ignore coupling along the magnetic field lines. In the following discussion, we follow the presentation by Fejer and Providakes (1988) quite closely.

In the lower E region (between about 90 and 100 km), the ions are unmagnetized ($v_i \ll \Omega_i$), and the electrons are collisionless to zero order ($\Omega_e \gg v_e$), so $V_D = E \times B/B^2$. Linear theory then is identical to the result in Chapter 4, and the oscillation frequency and growth rate, in the reference frame of the neutral wind, are given by

$$\omega_r = \mathbf{k} \cdot (\mathbf{V}_D + \Psi \mathbf{V}_{Di}) / (1 + \Psi) \tag{10.20}$$

$$\gamma = (1 + \Psi)^{-1} \left\{ (\Psi/v_i) \left[(\omega_r - \mathbf{k} \cdot \mathbf{V}_{Di})^2 - k^2 C_s^2 \right] \right. \\ \left. + (1/Lk^2) (\omega_r - \mathbf{k} \cdot \mathbf{V}_{Di}) + (v_i/\Omega_i) k_y \right\} - 2\alpha n_0 \tag{10.21}$$

where

$$\Psi = \Psi_0 \left[\left(k_\perp^2/k^2 \right) + \left(\Omega_e^2/v_e^2 \right) \left(k_\parallel^2/k^2 \right) \right]$$

and

$$\Psi_0 = v_e v_i / \Omega_e \Omega_i$$

In this expression $\mathbf{V}_{Di}$ is the ion drift velocity, which could be due either to neutral winds or to the ambient electric field. The assumptions used are $\omega_r \ll v_i$, or in other words that the wavelengths are much larger than the ion mean free path, and $|\gamma| \ll \omega_r$. In addition, we consider $k \gg k_0$, where $k_0 = (v_i/\Omega_i)[(1 + \Psi)L_N]^{-1}$, so the linear waves are nondispersive. These approximations are valid for wavelengths between a few meters and a few hundred meters. For shorter wavelengths, kinetic theories are needed. The first term in the right-hand side of (10.21) is the two-stream term, which includes a diffusive damping term of the form $k^2 C_s$. The second term describes the gradient drift instability, while the last term is due to recombinational damping. The two-stream term is dominant at short wavelengths ($1 \text{ m} \leq \lambda \leq 20 \text{ m}$) and yields instability when $\mathbf{k} \cdot \mathbf{V}_D > k C_s (1 + \psi)$. For reference a set of high-latitude parameters was chosen and the threshold drift velocity for instability evaluated as a function of wavelength. The results are shown in Fig. 10.31 for $k_\parallel = 0$ at a height of 105 km. The parameters used are $v_e = 4 \times 10^4 \text{ s}^{-1}$, $v_i = 2.5 \times 10^3 \text{ s}^{-1}$, $\Omega_e = 10^7 \text{ s}^{-1}$, $\Omega_i = 180 \text{ s}^{-1}$, $C_s = 360 \text{ m/s}$, and $2\alpha n_0 = 0.06 \text{ s}^{-1}$. Several different gradient scale lengths were used. Large-scale waves are easily excited only if the electron density gradient is destabilizing (L_N is negative). Note also that the two-stream threshold drift ($L_N = \infty$) increases rapidly with a wavelength for $\lambda \geq 20 \text{ m}$. Therefore, the two-stream instability mechanism is essentially restricted to short wavelengths.

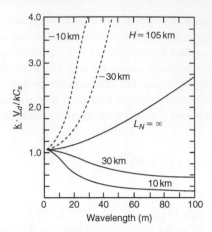

Figure 10.31 Variation of the normalized threshold drift velocity with wavelength and electron density gradient length for waves perpendicular to the magnetic field. The solid (dashed) curves correspond to destabilizing (stabilizing) electron density gradients. [After Fejer et al. (1984a). Reproduced with permission of the American Geophysical Union.]

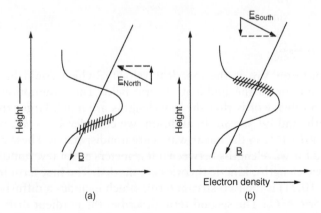

Figure 10.32 Sketch showing (a) unstable positive gradient in the vertical electron density profile in the presence of a poleward (northward in the northern hemisphere) electric field and (b) unstable negative gradient in the presence of an equatorial (southward in the northern hemisphere) electric field.

10.6.1   The Gradient Drift Instability

There are several ways to obtain density gradients at high latitudes. Figure 10.32 illustrates the case when a *vertical* density gradient exists in the presence of a dip angle which is large but is not equal to 90°. For example, this situation applies when solar illumination produces the E region plasma or when diffuse auroral precipitation is the source. Then a vertically upward gradient is unstable for a northward electric field, while a downward gradient is unstable for a southward field. The gradient scale length perpendicular to B is $L_\perp = L_N(\cos I)^{-1}$, where I is the dip angle. For the same value of L_N, $L_\perp$ is much larger at high latitudes than in the midlatitude or equatorial case. Even when classical sporadic E layers form at high latitudes, the $\cos I$ term reduces their importance. Steep horizontal gradients may arise due to very intense localized auroral arcs. However, such arcs have a high electron density, which drastically increases the recombinational damping. Thus, the gradient drift process is less important at high latitudes than elsewhere in the ionosphere.

That said, we now proceed to show that the gradients still cannot be totally ignored. In Fig. 10.33 the electron density profile and intermediate to short-wavelength wave data are plotted side by side for two rocket flights through the auroral electrojet. In the upper case, E had a northward component, while in the lower case E was southward. In each case the solid trace on the left shows the true electron density variation with altitude. In the upper plot the wave activity shows a local minimum exactly where the density gradient changes sign. In the lower example the wave activity increases when the vertical density gradient changes sign. The implication is that the electron density gradient controls or at least affects the intensity of electrojet turbulence but that the two-stream process is the dominant source of wave activity. In both of these events the electron drift speed was well over 1000 m/s, much higher than the typical sound speed.

One example which clearly seems to show an auroral event in which a dominant role is played by the gradient drift instability was shown earlier in Fig. 10.28. Here large-wavelength electric field fluctuations were observed on the bottom-side. The waves are quite intense ($\delta E \approx 25$ mV/m) and are polarized in the direction of the E $\times$ B drift (Pfaff, 1985). Of course, this low-apogee flight was one of the few rocket experiments capable of detecting large-scale waves on the bottomside and thus they may well exist much of the time when the electric field is northward and when the electron density, n_0, is not so high that recombination dominates.

This example is potentially quite interesting because it illustrates the altitude dependence of the wave amplitude. A detailed study of the linear theory for the event has been carried out and the results were plotted in panel c of Fig. 10.28. Indeed, the long-wavelength waves are observed at the same altitude as those where the growth rate in panel c shows positive values at long wavelengths. The observed longest-wavelength waves (panel e) seem to cut off just above 100 km altitude, although the intermediate to short-wavelength waves (panel d) remain strong between 100 and 110 km.

We now consider the extent to which these long-wavelength waves map to higher altitudes along the magnetic field lines. The theory of the mapping of electrostatic fields (which can include the fields associated with the long-wavelength waves) along magnetic field lines was reviewed in Chapter 2. Although the magnetic field lines may sometimes be considered equipotentials, in practice

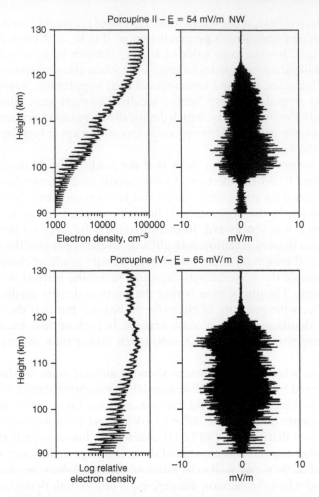

Figure 10.33 Wave electric field and fixed-bias Langmuir probe data from the traversal of the *Porcupine II* and *Porcupine IV* payloads through the electrojet region. The *Porcupine II* experiment was conducted during a northward electric field, whereas the *Porcupine IV* experiment was flown in the presence of a southward electric field. The solid line in the left-hand panel shows the density profile. The density decreases are due to probe end effects. [After Pfaff et al. (1984). Reproduced with permission of the American Geophysical Union.]

a nonzero attenuation will occur for wave electric fields. The parallel attenuation length resulting from the mapping of an electric field perpendicular to the magnetic field may be approximated by

$$\lambda_{\parallel} \cong (\sigma_0/\sigma_P)^{+1/2}\lambda_{\perp} \tag{10.22}$$

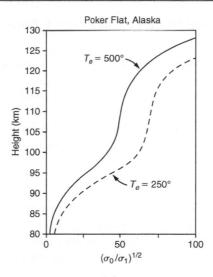

Figure 10.34 Curves representing $(\sigma_0/\sigma_P)^{1/2}$ in the auroral zone for constant temperature profiles corresponding to $T_e = 500\,\text{K}$ and $T_e = 250\,\text{K}$.

where σ_0 and σ_P are the parallel and Pedersen conductivities. In Fig. 10.34 $(\sigma_0\sigma_P)^{1/2}$ is plotted as a function of altitude using a model neutral atmosphere and the geomagnetic field at Poker Flat, Alaska. The computation of this ratio was very sensitive to the particular expression used for the collision frequencies and may vary quite a bit at these low altitudes. In addition, since these collision frequency values depend critically on the electron temperature, which also varies rapidly here, we have plotted two curves corresponding to $T_e = 500\,\text{K}$ and $T_e = 250\,\text{K}$. Notice that the conductivity ratio changes considerably in the region we are considering. For the sake of discussion, consider the value of this ratio at 96 km to be approximately 35. Thus, using (10.22), a 300 m wave would map through a distance of 10.5 km as it decays along the field lines. The calculated distance is somewhat larger than that given by the observations, but the basic idea seems correct. In any event, this calculation is consistent with the hypothesis that the source region of the waves was very narrow in height. In turn, this supports the claim that the waves were produced by a locally unstable electron density gradient and that the observed dominant wavelength may have corresponded to the peak in the linear growth rate for the gradient drift instability. The perturbation electric field mapped to a region where the waves were stable. In Chapter 5 we showed an example in which waves were detected on only one side of a midlatitude layer for which this condition held (Kelley and Gelinas, 2000). Seyler et al. (2004) developed a nonlocal theory for the midlatitude case, which can deal with such conditions.

10.6.2 The Two-Stream Instability and Type 4 Radar Echoes

Of course, the two-stream instability is very easily excited in the auroral case and it is probably responsible for most radar echoes. The possibility that such waves heat E region electrons to temperatures far in excess of that of the neutral gas makes their study very interesting, since such wave-induced processes are very important in space plasma physics in general.

The type 4 observations discussed earlier give evidence for local increases in the ion acoustic speed of up to 900 m/s. For a mean ion mass of 31 amu and an ion temperature of about 500 K, this corresponds to an electron temperature of about 2500 K for isothermal ions and electrons. The temperature would be reduced when electrons are nonisothermal, and slightly reduced for different specific heat ratios. Electron temperatures of this magnitude have been measured with incoherent radars during highly active periods (see Fig. 10.29). Cosmic noise absorption levels measured during events of this type are also consistent with large increases in the effective electron temperature in the locally heated region (Stauning, 1984).

Schlegel and St.-Maurice (1981) argued that the large electron temperature enhancements in the unstable electrojet layer cannot be explained by either particle precipitation or classical Joule heating. These enhancements maximize at the height (about 110 km) where two-stream waves are strongest. This led St.-Maurice et al. (1981) to suggest that these waves are responsible for the electron heating throughout the unstable region. Several theories were developed to explain the possible heating of the electron gas by strongly driven two-stream waves. Originally, St.-Maurice et al. (1981) considered a quasi-linear modified two-stream instability theory with the assumption that the wave energy is concentrated in a narrow band of short-wavelength ($\lambda \approx 20$ cm) waves where the linear growth rate is a maximum. However, the wave amplitudes necessary to generate the observed electron temperatures are much larger than measured by in situ probes in this wavelength range (e.g., Pfaff et al., 1984). St.-Maurice and Laher (1985) then suggested that heating of the electron gas is caused by parallel electric fields associated with long-wavelength gradient drift waves, while Primdahl and Bahnsen (1985) emphasized the role of an anomalous collision frequency, ν^*. Robinson (1986) also considered the heating of the electrons by the perpendicular gradient drift wave component in the presence of anomalous electron collisions. In this formulation, the electron heating rate can also be written

$$dW_e/dt \mid_{\text{waves}} = n_0 m \nu^* (V_D - C_s k_{\parallel}/k)^2$$

where ν^* is the anomalous electron collision frequency. Self-consistent calculations of anomalous electron collision frequencies necessary for the saturation of the two-stream waves and of the corresponding electron temperatures were in excellent agreement with the experimental results. (The concept of a ν^* was first proposed by Sudan (1983b), who used it to explain limitation of the wave phase velocity.)

A formula useful for studying electron heating by both collisional and wave-related phenomena has been given by St.-Maurice et al. (1986).

$$Q = N_e m_e \left(v_e + v^*\right) \left[\left(\frac{E}{B}\right)^2 + \left(\frac{\delta E}{B}\right)^2\right] + N_e m_e v_e \left(\frac{\Omega_e k_{||}}{v_e k_\perp}\right)^2 \left(\frac{\delta E}{B}\right)^2$$

In this expression $v_e + v^*$ is the sum of the collisional and anomalous electron collisions. The first term includes both the Joule heating due to the zero-order dc field and the Joule heating due to the wave field. Since $\delta E \leq E$, the first term is anomalous to the extent that $v^* > v_e$. The second term is entirely due to classical collisions, but the heating electric field in this case is due to the component of δE parallel to $\mathbf{B}$. This term shows that even a small $k_{||}$ can lead to very efficient heating due to waves.

Providakes et al. (1988) have argued that because radar experiments show that $k_{||}$ is large, the latter term is the most important. Furthermore, they point out that anomalous collisions could be included as well in this part of the expression which would increase the heating even more if $v^* > v_e$. Explaining the large value of $k_{||}$ is probably the single most important theoretical problem at this time.

10.6.3   Type 3 Radar Echoes: Are They Due to Ion Cyclotron Waves?

The type 3 observations seem to require additional flexibility in the theoretical analysis. Since there is evidence that such echoes come from the upper E region and are related to field-aligned currents, a general linear dispersion relation has been developed by Fejer et al. (1984a) for primary (directly excited) nearly field-aligned ($k_\perp >> k_{||}$) waves in the high-latitude E region, including cross-field and field-aligned drifts, ion inertia and magnetization, electron density gradients, and recombination. The general dispersion relation they derived describes the two-stream and gradient drift instability driven by Hall currents in the electrojet region, the magnetized two-stream ion cyclotron instability, which can be driven by either Hall or field-aligned currents between about 120 and 130 km, and the collisional electrostatic ion cyclotron instability driven by field-aligned currents above about 130 km.

In addition to the pulsed radar results shown in Fig. 10.24, observations with the University of Saskatchewan CW radar have shown type 3 spectra centered around +30 and +55 Hz during several periods of intense magnetic activity in the premidnight sector (Haldoupis et al., 1985). These resonant spectral peaks suggest the existence of cyclotron waves with frequencies the same as gyrofrequencies of the E region ion constituents (i.e., O^+, O_2^+, and NO^+). The different spectral components are probably generated in separate regions within the scattering volume. For field-aligned current-driven waves a positive Doppler shift corresponds to an upward current. These echoes are usually localized in space and time and are associated with substantial increases of the echo power but show very little variation in their Doppler shifts, implying highly coherent wave structures. One very significant feature of the observations is the large deviation of the scattering wave vector from the plane perpendicular to **B**. In other words, the waves seem to propagate at angles in the range of 82 to 85° from **B**. At such large angles the linear theory described previously indicated that the plasma should be stable. HF radar measurements show that the real part ω is in excellent agreement with ion cyclotron theory (Villain et al., 1987).

Interferometer observations generally show the region of type 3 echoes to be associated with horizontal shears in the cross-field plasma flow and with highly localized and irregular structures within the scattering volume. Providakes et al. (1985) reported type 3 echoes along the northern edge of a discrete visible auroral arc. These correspond to the darkened spectral peaks shown in Fig. 10.35. The type 3 echoes were associated with a negative velocity shear $W_z = -0.5\,\mathrm{s}^{-1}$, which, as noted earlier in this chapter, suggests the presence of large downward field-aligned currents. Rocket and satellite data show that the edges of auroral arcs are characterized by very large shears, which in turn are associated with intense field-aligned currents of many tens of microamperes per square meter.

One theory of electrostatic ion cyclotron (EIC) waves in the auroral ionosphere has been discussed by many investigators. In particular, Kindel and Kennel (1971) derived a kinetic linear dispersion relation for electrostatic waves in a weakly collisional $(\nu_e < k_{\parallel} V_e)$ uniformly magnetized plasma. We are interested here, however, in the excitation of EIC waves in the upper E region at wavelengths for which the electrons are strongly collisional, so the weak collision theory is not appropriate.

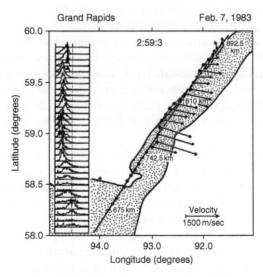

Figure 10.35 Fifty-megahertz Doppler spectra and radar velocity-position map during a discrete visual arc event in the morning sector. [After Providakes et al. (1985). Reproduced with permission of the American Geophysical Union.]

Kinetic studies of collisional EIC waves $(v_e > k_{||}V_e)$ in the lower auroral ionosphere were pursued by Providakes et al. (1985) and Satyanarayana and Chaturvedi (1985). These authors have shown that the destabilizing effect of electron collisions is necessary for the excitation of EIC waves in the upper E region, where ion-neutral collisions are stabilizing.

Satyanarayana and Chaturvedi (1985) showed that the maximum growth rates for collisional EIC waves in the bottomside ionosphere are comparable to those of collisionless EIC waves above the F region, where ion-neutral collisions are negligible and ion-ion collisions become an important damping mechanism. The critical electron drift velocity to excite EIC waves depends on electron density, on T_e/T_i and on the ion mass. Figure 10.36 shows three typical ionospheric electron density profiles and the corresponding critical drift velocities for generation of EIC waves. The critical drift velocity has a minimum between 150 and 200 km. This height range is consistent with VHF coherent radar observations. At 150 km the electron thermal speed (V_e) is about 100 km/s, which gives a threshold parallel drift velocity $V_D \cong 20$ km/s.

The theories discussed previously indicate that field-aligned drifts of about 30 km/s can excite collisional EIC waves at wavelengths of about 20 m in the upper E region. For an electron density of 5×10^4 cm^{-3}, this corresponds to a field-aligned current density of about $100 \, \mu$A/m^2. As noted previously, field-aligned currents up to $1000 \, \mu$A/m^2 and higher were observed during disturbed conditions in highly localized regions near the cusp on the Orsted satellite (Neubert and Christiansen, 2003). Since the currents along the geomagnetic field are alternatively going up and down, the field-aligned current density averaged over larger scales (as, for example, the spacial resolution of the EISCAT radar) are close to that observed with radar techniques and is about $10 \, \mu$A/m^2. These results provide considerable evidence that auroral field-aligned currents are

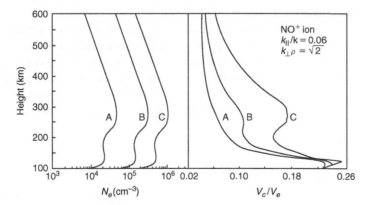

Figure 10.36 Three typical electron density profiles and the corresponding plot of the critical drift velocity for the generation of EIC. V_e is the electron thermal speed. (Figure courtesy of J. Providakes.)

large enough to generate collisional EIC waves in the upper E region at wave-lengths of 10–20 m. The most favorable altitude predicted by theory (about 150 km) is consistent with the observations, and the measured frequency range of 50–80 Hz is near the gyrofrequencies of the E-region ion constituents. How-ever, 3 m ion cyclotron waves can be directly excited only by unreasonably large field-aligned drifts (greater than the electron thermal speed). In addition, the theory predicts that for minimum threshold, $k_{||}/k = 0.007$ (aspect angle of 89.6°), whereas radar measurements indicate that aspect angles are as small as 80° ($k_{||}/k \cong 0.17$). The threshold velocity is much larger at these small aspect angles. It is possible that some nonlinear mechanism could generate coherent (resonant) 3 m waves from linearly unstable 20 m waves. In fact, Satyanarayana and Chaturvedi (1985) suggested that ion resonance broadening could provide such a cascade to short wavelengths. However, ion resonance broadening would generate broad spectra (spectral width much greater than Ω_i) at 3 m and, there-fore, cannot explain the radar data. Seyler and Providakes (1987) performed a numerical simulator that reproduced the type 3 echoes, but considerable work remains before we will finally understand these echoes.

10.6.4 Nonlinear Theories

The analytic (turbulence-based) gradient drift instability theory of Sudan and co-workers has already been discussed in Chapter 4, as have the simulation results of McDonald and co-workers. The theory should be applicable to the auroral case, although pure primary gradient drift waves are not as important at high altitudes as they are at the equator and at midlatitudes. In this approach the 3 m waves responsible for 50 MHz backscatter are due to a cascade of energy from unstable long-wavelength waves to stable short-wavelength modes. Because of the interest in electron heating at high latitudes we pursue the case of the nonlinear two-stream wave here.

Sudan (1983a, b) has suggested that momentum transfer between electrons and short-wavelength waves leads to an anomalous cross-field diffusion coefficient D^*. Since D^* is proportional to the collision frequency, there is a corresponding anomalous electron collision frequency v^* that increases with the wave ampli-tude. The nonlinear dispersion relation can be obtained by replacing v_e with v^*. The nonlinear two-stream oscillation frequency and growth rate are then given by

$$\omega_r^{NL} = \frac{\mathbf{k} \cdot \mathbf{V}_D}{1 + \Psi + \Psi^*} = kC_s \tag{10.23}$$

$$\gamma^{NL} = \frac{\Psi + \Psi^*}{v_i \left(1 + \Psi + \Psi^*\right)} \left[\left(\omega_r^{NL}\right)^2 - k^2 C_s^2 \right] \tag{10.24}$$

where ω_r^{NL} is the nonlinear oscillation frequency and $\Psi/\Psi^* = v/v^*$. In this theory the increasing value of Ψ^* in the denominator of (10.23) is what limits the phase velocity to values near marginal stability, $\omega_r^{NL}/k = C_s$, even though $\hat{k} \cdot \mathbf{V}_D$ is much larger than C_s. Large values of v^*/v_e are generally required to keep the two-stream phase velocity equal to C_s in the high-latitude electrojet region during periods of large plasma drifts, even though both C_s and v_e increase on these occasions. For example, for an altitude of 105 km, using $V_D = 1000$ m/s, $C_s = 420$ m/s, $v_e = 10.7 \times 10^4 \, \text{s}^{-1}$, $\Omega_e = 10^7 \, \text{s}^{-1}$, and $\Omega_i = 180 \, \text{s}^{-1}$, we must have $v^* \cong 18v_e \cong 1.5 \times 10^6 \, \text{s}^{-1}$. These magnitudes of v_e^*, however, require unreasonably large wave amplitudes.

A number of other investigations of the nonlinear stage of instability explain certain nonlinear features of the waves such as phase speed limitation, amplitude saturation, and certain other anomalous features (Hamza and St.-Maurice, 1993; St.-Maurice et al., 1994; Sahr and Fejer, 1996; Moorcroft, 2002; Brexler et al., 2002). The most impressive to date is the work of Otani and Oppenheim (1998) invoking three-wave coupling. This process was suggested previously to explain the strong vertical 50 MHz echoes in the region of equatorial pure two-stream instability (Kudeki et al., 1987), which have a broad spectrum with a mean Doppler shift of zero, and to explain strong, short-wavelength waves with vertical wave vectors in that same region as seen in a rocket experiment (Pfaff et al., 1987).

Otani and Oppenheim (1998) presented a simplified three-mode system reproducing many of the features seen in both the radar data and the particle simulations. The model followed the nonlinear time evolution of three spatial Fourier modes by integrating the two-fluid equations relevant to the Farley-Buneman instability. In equatorial electrojet geometry, the three modes used were a horizontally propagating mode, intended to represent the unstable primary mode driven by the electrojet vertical electric field, and two secondary modes traveling in oblique directions. The modes were chosen so as to satisfy the resonance condition $k_1 + k_2 = k_3$. For typical equatorial electrojet parameters, they found that the simple system generally reached a saturated state characterized by (1) a density perturbation of $\sim 5\%$, (2) vertically propagating secondary density perturbations propagating primarily along the density extrema of the primary wave, and (3) a slowing of the primary phase velocity from its linear value down to a value near the sound speed. These are all characteristics common to the saturated state observed in particle simulations (Oppenheim and Otani, 1996). The saturated state was also characterized by sinusoidal time dependences of all three modes, to be expected of this type of system. The frequencies of the modes obey the resonance condition $\omega_1 + \omega_2 = \omega_3$.

Computer simulations of the three-wave system produce steady states in close agreement with those obtained from the analytical method. Despite the fact that the system only contains three modes, a number of features also agree well with observation, including density fluctuation magnitudes ($|\delta n|/n_0 = 5\%$), propagation speeds (clustered around the sound speed), and power falloff as a function of elevation angle. The spatial distributions of the phases of the electron advection term and electron $\mathbf{E} \times \mathbf{B}$ velocity for the secondary modes lead to the partial cancellation of the destabilizing zero-order electron drift, thereby saturating the Farley-Buneman instability.

Built on these results, Bahcivan and Hysell (2006) proposed a theory for type 2 (broad) echoes they observed at an angle to the electron drift (Bahcivan et al., 2005). They propose secondary ion-acoustic structures driven by the electric fields in primary Farley-Buneman waves. In their model, the wave spectra broaden with a mean value equal to $C_s \cos\theta$.

10.7 Summary

In this chapter we only scratched the surface in the plasma physics of high-latitude phenomena. We have in some cases been able to export results from other latitudes, but, in general, the high-latitude regime has added features that complicate the physics. These problems are likely to keep ionospheric plasma physics an exciting discipline for the foreseeable future.

References

Bahcivan, H., and Hysell, D. L. (2006). A model of secondary Farley-Buneman waves in the auroral electrojet. *J. Geophys. Res.* 111, A01304, doi:10.1029/2005JA011408.

Bahcivan, H., Hysell, D. L., Larsen, M. F., and Pfaff, R. F. (2005). The 30 MHz imaging radar observations of auroral irregularities during the JOULE campaign. *J. Geophys. Res.* 110, A05307, doi:10.1029/2004JA010975.

Banks, P. M., and Kockarts, G. (1973). *Aeronomy,* Parts A and B. Academic Press, New York.

Basu, Su., Basu, Sa., S., C., Weimer, D., Neilsen, E., and Fougere, P. F. (1986). Velocity shears and sub-km scale irregularities in the nighttime auroral F region. *Geophys. Res. Lett.* 13, 101.

Bering, E. A., Kelley, M. C., and Mozer, F. S. (1975). Observations of an intense field-aligned thermal ion flow and associated intense narrow band electric field oscillations. *J. Geophys. Res.* 80, 4612.

Brexler, J., St.-Maurice, J.-P., Chen, D., and Moorcroft, D. R. (2002). New insights from a nonlocal generalization of the Farley-Buneman instability at high latitudes. *Ann. Geophys.* 20, 2003.

Burke, W. J., Kelley, M. C., Sagalyn, R. C., Smiddy, M., and Lai, S. T. (1979). Polar cap electric field structures with a northward interplanetary magnetic field. *Geophys. Res. Lett.* 6, 21.

Burke, W. J., Hardy, D. A., Rich, F. J., Kelley, M. C., Smiddy, M., Shuman, B., Sagalyn, R. C., Vancour, R. P., Wildman, P. J. L., and Lai, S. T. (1980). Electrodynamic structure of the late evening sector of the auroral zone. *J. Geophys. Res.* 85, 1179.

Burke, W. J., Gussenhoven, M. S., Kelley, M. C., Hardy, D. A., and Rich, F. J. (1982). Electric and magnetic field characteristics and discrete arcs in the polar cap. *J. Geophys. Res.* **87**, 2431.

Cerisier, J. C., Berthelier, J. J., and Beghin, C. (1985). Unstable density gradients in the high-latitude ionosphere. *Radio Sci.* **20**, 755.

Cole, K. D. (1965). Stable auroral red arcs, sinks for energy of DST main phase. *J. Geophys. Res.* **70**, 1689.

Costa, E., and Kelley, M. C. (1978). On the role of steepened structures and drift waves in equatorial spread F. *J. Geophys. Res.* **83**, 4359.

Cox, R. M., and Plane, J. M. C. (1998). An ion-molecule mechanism for the formation of neutral sporadic Na layers. *J. Geophys. Res.* **103**, 6349.

Crowley, G., and Hackert, C. L. (2001). Quantification of high latitude electric field variability. *Geophys. Res. Lett.* **28**(14), 2783.

Dyson, P. L., and Winningham, J. D. (1974). Topside ionospheric spread F and particle precipitation in the dayside magnetospheric clefts. *J. Geophys. Res.* **79**, 5219.

Dyson, P. L., McClure, J. P., and Hanson, W. B. (1974). In situ measurements of the spectral characteristics of F region ionospheric irregularities. *J. Geophys. Res.* **79**, 1497.

Earle, G. D. (1988). *Electrostatic Plasma Waves and Turbulence Near Auroral Arcs.* Ph.D. Thesis. Cornell University, Ithaca, New York.

Earle, G. D., and Kelley, M. C. (1993). Spectral evidence for stirring scales and two-dimensional turbulence in the auroral ionosphere. *J. Geophys. Res.* **98**(A7), 11,543–11,548.

Erukhimov, L. M., and Kagan, L. M. (1994). Thermo magnetic effects in ionospheric plasma. *J. Atmos. Terr. Phys.* **56**, 133.

Erukhimov, L. M., Kagan, L. M., and Myasnikov, E. N. (1982). Heating mechanism of irregularity generation in ionospheric F-layer. *Geomag. Aeron.* **22**(5), 597 (translated from Russian).

Farley, D. T. (1959). A theory of electrostatic fields in a horizontally stratified ionosphere subject to a vertical magnetic field. *J. Geophys. Res.* **64**, 1225.

Fejer, B. G., and Kelley, M. C. (1980). Ionosphere irregularities. *Rev. Geophys. Space Phys.* **18**, 401.

Fejer, B. G., and Providakes, J. (1988). High latitude E region irregularities: New results. *Phys. Scr.* **T18**, 167.

Fejer, B. G., Providakes, J. and Farley, D. T. (1984a). Theory of plasma waves in the auroral E region. *J. Geophys. Res.* **89**, 7487.

Fejer, B. G., Reed, R. W., Farley, D. T., Swartz, W. E., and Kelley, M. C. (1984b). Ion cyclotron waves as a possible source of resonant auroral radar echoes. *J. Geophys. Res.* **89**, 187.

Ganguli, G., Palmadesso, P., and Lee, C. Y. (1985). Electrostatic ion cyclotron instability caused by a nonuniform electric field perpendicular to the external magnetic field. *Phys. Fluids* **28**, 761.

Goldman, S. R., Baker, L., Ossakow, S. L., and Scannapieco, A. J. (1976). Striation formation associated with barium clouds in an inhomogeneous ionosphere. *J. Geophys. Res.* **81**, 5097.

Greenwald, R. A., Baker, K. D., and Villain, J. P. (1983). Initial studies of small-scale F region irregularities at very high latitudes. *Radio Sci.* **18**, 1122.

Gussenhoven, M. S., Hardy, D. A., and Heinemann, N. (1983). Systematics of the equatorward diffuse auroral boundary. *J. Geophys. Res.* 88, 5692.

Haldoupis, C. I., Prikryl, P., Sojko, G. J., and Koehler, D. J. (1985). Evidence for 50 MHz bistatic radio observations of electrostatic ion cyclotron waves in the auroral plasma. *J. Geophys. Res.* 90, 10,983.

Hamza, A. M., and St.-Maurice, J.-P. (1993). A self-consistent fully turbulent theory of auroral E region irregularities. *J. Geophys. Res.* 98, 11,601.

Heelis, R. A., Vickrey, J. F., and Walker, N. B. (1985). Electrical coupling effects on the temporal evolution of F layer plasma structure. *J. Geophys. Res.* 90, 437.

Heinselman, C. J., Thayer, J. P., and Watkins, B. J. (1998). A high-latitude observation of sporadic sodium and sporadic E-layer formation. *Geophys. Res. Lett.* 25, 3059.

Hudson, M. K., and Kelley, M. C. (1976). The temperature gradient drift instability at the equatorial edge of the ionospheric plasma trough. *J. Geophys. Res.* 81, 3913–3918.

Kagan, L. M., and Kelley, M. C. (2000). A thermal mechanism for generation of type 2 small-scale irregularities in the ionospheric E region. *J. Geophys. Res.* 105, 5391.

Kagan, L. M., and St.-Maurice, J.-P. (2005). Origin of type-2 thermal-ion upflows in the auroral ionosphere. *Ann. Geophys.* 23(1), 13–24.

Kagan, L. M., Myasnikov, E. N., Kosolapenko, V. I., Kryazhev, V. A., Cheremnyj, V. A., and Persson, M. A. L. (1995). F-layer irregularities' formation at auroral latitudes: Radio wave scintillation and EISCAT observations. *J. Atmos. Terr. Phys.* 57, 917.

Kelley, M. C. (1986). Intense sheared flow as the origin of large-scale undulations of the edge of diffuse aurora. *J. Geophys. Res.* 91, 3225.

Kelley, M. C., and Carlson, C. W. (1977). Observation of intense velocity shear and associated electrostatic waves near an auroral arc. *J. Geophys. Res.* 82, 2343.

Kelley, M. C., and Gelinas, L. J. (2000). Gradient drift instability in midlatitude sporadic E layers: Localization of physical and wave number space. *Geophys. Res. Lett.* 27, 457.

Kelley, M. C., Baker, K. D., and Ulwick, J. C. (1979). Late time barium cloud striations and their relationships to equatorial spread F. *J. Geophys. Res.* 84, 18,910.

Kelley, M. C., Bering, E. A., and Mozer, F. S. (1975). Evidence that the ion cyclotron instability is saturated by ion heating. *Phys. Fluids* 18, 1590.

Kelley, M. C., Livingston, R. C., Rino, C. L., and Tsunoda, R. T. (1982a). The vertical wave number spectrum of topside equatorial spread F: Estimates of backscatter levels and implications for a unified theory. *J. Geophys. Res.* 87, 5217.

Kelley, M. C., Vickrey, J. F., Carlson, C. W., and Torbert, R. (1982b). On the origin and spatial extent of high-latitude F region irregularities. *J. Geophys. Res.* 87, 4469.

Kelley, M. C., Baker, K. D., Ulwick, J. C., Rino, C. L., and Baron, M. J. (1980). Simultaneous rocket probe scintillation and incoherent scatter radar observations of irregularities in the auroral zone ionosphere. *Radio Sci.* 15, 491.

Keskinen, M. J., Mitchell, H. G., Fedder, J. A., Satyanarayama, P., Zalesak, S. T., and Huba, J. D. (1988). Nonlinear evolution of the Kelvin-Helmholtz instability in the high latitude ionosphere. *J. Geophys. Res.* 93, 137.

Kindel, J. M., and Kennel, C. F. (1971). Topside current instabilities. *J. Geophys. Res.* 76, 3055.

Kintner, P. M., and Seyler, C. E. (1985). The status of observations and theory of high latitude ionospheric and magnetospheric plasma turbulence. *Space Sci. Rev.* 41, 91.

Kintner, P. M., Kelley, M. C., and Mozer, F. S. (1978). Electrostatic hydrogen cyclotron waves near one earth radius in the polar magnetosphere. *Geophys. Res. Lett.* 5, 139.

Knudsen, D. T. (1996). Spacial modulation of electron energy and density by nonlinear stationary Alfvén waves. *J. Geophys. Res.* **101**(A5), 10,761–10,772.

Knudsen, D. J., Kelley, M. C., and Vickrey, J. F. (1992). Alfvén waves in the auroral ionosphere: A numerical model compared with measurements. *J. Geophys. Res.* **97**(A1), 77–90.

Kudeki, E., Fejer, B. G., Farley, D. T., and Hanuise, C. (1987). The CONDOR equatorial electrojet campaign: Radar results. *J. Geophys. Res.* **92**, 13,451.

LaBelle, J., Sica, R. J., Kletzing, C., Earle, G. D., Kelley, M. C., Lummerzheim, D., Torbert, R. B., Baker, K. D., and Berg, G. (1989). Ionization from soft electron precipitation in the auroral F region. *J. Geophys. Res.* **96**(A4), 3791–3798.

Lassen, K., and Danielsen, C. (1978). Quiet time pattern of auroral arcs for different directions of the interplanetary magnetic field in the y-z plane. *J. Geophys. Res.* **83**, 5277.

Lewis, W. S., Burch, J. L., Goldstein, J., Horton, W., Perex, J. C., Frey, H. U., and Anderson, P. C. (2005). Duskside auroral undulations observed by IMAGE and their possible association with large-scale structures on the inner edge of the electron plasma sheet. *Geophys. Res. Lett.* **32**, L24103, doi:10.1029/2005GL024390.

Lui, A. T. Y., Meng, C.-I., and Ismail, S. (1982). Large amplitude undulations on the equatorward boundary of the diffuse aurora. *J. Geophys. Res.* **87**, 2385.

MacDougall, J. W., Plane, J. M. C., and Jayachandran, P. T. (2000a). Polar cap sporadic-E: Part 1, observations. *J. Atmos. Solar-Terr. Phys.* **62**, 1155–1167.

———. (2000b). Polar cap sporadic-E: Part 2, modeling. *J. Atmos. Solar-Terr. Phys.* **62**, 1169–1176.

Mozer, F. S., and Lucht, P. (1974). The average auroral zone electric field. *J. Geophys. Res.* **79**, 1001.

Neubert, T., and Christiansen, F. (2003). Small-scale, field-aligned currents at the top-side ionosphere. *Geophys. Res. Lett.* **30**(19), doi:10.1029/2003GL017808, 2010–2013.

Nicolls, M. J., Kelley, M. C., and Erdogen, C. (2005). Small-scale structure on the poleward edge of a stable auroral red arc. *IEEE Trans. Plasma Sci.* **33**(2), 412–413.

Nielsen, E., and Schlegel, K. (1985). Coherent radar Doppler measurements and their relationship to the ionospheric electron drift velocity. *J. Geophys. Res.* **90**, 3498–3504.

Oppenheim, M., and Otani, N. (1996). Spectral characteristics of the Farley-Buneman instability: Simulations vs. observations. *J. Geophys. Res.* **101**, 24,5731.

Ossakow, S. L., and Chaturvedi, P. K. (1979). Current convective instability in the diffuse aurora. *Geophys. Res. Lett.* **6**, 332.

Otani, N., and Oppenheim, M. (1998). A saturation mechanism for the Farley-Buneman instability. *Geophys. Res. Lett.* **25**(11), 1833.

Pfaff, R. F. (1985). *Rocket Measurements of Plasma Turbulence in the Equatorial and Auroral Elements*. Ph.D. Thesis. Cornell University, Ithaca, New York.

Pfaff, R. F., Kelley, M. C., Kudeki, E., Fejer. B. G., and Baker, K. D. (1987). Electric field and plasma density measurements in the strongly driven daytime equatorial electrojet. 2. Two-stream waves. *J. Geophys. Res.* **92**, 13,597.

Pfaff, R. F., Kelley, M. C., Fejer, B. G., Kudeki, E., Carlson, C. W., Pedersen, A., and Haüsler, B. (1984). Electric field and plasma density measurements in the auroral electrojet. *J. Geophys. Res.* **89**, 236.

Primdahl, F., and Bahnsen, A. (1985). Auroral E region diagnosis by means of nonlinearly stabilized plasma waves. *Ann. Geophys.* **3**, 57.

Providakes, J. F., Farley, D. T., Swartz, W. E., and Riggin, D. (1985). Plasma irregularities associated with a morning discrete auroral arc: Radar interferometer observations and theory. *J. Geophys. Res.* 90, 7513.

Providakes, J. F., Kelley, M. C., Swartz, W. E., Mendillo, M., and Holt, J. (1989). Radar and optical measurements of ionospheric processes associated with large electric fields near the plasmapause. *J. Geophys. Res.* 94, 5350.

Providakes, J. F., Farley, D. T., Fejer, B. G., Sahr, J., Swartz, W. E., Häggström, I., Hedberg, Å. and Nordling, J. A. (1988). Observations of auroral E region plasma waves and electron beating with EISCAT and a VHF radar interferometer. *J. Atmos. Terr. Phys.* 50, 339.

Rich, F. J., Burke, W. J., Kelley, M. C., and Smiddy, M. (1980). Observations of field-aligned currents in association with strong convection electric fields at subauroral latitudes. *J. Geophys. Res.* 85, 2335.

Rishbeth, H., and Garriott, O. K. (1969). *Introduction to Ionospheric Physics.* Int. Geophys. Ser. 14, Academic Press, New York.

Robinson, R. M., Tsunoda, R. T., Vickrey, J. F., and Guerin, L. (1985). Sources of F region ionization enhancements in the nighttime auroral zone. *J. Geophys. Res.* 90, 7533.

Robinson, T. R. (1986). Towards a self-consistent nonlinear theory of radar auroral backscatter. *J. Atmos. Terr. Phys.* 48, 417.

Roble, R. G., and Rees, M. H. (1977). Time-dependent studies of the aurora: Effects of particle precipitation on the dynamic morphology of ionospheric and atmospheric properties. *Planet. Space Sci.* 25, 991.

Ruohoniemi, J. M., Greenwald, A., Baker, K. D., Villain, J. P., and McCready, M. A. (1987). Drift motions of small scale irregularities in the high-latitude F region: An experimental comparison with plasma drift motions. *J. Geophys. Res.* 92, 4553.

Sagalyn, R. C., Smiddy, M., and Ahmed, M. (1974). High-latitude irregularities in the topside ionosphere based on Isis I thermal probe. *J. Geophys. Res.* 79, 4252.

Sahr, J. D., and Fejer, B. G. (1996). Auroral electrojet plasma irregularity theory and experiment: A critical review of present understanding and future directions. *J. Geophys. Res.* 101, 26,893.

St.-Maurice, J.-P., and Laher, R. (1985). Are observed broadband plasma wave amplitudes large enough to explain the enhanced electron temperatures of the high-latitude E region. *J. Geophys. Res.* 90, 2843.

St.-Maurice, J.-P., Hanuise, C., and Kudeki, E. (1986). Anomalous heating in the auroral electrojet. *J. Geophys. Res.* 91, 13,493.

St.-Maurice, J.-P., Schlegel, K., and Banks, P. M. (1981). Anomalous heating of the polar F region by unstable plasma waves. 2. Theory. *J. Geophys. Res.* 86, 1453.

St.-Maurice, J.-P., Prikryl, P., Danskin, D., Hamza, A. M., Sofko, G. J., Koehler, J. A., Kustov, A., and Chen, J. (1994). On the origin of narrow non-ion acoustic coherent radar spectra in the high latitude E region. *J. Geophys. Res.* 99, 6447.

Satyanarayana, P., and Chaturvedi, P. K. (1985). Theory of the current-driven ion cyclotron instability in the bottomside ionosphere. *J. Geophys. Res.* 90, 12,209.

Schlegel, K., and St.-Maurice, J. P. (1981). Anomalous heating of the polar E region by unstable plasma waves. 1. Observations. *J. Geophys. Res.* 86, 1447.

Schunk, R. W., and Sojka, J. J. (1987). A theoretical study of the lifetime and transport of large ionospheric density structures. *J. Geophys. Res.* 92, 12,343.

Schunk, R. W., Banks, P. M., and Raitt, W. J. (1976). Effects of electric fields and other processes upon the nighttime high latitude F layer. *J. Geophys. Res.* **81**, 3271.

Seyler, C. E., and Providakes, J. (1987). Particle and fluid simulations of resistive current-driven electrostatic ion cyclotron waves. *Phys. Fluids* **30**, 3113.

Seyler, C. E., Rosado-Roman, J. M., and Farley, D. T. (2004). A nonlocal theory of the gradient-drift instability in the ionospheric E region plasma at midlatitudes. *J. Atmos. Solar-Terr. Phys.* **66**, 1627–1637.

Smiddy, M., Kelley, M. C., Burke, W., Rich, R., Sagalyn, R., Shuman, B., Hays, R., and Lai, S. (1977). Intense poleward-directed electric fields near the ionospheric projection of the plasmapause. *Geophys. Res. Lett.* **4**, 543.

Spiro, R. W., Heelis, R. A., and Hanson, W. B. (1979). Rapid sub-auroral ion drifts observed by Atmosphere Explorer-C. *Geophys. Res. Lett.* **6**, 657.

Stauning, P. (1984). Absorption of cosmic noise in the E region during electron heating events. A new class of riometer absorption events. *Geophys. Res. Lett.* **11**, 1184.

Sudan, R. N. (1983a). Unified theory of type I and type II irregularities in the equatorial electrojet. *J. Geophys. Res.* **88**, 4853.

Sudan, R. N. (1983b). Nonlinear theory of type I irregularities in the equatorial electrojet. *Geophys. Res. Lett.* **10**, 983.

Tanskanen, P. J., Hardy, D. A., and Burke, W. J. (1981). Spectral characteristics of precipitation electrons associated with visible aurora in the premidnight oval during periods of substorm activity. *J. Geophys. Res.* **86**, 1379.

Tsunoda, R. T. (1988). High latitude F region irregularities: A review and synthesis. *Rev. Geophys. Space Phys.* **26**, 719.

Tsunoda, R. T., Häggerström, I., Pellinen-Wannberg, A., Steen, Å., and Wannberg, G. (1985). Direct evidence of plasma density structuring in the auroral F region ionosphere. *Radio Sci.* **20**, 762.

Vickrey, J. F., and Kelley, M. C. (1982). The effects of a conducting E layer on classical F region cross-field plasma diffusion. *J. Geophys. Res.* **87**, 4461.

Vickrey, J. F., Rino, C. L., and Potemra, T. A. (1980). Chatanika/TRIAD observations of unstable ionization enhancements in the auroral F region. *Geophys. Res. Lett.* **7**, 789.

Vickrey, J. F., Kelley, M. C., Pfaff, R., and Goldman, S. R. (1984). Low-altitude image striations associated with bottomside equatorial spread F: Observations and theory. *J. Geophys. Res.* **89**, 2955.

Vickrey, J. F., Livingston, R. C., Walker, N. B., Potemra, T. A., Heelis, R. A., Kelley, M. C., and Rich, F. J. (1986). On the current-voltage relationship of the magnetospheric generator at intermediate spatial scales. *Geophys. Res. Lett.* **13**, 495.

Villain, J.-P., Caudal, G., and Hanuise, C. (1985). A SAFARI-EISCAT comparison between the velocity of F region small-scale irregularities and the ion drift. *J. Geophys. Res.* **90**, 8433.

Villain, J.-P., Greenwald, R. A., Baker, K. B., and Ronhoniemi, J. M. (1987). HF radar observations of E region plasma irregularities produced by oblique electron streaming. *J. Geophys. Res.* **92**, 12,327.

von Zahn, U., von der Gathen, P., and Hansen, G. (1987). Forced release of sodium from upper atmospheric dust particles. *Geophys. Res. Lett.* **14**, 76.

Weber, E. J., Tsunoda, R. T., Buchau, J., Sheehan, R.-E., Strickland, D. J., Whiting, W., and Moore, J. G. (1985). Coordinated measurements of auroral zone plasma enhancements. *J. Geophys. Res.* **90**, 6497.

Weber, E. J., Klobuchar, J. A., Buchau, J., Carlson, H. C., Jr., Livingston, R. C., de La Beaujardiere, O., McCready, M., Moore, J. G., and Bishop, G. J. (1986). Polar cap F layer patches: Structure and dynamics. *J. Geophys. Res.* 91, 12,121.

Weber, E. J., Kelley, M. C., Ballenthin, J. D., Basu, S., Carlson, H. C., Fleischman, J. R., Hardy, D. A., Maynard, N. C., Pfaff, R. F., Rodriguez, P., Sheehan, R. E., and Smiddy, M. (1989). Rocket measurements within a polar cap arc: Plasma, particle and electric circuit parameters. *J. Geophys. Res.* 94, 6692.

Yamamoto, T., Ozaki, M., Inoue, S., Makita, K., and Meng, C.-I. (1994). Convective generation of "giant" undulations on the evening diffuse auroral boundary. *J. Geophys. Res.* 99, 19,499.

Index

Note: Page numbers followed by "f" indicates figures and "t" indicates tables.

International Geophysics Series

Edited by

Renata Dmowska
School of Engineering and Applied Sciences
Harvard University
Cambridge, Massachusetts

Dennis Hartmann
Department of Atmospheric Sciences
University of Washington
Seattle, Washington

H. Thomas Rossby
Graduate School of Oceanography
University of Rhode Island
Narragansett, Rhode Island

Volume 37 J. A. Jacobs. *The Earth's Core*, 2nd Ed. 1987*

Volume 38 J. R. Apel. *Principles of Ocean Physics*. 1987

Volume 39 Martin A. Uman. *The Lightning Discharge*. 1987*

Volume 40 David G. Andrews, James R. Holton and Conway B. Leovy. *Middle Atmosphere Dynamics*. 1987

Volume 41 Peter Warneck. *Chemistry of the National Atmosphere*. 1988*

Volume 42 S. Pal Arya. *Introduction to Micrometeorology*. 1988*

Volume 43 Michael C. Kelley. *The Earth's Ionosphere*. 1989*

Volume 44 William R. Cotton and Richard A. Anthes. *Storm and Cloud Dynamics*. 1989

Volume 45 William Menke. *Geophysical Data Analysis: Discrete Inverse Theory*, Revised Edition. 1989

Volume 46 S. George Philander. *El Niño, La Niña, and the Southern Oscillation*. 1990

Volume 47 Robert A. Brown. *Fluid Mechanics of the Atmosphere*. 1991

Volume 48 James R. Holton. *An Introduction to Dynamic Meteorology*, 3rd Ed. 1992

Volume 49 Alexander A. Kaufman. *Geophysical Field Theory and Method*.

　Part A: Gravitational, Electric, and Magnetic Fields. 1992*

　Part B: Electromagnetic Fields I. 1994*

　Part C: Electromagnetic Fields II. 1994*

Volume 50 Samuel S. Butcher, Gordon H. Orians, Robert J. Carlson, and Gordon V. Wolfe. *Global Biogeochemical Cycles*. 1992

Volume 51 Brian Evans and Teng-Fong Wong. *Fault Mechanics and Transport Properties of Rocks*. 1992

Volume 52 Robert E. Huffman. *Atmospheric Ultraviolet Remote Sensing*. 1992

Volume 53 Robert E. Houze, Jr. *Cloud Dynamics*. 1993

Volume 54 Peter V. Hobbs. *Aerosol-Cloud-Climate Interactions*. 1993

Volume 55 S. J. Gibowicz and A. Kijko. *An Introduction to Mining Seismology*. 1993

Volume 56 Dennis L. Hartmann. *Global Physical Climatology*. 1994

Volume 57 Michael P. Ryan. *Magmatic Systems*. 1994

Volume 58 Thorne Lay and Terry C. Wallace. *Modern Global Seismology*. 1995

Volume 59 Daniel S. Wilks. *Statistical Methods in the Atmospheric Sciences*. 1995

Volume 60 Fredrik Nebeker. *Calculating the Weather*. 1995

*Out of Print.

Printed and bound by CPI Group (UK) Ltd, Croydon, CR0 4YY

03/10/2024

01040415-0016